MINERAL EXPLORATION:
BIOLOGICAL SYSTEMS
AND ORGANIC MATTER

W. W. Rubey (1898–1974)

DONALD CARLISLE

WADE L. BERRY

ISAAC R. KAPLAN

JOHN R. WATTERSON

Editors

MINERAL EXPLORATION: BIOLOGICAL SYSTEMS AND ORGANIC MATTER

Rubey Volume V

Prentice-Hall, Englewood Cliffs, New Jersey 07632

Library of Congress Cataloging-in-Publication Data

Mineral exploration.

    (Rubey volume ; 5)
    Papers from the Rubey Colloquium held Feb. 14-18,
1983 at the University of California, Los Angeles,
conducted by its Dept. of Earth and Space Sciences.
    Includes bibliographies and index.
    1. Biogeochemical prospecting—Congresses.
2. Geobotanical prospecting—Congresses. 1. Carlisle,
Donald, date. II. University of California,
Los Angeles. Dept. of Earth and Space Sciences.
III. Rubey Colloquium (1983: University of California,
Los Angeles) IV. Series.
TN270.A1M48 1986   6221.1   86-3177
ISBN 0-13-583634-4

Editorial/production supervision and
   interior design: *Virginia Huebner*
Cover design: *Edsal Enterprises*
Manufacturing buyer: *John B. Hall*

© 1986 by Prentice-Hall
A Division of Simon and Schuster, Inc.
Englewood Cliffs, New Jersey 07632

Printed in the United States of America

10 9 8 7 6 5 4 3 2 1

ISBN 0-13-583634-4 01

Prentice-Hall International (UK) Limited, *London*
Prentice-Hall of Australia Pty. Limited, *Sydney*
Prentice-Hall Canada Inc., *Toronto*
Prentice-Hall Hispanoamericana, S.A., *Mexico*
Prentice-Hall of India Private Limited, *New Delhi*
Prentice-Hall of Japan, Inc., *Tokyo*
Prentice-Hall of Southeast Asia Pte. Ltd., *Singapore*
Editora Prentice-Hall do Brasil, Ltda., *Rio de Janeiro*
Whitehall Books Limited, *Wellington, New Zealand*

CONTENTS

PREFACE

The Department of Earth and Space Sciences of the University of California Los Angeles (UCLA) conducts an annual symposium of several days length or a lecture series known as "The Rubey Colloquium." Specialists from research institutions, industry and government organizations around the world are invited to present lectures on a central theme of broad interest to earth and/or space scientists. Written summaries or elaborations of these presentations have been gathered together as a series of books known collectively as the Rubey Volumes.

Both the colloquia and the book series are named in honor of the late W. W. Rubey (1898-1984), career geologist with the U.S. Geological Survey and Professor of Geology and Geophysics at UCLA. A brief sketch of Rubey's geological accomplishments is contained in the preface to Rubey Volume I, *The Geotectonic Development of California* (W. G. Ernst, ed., 1981).

The colloquium *Organic Matter, Biological Systems and Mineral Exploration* on which the present Rubey Volume V is based, was held at UCLA on February 14–18, 1983, and attended by just over 100 participants from nine countries, in addition to students. Its purpose was to bring together explorationists and scientists in several disciplines interested in innovative techniques for the discovery of hidden mineral deposits. The focus was on biogeochemical exploration, a methodology that has had variable acceptance and success but that may have recognizable new potential because of recent developments in technology and concepts. An underlying question was whether the customary methods of biogeochemical exploration using higher plants as the sampling medium may not have suffered from too little input from plant physiologists. In addition to that, there were some tantalizing suggestions, new data, and techniques from the rapidly growing field of microbiology and from research on soil organic matter.

The subject matter was introduced by means of a short course (Brooks, Peterson, and Watterson) on state-of-the-art biogeochemical exploration and by presentation and critical evaluation of several case histories. A dozen of these are included in the volume either as full papers or as abstracts. This was followed by the symposium itself lasting three days, with papers by internationally known research scientists in three areas:

1. Plant physiology, plant-soil interactions, and the use of larger plants in geochemical exploration
2. Microbiology and its potential use in mineral exploration
3. Mull and soil organic matter in relation to trace metals

The volume is divided into these same three subject areas.

This volume, then, is an attempt to identify new directions, to summarize recent basic research and to review the state of knowledge about underlying principles that may influence the success of biogeochemical exploration. We hope that in this sense it will be useful. It is, however, neither a handbook nor a comprehensive treatment of biogeochemical exploration (Brooks, 1983; Kovalevskii, 1979). Remote sensing of vegetation, aerial sampling techniques, and geobotanical methods that are not dependent upon chemical analysis, for example, are reviewed only briefly (see Chapter 7, a present status report by Robert Brooks, and abstracts by Olwell and by Collins and Chang). However, fundamental biological considerations that underlie these techniques are brought out in several of the chapters in Part I, *Larger Plants in Geochemical Exploration.* The extensive work on uranium biogeochemistry, although illustrated in an important case history by Dunn, was purposely omitted partly because of limitations of time and space and partly because of an ongoing project on this topic under the auspices of the International Atomic Energy Agency (Dunn *et al.,* 1985). Processing of numerical data is not discussed at all. On the other hand, the entire discussion of microbiology as it might apply to geochemical exploration for minerals is new and, at the time of the symposium, quite speculative. So, too, is much of the discussion on humic substances in relation to exploration. Both have been given full coverage.

We would like to express our appreciation to the many distinguished research scientists, not all represented by papers in this volume, who contributed their time and their thoughts and their enthusiasm to the colloquium. We especially thank the industry representatives who joined the fray with such good humor, including in particular those who led the final all-participants forum "Exploration Managers Fight Back." Catherine Elkins, Victoria Doyle Jones, Diane Hunter, Susanne Luera, and Gail Marshall have been enormously helpful throughout the effort.

DONALD CARLISLE
Department of Earth and Space Sciences
University of California Los Angeles

WADE L. BERRY
Laboratory of Biomedical and
Environmental Sciences
University of California Los Angeles

ISAAC R. KAPLAN
Department of Earth and Space Sciences
University of California Los Angeles

JOHN R. WATTERSON
U.S. Geological Survey
Denver Federal Center, Denver, Colorado

REFERENCES

Brooks, R. R., 1983, *Biological methods of prospecting for minerals:* Wiley & Sons, New York, 322 p.

Dunn, C. E., Ek, J. and Byman, J., 1985, *Uranium biogeochemistry: A bibliography and report on the state of the art:* International Atomic Energy Agency, IAEA-TECDOC-327, Vienna, Austria, 83 p.

Kovalevskii, A. L., 1979, *Biogeochemical exploration for mineral deposits:* (Translated from the Russian), Oxonian Press Pvt. Ltd., New Delhi, 136 p.

Harry V. Warren
Honorary Professor
Department of Geological Sciences
University of British Columbia
Vancouver, British Columbia, Canada

PROLOGUE

BIOGEOCHEMISTRY
IN BRITISH COLUMBIA
1944-1982
A PERSONAL ACCOUNT

Editorial Note: The use of biological materials as media for geochemical exploration has evolved at different rates and often independently on several continents. The earliest case history and the largest number of reports have come from the Soviet Union (see Brooks, Chapter 6, this volume). Work in Scandinavia was soon to follow and then, within a few years, in Canada, the United States, the United Kingdom, New Zealand, and more recently in several other countries. Harry V. Warren of the University of British Columbia has been a pioneer and a leader in biogeochemistry in North America. He began his studies essentially as a prospector in glaciated and forested mountainous terrain and has pursued them energetically ever since. Here, as a prologue to our volume, is a personal account of his pioneering efforts and those of his associates.

DC

ABSTRACT

In British Columbia the idea of using biogeochemistry as a prospecting tool evolved during the summer of 1944. Published data first appeared in 1947 but dealt only with copper and zinc. By 1950 a paper described the successful recovery of gold and silver beads from vegetal matter. Thanks to the help of several collaborators, many biogeochemical papers were published during the years 1952–1973 which dealt in turn with the elements iron, manganese, molybdenum, nickel, cobalt, lead, arsenic, mercury, and tellurium.

By 1980 improved analytical techniques had added greatly to the practicability of using biogeochemistry as a prospecting tool. Gold and silver determinations could be made on 1-g samples of dry vegetal matter. Between 1980 and 1982 it was discovered that 1-g samples of pollen might be used to obtain results comparable to those obtained on vegetal matter, and this held even for gold and silver.

Coincident with the investigations noted above, it became apparent that trace elements had an important part to play in many forestry and epidemiological problems.

BEGINNINGS

Picture if you will a geologist trying to get a look at bedrock by removing some 3 to 5 ft of overburden by ground sluicing. The lower part of this overburden consisted of roots lying on a layer of glacial clay or "hardpan." Water alone could not achieve the desired results. Pick and mattock had to be enlisted to do the job. Picture as well swarms of mosquitos and a miserable light rain and you may well sympathize with an assistant who complained that he had done more shoveling since prospecting with me than in all the time he had worked on a railroad gang to help put himself through university!

Lying in my sleeping bag after a particularly exhausting day the idea came that maybe those "pesky" roots might have some use besides supplying moisture and anchoring trees. Might they not reflect the presence or absence of mineralization in the underlying ground? Further weeks of labor at the end of pick and shovel served to enhance this idea and when autumn came I broached the idea to some of my colleagues.

The reaction of my colleagues could never be described as enthusiastic. Their scepticism centered around three main points, all of which had varying degrees of merit. First, I could in no way claim to be knowledgeable in either chemistry or botany. If, botany and subsurface geology, as this subject was then identified, was to be further developed, assistance from these sister disciplines would have to be enlisted. This would add materially to the cost of any research project. It was seriously suggested that should the idea have any value it would be taken up by investigators in the United States who, with greater resources, would better be able to assess the merits of the idea. Only if early results were satisfactory would we in Canada be justified in initiating further programs in this new and unproven field. Second, as one of our leading botanists pointed out, everywhere you saw trees you knew that there must be copper and zinc present; they could not exist without these elements. However, the trees might be expected to absorb only what they needed for metabolic processes and could not be expected to take up what they did not need. Because trees had no need to have lead, there would be no reason to expect them to absorb this element. Finally, as a geological colleague pointed out, trees would primarily be sampling the soil on which they were growing and not necessarily the underlying bedrock.

All the opinions noted above demanded serious consideration, although it was obvious from the beginning that any biogeochemical program, to be effective, would require an interdisciplinary approach. It was soon pointed out, too, that Canadians should not wait until others did the research for them. Fortunately, the biogeochemical studies, once begun, soon showed that the difficulties suggested by the botanists were not fatal.

The third point raised warrants elaboration. Boulder tracing is a recognized form of prospecting in glaciated terranes. One does not expect to find ore under a mineralized boulder. In glaciated country one locates as many mineralized boulders as possible and when all relevant data have been assembled one attempts to discover the source of the boulders. Similarly, in glaciated country one should not necessarily look directly under anomalous vegetation for an ore body. One merely adds any biogeochemical data to all the other relevant information and then attempts to locate the source of mineralization.

In retrospect, it would appear that there were answers to all the objections that had been raised in connection with biogeochemistry—although some regions are still refractory to this method. Nevertheless, only a series of unanticipated and fortuitious events actually prevented biogeochemistry in Canada from being stillborn, or at least delayed.

EARLY DEVELOPMENTS

Dorothy and William "Dorbils" offered a prize to the University of British Columbia for the best essay written by an undergraduate on the subject "Botany and Geology." The "Dorbils" came from Cornwall, where they had heard that a particular plant betrayed the presence of tin mineralization. The late university president Walter Gage, who at that time looked after bequests and scholarships, asked the Botany Department if they would accept this offer and, when it was declined, asked if I would. Unhesitantly I said "yes," but wondered how I could take advantage of the offer. Most opportunely, Charles Howatson, a former student of mine and a veteran of World War II, had asked if he could take some courses with me: He could use a little extra cash. How would he like to write an essay on "Botany and Geology"? If he could have some guidance, the project would appeal to him. The result was a paper (Warren and Howatson, 1947) which many of you have probably read.

The University of British Columbia's Board of Governors, with persuasive support from the late Dean Daniel Buchanan of the Faculty of Arts and Science, alloted sufficient funds for us to employ J. W. Warden, a competent and experienced botanist, to collect some 250 samples. These samples were collected under the guidance of the staff of various mines. Warden was not told whether the sample areas were over mineralized or barren ground. Looking back over 35 years our results must appear somewhat crude; nevertheless, they were sufficient to justify further work in the field of biogeochemistry. In essence, this initial program established beyond doubt that when trees and plants were growing over copper or zinc mineralization they reported anomalous concentrations of these metals.

Now for some time fortune smiled on this project. The late M. Y. Williams, then head of the Department of Geology and Geography at the University of British Columbia, supported our project almost from its inception: He had been impressed by the differences he had observed in the vegetation growing over plutonic and over sedimentary rocks in southwestern Ontario. We asked for support from the Penrose Fund of the Geological Society of America for the further development of biogeochemistry. In assessing our request the Society conferred with W. H. Bradley, Director of the U.S. Geological Survey, who in turn consulted H. E. Hawkes and M. F. Fleischer. Their responses were positive and I will always be grateful to all concerned for the generous response to our appeal for assistance in our efforts to establish what, at that time, was a comparatively little known field of study.

Coincident with receiving Penrose funds from the Geological Society of America there happened another event which was significantly to enhance the development of biogeochemistry. An introduction to Robert E. Delavault provided what had so far been sadly lacking: sound chemical knowledge. Delavault had been born in Edmonton but had returned to Paris, where at the Sorbonne he initially studied medicine before turning his attention to botany and crystallography. During the next three years our joint studies resulted in several papers (Warren and Delavault, 1948; 1949a,b; 1950a,b).

At about this time we had two setbacks. After presenting a paper at the Annual Meeting of the Canadian Institute of Mining and Metallurgy in Toronto in 1950, a geologist from a well-known North American company rose from the audience and stated that his company had tried biogeochemistry and found it to be of little help. To cap this discouragement an institution to whom we had appealed for help informed us that because we were using dithizone, which they considered to be an unreliable chemical, they would not support our project. We had found dithizone to be a most useful and, indeed, an almost indispensable chemical for nearly three years.

AN EARLY TEST

However, at this juncture another event turned the tides of fortune in our favor. In 1948, O. S. Lovekin, a dentist from the United States, called on me and suggested that we have our claims for biogeochemistry put to a practical test. We accepted this challenge.

Lovekin had a son at the University of Arizona and between them they arranged to have sets of vegetation samples sent to us. Some samples were to be taken from over mineralized ground and some from unmineralized areas: all samples were to come from the general area of the San Manuel copper mine in Arizona. We carried out analyses on these samples without having any idea where the samples had been taken.

Seven suites of samples were sent. Two we pronounced positive, four negative, and one indeterminate. We were told that we were correct in all except the last named one, which should have been negative. However, a few weeks later another letter informed us that they had been in error and that the suite in question had indeed revealed sub-economic mineralization.

A year previously, Lovering *et al.*, (1950) of the U.S. Geological Survey had published a paper on "Dispersion of Copper from the San Manual Copper Deposit." Although they had concentrated on soil sampling their vegetal results were in the same "ballpark" as ours. H. E. Hawkes, at that time with the U.S. Geological Survey, discussed the general problems of San Manuel with us and has, throughout all our biogeochemical investigations, afforded us every kindness and consideration. This set of circumstances put biogeochemistry on its feet (Warren *et al.*, 1951).

Neither time nor space allow me to do more than report that during the next 16 years we published many papers on the subject of biogeochemistry, all of which are listed in the References and Bibliography. A few of these investigations provided stories in themselves. Among these are the early detection of molybdenum at the Endako mine prospect (Warren *et al.*, 1953; Warren and Delavault, 1965), of copper in sagebrush (*Artemesia tridentata*) at the Afton mine and in Douglas fir (*Pseudotsuga menziesii*), alder (*Alnus* s.) and Englemann spruce (*Picea engelmanni*) at the Bethlehem mine properties (Warren *et al.*, 1957). Of special interest to human health was the, at that time, problematical detection of lead in several commonly occurring trees from various sites across British Columbia (Warren and Delavault, 1960). Some high values of lead were conceivably derived from automobile exhaust, as subsequently shown in the United States by Helen Cannon and Jessie Bowles (1962) and again by Everett *et al.* (1967) in English privet.

RECENT DEVELOPMENTS

There is an old saying that goes "what the old cock crows the young ones learn." Some 60 years ago the United States initiated the exploitation of disseminated copper and molybdenum occurrences, commonly referred to as prophyry deposits. The late S. J. Schofield, Professor of Economic Geology at the University of British Columbia, constantly drew attention to the many similarities that exist between the geology north and south of our common border. Had he lived just a little longer, doubtless he would have been delighted to follow the development of the many porphyry deposits that have since been found north of the 49th parallel.

The production of gold at Carlin in 1965 also represented a radical new development. Indeed, the exploitation of the finely disseminated gold at Carlin was every bit as novel and exciting as the earlier development of disseminated copper and molybdenum.

Remembering the historical significance of the development of the porphyry deposits in British Columbia, and inspired in no small measure by the example of Carlin and related mines, we felt that once again we might take advantage of the discoveries and developments initiated by our good neighbors to the south.

Unfortunately, our problem is greatly complicated by the fact that much of British Columbia is covered with a mantle of glacial drift. This fact suggested that new techniques would be welcome in searching for buried disseminated gold deposits. It is well to remember that in prospecting for these disseminated gold deposits the use of the gold pan has not been effective. Obviously, biogeochemical techniques should be considered. During the last 10 years much of the attention has been focused on discovering the most appropriate techniques applicable to finding Carlin-type deposits in British Columbia.

Carlin-type deposits are associated with modest amounts of arsenic, tellurium, antimony, and mercury. Routine analytical techniques are now available for discovering anomalous amounts of any of these elements in plants. Douglas fir (*Pseudotsuga menziesii*) had already been found to be particularly useful where arsenic is concerned (Warren *et al.*, 1964, 1968). Already one Carlin type of deposit has been described (Warren and Hajek, 1973).

In British Columbia many plants have demonstrated their ability to reflect anomalous concentrations of gold in overburden. Particularly interesting has been the discovery that at least two plants, mountain phacelia (*Phacelia sericea*) and mountain avens (*Dryas drummondii*), may pick up 40 to 50 times their background content of gold. Both these plants are cyanogens, which presumably accounts for their ability to do this (Girling *et al.*, 1979). How gold enters a plant and is distributed from roots to flowers has been beautifully demonstrated by an autoradiogram (Girling and Peterson, 1980).

Extrapolating from the phacelia studies we hypothesized that when these plants died, their cyanide might break down and their contained gold might be in a chemical form suitable for precipitation. After careful searching, gold crystals and fragments of crystals, always in the form of octahedra, have in fact been found in overburden below concentrations of phacelia plants (Warren, 1982).

We are still engaged in attempting to determine what are normal and anomalous concentrations of gold in the various organs of the several species of plants. However, as a general indication it may be said, that, in terms of dry plant material, a majority of plants carry background amounts of gold between 1 and 2 parts per billion. Anomalous material may run between 3 and 25 parts per billion, and occasionally considerable more (Warren and Barakso, 1982).

Girling found why some of our early work, notably on phacelia, failed. More than 90 percent of the gold in phacelia is lost when this plant is ashed at a dull red heat (550°C). Presumably, this is because the gold in this plant is present as gold cyanide, which is volatile at about 300°C.

POLLEN ANOTHER BIOGEOCHEMICAL TOOL

Another possible use of biogeochemistry involves the use of pollen as a tool with which broad geochemical features may be outlined. In a general way pollen reflects the trace element relationships found in the plants from which the pollen was derived. We have ascertained that pollen in several areas may admirably reflect the presence of anomalous copper, zinc, lead, and molybdenum. The potential value of this medium is further exemplified by a collection of five different pollens from the Ashloo property near Squamish, a short distance north of Vancouver. These pollens ran 0.3, 0.7, 0.2, 0.4, and 0.4 ppm *gold*. They carried from twice to five times as much gold as silver corresponding with the relative precious metal contents of the ore at this property (Warren and Horsky, 1982).

CONCLUSION

During the past 38 years much has been accomplished in biogeochemistry. Advances in analytical techniques have made it possible to apply this relatively new prospecting tool to a wide variety of elements. Until a few years ago mineral prospects in British Columbia had been discovered by prospectors who had found mineralized outcrops. However, geological, geophysical, and geochemical skills did lead to many of these discoveries becoming mines. I know of no mines operating today in British Columbia that owe their discovery solely to biogeochemistry. However, biogeochemistry has played a significant part not only in furthering the development of some prospects but also in directing attention to some problems of environmental significance. Perhaps it would be appropriate to repeat some remarks I made more than 30 years ago.

"Biogeochemistry is not a rival of geophysics, geology, or hard work as a means of finding ore. It is another tool which, employed intelligently under appropriate, but not necessarily ideal conditions, and in conjunction with other suitable tools, can do much to assist in the search for hidden ore."

ACKNOWLEDGMENTS

A perusal of the list of coauthors will indicate that biogeochemical studies at the University of British Columbia represent the work of many persons. The accompanying References and Bibliography list many other people, to each of whom the author gratefully acknowledges his indebtedness.

None of these studies would have occurred had we not been supported financially by several organizations, notably the Geological Society of America, the Geological Survey of Canada, the National Research Council of Canada, the Defence Research Board of Canada, Kennco Explorations (Western) Ltd., and Canada's Natural Science and Engineering Research Council. Subsidiary funds were supplied by several mining companies and numerous personal contributors.

REFERENCES AND BIBLIOGRAPHY

Warren, H. V., and Howatson, C. H., 1947, Biogeochemical prospecting for zinc: *Bull. Geol. Soc. Am.*, v. 58, p. 803–820.

Warren, H. V., and Delavault, R. E., 1948, Biogeochemical investigation in British Columbia: *Geophysics*, v. 13, p. 609–624.

Warren, H. V., and Delavault, R. E., 1949a, Further studies in biogeochemistry: *Bull. Geol. Soc. Am.*, v. 60, p. 531–559.

Warren, H. V., and Delavault, R. E., 1949b, Biogeochemical researches on copper in British Columbia: *Trans. R. Soc. Can.*, Third Series, v. 43, Sec. 4, p. 119–137.

Warren, H. V., and Delavault, R. E., 1950a, Gold and silver content of some trees and horsetails in British Columbia: *Bull. Geol. Soc. Am.*, v. 61, p. 123–128.

Warren, H. V., and Delavault, R. E., 1950b, History of biogeochemical investigations in British Columbia: *Can. Inst. Min. Metall. Trans.*, v. 53, p. 236–242.

Lovering, T. S., Huff, L. C., and Almond, H., 1950c, Dispersion of copper from the San Manuel copper deposit, Pinal County, Arizona: *Econ. Geol.*, v. 45, p. 493–514.

Warren, H. V., Delavault, R. E., and Irish R. I., 1951, Further biogeochemical data from the San Manuel copper deposit, Pinal County, Arizona: *Bull. Geol. Soc. Am.*, v. 62, p. 919–929.

Warren, H. V., Delavault, R. E., and Irish, R. I., 1952a, Preliminary studies on the biogeochemistry of iron and manganese: *Econ. Geol.*, v. 47, p. 131–145.

Warren, H. V., Delavault, R. E., and Irish, R. I., 1952b, Biogeochemical investigations of the Pacific Northwest: *Bull. Geol. Soc. Am.*, v. 63, p. 435–484.

Warren, H. V., Delavault, R. E., and Routley, D. G., 1953, Preliminary studies of the biogeochemistry of molybdenum: *Trans. R. Soc. Can.*, Third Series, v. 47, Sec. 4, p. 71–75.

Warren, H. V., and Delavault, R. E., 1954, Variations in the nickel content of some Canadian trees: *Trans. R. Soc. Can.*, Third Series, v. 48, Sec. 4, p. 71–74.

Warren, H. V., Delavault, R. E., and Fortescue, J. A. C., 1955, Sampling in biogeochemistry: *Bull. Geol. Soc. Am.*, v. 66, p. 229–238.

Warren, H. V., and Delavault, R. E., 1955a, Some biogeochemical investigations in eastern Canada: *Can. Min. J.*, Part 1, v. 76, no. 7, p. 49–54; Part 2, v. 76, no. 8, p. 58–63.

Warren, H. V., and Delavault, R. E., 1955b, Biogeochemical prospecting in northern latitudes: *Trans. R. Soc. Can.*, Third Series, v. 49, Sec. 4, p. 111–115.

Warren, H. V., and Delavault, R. E., 1957, Prospecting for cobalt: *Trans. R. Soc. Can.*, Third Series, v. 51, Sec. 4, p. 33–37.

Warren, H. V., Delavault, R. E., and Cross, C. H., 1957, Geochemical anomalies related to some British Columbia copper mineralization, p. 277–282 in *Methods and Case Histories in Mining Geophysics:* 6th Commonwealth Min. Metall. Congr. (published in 1959).

Warren, H. V., and Delavault, R. E., 1960, Observations on the biogeochemistry of lead in Canada: *Trans. R. Soc. Can.*, Third Series, v. 54, Sec. 4, p. 11–20.

Warren, H. V., 1962, Background data for biogeochemical prospecting in British Columbia: *Trans. R. Soc. Can.*, Third Series, v. 56, Sec. 3, p. 21–30.

Cannon, H. L., and Bowles, J. M., 1962, Contamination of vegetation by tetraethyl lead: *Science*, Sept. 7, v. 137, no. 3532, p. 765–766.

Warren, H. V., Delavault, R. E., and Barakso, J., 1964, The role of arsenic as a pathfinder in biogeochemical prospecting: *Econ. Geol.*, v. 59, p. 1381–1383.

Warren, H. V., and Delavault, R. E., 1965, Further studies of the biogeochemistry of molybdenum: *West. Miner*, v. 38, p. 64–72.

Warren, H. V., Delavault, R. E., and Barakso, J., 1966, Some observations on the geochemistry of mercury: *Econ. Geol.*, v. 61, p. 1010–1028.

Everett, J. L., *et al.*, 1967, Comparative survey of lead at selected sites in the British Isles in relation to air pollution: *Food Cosmet. Toxicol.*, v. 5, p. 29–35.

Warren, H. V., Delavault, R. E., and Barakso, J., 1968, The arsenic content of Douglas fir as a guide to some gold, silver and base metal deposits: *Can. Min. Metall. Bull. 61,* p. 860-866.

Radtke, A. S., Heropoules, C., Fabbi, B. P., Scheiner, B. J., and Essington, M., 1972, Data in major and minor elements in host rocks and ores, Carlin gold deposit, Nevada: *Econ. Geol.,* v. 67, no. 7, p. 975-978.

Warren, H. V., and Hajek, J. H., 1973, An attempt to discover a Carlin-Cortez type of deposit in B.C.: *West. Miner,* v. 46, p. 124-134.

Girling, C. A., Peterson, P. J., and Warren, H. V., 1979, Plants as indicators of gold mineralization at Watson Bar, British Columbia: *Econ. Geol.,* v. 74, p. 902-907.

Girling, C. A., and Peterson, P. J., 1980, Gold in plants: *Gold Bull.,* v. 13, no. 4, p. 151-157.

Warren, H. V., 1982, The significance of a discovery of gold crystals in overburden, p. 45-51 *in* Levinson, A. A. (ed.), *Precious Metals in the Northern Cordillera:* Association of Exploration Geochemists, Rexdale, Ont.

Warren, H. V., and Barakso, J., 1982, The development of biogeochemistry as a practical prospecting tool for gold: *West. Miner,* v. 55, p. 27-32.

Warren, H. V., and Horsky, S. J., 1982, Further notes on the use of pollen as an exploration tool: *West. Miner,* v. 55, p. 42-43.

Warren, H. V., Horsky, S. J., Kruckeberg, A., Towers, G. H. N., and Armstrong, J. E., 1983, Biogeochemistry, a prospecting tool in the search for mercury mineralization: *J. Geochem. Explor.,* v. 18, p. 169-173.

Warren, H. V., Horsky, S. J., and Barakso, J. J., 1984, Preliminary studies of the biogeochemistry of silver in British Columbia: *CIM Bull. 77*(863), p. 95-98.

Warren, H. V., Horsky, S. J., and Lipp, C., 1984, Biogeochemistry indicates mineral anomalies along southern extensions of the Pinchi fault: *West. Miner,* v. 57, No. 6, June, p. 31-34.

*Entries are listed in historical sequence.

Donald Carlisle
Department of Earth and Space Sciences
University of California, Los Angeles
Los Angeles, California

Wade L. Berry
Laboratory of Biomedical and Environmental Sciences
University of California, Los Angeles
Los Angeles, California

I

AN INTRODUCTION TO LARGER PLANTS IN GEOCHEMICAL EXPLORATION

The notion that a mineral explorationist might collect and analyze parts of living trees or shrubs in the search for economic mineral deposits was introduced less than 50 years ago. As Harry Warren has recounted, it met with enthusiasm from some and audible skepticism from others. Since then, the conept has been developed in several universities, geological surveys, and by a substantial number of enterprising mining companies. The investigators have been geologists primarily, sometimes in collaboration with botanists. In only a few research projects have plant physiologists been directly involved. The results indicate that measurement of element concentration by plants can indeed be one of several useful and mutually supportive tools in mineral exploration but that there are many unresolved problems and many unexplained failures to detect hidden mineralization. Can the technique be improved by incorporating a greater knowledge of plant physiology, biochemistry, and ecology and, in conjunction with this, by the application of new rapid, inexpensive methods of multielement analysis to increase the data base? Part I of the volume addresses this question.

Because the plant samples collected are the keystone of the biogeochemical method, the first chapter, by Wade L. Berry, a plant nutritionist, examines criteria for choosing the most desirable tissue for sampling and the effect that this choice has on the interpretation of analytical data. The yield–dose response curve and its analog, the tissue concentration–dose response curve, are new, in this form, to geological literature and fundamental to an understanding of biogeochmical exploration. The role of multielement analysis, the phenomenon of "tolerance," and the meanings of such terms as "accumulator," "excluder," "essential," and "nonessential" are introduced.

The second chapter, by Larry P. Gough, a geobotanist with the U.S. Geological Survey, builds on Chapter 1 and emphasizes that multiple stresses are the rule in nature. There are reasons why boundaries between kinds of plants may be much more abrupt in the field than one might expect from the laboratory-determined yield–dose response curve, why stress factors other than soil metal anomalies may be difficult to detect, or why the anomalies themselves might not be recognizable in the vegetation.

In Chapter 3, Iain Thornton, a biogeochemist, draws on British examples to summarize parameters, in addition to the natural absolute metal content of soils, which can influence the analyses obtained from plant tissue. These are particularly important in areas where several different soil series occur.

The practical value of plant physiology and biochemistry are well illustrated in Chapter 4, *Tolerance: A Plant's Response to Metal Stress* by Peter J. Peterson. The author examines, in some detail, three categories of mechanisms that plants have developed to adapt to high metal concentrations. Some of the mechanisms, such as those exhibited by specific metal hyperaccumulators, can be used to advantage in biogeochemical exploration and evaluation. Others, such as exclusion, simply make things more difficult.

Chapter 5, the last one with a biological emphasis, is by J. H. Richards, a plant physiologist with special interests in plant root systems. It is commonly stated that one of the unique advantages of larger plants as sampling media is the fact that their root systems extend much more deeply into the subsurface and encompass a larger volume than is practical with normal soil or rock geochemical sampling. Beyond that the mineral explorationist may have precious little information on the subject. Not only are root form and depth determined by many factors, but in addition, as pointed in Chapter 1, roots commonly have avoidance and evasion strategies in the presence of abnormal metal concentrations.

The final chapter in this conceptual section is by Robert R. Brooks, biogeochemist and director of the short course that introduced the symposium. His extensive experience in both geobotanical and biogeochemical research provides him with a unique and unusually strong background from which to evaluate the status and recent developments in applied biogeochemical exploration.

Finally, in this first part of the volume, there are five chapters and six abstracts recounting case histories and one abstract on airborne multispectral techniques, all of which illustrate principles as well as data and methodology under a range of conditions. Gold is the metal of interest in five of the case histories, platinum in one that may be the first such study published, uranium in one, and silver and several base metals in others. Of unusual interest are the very substantial quantities of gold detected by W. E. Baker in the ash of trees growing over alluvial gold deposits in Tasmania. The value of forest litter or "mull" as a sampling medium for gold exploration is illustrated in two of the chapters (see also Chapter 22). Two of the chapters report anomalous amounts of gold in larger trees.

Donald Carlisle   Wade L. Berry

Wade L. Berry
Laboratory of Biomedical and Environmental Sciences
University of California, Los Angeles
Los Angeles, California

1

PLANT FACTORS INFLUENCING THE USE OF PLANT ANALYSIS AS A TOOL FOR BIOGEOCHEMICAL PROSPECTING

ABSTRACT

The composition of plant tissues reflects the composition of the soil on which the plant grows. This reflection, however, is modified in many plants by tolerance mechanisms. It is these mechanisms that provide certain plants with the ability to grow in the presence of unusually high concentrations of specific elements. Accordingly, biogeochemical prospecting can be effective only when these mechanisms are considered in the interpretation of plant analysis data. On the other hand, plants used for biogeochemical prospecting must grow on a wide range of elemental concentrations in the soil. Those with no tolerance will be limited to soils with only background levels and those with high tolerance will flourish only in the presence of high concentrations. Thus the appropriate plant for biogeochemical prospecting will have a limited degree of tolerance and will grow over the range of soil metal concentrations of interest.

The principal plant factors to consider are (1) that element uptake by the plant is typically not linear over the entire range of concentrations of interest; (2) that normal plants respond to each element in a characteristic manner although there are groupings of elements that may have common parameters (e.g., divalent cations, polyvalent ions, essential and nonessential elements); (3) the plant is a multitissue organism, specific responses to metal and nonmetal stress must be defined on an individual tissue basis, and plant analysis of the total plant is very difficult to interpret; (4) elemental concentrations in tissues are age dependent, and both chronological and physiological age must be considered; (5) the overall health of the plant must be considered because nonhealthy plants do not respond in a characteristic manner; and (6) tissue element concentrations can be influenced by the presence of high concentrations of other elements; the greater the number of elements whose concentration is known in the plant tissue, the higher will be the degree of confidence in interpretation.

INTRODUCTION

As plants depend on the soil for essential minerals, the mineral content of their tissues reflects soil composition. During evolution and colonization of new areas some adaptations occur that allow the plant to tolerate new conditions or to be able to modify the environment to meet their needs. Thus the composition of a plant will reflect the soil composition, but the reflection may be modified by the plant's own physiology. The possibility then exists that if the nature and extent of the modifications are known, it should be possible to evaluate the soil composition by analyzing the tissue of the plants growing on the soils. This is the basic premise of biogeochemical prospecting through plant analysis.

Historically, plant analysis was developed as a tool to evaluate the nutrient status of agronomic plants (Ulrich, 1952; Chapman, 1966; Bates, 1971; van den Driessche, 1974). The purpose was to determine if nutrient elements present in the soil were at desirable concentrations for plant growth and to estimate the total amount of fertilizer to apply to correct deficiencies where they occurred. The principal soil factors considered were (1) the total soil nutrient concentration and (2) the proportion of the total soil nutrient pool available to the plant. The mineral elements of most concern have been N, K, and P because these are the elements in high concentration in plant tissues and are usually limiting in the soil for plant growth and thus comprise the bulk of the applied chemical fertilizers both in weight and cost. After the recognition in the first half of the twentieth century of a number of trace metals, such as Mo, Zn, and Cu, as essential elements for plant growth, these elements have also become an important part of agronomic plant analysis (Bowen, 1966). However, it was not until the 1970s, when society as a whole became highly concerned about environmental problems, that significant effort was directed toward understanding the effects and interactions of metals and elements other than the essential nutrients (Lisk, 1972; Furr et al., 1976; Cunningham et al., 1975). The recent concern about environmental pollution has resulted in considerable effort toward understanding the extent and mechanism of plant absorption and translocation for a wide range of metal elements, including such metals as Ni, Cd, Co, Cr, and V (Antonovics et al., 1971; Foy et al., 1978; Berry, 1977, 1978; Beckett and Davis, 1977; Davis and Beckett, 1978). Although incomplete, this recently developed information on the absorption and translocation of many nonessential trace metals along with the basic information from plant physiology and agronomic plant analyses could well provide the foundation for a resurgence of interest and use of plant analyses for biogeochemical prospecting (Warren and Delavault, 1948; Webb and Millman, 1951; Vinogradov, 1959; Carlisle and Cleveland, 1958; Cannon, 1960; Malyuga, 1964; Brooks, 1972).

USE OF INDICATOR PLANTS

The study of soils and plants can provide a wide range of information for the geologist (Mason and Stout, 1954; Jenny, 1980). This paper is concerned primarily with the use of plant nutrition information to assess soil composition by analyzing the mineral content of plant tissues (Allaway, 1968; Chapman, 1966; Clarkson and Hanson, 1980; Chapin, 1980). However, it was from the field of plant geography and the use of indicator plants that the first interface between botany and geology was developed to pinpoint potential ore zones (Jowett, 1959; Kruckeberg, 1954; Peterson, 1976; Wainwright and Woolhouse, 1976). Geobotanical prospecting, in contrast to biogeochemical prospecting, employs indicator plants that naturally occur and grow or have specific morphological variations only in the presence of an abnormally high concentration of certain metals (Brooks, 1979). These plants are often metal specific and thus provide an instantaneous field verification of the presence of high metal concentrations in the surface soil (Jowett, 1964). They can be very useful when available, but unfortunately there is no universal indicator plant that grows in all climates or places in which a given metal occurs.

The very fact that only a very restricted group of plants can grow in soil with high concentrations of specific trace elements suggests that some very special physiological adaptations were required by plants before they could survive in these metal-stressed environments. There is no reason to believe that there is only one strategy that plants utilize to colonize the high metal soils found over ore bodies and mine tailings. Nor is there reason to believe that these strategies are an all-or-none adaptation to growth under metal-stressed conditions (Gerloff, 1963; Antonovics et al., 1971; Snaydon and Bradshaw, 1962).

ECOLOGICAL RELATIONSHIPS

Although the emphasis of this chapter is on plant analysis, any biogeochemical survey must take into account the overall ecology of a given area (Fortescue, 1979). There are many plants in which tissue metal concentration is well correlated with the soil metal concentration, but if the ecological relationships are such that the plant does not grow widely in the area of concern, it is of little or no practical use in biogeochemical prospecting. There are many ecological conditions, such as drought competition, temperature, isolation, and general climate conditions that could limit plant distribution in addition to those related to metal concentrations. The first priority, therefore, is to select plants that grow under a range of metal concentrations, including both background conditions and those conditions found in the halos of ore deposits.

PLANT RESPONSE
TO ABNORMAL METAL CONCENTRATIONS

Dependence on Concentration:
the Yield–Dose Response Curve

A plant's response to a given change in element concentration is concentration dependent; that is, the degree of response of a plant to a change in concentration of a specific element may not be the same when the concentration is high as where there is only a low background concentration of the same element (Asher and Edwards, 1978). A dose-response curve (Fig. 1-1) is a continuum having multiple distinct phases starting at or near zero concentration and terminating when lethal concentrations are reached. The yield-dose response as shown in the response curve is composed of three principal zones which define the nutrient status of the overall plant: deficiency, tolerance, and toxicity (Berry and Wallace, 1981). The deficient zone or low end of the concentration range can be of interest in biogeochemical prospecting because it is usually the concentration range of the normal background and it can also provide significant information about which elements may be depleted in an area. This knowledge can be useful in the recognition of a geochemical contrast in the substrate soil or rock and its precise boundary. In some circum-

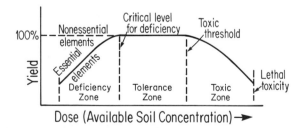

Fig. 1-1   Generalized yield–dose response curve. Diagram not to scale, thus shows relationships only; the degree of expression of each zone is a function of the specific element being considered.

stances a depletion of a specific element can be as useful an indicator as an enrichment, for both provide information about the geochemistry of the area. The high-concentration end of the yield–dose response curve is particularly useful as a source of information about the possible presence of mineral anomalies. Two typical examples are Cu and Ni, as shown in Figures 1-2 and 1-3. Both figures show that there is a marked break in tissue metal concentration as the external metal concentration exceeds tolerable levels. From these figures it can be seen that there are two phases of tissue metal concentration or metal uptake. In the low range of soil or external metal concentration the rate of uptake increases only slowly as external metal concentration increases. But if the external metal concentration exceeds this low concentration range, uptake enters a new phase and the uptake rate increases rapidly with increasing external metal concentration. In this high range of external metal concentration, plant tissue analysis is very sensitive to changes in soil metal concentration. Unfortunately, this is generally a limited range because when the lethal concentration is reached, the roots are killed and can no longer function as part of an organized transport system to transport metal to the tops of the plant.

The slope of the yield–dose response curve in this "toxic zone" is a measure of the unit toxicity of the metal involved. Unit toxicity is defined as the amount of yield decrease per unit of added toxicant after the toxic threshold is exceeded. Those metals with a high unit toxicity, such as Cu, will have a limited range between threshold and lethal toxicity, while metals with a low unit toxicity, such as Cr or Mn, will have a broad range (Berry, 1978). A high unit toxicity would be indicated by a steep slope in the toxic zone of Figure 1-1. The toxic threshold or upper limit of the tolerance zone is the lowest con-

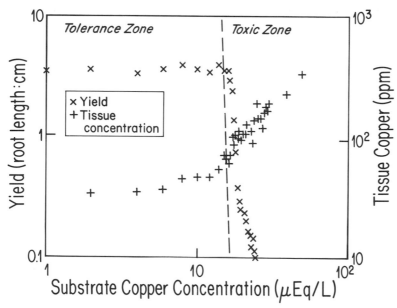

Fig. 1-2 Dose-response curve of lettuce for copper, showing yield as a function of available copper in the rooting zone. Superimposed on this is the leaf tissue copper concentration, showing a change in absorption made associated with the change in nutrient status from the tolerant zone to the toxic zone. (From Berry, 1978.)

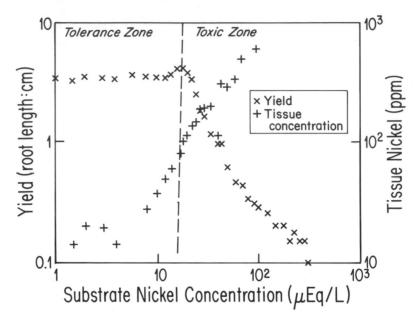

Fig. 1-3   Dose-response curve of lettuce for nickel, showing yield as a function of available nickel in the rooting zone. Superimposed on this is the leaf tissue nickel concentration, showing a change in absorption made associated with the change in nutrient status from the tolerant zone to the toxic zone. (From Berry, 1978.)

centration of an element that causes a direct reduction in yield. These two parameters of toxicity, unit toxicity and toxic threshold, are independent of each other. Thus an element can have both a high toxicity threshold and a high unit toxicity. Such an element would not cause any toxicity until high concentrations were reached, but the difference in concentration between no response and death would be small. An element with a low toxicity threshold and low unit toxicity would be inhibiting at low concentrations, but each additional unit amount of the toxic element would cause relatively little damage.

Multielement Interactions

The plant's response to soil nutrient concentrations also may be influenced by the amount of other elements present (Wallace *et al.*,1977). For example, a number of plants will actively secrete protons and chelating agents in response to Fe deficiency (Landsburg, 1981). The soil around the plant roots will become more acid and many metals, in addition to Fe, will become more available for absorption by the plant. Thus plants interact with their environment and changes in one nutrient can result in changes in tissue concentration of many other elements (Chaney *et al.*, 1976; Miller *et al.*, 1977). This is in addition to those interactions between elements in the soil itself which are the direct result of soil geochemistry. Because of this, biogeochemical surveys must utilize a multi-element approach in order to distinguish soil and plant anomalies.

Root Evasion

A plant's general response to small localized areas of high elemental concentration (the nugget effect as seen by plants) is avoidance. When the metal is evenly distributed in the soil, all roots are affected in the same way and tissue analysis will reflect the soil concentration. However, when the high-concentration areas are localized, root growth in these toxic areas will be inhibited while the root growth in the rest of the soil will continue at a normal rate. Within a short time there will be an uneven distribution of roots, with very few roots in the localized high-concentration areas. Under these circumstances tissue analysis will reflect the concentration of the soil where most of the roots occur (Wallace, 1980). However, if the root system is severely restricted by only a limited area of favorable conditions, the plant will be more susceptible to another stress, such as drought, which will indicate the presence of the localized toxic conditions.

TOLERANCE STRATEGIES
AND THEIR CONSEQUENCES FOR EXPLORATION

There are at least three distinct types of strategies used by plants that grow in areas of elevated elemental concentrations. The first is *avoidance*, the second is *detoxification*, and the third is *biochemical tolerance*. Each of these strategies will alter the tissue metal concentration in different ways. For instance, a small increase in the tissue concentration of a plant using the avoidance strategy could indicate a geological anomaly, while an increase in concentration of the same magnitude in a plant utilizing the detoxification strategy may not be significant. It is the amount of change in relation to the overall physiological range that is diagnostic rather than the absolute change in tissue concentration.

Plants that use the *avoidance* strategy typically employ biochemical mechanisms in the roots which limit uptake of certain toxic materials into the root or mechanisms at the root shoot interface which limit the translocation of the elements to the tops (Brown *et al.*, 1972). Some of the most highly studied examples of this strategy are sodium excluders, such as beans, potatoes, and buckwheat, which achieve salinity tolerance by excluding Na from their actively growing tissues. The same type of mechanism is used by many metal-tolerant plants, as exemplified by the fact that they can grow in areas of high soil metal concentration and still have low tissue metal concentrations (Nicolls *et al.*, 1965).

The *detoxification* strategy is conceptually similar to the avoidance strategy except that avoidance is accomplished by internal mechanisms. The toxic element is either partitioned into a nonmetabolically active structure in the plant, such as the cell wall, or is chemically bound in such a way that it can not interfere with the plant's metabolism (Clark, 1983, Hecht-Buchholz, 1983; Turner and Marshall, 1972). In this way the plant avoids toxic metal concentrations in the functioning cytoplasm. It is therefore possible to find tissue for analysis from such a tolerant plant that is enriched in the element of concern (Collins *et al.*, 1976; Cunningham *et al.*, 1976). It is this basic strategy that results in the older (second year) leaf and stems becoming high in metal concentration, for these tissues are no longer the principal metabolic tissue and will soon be lost through senescence.

The third strategy is that of pure *biochemical tolerance*, where the plant finds and utilizes alternative metabolic paths which are less sensitive to high elemental concentrations (Ernst, 1976).

The extreme case of all these tolerance strategies is exemplified by indicator and hyperaccumulator plants, which grow in the presence of high elemental concentration (Malaisse *et al.*, 1978; Severne, 1974). Plants that colonize outcrops and mine tailings are an example of an extreme case in which the element of high tolerance is a metal of geological concern. Nonmetals also can be tolerated and accumulated by plants to a high degree. An example is Se, which is accumulated by several plants and has been used as a pathfinder element for U (Cannon, 1971; Rosenfeld and Beath, 1965). The mere presence of large numbers of such plants generally indicates a high concentration of the element in the soil and the use of plant analysis to confirm this would, in most cases, be superfluous. These plants are the plants used for geobotanical prospecting and represent the end member of a family of physiological adaptations to metal tolerance (Cannon, 1971).

There are many plants that possess one or more of these tolerances but to a lesser degree (Wu *et al.*, 1975). Knowing the physiology of these plants facilitates understanding their response to unusually high concentrations of metals in the environment. The response in elemental uptake is not always linear, nor is one mechanism necessarily operative over the entire concentration range. Consequently, the way in which tissue concentration changes provides information that can enhance any conclusion drawn from simple concentration data. This knowledge can then be used to develop efficient procedures for identifying and using these plants for biogeochemical prospecting.

METAL CONTENT OF PLANT TISSUE

Plant Parameters

Although it is a truism that high tissue metal content has to have its source in the soil, in only relatively few soils and under only a limited range of concentration is there a direct linear relationship between total soil concentration and tissue concentration. The important plant parameters other than tolerance mechanisms that modify and govern tissue concentration are (1) the physiology of a particular plant species, (2) the particular plant tissue, (3) tissue age, and (4) the growth rate of the plant. If these parameters are known or remain constant, most plant species could theoretically be useful for biogeochemical prospecting. In fact, only a limited number of plants are presently useful for locating mineral anomalies. The unsuitability of many plants is mainly the result of insufficient precision in the overall analytical method or the limited time during which suitable tissue can be obtained each year. When there is only a small gradient in tissue concentration between background and anomalous areas, both sampling and chemical analysis must have a higher degree of precision than is often presently feasible for routine operations. This is further complicated by the fact that many native plant species have a high degree of genetic variability, which would make it necessary to have a relatively large sample size to determine small differences. However, because of the number of plant species normally found in most areas, it is not unreasonable to expect to find some species with characteristics that can be used for biogeochemical prospecting. Owing to the relative immaturity of this field, there is presently a limited amount of background information covering only a few specific plants with tolerance to certain metals that are well documented. This simply means that with continued use of biogeochemical prospecting, the effort required to develop the necessary background information for a given site will decrease.

Uptake Physiology

The first factor to consider when deciding which plant to use for a biogeochemical survey is the general biology and physiology of the plant itself. The mineral uptake and accumulation characteristics of a plant with respect to a given element at a given soil concentration reflect one of three general conditions: accumulation, exclusion, or passive uptake. With respect to the essential elements, most plants function as accumulators at low background concentrations. They are excluders at moderate concentrations and exhibit passive uptake when stressed with toxic concentrations of an element. Both the plant and the element are important in determining the conditions under which an element is accumulated. There is no plant that is an accumulator for all elements, nor is there an element accumulated in excessive quantities in all plants. The description of a plant as an accumulator is thus insufficient, for the plant itself is only one half of a functioning system, the element being the other half; an appropriate description would be a Cu accumulator or a Ni accumulator. Even this description is not complete if the range in metal concentration considered is not given.

Accumulator Plants

A plant described as an accumulator for a metal has tissue concentrations of that metal greater than that available in the substrate on which it is grown (i.e., the plant concentrates the element in its tissues). This concentration is achieved through an active biological mechanism. Such plants may exhibit very large changes in tissue concentration in relation to relatively small changes in soil concentration. However, the uptake capacity of a plant for a given element is not always a simple linear or a continuous function over an extended dose range. This makes it difficult to evaluate the changes that occur over a wide range of soil concentrations. A plant can be an accumulator at low element concentrations but be an excluder at higher concentrations.

The concept of an element accumulator plant is relatively simple and is based on the tissue concentration found at the time of sampling in relation to soil concentration. Expressed as a ratio, this is the *biological absorption coefficient* (BAC). Most calculated BACs are useful only for comparison purposes because of the many inherent problems in comparing elemental concentrations in two substances as dissimilar as soil and plant tissue. Thus, although the concept of accumulators and excluders is discussed extensively in the literature, the exact distinction between these two categories depends on how concentrations are measured and expressed. For biogeochemical prospecting the distinction is not important. As long as the BAC does not change as a function of soil concentration, the interpretation is straightforward. The principal problem with low BACs is that the change in tissue concentration between background and enriched areas will be small and therefore difficult to determine. However, if the BAC changes with soil concentration, the nature of the change also must be known before plant analysis data can be interpreted.

There is another complicating factor. Because of the interdisciplinary nature of biogeochemistry, tissue concentrations are not always calculated on the same basis. Plant physiologists and agronomists will generally use dry weight, whereas geochemists often use ash weight as the basis for calculating tissue concentration. Because of cellulose and storage materials in many plant tissues, ash weights range from around a low of 0.1 percent in some woody tissues to a high of 10 percent in some leaves. Thus the use of these two bases of reference can cause an apparent difference in calculated tissue concentration of a factor of 10 or more. In addition, the methods of characterization of soil metal concentration are not standardized. This must also be dealt with when utilizing information from the literature.

Excluder Plants

Plants that respond as excluders of an element tend to have a very low concentration of the excluded element in their tissue, even in the presence of high soil concentrations (Foster, 1977). These plants have a biological mechanism that actively excludes the element, but in many plants the mechanism may function only over a limited range of elemental concentrations. Thus plants collected from different soils with different elemental concentrations may show different metal accumulation characteristics. The generally low tissue concentration associated with plant excluders does not in itself make this type of plant metal system unsuited for exploration use, but does impose a very severe challenge to any tissue analysis procedure, because very small differences must be determined, which means that extreme care must be taken in both sampling and analysis. For most cases the analytical problem could prove too great a problem for the practical use of excluders when other options are available.

Passive Uptake

The plants responding in the passive mode of uptake for a given element generally are very suitable for biogeochemical exploration. Many plants show this relationship over a broad concentration range for many of the nonessential trace metals. In plants functioning in this manner, the element of concern is passively taken up along with the water that the plant utilizes in transpiration. Tissue concentration of these plants usually bears a simple relationship to soil concentration and is often uniform over a wide range of concentrations. The passive mode of uptake may continue even into the toxic range of metal concentration.

Uptake characteristics of plants unfortunately cannot be generalized for all elements; some plants will accumulate element x and exclude element y, while other plants will have a completely different profile of elemental accumulation (Foy et al., 1973). There are few generalizations that can be made about which plant will take up which element. The major generalization may be that for any element there will be some plant somewhere that will accumulate it to very high concentrations (Peterson, 1971; Peterson and Girling, 1981; Brooks et al., 1974; Brooks, 1977).

Essential Elements

The uptake characteristics of the essential plant nutrients change as a function of the nutrient status of the plant (Berry and Ulrich, 1979). The essential elements of most direct value biogeochemically are B, Cu, Zn, and Mo. However, the nutrient status of other essential nutrients, which may be of limited value in relation to mining, can provide valuable information about the geologic province in which the plants are growing (Goldschmidt, 1954). When the essential nutrient trace metals are deficient for plant growth, they are actively accumulated by the plant (Ulrich and Berry, 1961). Once the tissue concentration has reached the critical concentration required for normal biological functioning, the uptake rate is reduced under homeostatic control and the plant in effect becomes an excluder in regard to additional increases in soil concentration. When the tissue concentration reaches the toxic threshold, metabolic control over uptake is lost and uptake increases as a function of external concentration (Figs. 1-2 and 1-3). In this high concentration range (toxic), the uptake of such essential nutrients as Zn and Cu is essentially passive; hence they may be very amenable for use in biogeochemical prospecting in their toxic but sublethal concentration range (Timperley et al., 1970).

Nonessential Elements

The uptake of the nonessential elements is also influenced by the nutrient status of the plants. However, for the nonessential elements only two nutrient states exist, tolerance and toxicity (see Fig. 1-1). Because these elements are nonessential, there is not necessarily an inherent biological mechanism for accumulation. However, in some cases the chemistry of the nonessential elements is sufficiently similar to that of an essential element that they are absorbed together with the essential element. Examples of this are Sr for Ca, Rb for K, and As for P. Most nonessential elements tend to be taken up passively or excluded to some degree. At low concentrations there is often a nonmetabolic accumulation by absorption of charged elements on exchange sites both on and in plant tissues. This can give the appearance of active accumulation but is of limited degree and not related to metabolic accumulation. Once the tissue concentration has reached the toxic threshold, biological control, if present, is reduced and uptake becomes passive. Passive uptake induced by toxicity will be modified to the degree that the roots lose integrity as a function of toxicity.

Plant Age

As with most living organisms the age of a plant can be described either in chronological or physiological terms. The nutrient status of the plant is related to physiological age rather than chronological age. The plant, as a whole, passes through a series of developmental phases. Within each phase each of its tissues has its own independent physiological age (Woodwell, 1974). The physiological age of the tissue determines its metabolic requirements and therefore its nutritional needs in relationship to the rest of the plant. Young plants may have only young leaves; however, old plants will have young, mature and old leaves. Certain plants with an indeterminate growth pattern (where the growing point grows continuously during the growth season) will have leaves of all ages at all times. Other plants which have only a flush of growth during the spring will have young leaves in the spring and no old leaves. This situation will be reversed in the fall when the plant that produces growth flushes will have only old leaves. Plants with the flush type of growth can be sampled only at comparable times in term of flush growth (i.e., the same time each year if it is desired that the results be comparable with tissue sampled in other years). Plants with an indeterminate growth pattern can be sampled at any time of the year when they are actively growing. Young plants and plants early in the growing season generally have a higher nutrient content than later in the season. However, there are certain elements, such as Ca, which continue to increase in leaf tissues as a function of chronological age (Loneragan and Snowball, 1969).

Plant Tissues

Each plant tissue has a characteristic mineral profile. For the purposes of biogeochemical prospecting these tissues can be grouped into a relatively small number of broad divisions such as the fibrous roots, stems, bark, petioles, leaves, and fruits or other storage organs. For many of the common nonessential minerals and for Zn and Cu under nondeficient conditions, the relative tissue concentration follows the following order: roots $>>>$ stems $>$ petioles $>$ leaves $>$ fruits (Chapman, 1966; Tiffin, 1977). The major concentration difference for metals is between roots and stems, where the difference is maybe an order of magnitude, while the difference between petioles and leaves is something closer to a factor of 2 (Berry, 1971). This relative order of tissue concentration can be altered when a metal-tolerant plant partitions stressful concentrations of a metal into a specific tissue such as bark. In this example, analysis of the bark would indicate a passive type of metal accumulation whenever the metal reaches stressful levels, but it would be the result of an active tolerance mechanism rather than passive uptake. The tissue is in effect functioning as a sink for excess metal; hence the concentration will be highly time dependent. This tolerance response can significantly increase the concentration range in which the plant can grow and thus its usefulness for biogeochemical prospecting. Elements such as B, Mo, and Ca are concentrated in the leaves rather than the roots. In the case of an elevated supply of B, its concentration is further biomagnified in the margins of the leaf, where concentrations as high as 3000 ppm can occur in comparison to petiole concentrations of only 30 ppm (Oertli, 1960). Many elements that move in the transpiration stream as uncharged molecules are commonly concentrated in the leaves. For many metals, the high metal concentration in the small fibrous roots would make them the most desirable tissue to sample; however, for roots there are some major sampling difficulties. Not only are roots hard to extract from the soil, it is often very difficult even to tell from which plant they have originated. Once the roots are extracted from the soil, it is very difficult to wash them free of adhering and absorbed soil and nutrients. The ease of sampling of both the leaf and stem tissue usually more than makes up for any added analytical problems as a result of their lower tissue concentration.

Chronological Tissue Age

In addition to physiological differences between tissues and tissues of different physiological age there is also a difference in the nutrient content of tissues as a function of chronological age (Morrison, 1972). The kind of difference associated with chronological age is strongly correlated with the difference in the function of the elements and, in the case of nonessential elements, their chemistry. Certain elements, such as Ca, Si, and Fe, increase in leaf tissue as a function of age. These are the elements that are relatively nonmobile in the plant and are often related to plant structure. Other elements, such as K, N, and S, whose functions are more closely related to metabolism are more subject to retranslocation within the plant, especially during active growth, when they are moved into new growth, or during senescence, when they are removed from the senescent tissue, and therefore have little tendency to build up in the tissue as a function of time.

Growth Rate

The growth rate of the plants will have an important effect on the nutrient content of the plant. Plants whose growth rate has been limited by some environmental factor other than nutritional deficiency will continue to absorb nutrients and the overall tissue concentration will increase to high levels. Thus, if a plant is limited in growth because of some factor such as low light it will have a higher tissue metal concentration than similar plants growing rapidly in the sun on the same soil provided that there is no change in the degree of succulence (Jarrell and Beverly, 1981).

Nutrient Deficiencies

The growth rate of plants under natural conditions is often limited by the deficiency of one of the essential nutrients or other environmental conditions. There are known examples of natural deficiencies of all the essential mineral nutrients with the exception of Cl. Deficiencies of N, Fe, and the trace elements are especially common in agronomic plants. and N is chronically deficient in all well-weathered natural ecosystems. Plants growing under a particular deficiency will appear enriched in the other mineral nutrients. Plants growing under these conditions can be utilized in biogeochemical work, provided that the same limiting condition is maintained throughout the entire survey. In plants growing under favorable conditions there are reciprocal relationships between certain elements such that if one is present at low concentration, the other will be higher than expected for that environment. These elemental pairs are Fe-Mn, Cu-Zn, Zn-Cd, Mo-W, Ca-Mg, Na-K, and most heavy metals-Fe (Berry and Ulrich, 1979). Not only are certain pairs of elements found associated in plants, but certain elements, such as Au and As are geochemically associated (Warren *et al.*, 1964, 1968). Only one element of such a pair needs to be amenable to plant analysis to provide a feasible biogeochemical prospecting system. Geochemical parameters such as pH, Eh, cation exchange complex, and salinity are all important in determining the availability of soil metals to plants (Lindsay, 1979; Loneragan, 1975; Hodgson, 1961). The plant can utilize only that fraction of soil nutrients which are available. Under most circumstances availability increases with increased solubility. If a biogeochemical survey crosses a contact zone where there is a shift in soil pH, the heavy metal concentration of the plant tissues would be expected to be higher on the low-pH side of the contact zone if the soil metal concentration remained the same. Any biogeochemical survey utilizing plant analysis must take these chemical functions into account.

Overview of Factors Affecting Tissue Concentration: the Tissue Concentration–Dose Response Curve

Full use of plant analysis as a biogeochemical prospecting tool can be made only if the evaluation is made at the level of a plant's reaction to its environment. The plant's basic growth response to increasing concentrations of an element is described diagrammatically by the yield–dose response curve shown in Figure 1-1. The type of response shown by the plant is determined by the nutrient status of the plant, which, in turn, is governed by tissue concentration. There are three nutrient states corresponding to the three zones: (1) deficiency, (2) tolerance, and (3) toxicity. All three zones are expressed for the essential nutrients, with only the latter two expressed for the nonessential elements with the tolerance zone extending down to zero concentration. Tissue concentrations of all elements are a product of the plant's response to available elemental concentrations. Tissue concentration also can be generalized and diagrammed using the same generalized nutrient status format used for the yield–dose response curve. In Figure 1-4 a generalized tissue concentration–dose response curve is shown with the three zones—deficiency, tolerance, and toxicity—indicated.

The plant response to the essential nutrients is governed by the nutrient status of the plant. In the deficient zone, tissue concentrations remain constant at the minimal tissue level, for as more nutrients become available, more tissue is produced. When the supply of the nutrient exceeds that required for growth, tissue concentration increases and the nutrient status shifts to that of the tolerant zone, where tissue concentration increases slowly under homeostasis control, a metabolic feedback mechanism for limiting change in tissue concentration. When the external concentration reaches the toxic threshold, metabolic control over elemental uptake is lost, which results in passive uptake for most elements. Passive uptake refers to uptake that is coincidental with water use. Thus, if growth is limited and water use remains high, tissue concentration can become large in relation to soil concentration.

In the generalized tissue concentration–dose response curve, the plant responds to the nonessential elements as if there were only two nutrient states, tolerance and toxic. For these elements, the tolerance zone extends to zero concentration. However, because of the diverse chemistry of the nonessential elements, the generalizations that can be made are more limited and less precise. In the case of elements, such as the heavy metals, which are strongly adsorbed by plant tissues, large amounts can be adsorbed by the roots before much is translocated to the tops. Elements that form noncharged soluble complexes with plant material, such as B and Mo, will commonly concentrate in the leaves after moving in the transpiration stream. There are also certain elements, such as Ba, Sr, W, As, Cd, and Rb, whose distribution parallels that of chemically similar essential elements. The big advantage of nonessential elements for biogeochemical prospecting is that

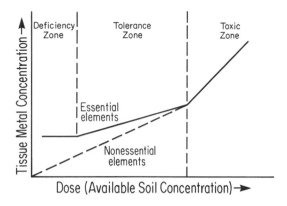

Fig. 1-4   Generalized tissue concentration-response curve. This curve represents metabolically active tissue, that is, nonstructural and nonstorage tissue of plants that do not have active tolerance mechanisms. In plants possessing active tolerance mechanisms, the pool of an element that is metabolically active may be only a fraction of the total tissue concentration. Diagram not to scale, thus shows relationships only; the degree of expression of each zone is a function of the specific element being considered.

Deficient Zone
1. Curve for essential elements does not go to zero; in all plant tissue there is minimum tissue concentration.
2. Deficient zone not applicable to nonessential elements.

Tolerance Zone
1. Tissue concentration of essential elements limited through homeostasis control.
2. Tissue concentration increases at a slower rate than available supply.
3. In the case of nonessential elements, the tolerance zone and the curve both extend to zero.

Toxic Zone
1. Passive uptake, metabolic control limited by toxicity.
2. Plants that are excluders can have their roots killed by toxic concentrations before their tops show a concentration buildup.

from a low background to toxicity their uptake is generally described by only one function rather than by the two functions normally found for the essential elements. Thus it is simpler to describe tissue concentration of an individual nonessential element but more difficult to make generalizations about them as a group. The generalized curve depicts the plant's response to increased elemental concentration assuming normal metabolism and the presence of no specific tolerance mechanism. As discussed earlier, many of the tolerance strategies involve partitioning toxic elements into tissues such as wood and bark which have limited metabolic functions. In such a scenario, these tissues will contain relatively high concentrations of an element because these plants have a tolerance mechanism rather than the normal metabolism considered in the generalized curve. If a second scenario is considered, where the tolerance is the result of exclusion in the roots, all top tissues will have very limited concentration.

ADDITIONAL IMPLICATION FOR BIOGEOCHEMICAL EXPLORATION

Use of Tissue–Soil Concentration Ratios

When tissue-metal concentration is essentially proportional to soil concentration, the tissue is considered to have no barrier to uptake and has been termed a *barrier-free tissue* (Kovaleskii, 1979; Brooks, Chapter 6, this volume). When tissue concentration is less than proportional to soil concentration over at least part of the soil concentration range, the tissue is termed a *barrier tissue*. From Figure 1-4 and the discussion above it would appear that the barrier category usually includes some tissues other than, or in addition to, metabolically active tissue. Such barrier tissues can include, for example, bark, woody structural tissue, fruits, seeds, and other storage tissues. Differential partitioning of elements between barrier and nonbarrier tissues due to tolerance mechanisms, adsorption of elements within tissues, and changes in availability of elements in the soil with changing soil concentrations can alter the ratio of tissue concentration to soil concentration (i.e., barrier–nonbarrier relationships). Thus barrier-type relationships are empirical and operationally defined. Tissue assignment to specific barrier categories can be influenced by variation in both biological and site-specific parameters. Prospecting criteria developed from a program based on the barrier concept could be useful for the specific project where it was developed but not have a high probability of being useful when applied to another project in a different ecosystem or location.

The usefulness of the interpretation of tissue analysis data from a series of plant samples is increased as more information is known about the soil–plant system. When both the physiology of the plant and soil conditions are known, multielement tissue analysis is a very powerful tool for biogeochemical prospecting. However, as with most exploration tools, plant analysis, even with limited data for interpretations, can add valuable knowledge to the overall program. Biogeochemistry is not intended to be a self-contained exploration method, but rather a tool to be used in conjunction with other information developed within a comprehensive exploration program.

ACKNOWLEDGMENTS

Although the research described in this article has been funded in part by the U.S. Environmental Protection Agency under assistance agreement number R809578-01-1, it has not been subjected to the agency's prescribed peer and administration review and therefore does not necessarily reflect the view of the agency and no official endorsement should be inferred. It has, however, been subjected to peer review for the publication process.

REFERENCES

Allaway, W. H., 1968, Agronomic controls over the environmental cycling of trace elements: *Adv. Agron.*, v. 20, p. 235–274.

Antonovics, J., Bradshaw, A. D., and Turner, R. G., 1971, Heavy metal tolerance in plants: *Adv. Ecol. Res.* v. 7, p. 1–85.

Asher, C. J., and Edwards, D. G., 1978, Critical external concentrations for nutrient deficiency and excess: *Proc. 8th Int. Colloq. Plant Anal. Fertil. Prob.*, Auckland, New Zealand, Aug. 28–Sept. 1, 1978. DSIR Inf. Series No. 134, p. 13–28.

Bates, T. E., 1971, Factors affecting critical nutrient concentrations in plants and their evaluation: a review: *Soil Sci.* v. 112, p. 116–130.

Beckett, P. H. T., and Davis, R. D., 1977, Upper critical levels of toxic elements in plants: *New Phytol.*, v. 79, p. 95–106.

Berry, W. L., 1971. The nutrient status of zinc in lettuce evaluated by plant analysis: *J. Am. Soc. Hort. Sci.*, v. 96, p. 412–414.

———, 1977, Dose-response curves for lettuce subjected to acute toxic levels of copper and zinc. *in Biological Implications of Metals in the Environment:* Proc. 15th Annu. Hanford Life Sci. Symp., Richland, Wash., Sept. 29–Oct. 1, 1975, TIC CONF-750929.

———, 1978, Comparative toxicity of VO_3, CrO^{2-}, Mn^{2+}, Co^{2+}, Ni^{2+}, Cu^{2+}, Zn^{2+}, and Cd^{2+} to lettuce seedlings, *in Environmental Chemistry and Cycling Processes:* Proc. Symp., Augusta, Ga., Apr. 28–May 1, 1976. TIC CONF-760429, p. 582–589.

———, and Ulrich, A., 1979, Interactions of environmental factors, p. 369–379 in Tibbitts, T. W. (Ed.), *Controlled Environment Guidelines for Plant Research:* University of Wisconsin, Madison, Wis.

———, and Wallace, A., 1981, Toxicity: the concept and relationship to the dose response curve: *J. Plant Nutr.*, v. 3, p. 13–19.

Bowen, H. J. M., 1966, *Trace Elements in Biochemistry:* Academic Press, London, New York, 241 p.

Brooks, R. R., 1972, *Geobotany and Biogeochemistry in Mineral Exploration:* Harper & Row, New York, 292 p.

———, 1977, Copper and cobalt uptake by *Haumaniastrum* species: *Plant Soil*, v. 48, p. 541–545.

———, 1979. Indicator plants for mineral prospecting–a critique: *J. Geochem. Explor*, v. 12, p. 67–78.

———, Lee, J., and Jaffré, T., 1974, Some New Zealand and New Caledonian plant accumulators of nickel: *J. Ecol.*, v. 62, p. 493–499.

Brown, J. C., Ambler, J. E., Chaney, R. L., and Foy, C. D., 1972. Differential responses of plant geotypes to micronutrients, p. 389–418 in Mortvedt, J. J., Giordano, P. M., and Lindsay, W. L. (Eds.), *Micronutrients in Agriculture:* Soil Sci. Soc. Am., Madison, Wis.

Cannon, H. L., 1960, Botanical prospecting for ore deposits: *Science*, v. 132, p. 591–598.

———, 1971. Use of plant indicators in ground water surveys, geologic mapping, and mineral prospecting: *Taxon*, v. 20, p. 227–256.

Carlisle, D., and Cleveland, G. B., 1958, Plants as a guide to mineralization: *Calif. State Div. Mines, Spec. Rep. 50.*

Chaney, R. L., White, M. C., and van Tienhoven, M., 1976. Interaction of Cd and Zn in phytotoxicity and uptake by soybean: *Agron. Abstr.*, 1976, p. 21.

Chapin, F. Stuart, III, 1980, The mineral nutrition of wild plants: *Annu. Rev. Ecol. Syst.* v. 11, p. 233–260.

Chapman, H. D., 1966, *Diagnostic Criteria for Plants and Soils:* University of California Division of Agricultural Science, Riverside, Calif., 793 p.

Clark, R. B., 1983, Plant genotype differences and the uptake, translocation, accumulation, and use of mineral elements required for plant growth: *Plant Soil,* v. 72, p. 175–196.

Clarkson, D. T., and Hanson, J. B., 1980, The mineral nutrition of higher plants: *Annu. Rev. Plant Physiol.,* v. 31, p. 239–298.

Collins, F. W., Cunningham, L. M., and Hutchinson, T. C., 1976, Physiological and biochemical aspects of cadmium toxicity in soybean. Part II. Toxicity, bioaccumulation, and subcellular fractionation of cadmium in soybean grown at subchronic to acute cadmium levels, *in* Hemphill, D. D. (Ed.), *Trace Substances in Environmental Health,* Vol. 10: University of Missouri, Columbia, Mo. p. 145–166.

Cunningham, J. D., Keeney, D. R., and Ryan, J. A., 1975, Phytotoxicity and uptake of metals added to soil as inorganic salts or in sewage sludge: *J. Environ. Qual.,* v. 4, p. 460–462.

Cunningham, L. M., Collins, F. W., and Hutchinson, T. C., 1976, Physiological and biochemical aspects of cadmium toxicity in soybean. Part I. Toxicity symptoms and autoradiographic distribution of cadmium in roots, stems, and leaves: *Proc. Int. Conf. Heavy Met. Environ.,* v. 2, p. 97–120.

Davis, R. D., and Beckett, P. H. T., 1978, Upper critical levels of toxic elements in plants. II. Critical levels of copper in young barley, wheat, rape, lettuce and ryegrass, and of nickel and zinc in young barley and ryegrass: *New Phytol.,* v. 80, p. 23–32.

Ernst, W. H. O., 1976, Physiology of heavy metal resistance in plants: *Proc. Int. Conf. Heavy Met. Environ.,* v. 2, p. 121–136.

Fortescue, J. A. C., 1979, *Environmental Geochemistry, a Holistic Approach:* Springer-Verlag, New York, 347 p.

Foster, P. L., 1977, Copper exclusion as a mechanism of heavy metal tolerance in a green alga: *Nature,* v. 269, p. 322–323.

Foy, C. D., Fleming, A. L., and Schwartz, J. W., 1973, Opposite aluminum and manganese tolerances in two wheat varieties: *Agron. J.,* v. 65, p. 123–126.

——, Chaney, R. L., and White, M. C., 1978, The physiology of metal toxicity in plants: *Annu. Rev. Plant Physiol.,* v. 29, p. 511–566.

Furr, A. K., Kelly, W. C., Bache, C. A., Gutenmann, W., II, and Lisk, D. J., 1976, Multi-element absorption by crops grown in pots on municipal sludge-amended soil: *J. Agric. Food Chem.,* v. 24, p. 889–892.

Gerloff, G. C., 1963, Comparative mineral nutrition of plants: *Annu. Rev. Plant Physiol.,* v. 14, p. 107–124.

Goldschmidt, U. M., 1954, *Geochemistry:* Clarendon Press, Oxford, p. 730.

Hecht-Buchholz, C. H., 1983, Light and electron microscope investigations of the reactions of various genotypes to nutritional disorders: *Plant Soil,* v. 72, p. 151–165.

Hodgson, J. F., 1961, Chemistry of the micronutrient elements in soil: *Adv. Agron.,* v. 15, p. 119–159.

Jarrell, W. M., and Beverly, R. B., 1981, The dilution effect in plant nutrition studies: *Adv. Agron.,* v. 34, p. 197–224.

Jenny, H., 1980, *The Soil Resource, Origin and Behavior:* Springer-Verlag, New York, 377 p.

Jowett, D., 1959, Adaptation of a lead tolerant population of *Agrostis tenuis* to low soil fertility: *Nature (Lond),* v. 184, p. 43.

———, 1964, Population studies on lead tolerant *Agrostis tenuis: Evolution,* v. 18, p. 70–80.

Kovalevskii, A. L., 1979, *Biogeochemical Exploration for Mineral Deposits:* Amerind, New Delhi, 136 p.

Kruckeberg, A. R., 1954, The ecology of serpentine soils. III. Plant species in relation to serpentine soils: *Ecology,* v. 35, p. 267–274.

Landsberg, F-Ch., 1981, Organic acid synthesis and release of hydrogen ions in response to Fe deficiency stress of mono- and dicotyledonous plant species: *J. Plant Nutr.,* v. 3, p. 579–591.

Lindsay, W. L., 1979, *Chemical Equilibria in Soils:* Wiley-Interscience, New York.

Lisk, D. J., 1972, Trace metals in soils, plants, and animals: *Adv. Agron.,* v. 24, p. 267–325.

Loneragan, J. F., 1975, The availability and absorption of trace elements in soil-plant systems and their relation to movement and concentrations of trace elements in plants, p. 109–134 *in* Nicholas, D. J. D., and Egan, A. R. (Eds.), *Trace Elements in Soil-Plant-Animal Systems:* Academic Press, New York.

———, and Snowball, K., 1969, Calcium requirements of plants: *Aust. J. Agric. Res.,* v. 20, p. 465–478.

Malaisse, F., Gregoire, J., Brooks, R. R., Morrison, R. S., and Reeves, R. D., 1978, *Aeolanthus biformifolius:* a hyperaccumulator of copper from Zaire: *Science,* v. 199, p. 887–888.

Malyuga, D. P., 1964, *Biogeochemical Methods of Prospecting:* Consultants Bureau, New York, 205 p.

Mason, H. L., and Stout, P. R., 1954, The role of plant physiology in plant geography: *Annu. Rev. Plant Physiol.,* v. 5, p. 249–270.

Miller, J. E., Hassett, J. J., and Koeppe, D. E., 1977, Interactions of lead and cadmium on metal uptake and growth of corn plants: *J. Environ. Qual.,* v. 6, p. 18–20.

Morrison, I. K., 1972, Variation with crown position and leaf age in content of seven elements in leaves of *Pinus banksiana* Lamb: *Can. J. For. Res.,* v. 2, p. 89–94.

Nicolls, O. W., Provan, D. M. J., Cole, M. M., and Tooms, J. S., 1965, Geobotany and geochemistry in mineral exploration in the Dugald River Area, Cloncurry District, Australia: *Trans. Inst. Min. Metall.,* v. 74, p. 695–799.

Oertli, J. J., 1960, The distribution of normal and toxic amounts of boron in leaves of rough lemon: *Agron. J.,* v. 52, p. 530–532.

Peterson, P. J., 1971, Unusual accumulations of elements by plants and animals: *Sci. Prog.* (Oxford), v. 59, p. 505–526.

———, 1976, Element accumulation by plants and their tolerance of toxic mineral soils: *Proc. Int. Conf. Heavy Met. Environ.,* v. 2, p. 39–54.

———, and Girling, C. A., 1981, Other trace metals, p. 213–278 *in* Lepp, H. W. (Ed.), *Effects of Heavy Metal Pollution on Plants,* Vol. 1: Applied Science Publishers, Barking, Essex, England.

Rosenfeld, I., and Beath, O. A., 1965. *Selenium: Geobotany, Biochemistry, Toxicity and Nutrition:* Academic Press, New York.

Severne, B. C., 1974, Nickel accumulation by *Hybanthus floribundus: Nature (Lond),* v. 248, p. 807–808.

Snaydon, R. W., and Bradshaw, A. D., 1962, Differences between natural populations of *Trifolium repens* in response to mineral nutrients. I. Phosphate: *J. Exp. Bot.,* v. 13, p. 422–434.

Tiffin, L. O., 1977, The form and distribution of metals in plants: An overview,

p. 315–334 in *Biological Implications of Metals in the Environment:* Proc. 15th Annu. Hanford Life Sci. Symp., Sept. 29–Oct. 1, 1975. ERDA-TIC-CONF-750929. Oak Ridge, Tenn.

Timperley, M. H., Brooks, R. R., and Peterson, P. J., 1970, The significance of essential and non-essential trace elements in plants in relation to biogeochemical prospecting: *J. Appl. Ecol.*, vo. 7, p. 429–439.

Turner, R. G., and Marshall, C., 1972, The accumulation of zinc by subcellular fractions of roots of *Agrostis tenuis* Sibth. in relation to zinc tolerance: *New Phytol.*, v. 72, p. 671–676.

Ulrich, A., 1952, Physiological bases for assessing the nutritional requirements of plants: *Annu. Rev. Plant Physiol.*, v. 3, p. 207–228.

——, and Berry, W. L., 1961, Critical phosphorus levels for lima bean growth: *Plant Physiol.*, v. 36, p. 626–632.

van den Driessche, R., 1974, Prediction of mineral nutrient status of trees by foliar analysis: *Bot. Rev.*, v. 40, p. 347–394.

Vinogradov, A. P., 1959, *Geochemistry of Rare and Dispersed Chemical Elements in Soils* (translated from the Russian): Consultants Bureau, New York, 209 p.

Wainwright, S. J., and Woolhouse, H. W., 1976, Physiological mechanisms of heavy metal tolerance, in Chadwicks, M. J., and Goodman, G. T. (Eds.), *The Ecology of Resource Degradation and Renewal*, Br. Ecol. Soc. Symp., Vol. 15: Blackwell, Oxford, p. 231–257.

Walker, R. B., 1954, The ecology of serpentine soils. II. Factors affecting plant growth on serpentine soils: *Ecology*, v. 35, p. 259–266.

Wallace, A., 1980, Trace metal placement in soil on metal uptake and phytotoxicity: *J. Plant Nutr.*, v. 2, p. 35–38.

——, El-Gazzar, A., and Soufi, S. M., 1968, The role of calcium as a micronutrient and its relationship to other micronutrients: *Proc. 9th Int. Congr. Soil Sci. Adelaide*, v. 2, p. 357–366.

——, Romney, E. M., Alexander, G. V., Soufi, S. M., and Patel, P. M., 1977, Some interactions in plants among cadmium, other heavy metals, and chelating agents: *Agron. J.*, v. 69, p. 18–20.

Warren, H. V., and Delavault, R. E., 1948, Biochemical investigations in British Columbia: *Geophysics*, v. 13, p. 609–624.

——, Delavault, R. E., and Barakso, J., 1964, The role of arsenic as a pathfinder in biogeochemical prospecting: *Econ. Geol.*, v. 59, p. 1381–1389.

——, Delavault, R. E., and Barakso, J., 1968, The arsenic content of Douglas fir as a guide to some gold, silver and base metal deposits: *Can. Min. Metall. Bull. 61*, p. 860–866.

Webb, J. S. and Millman, A. P., 1951, Heavy metals in vegetation as a guide to ore: *Trans. Inst. Min. Metall.*, v. 60, p. 473–504.

Woodwell, G. M., 1974, Variation in the nutrient content of leaves of *Quercus alba, Quercus coccinea,* and *Pinus rigida* in the Brookhaven forest from budbreak to abscission: *Am. J. Bot.*, v. 61, p. 749–753.

Wu, L., Bradshaw, A. D., and Thurman, D. A., 1975, The potential for evolution of heavy metal tolerance in plants. III. The rapid evolution of copper tolerance in *Agrostis stolonifera: Heredity*, v. 34, p. 165–187.

Larry P. Gough
U.S. Geological Survey
Denver Federal Center
Denver, Colorado

2

VEGETATION ZONATION, METAL STRESS, AND MINERAL EXPLORATION

ABSTRACT

Causative factors that influence vegetation zonation are presented with special emphasis on the physiological ecology of metal stress. The tolerance of plants to environmental stress (including high levels of available toxic metals) is usually quite broad. Where anomalously high levels of a metal occur in the substrate, sensitive species are generally absent from the community because of their inability to compete with metal-tolerant species; only rarely are the classic laboratory physiological symptoms of metal stress observed in the field. Geobotanists have shown that indicator species and morphologically distinguishable ecotypes of species can be beneficial in the delineation of metalliferous areas. There are many environmental and biological factors, however, that can limit the occurrence of an indicator species; its absence from a community does not necessarily mean the absence of the target element. The "geochemical ecologist" may find it more advantageous to map the abrupt disappearance in a community or an otherwise commonly encountered species.

Metal stress causes abrupt changes in plant communities, but other physical and chemical features of soils can also cause such changes. This chapter emphasizes caution when interpreting plant community structure and gives as an example the multifaceted and complex nature of vegetation zonation at Eagle Bluff, Alaska. Although metal stress influences this community, the very unusual vegetation results from a combination of factors: absence of glaciation, soils that are serpentine in character, southwest exposure, unstable substrate, and extreme fluctuations in soil moisture and temperature.

INTRODUCTION

Vegetation zonation, on both local and regional scales, is easily observed and has long been used for practical purposes. Eighteenth- and nineteenth-century settlers in the midwestern United States knew that certain calciphilous vegetation types indicated fertile soils and they located their farms accordingly. Conversely, they avoided areas whose vegetation indicated infertile soils (usually a consequence of low base-exchange capacity). In semiarid portions of the Great Plains, halophytic vegetation delineated infertile zones in otherwise prime dryland wheat-growing soils. Cannon (1979) discusses the history of the use of plant communities in the delineation of metalliferous soils and notes that its importance was recognized more than 300 years ago. The problem for the geologist interested in mineral exploration is in the correct interpretation of plant communities as they may relate to substrate chemistry. Such an oblique or indirect approach to resource discovery and appraisal is justified only if community structure can be accurately interpreted as being associated with metal stress.

The causes of vegetation zonation are most often multifaceted (Jenny, 1980) and include climate, aspect (topography), soil physical and chemical properties, biota, and the presence or absence of anthropogenic and natural disturbances. Figure 2-1 illustrates a hypothetical vegetation zonation pattern in Alaska that can be explained in five different ways: Figure 2-1A shows that the change from a sedge meadow to a spruce forest to an aspen forest (right to left in the figure) occurs because of the contact between lake bed sediments and glacial till; in Figure 2-1B the change is a response to differences in water table depth (geology unchanged); in Figure 2-1C it is caused by the distribution of permafrost; vegetation zonation in Figure 2-1D reflects the occurrence of different

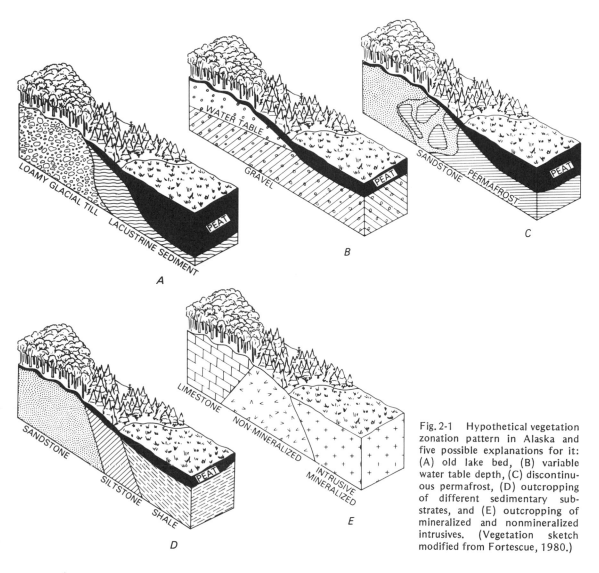

Fig. 2-1 Hypothetical vegetation zonation pattern in Alaska and five possible explanations for it: (A) old lake bed, (B) variable water table depth, (C) discontinuous permafrost, (D) outcropping of different sedimentary substrates, and (E) outcropping of mineralized and nonmineralized intrusives. (Vegetation sketch modified from Fortescue, 1980.)

sedimentary outcrops, and the pattern in Figure 2-1E is a response to the occurrence of metal stress from a mineralized intrusive substrate. In reality, the example given in Figure 2-1E is an unusual occurrence. Foy *et al.* (1978, p. 535) made the following general statement:

> Although any of the heavy metals can be toxic to plants at some level of solubility only a few have been generally observed to cause phytotoxicity in soils. In soils, most heavy metals occur as inorganic compounds or are bound to organic matter, clays, or hydrous oxides of Fe, Mn, and Al. Because of this precipitation and sorption of most metals by soils, only Zn, Cu, and Ni toxicities have occurred frequently. Toxicities of Pb, Co, Be, As, and Cd occur only under very unusual conditions. Other elements may be toxic in solution cultures but are not phytotoxic in soils, even at very high levels (e.g., Cr, Ag, Sn, Ga, and Ge).

Although direct toxicity to many of the metals is unusual under field conditions, excessive levels of Cr, Co, Cu, Mn, Ni, and Zn have been shown to inhibit Fe assimilation and cause chronic Fe deficiency (Brooks, 1983). Induced deficiency is therefore an indirect form of metal stress that affects plant vigor.

RESPONSE OF SPECIES TO METAL STRESS

Many plant species are capable of tolerating broad fluctuations in environmental parameters. When the tolerance of a particular parameter is especially great and apparently genetically fixed, we speak of the occurrence within the species of an *ecotype*. An ecotype has, therefore, limits of tolerance that are adjusted to local conditions. If, in turn, the ecotype is morphologically distinctive (taxonomically unique) and composed of an aggregate of local populations of a species, it is referred to as a *subspecies*. Much confusion exists in the geochemical ecology literature over the exact meaning and therefore the proper use of the terms *ecotype, ecospecies, coenospecies, subspecies, race,* and *variety,* but for purposes of geobotanical prospecting, the differences are unimportant. [The student of ecological etymology should consult Odum (1971) and Sukachev and Dylis (1964).] The geochemical ecologist should realize that a metal-tolerant ecotype may be morphologically indistinguishable from its metal-intolerant parent stock.

Ecotypes can also evolve which are less tolerant to a particular stress than the species as a whole. Much less frequently, adaptation to stress can lead to the development of a new species that requires the stress-producing parameter. Species of this type (called the *indicator species*) have become important mineral exploration tools (Malyuga, 1964; Cannon, 1960, 1971; Ernst, 1974; Brooks, 1979). It is not the purpose of this chapter to repeat the work of past reviews on this subject (Antonovics *et al.,* 1971; Cannon, 1979; Brooks, 1983). It must be emphasized, however, that of the 100 or so species that have been identified as being characteristic of a particular type of mineral deposit, only a very few are truly obligate and require the presence in the substrate of the mineral or of an available form of the particular metal. A recognition of the oblique relations between plant species and various metals has led to several different classifications of "metallophytes" and of their associations (Malyuga, 1964; Antonovics *et al.,* 1971; Ernst, 1974; Birse, 1982).

Figure 2-2A illustrates the general response (yield) of a plant population to an increase in the dose of either an essential or a nonessential metal. An essential metal must be in the substrate at a specific concentration before a plant will grow. In the deficiency range of the growth curve for an essential metal, yield increases as dose increases. Above the critical level for deficiency is the tolerance plateau, within which increasing the dose of a metal does nothing to the yield. Toxicity of a metal will manifest itself by a decrease in yield above some critical level. Both the position of the toxicity threshold with respect to dose and the intensity of the toxicity response (slope of the curve) will vary depending on the species (or ecotype) and the interaction of substrate parameters (Berry and Wallace, 1981; Berry, Chapter 1, this volume).

The curve that is diagrammed in Figure 2-2A is theoretical. It is the response of an individual population of a species grown under otherwise ideal greenhouse conditions. Controlled conditions like these permit observations of parameters such as decreased root

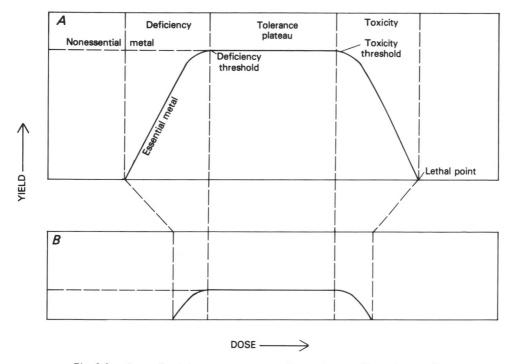

Fig. 2-2   Generalized dose-response curves illustrating the effect of the availability of an essential or a nonessential metal on plant growth under laboratory conditions (A) and in the field (B). Because of competition, the deficiency and toxicity zones for a plant population in the field (B) are more restricted than the zones for a population growing under ideal conditions in the greenhouse (A). (Part A modified from Berry and Wallace, 1981.)

growth, leaf chlorosis or necrosis, or internode stunting—all of which are manifestations of chronic and acute toxicity—in addition to simply a decrease in yield.

The controlled response described above seldom occurs under field conditions. Figure 2-2B shows that the yield of a species, relative to an increase in the dose of a metal (such as would occur along a transect that bisects a mineral deposit), remains essentially unchanged within the "tolerance plateau" part of the curve and falls off abruptly beyond either threshold. This is because as the toxicity (or deficiency) threshold is reached, an individual organism is less likely to effectively compete in a community and is therefore eliminated from the community. A narrow zone may exist immediately below the deficiency threshold and immediately above the toxicity threshold within which certain individuals of a species may be found, but, in general, one simply will not find a given species in a community unless the substrate possesses metal levels that are adequate for normal growth, development, and reproduction.

Simply stated, the phytotoxic response observed in the laboratory, or under controlled agronomic conditions in cultivated plots, probably will not be of much assistance to the geochemical ecologist in the field. Classic toxicity symptoms indicating metal stress can occur in the field (Brooks, 1983) but commonly occur as the result of some unusual physical or chemical event. For example, chlorosis of leaves of a deciduous tree species could theoretically occur following a decrease in soil Eh and pH resulting from waterlogging, which, in turn, would mobilize divalent metal ions in soils anomalously high in total metal but not high in available metal (see, e.g., Fraser, 1961). The very presence of a mineralized area implies that at this point in geologic time, the constituent metals are immobile. Antonovics *et al.* (1971, p. 49) state: "The majority of naturally occurring metallic ores are usually found as mixtures of the sulphides which are virtually insoluble: thus the amounts of metal available to plant life are very small."

RESPONSE OF COMMUNITIES TO METAL STRESS

The dose-response relationship depicted in Figure 2-2B indicates that metal stress will result in the discontinuous distribution of species if the metabolic tolerance limits for a metal are exceeded. The idealized "bull's-eye" appearance of a plant community growing over a toxic mineralized zone is, therefore, a manifestation of several successive abrupt changes from one community to another, not the result of progressive stunting (or other gradual morphological expression) of individuals of a particular community. Figure 2-3

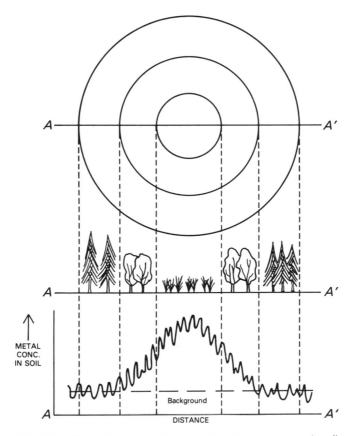

Fig. 2-3   Generalized vegetation zonation along a transect (A-A'), showing the relation between the available toxic metal concentration in soil (bottom) and abrupt changes in community structure (top).

illustrates this concept. The community structure changes along transect A-A′ in relation to a change in the concentration of some available toxic metal. A background metal concentration in soil has an associated natural variability to which particular species in a community are tolerant. As generally larger concentrations of the metal occur in the soil, the variability begins to exceed the tolerance limits of these species and the community structure changes. In extreme instances, metal toxicity can be so severe as to result in an absence of plant cover altogether (Bølviken *et al.*, 1977).

Figures 2-4 and 2-5 demonstrate vegetation zonation as it relates to the relative tolerance of individual species to high or low levels of an available metal. In Figure 2-4, species A possesses a much broader tolerance plateau than species B for a particular limiting metal and is therefore found both off and on the metal-rich outcrop. An abrupt zonation occurs because of the metal intolerance of species B. Figure 2-5 also shows the intolerance of species B for the metal-rich outcrop, but in this instance species A is truly an "indicator species": it is found exclusively on the outcrop because of its requirement for large available concentrations of the essential metal.

Although the abruptness of the changes in Figures 2-3, 2-4, and 2-5 is caused by

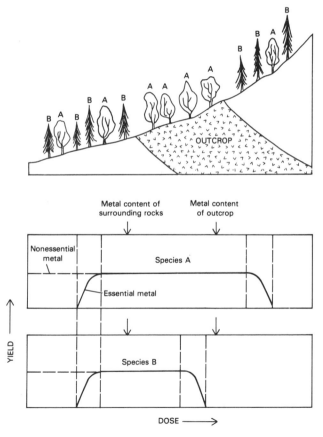

Fig. 2-4   Generalized vegetation zonation over a metalliferous outcrop where one species (A) has greater tolerance than another (B) for high available levels of an essential or nonessential metal.

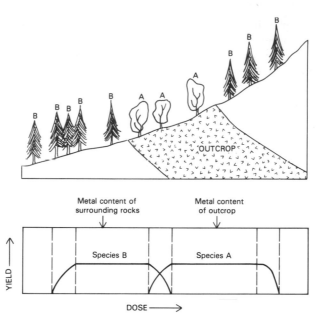

Fig. 2-5   Generalized vegetation zonation over a metalliferous outcrop where one species (A) requires high available levels of an essential metal and another species (B) cannot tolerate such high levels.

the dose-response relationship illustrated in Figure 2-2, certain environmental disruptions serve to perpetuate the community boundaries. For example, community composition tends to simplify as metal stress increases, resulting in increases in soil surface insolation, temperature, and air movement and decreases in soil moisture, structure, and organic matter. This situation in turn affects plant competition, seed germination potential, and seedling establishment. Together, these factors serve to compound the impact of metal stress on community composition.

EAGLE BLUFF, AN EXAMPLE OF THE MULTIFACETED NATURE OF VEGETATION ZONATION

Eagle Bluff is 1.5 km west of the village of Eagle, Alaska, which is on the Yukon River 10 km west of the Yukon Territory–Alaska boundary. Physiographically, Péwé (1975) described the area as part of the rolling hills and low mountains of the Yukon–Tanana Highlands that essentially escaped glaciation during both the Pleistocene and Holocene Epochs.

Eagle Bluff is an outcrop of relatively uncommon fine-grained basaltic greenstone (Foster, 1972) possibly of Precambrian age. Except for the unusually large concentrations of Ca, whose source may be limestone formations in the immediate vicinity, the residual soils at the site are "serpentine" (the term is used in its broad ecological sense,

Shacklette, 1966). Table 2-1 lists the total concentration of selected alkaline-earth and transition metals in channel samples of an Eagle Bluff soil as well as their concentrations in other Alaskan mineral soils (Usibelli, near Healy; Capps Creek, near Tyonek). The unusually large levels of Mg, Cr, and Ni in the Eagle Bluff sample are apparent. In separate studies, Clark and Foster (1969) and Foster and Yount (1971) found that the greenstone, which is the soil parent material for Shacklette's samples, also contained anomalously high levels of Cr and Ni ($\geq$ 200 and 100 ppm, respectively). Apparently unlike the soils, however, the rock samples also possessed anomalous levels of the potentially toxic metals Cu, Pb, and Zn.

TABLE 2-1   Total Concentration of Metals in Selected Mineral Soils, Alaska

	Eagle Bluff[a]	Usibelli[b]	Capps Creek[c]
Alkaline-earth metals			
Mg (%)	5.2	0.31	1.5
Ca (%)	2.1	0.68	2.7
Sr (ppm)	250	150	390
Ba (ppm)	1300	1300	450
Transition metals			
Cr (ppm)	80	47	26
Fe (%)	2.0	1.6	3.8
Co (ppm)	9.0	7.4	13
Ni (ppm)	58	26	11
Cu (ppm)	37	14	30
Zn (ppm)	42	38	64

[a]From Shacklette (1966). Arithmetic mean of three samples. Soils coarse in texture, derived from basaltic greenstone.
[b]Arithmetic mean of three samples. Soils fine in texture, derived from schist.
[c]Geometric mean of 100 samples. Soils fine sand, derived from volcanic tephra.

Close inspection of the vegetation of the southwest flank of the bluff reveals an unusual assemblage of plant species. Shacklette (1966, p. F22) gives the following interpretation:

> The development of the plant association named in this report is controlled by both physical and chemical characteristics of the site and . . . is dependent on soils derived from parent mafic and ultramafic material. Very different plant associations develop at other sites in the region that have similar physical features but that overlie schist, slate, or limestone. The combination of plant species found at Eagle Bluff appears to relate to the tolerance of the species to toxic elements ultimately derived from bedrock.

The situation at Eagle Bluff, therefore, is not unlike the one depicted in Figure 2-5.

In addition to metal stress and the direct influence it has on the community structure at Eagle Bluff, this site is an example of the compounding influence of historical, physical, and chemical factors on vegetation zonation and species composition. The site was apparently a refugium, and the absence of glaciation meant that species already established could theoretically survive and evolve in place. In addition to the site's elevation (500 to 610 m), the steep (20 to 40 percent), rocky, southern exposure and the element content of the soils have contributed to a unique community structure that includes one vascular species never before reported (*Cryptantha shackletteana*, Table 2-2) and two species not previously known from Alaska.[1]

TABLE 2-2 Unusual Members of a Vegetation Community from Eagle Bluff, Alaska[a]

Flowering Species
***Cryptantha shackletteana* L. C. Higgins
**Eriogonum flavum var. *aquilinum* J. Reveal
*Erysimum asperum var. *angustatum* (Rydb.) B. Boivin
*Montia bostockii (A. E. Porsild) S. L. Welsh
Phacelia mollis Macbr.
**Podistera yukonensis* Math. & Const.

Lichen Species
Growing on greenstone
Caloplaca flavovirescens (Wulf.) D. T. & Sarn.
Lecanora cenisia Ach.
Phaeophyscia constipata (Norrl. & Nyl.) Moberg
Growing on greenstone-derived soils
Xanthoparmelia wyomingica (Gyeln.) Hale

[a]According to the classification of Murray (1980), **means "endangered," * means "threatened."

In 1979, I visited Eagle Bluff and examined the lichen (nonvascular) vegetation. It, too, proved to be floristically unusual. One of the species, *Xanthoparmelia wyomingica* (Table 2-2), was found growing among clumps of the grass *Calamagrostis purpurescens*. This lichen is commonly found on soils and rocks in the cool deserts, plains, and mountains of Europe and North America. The presence of *Xanthoparmelia* reflects the environmental parameters characteristic of the steep, southwest-facing slope and the unstable soil conditions more than the chemical characteristics of the substrate. I have collected species of this genus at sites that have similar exposures in the Copper River valley of south-central Alaska.

Table 2-2 lists, in addition to *Xanthoparmelia*, three other unusual lichen species that are not given in the species checklists for Alaska (Thomson, 1979; Murray, 1974; Murray and Murray, 1978) and probably represent species not previously found in the

[1]Shacklette (1966, p. F23–F24) originally described *Cryptantha sobolifera, Eriogonum flavum*, and *Erysimum angustatum* as new to Alaska. A subsequent reevaluation of the taxonomy of these species (Murray, 1980) has resulted in the names *Cryptantha shackletteana, Eriogonum flavum* var. *aquilinum*, and *Erysimum asperum* var. *angustatum*, respectively. A fourth species, *Phacelia sericea*, reported by Shacklette as new to Alaska, has since been reclassified as *P. mollis* (Hultén, 1974). Although not new to Alaska, *P. mollis* is, nevertheless, an endemic of the Alaska–Yukon Territory flora.

Alaska flora. One of these, *Phaeophyscia constipata*, has occasionally been collected but not yet reported as a new addition (B. M. Murray, oral communication, 1983).

It is apparent that the unique community structure and species composition of Eagle Bluff reflects the physical and chemical characteristics of this basaltic greenstone habitat. The abrupt occurrence of vegetation zonation, like those depicted in Figure 2-3, are easily noted. Whether the species found here exclusively reflect the examples depicted in Figure 2-4 or 2-5 or whether their distribution is also governed by a combination of factors found at this site awaits additional study. The exploration geologist, however, is interested in the end result—an easily observable change in flora.

CONCLUSION

1. The absence of an indicator species (or of a metallophilic ecotype of a species) does not necessarily mean the absence of the target mineral. Even in otherwise favorable environments, a plant may not be found because of restrictions in the dissemination of its diaspores (seeds, fruits, spores, fragments, etc.).

2. The abrupt absence in a community of a commonly encountered species (Fig. 2-4, species B) usually has more significance in identifying metal stress than the abrupt occurrence of an unusual species.

3. High metal content in soil can cause very abrupt changes in plant community composition (Fig. 2-3). It should be emphasized, however, that other soil physical and chemical features (pH, Eh, salinity, etc.) can also cause easily observable zonation.

4. The example of vegetation zonation at Eagle Bluff, Alaska, illustrates the multi-faceted and complex nature of abrupt vegetation changes. The area was probably a plant refugium that escaped glaciation; except for high calcium levels, the soil is similar to serpentine in chemical composition and is derived from the weathering of an unusual basaltic greenstone; the steep, southwest-facing slope of the site results in an unstable substrate (downslope soil movement) and extremes in some environmental parameters such as surface temperature and soil moisture. Although metal stress is involved to some unknown degree in the characterization of this community, the combination of other factors, listed above, has influenced the vegetation composition of this unique community.

5. The subjective evaluation of the vegetation of an area by a competent geochemical ecologist can be a significant aid to exploration for a given type of mineral deposit.

RESEARCH NEEDS

Research in the field of vegetation zonation, as it relates to metal stress and the identification of metalliferous substrates, can be approached using both laboratory and field methods. Of the numerous areas in need of study, the following broad research categories should prove rewarding:

1. *Allelopathy.* There may be allelopathic agents that function by mobilizing otherwise unavailable toxic elements. Mechanisms may exist in a given species that increase, directly or indirectly, the stress on the establishment and growth of another species

through its ability to make available otherwise immobile metals. (For a recent review of allelopathy, see Rice, 1979.)

2. *Metal synergisms.* Not enough is known concerning the additive effect, on vegetation zonation, of various combinations of available toxic metals. For example, large concentrations of available Cu and Zn can depress Fe uptake, causing Fe-deficiency chlorosis (not, necessarily, Cu or Zn toxicosis). Laboratory studies using nutrient solutions (Wallace *et al.*, 1981) and field studies employing multiple regression and other multivariate statistical techniques (Simon, 1978; Gough *et al.*, 1980) are needed.

3. *Remote sensing.* Although problems remain in the areas of instrument sensitivity and background waveform definition, recent advances in airborne spectroradiometry (F. C. Canney and N. M. Milton, oral communication, 1982) show promise in identifying subtle spectral indicators of metal stress in vegetation (see, e.g., Raines and Canney, 1980).

4. *Metal availability and mineral deposits.* The weathering and mobilization of metals that result in secondary mineral deposits is well studied (Levinson, 1980). What are needed, however, are studies that define and generalize the processes of metal uptake by plants growing over mineralized regions, so that predictive models of metal stress can be made for specific, widespread vegetation-type communities, such as the black spruce woodlands of Alaska or the big sagebrush shrublands of the Great Basin.

5. *Indicator species.* The identification and cataloguing of plant ecotypes, species, and general plant groups that are adapted to high levels of available metals should continue. It must be emphasized, however, that the use of indicator species is strongly localized, and their utility over a broad geographic area is limited.

ACKNOWLEDGMENTS

J. L. Peard (U.S. Geological Survey), J. W. Thomson (University of Wisconsin), and R. S. Egan (University of Nebraska, Omaha) examined and verified the identification of the lichen specimens collected at Eagle Bluff, and B. M. Murray (University of Alaska) assisted in identifying their phytogeographic distribution.

REFERENCES

Antonovics, J., Bradshaw, A. D., and Turner, R. G., 1971, Heavy metal tolerance in plants: *Adv. Ecol. Res.*, v. 7, p. 1–85.

Berry, W. L., and Wallace, A., 1981, Toxicity: the concept and relationship to the dose response curve: *J. Plant Nutr.*, v. 3, p. 13–19.

Birse, E. L., 1982, Plant communities on serpentine in Scotland: *Vegetatio*, v. 49, p. 141–162.

Bølviken, B., Honey, F., Levine, S. R., Lyon, R. J. P., and Prelat, A., 1977, Detection of naturally heavy-metal-poisoned areas by LANDSAT-1 digital data: *J. Geochem. Explor.*, v. 8, p. 457–471.

Brooks, R. R., 1979, Indicator plants for mineral prospecting—a critique: *J. Geochem. Explor.*, v. 12, p. 67–78.

——— , 1983, *Biological Methods of Prospecting for Minerals:* Wiley, New York, 322 p.

Cannon, H. L., 1960, Botanical prospecting for ore deposits: *Science,* v. 132, p. 591–598.

——— , 1971, The use of plant indicators in ground water surveys, geologic mapping, and mineral prospecting: *Taxon,* v. 20, p. 227–256.

——— , 1979, Advances in botanical methods of prospecting for minerals. Part I. Advances in geobotanical methods, p. 385–395 *in* Hood, P. J. (Ed.), *Geophysics and Geochemistry in the Search for Metallic Ores:* Geol. Surv. Can., Econ. Geol. Rep. 31.

Clark, S. H. B., and Foster, H. L., 1969, Analyses of stream-sediment, rock, and soil samples from a part of the Seventymile River area, Eagle quadrangle, Alaska: *U.S. Geol. Surv. Open-File Rep.,* 129 p.

Ernst, W., 1974, *Schwermetallvegetation der Erde:* Gustav Fischer, Stuttgart, 194 p.

Fortescue, J. A. C., 1980, *Environmental Geochemistry–A Holistic Approach:* Springer-Verlag, New York, 347 p.

Foster, H. L., 1972, Preliminary geologic map of the Eagle quadrangle, Alaska: *U.S. Geol. Surv. Misc. Field Stud. Map MF-358.*

——— , and Yount, M. E., 1971, Maps showing distribution of anomalous amounts of selected elements in stream-sediment and rock samples from the Eagle quadrangle, east-central Alaska: *U.S. Geol. Surv. Open-File Rep.,* 6 p.

Foy, C. D., Chaney, R. L., and White, M. C., 1978, The physiology of metal toxicity in plants: *Annu. Rev. Plant Physiol..* v. 29, p. 511–566.

Fraser, D. C., 1961, A syngenetic copper deposit of recent age: *Econ. Geol.,* v. 56, p. 951–962.

Gough, L. P., McNeal, J. M., and Severson, R. C., 1980, Predicting native plant copper, iron, manganese, and zinc levels using DTPA and EDTA soil extractants, northern Great Plains: *Soil Sci. Soc. Am. J.,* v. 44, p. 1030–1036.

Hultén, E., 1974, *Flora of Alaska and Neighboring Territories:* Stanford University Press, Stanford, Calif., 1008 p.

Jenny, H., 1980, *The Soil Resource, Origin and Behavior:* Springer-Verlag, New York, 377 p.

Levinson, A. A., 1980, *Introduction to Exploration Geochemistry:* Applied Publishing, Wilmette, Ill., 924 p.

Malyuga, D. P., 1964, *Biogeochemical Methods of Prospecting:* Consultants Bureau, New York, 205 p.

Murray, B. M., 1974, Catalog of bryophytes and lichens of the central Brooks Range, Alaska—a literature review: *Univ. Alaska Mus. Publ., Fairbanks,* (no number), 46 p.

——— , and Murray, D. F., 1978, Appendix: Checklist of vascular plants, bryophytes, and lichens for the Alaskan U.S. IBP Tundra Biome study areas—Barrow, Prudhoe Bay, Eagle Summit, p. 647–677 *in* Tieszen, L. L. (Ed.), *Vegetation and Production Ecology of an Alaskan Arctic Tundra:* Springer-Verlag, New York, 677 p.

Murray, D. F., 1980, Threatened and endangered plants of Alaska: *USDA For. Serv./ USDI Bur. Land Manage. Joint Publ.* (no number), 59 p.

Odum. E. P., 1971, *Fundamentals of Ecology:* W.B. Saunders, Philadelphia, 574 p.

Péwé, T. L., 1975, Quaternary geology of Alaska: *U.S. Geol. Surv. Prof. Pap. 835,* 145 p., 3 maps.

Raines, G. L., and Canney, F. C., 1980, Vegetation and geology, p. 365–380 *in* Siegal, B. S., and Gillespie, A. R. (Eds.), *Remote Sensing in Geology:* Wiley, New York, 702 p.

Rice, E. L., 1979, Allelopathy—an update: *Bot. Rev.,* v. 45, p. 15–109.

Shacklette, H. T., 1966, Phytoecology of a greenstone habitat at Eagle, Alaska: *U.S. Geol. Surv. Bull. 1198-F,* 36 p.

Simon, E., 1978, Heavy metals in soils, vegetation development and heavy metal tolerance in plant populations from metalliferous areas: *New Phytol.,* v. 81, p. 175–188.

Sukachev, V. N., and Dylis, N. V. (Eds.), 1964, *Fundamentals of Forest Biogeocoenology* (in Russian): Bot. Inst. and Lab. For. Sci., Moscow, 474 p.

Thomson, J. W., 1979, *Lichens of the Alaskan Arctic Slope:* University of Toronto Press, Toronto, 314 p.

Wallace, A., Romney, E. M., and Alexander, G. V., 1981, Multiple trace element toxicities in plants: *J. Plant Nutr.,* v. 3, p. 257–263.

Iain Thornton
Applied Geochemistry Research Group
Department of Geology
Imperial College,
London, England

3

SOIL AND PLANT FACTORS THAT INFLUENCE ELEMENT AVAILABILITY AND UPTAKE: IMPLICATIONS FOR GEOCHEMICAL PROSPECTING

ABSTRACT

Concentrations of elements in soils reflect (1) natural geochemistry of the parent material, (2) soil-forming factors such as podzolization, and (3) anthropogenic inputs from industry, sewage sludge, and so on. Availability to plant roots is influenced both by the total concentration and by the forms (or "species") of elements present in soil together with a number of soil factors, including pH, drainage, organic status, and amounts of other interacting ions. The chemical composition of plant roots and shoots may vary markedly between species and between different seasons of the year. These factors must be taken into account in planning and interpreting programs of biogeochemical exploration.

INTRODUCTION

Hawkes and Webb (1962) define geochemical prospecting for minerals as including "any method of mineral exploration based on systematic measurement of one or more chemical properties of a naturally occurring material," with the purpose of discovering "abnormal chemical patterns, or geochemical anomalies, related to mineralization." The materials sampled may include weathered or unweathered surficial materials and/or vegetation. Anomalies determined by the chemical composition of plants form the basis for biogeochemical prospecting results reviewed by Warren (1980), in contrast to geobotanical methods, which depend on the distribution of plant species or of unusual variations in plant morphology. The purpose of this contribution is to provide some basic information and current thoughts on the principal factors that influence the relationships between chemical elements, and particularly metals, in the soil–plant system, by illustrating some of the problems of interpretation of data based on the composition of plants for the identification of mineral deposits.

SOURCES OF METALS IN SOILS

The primary sources of metals in soils are the parent materials from which they are derived. These are usually weathered bedrock or overburden transported by wind, water, or glaciation. The chemical composition of different rock types may vary widely, and in fact the composition of a single rock type may also show significant variation (Table 3-1). Coarse-grained sedimentary rocks, such as sands and sandstones, together with acid-igneous rocks, including unmineralized granites and rhyolites, tend to contain relatively small amounts of trace elements. Fine-grained sediments, such as clays and shales, however, contain larger amounts. In particular, organic-rich black shales may be enriched in Cu, Cd, Pb, Zn, Mo, and Hg (Table 3-1).

The weathering of igneous and sedimentary rocks in the surface environment has been discussed in detail by Mitchell (1974) and West (1981) and does not fall within the scope of this chapter. However, the complex pedological, biological, and microbiological processes involved in soil formation are emphasized; in addition, the contributions of farmers and urban and industrial developers must be recognized.

TABLE 3-1 Range and Mean Concentrations of Some Metals and Metalloids in Igneous and Sedimentary Rocks (ppm)

	Basaltic Igneous	Granitic Igneous	Shales and Clays	Black Shales	Limestones	Sandstones
As	0.2–10	0.2–13.8	–	–	0.1–8.1	0.6–9.7
	2.0	2.0	10	–	1.7	2.0
Cd	0.006–0.6	0.003–0.18	0–11	<0.3–8.4	–	–
	0.2	0.15	1.4	1.0	0.05	0.05
Cr	40–600	2–90	0–590	?6–1000	–	–
	220	20	120	100	10	35
Co	24–90	1–15	5–25	7–100		
	50	5	20	10	0.1	0.3
Cu	30–160	4–30	18–120	20–200	–	–
	90	15	50	70	4	2
Hg	0.002–0.5	0.005–0.4	0.005–0.51	0.03–2.8	0.01–0.22	0.001–0.3
	0.05	0.06	0.09	0.5	0.04	0.05
Pb	2–18	6–30	16–50	7–150	–	<1–31
	6	18	20	30	9	12
Mo	0.9–7	1–6	–	1–300	–	–
	1.5	1.4	2.5	10	0.4	0.2
Ni	45–410	2–20	20–250	10–500	–	–
	140	8	68	50	20	2
Se	–	–	–	–		
	0.05	0.05	0.6	–	0.08	0.05
Zn	48–240	5–140	18–180	34–1500	–	2–41
	110	40	90	100	20	16

Source: Adapted from a table compiled by M. Fleisher and H. L. Cannon (Cannon et al., 1978).

Anomalously high levels of metals in soils may be derived not only from mineralized materials but also from metal-rich source rocks (Table 3-2). For example, in Scotland, soils derived from ultrabasic rocks containing Ni-rich ferromagnesium minerals may, under poorly drained conditions, contain sufficient metal in an available form to result in Ni toxicity in cereal and other crops (Mitchell, 1974). Soils derived from Carboniferous black shales and dolomitic conglomerate may contain as much as 20 to 30 ppm Cd (normally soils contain less than 2 ppm), and those weathered from marine black shales may contain 10 to 100 ppm Mo. These large variations in the metal content of soils derived from unmineralized bedrock and overburden serve to emphasize the importance of obtaining reliable background information in any soil–plant sampling program aimed at identifying mineralized or polluted areas.

Where soil parent materials have been redistributed or mixed as a result of wind, glacial activity, or alluvial transport, the influence of the underlying rock on the composition of surficial materials and soils may be modified or even completely masked. The influence of parent materials on the trace metal content of soils is further modified by soil-forming processes, such as gleying, leaching, surface organic matter accumulation, and podzolization, which may lead to the mobilization and redistribution of metals both within the soil profile and between neighboring soils (Swaine and Mitchell, 1960; Mitchell, 1964). The solubility and mobility of individual metals varies considerably. For example, Mn, Co, Cd, Zn, and Fe are relatively mobile compared with Pb and Mo and are sometimes redistributed in the course of soil formation. These redistribution processes may lead to

TABLE 3-2   Trace Elements in Soils Derived from Normal and Geochemically Anomalous Parent Materials

Normal Range in Soil (ppm)	Metal-Rich Soils (ppm)	Source	Possible Effects
As <5–100	Up to 2500	Mineralization	Toxicity in plants and live-stock; excess in food crops
	Up to 250	Metamorphosed rocks around Dartmoor	
Cd <1–2	Up to 30 (Up to 800)	Mineralization (Shipham, Somerset)	Excess in food crops
	Up to 20	Carboniferous black shale	
Cu 2–60	Up t0 2000	Mineralization	Toxicity in cereal crops
Mo <1–5	10–100	Marine black shales of varying age	Molybdenosis or molybdenum-induced hypocuprosis in cattle
Ni 2–100	Up to 8000	Ultrabasic rocks in Scotland	Toxicity in cereal and other crops
Pb 10–150	1% or more	Mineralization	Toxicity in livestock; excess in foodstuffs
Se <1–2	Up to 7	Marine black shales in England and Wales	No effect
	Up to 500	Namurian shales in Ireland	Chronic selenosis in horses and cattle
Zn 25–200	1% or more	Mineralization	Toxicity in cereal crops

Source: Thornton (1980).

a depletion of one or more metals in the rooting zone of many plants, as in the case of the surface layers of podzols. For example, Se and Fe may be removed and subsequently enriched in the B horizon (Smith, 1983). For prospecting purposes, the sampling of deep-rooted species, such as trees, is obviously preferable to shallow-rooting shrubs or grasses under these conditions.

Biogeochemical prospecting is based on the thesis that soils and plants in mineralized areas are often enriched in ore-forming metals. In the United Kingdom both soil and plant materials may contain high concentrations of one or more of the elements Cu, Pb, Zn, Cd, and As. A detailed discussion of the factors controlling metal disperson from mineralized sources into residual soils, glacial drift, alluvial soils, and organic soils is outside the scope of this contribution; the subject has been previously reviewed by Hawkes and Webb (1962), Levinson (1974), and Siegal (1974). In those areas where mineral deposits have been worked, the mining and processing of ore minerals has often led to appreciable soil contamination usually affecting much larger areas than the original mineral deposits (Davies, 1980). It is at times extremely difficult to differentiate between soils directly weathered from mineralized material and those contaminated by mining operations.

The soil–plant system may be affected by several other sources of metals arising from human activities. These include (1) industrial emissions and effluents; (2) contaminated dusts and rainfall; (3) agricultural fertilizers, soil ameliorants, and pesticides; (4) waste materials, such as sewage sludge, pig slurry, and composted town refuse; and (5) urban development and vehicle traffic. One or more of these sources may locally give rise to metal anomalies in surface soils comparable with those resulting from mineralization.

The availability of metals in the soil to plants is related both to their total concentration and to their chemical forms. They may be present as (1) insoluble salts such as oxides or hydroxides, (2) free cations (Zn^{2+}) and anions (MoO_4^{2-}) in solution, (3) complexed with organic ligands coarsely termed "humic" and "fulvic" acids, or (4) sorbed onto reactive exchange surfaces on soil constituents such as Fe and Al oxides, clays, and organic matter. Trace metals occluded in precipitated oxides, carbonates, or in the lattices of secondary minerals, or that are bound in insoluble organic species, are largely unavailable to plants but can replenish the soil solution or exchangeable pool; those locked within primary minerals and some secondary minerals can be released for plant uptake only through weathering (West, 1981).

The categories constituting the potential supply of trace elements to plants, and the time involved in attaining equilibria between forms, have been summarized by West (1981) and are shown below. The intensity factor represents those forms of elements in the soil solution that are available for immediate plant uptake; the quantity factor comprises those forms that are potentially available to the soil solution.

(1)		(2)		(3)		(4)
	k_1		k_2		k_3	
unavailable	$\xrightleftharpoons{}$	intermediate	$\xrightleftharpoons{}$	labile	$\xrightleftharpoons{}$	soil solution
(inert)						
primary	(years)	secondary	(months)	adsorbed	(days)	simple ions,
minerals		minerals		ions and		soluble
				compounds		complexes

$$\longleftarrow \text{quantity factor } (Q) \longrightarrow \text{intensity factor } (I)$$

The actual chemical forms or "species" of trace metals in soils are still not fully understood. A detailed review by Sposito (1983) outlines present-day knowledge of solid, adsorbed, and aqueous "species" and describes computer models based on thermodynamic principles that have recently been developed to calculate trace metal speciation in soils. One such model, GEOCHEM, has also been described by Mattigod et al. (1981) and Mattigod and Page (1983).

AVAILABILITY OF METALS TO PLANTS

The determination of the "plant-available" fraction of an element in the soil has usually been undertaken using empirical methods based on chemical soil extractants such as ammonium acetate, EDTA, and DTPA. However, techniques based on this approach were initially developed with the aim of diagnosing potential deficiency situations for agricultural and horticultural crops and have not yet been fully tested over the elevated ranges of soil metal concentrations found in mineralized and contaminated areas. Davies (1978) has successfully used extraction of "available" Pb with 0.05 M diammonium EDTA from garden soils as a prediction of uptake by vegetables. However, results of partial extraction may be difficult to interpret and should be used with caution. Results applicable to one soil type may not be applicable to another, in particular if the soils to be compared cover a wide range of pH values, drainage status, or organic matter content. It is of interest that the use of the selective chemical extractants N ammonium acetate (pH 7) and DTPA to estimate available Cd and Zn in soils of an old Zn-mining area at Shipham (southwest England) showed that a far greater proportion of Cd could be extracted from these soils than Zn (Matthews and Thornton, 1982; Table 3-3). This was not reflected in a high proportional uptake of Cd to Zn in the pasture plants studied and is probably the result of a variety of other factors discussed below.

TABLE 3-3 Comparison of the Zn/Cd Ratios Obtained Using Total and Partial Extractants on Shipham Soils

	Zn/Cd Ratios	
Extractant	Mean	Range
Nitric acid	88	68–112
Ammonium acetate	18	9–27
DTPA	22	13–30

FACTORS INFLUENCING THE UPTAKE OF METALS INTO PLANTS

Nye and Tinker (1977) have expressed the mean concentration of any element in a plant in relation to the simultaneous processes of growth and uptake by using the formula

$$X = (S/R_w)(M_R/M)$$

where X is the fraction of the element to dry matter, which is governed by the specific nutrient absorbing power of the root (S), the relative growth rate of the plant (R_w), and the root/plant mass ratio (M_R/M). Tinker (1981) discusses the interpretation of this formula:

> For elements essential to plant growth, there is a functional relation between X and growth rate, in which growth increases rapidly with X until the latter reaches a so-called "critical level"; if X rises still further, growth is not affected at first, then declines as toxic levels are reached. This relation is idealized, and the "critical level" may vary quite widely with environmental conditions, cultivar or species of plant, and with the level of supply of other nutrients. The elemental composition may thus vary over a wide range without affecting plant health. For non-essential elements, there is no initial increase in growth, and the only effect is the possible development of plant toxicity. With tolerant strains, and even more with accumulator species (for example, Astragalus [spp.], for Se), the horizontal part of the curve is extended to extremely high levels.

The distinction between passive and active uptake of trace elements is discussed by Tinker (1981) and is complicated by the strong physicochemical adsorption of ions, including Zn and Cu, on the root cell walls (Loneragan, 1975).

Uptake of metals into plants is influenced by a number of soil and plant factors, discussed below.

Soil Factors

Under temperate conditions in the United Kingdom, the pedological drainage status of the soil is the most significant factor in influencing trace element supply from a particular soil type (West, 1981). For example, pasture plants growing on soils derived from argillaceous schist in Scotland contained seven times the Co, three times the Mn, and 2.5 times the Ni under poorly drained (compared with freely drained) conditions (Mitchell *et al.*, 1957).

Acidity is probably the second most important factor. An increase of soil pH of one unit in the region 5 to 6 will decrease the availability of most cations by half (with the exception of Cu) and considerably increase the availability of Mo and Se (which are taken up as anions, West, 1981). Thus raising soil pH by one unit from 5.4 to 6.4 with lime reduced the content of Zn in red clover from 61 ppm to 51 ppm dry matter and Mn from 56 ppm to 25 ppm; at the same time the Mo content was raised from 0.3 ppm to 1.9 ppm (Mitchell and Burridge, 1979).

Tinker (1981) draws attention to the ease with which trace metals form complexes with soluble and insoluble ligands in the soil, with the root cell wall, and with other tissues of the plant; he also discusses in detail the uptake of simple ions in solution and the uptake from chelating solutions. There is no doubt that the nature of organic complexes with metals both in the soil and in the plant are still far from fully understood. For example, in organic soils the availability of Cu to plants depends not only on the concentration of Cu in the soil solution but also on the form in which the Cu is present (Mercer and Richmond, 1970); Cu complexes of molecular weight 1000 were found to be more available to plants than those exceeding 5000. The role of soil organic matter in controlling the release of some essential elements and immobilizing others is important. For instance, Pb added to soil from smelting and other pollution sources is closely held by organic matter in surface soils, probably absorbed as Pb^{2+} on the surfaces of clay minerals and organic colloids, and by the formation of insoluble lead chelates with organic matter (Lagerwerff, 1972). As a result, Pb is relatively unavailable for uptake into plants and is not noticeably leached down the soil profile into groundwaters.

Both solubility and uptake of specific elements may also be influenced by the presence of other elements in the soil. For example, uptake of Mo as the MoO_4^{2-} ion may be modified by PO_4^{3-} and SO_4^{2-}, although the evidence is conflicting. The presence of large amounts of Zn and Ca has been shown by solution culture techniques to depress Cd uptake into ryegrass, through competition for exchange sites at the root surface (Jarvis and Jones, 1978). Both antagonistic and synergistic effects are probably significant in the majority of situations, and interactions between trace metals themselves and with other major soil constituents need further evaluation under a wide range of soil conditions.

Finally, it is important to emphasize the potential importance of soil microorganisms on metal availability. Recently reviewed by Olson (1983), this is a complicated area and is far from being clearly understood. It is obvious that with as many as 10^9 bacteria in each gram of soil at any moment in time, a significant proportion of some metals will be in or adsorbed onto bacterial cells or associated with bacterial exudates such as muco-polysaccharides.

Plant Factors

Root excretions, including citric and malic acids and amino acids, may increase the concentration of chelated metals in the soil and thus influence uptake by the plant (Tinker, 1980; Elgawharz *et al.*, 1970). In addition, fungal hyphae of mycorrhizae infecting plant roots have been shown to increase the supply of Zn and Cu (Tinker, 1978).

Research based on solution culture techniques has shown that 88 percent and 57 to 80 percent of the Cd and Pb, respectively, taken up by ryegrass was localized in the root material and not translocated into the aerial tissues of the plant (Jarvis *et al.*, 1977; Jarvis and Jones, 1978). This is illustrated in Figure 3-1 using the example of *Holcus*

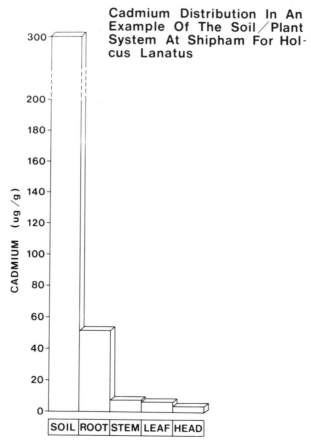

Cadmium Distribution In An Example Of The Soil⁄Plant System At Shipham For Holcus Lanatus

Fig. 3-1   Cadmium distribution (dry weight basis) in *Holcus lanatus* from Shipham as an example of the soil–plant system. (From Matthews and Thornton, 1982.)

lanatus (Yorkshire fog), a common pasture grass in the United Kingdom, growing on heavily contaminated mineralized land at Shipham (Matthews and Thornton, 1982). Differences shown between the concentrations of Cd in the stem, leaf, and head confirm many earlier findings, including those of Fleming (1963, 1965) for Co, Ni, Cr, Mn, Fe, Cu, Pb, Zn, and V. These differences are found in most plants and serve to highlight the importance of careful tissue selection in biogeochemical prospecting sampling programs.

The influence of plant species on metal uptake from the soil is often large (Fleming, 1965; Archer, 1971). Different species, and even different cultivars, regulate metal uptake at both the soil–root and root–shoot interfaces to varying degrees. Individual metals behave differently depending on their solubility, tendency to form complexes, ionic size, and essential or nonessential nature. For example, clover tends to contain larger amounts of most trace elements—particularly Mo—than do grasses. Amounts of Cd in the leaves of pasture species at Shipham varied markedly between the species samples (Fig. 3-2). Accumulator plants, such as *Astragalus* spp. found in some arid areas of the United States, may contain several thousand ppm Se, compared with only a few ppm in grass species growing in the same soil. Arsenic accumulator plants have been found on mine waste in southwest England (Nye and Peterson, 1975). Warren (1980) draws attention to *Pseudotsuga menziesii* (Douglas fir) as an As accumulator and a useful pathfinder in relation to Au mineralization. He reports a concentration of 3500 ppm As in the first-year stems of this tree compared with 570 ppm As in first-year needles. *Betula* spp. (birch) can accumulate Zn, and pine and spruce can accumulate Cu. Metal-tolerant ecotypes of grasses that have adapted to metal-rich soils and waste materials have been shown to take up different amounts of Pb and Zn when compared to different cultivars growing in uncontaminated land (Antonovics *et al.*, 1971). These metal-tolerant plants are also found on undisturbed mineral soils.

Stage of growth is another important factor (Warren, 1980). In general, the trace element concentration in the whole plant tends to decrease with increasing maturity (Fleming, 1965; Antonovics *et al.*, 1971; Thomas *et al.*, 1972). Seasonal influence on the heavy metal content of plant tissue is exemplified by a marked increase in the Pb content

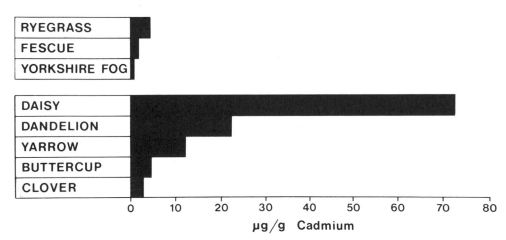

Fig. 3-2 Variations in cadmium concentration (dry weight basis) of common pasture species at Shipham. (From Matthews and Thornton, 1982.)

of grass in winter months (Mitchell and Reith, 1966; Thornton and Kinniburgh, 1978). This seasonal effect is illustrated in Figure 3-3 for *Holcus lanatus* growing on metal-contaminated land at Shipham, where the increase in Pb content from low summer values to a winter maximum is around 4-fold (Matthews and Thornton, 1982). Possible reasons proposed include (1) changes in the amount of plant dry matter due to plant senescence in winter (not substantiated by the small variation in Cu in Figure 3-3) (Rains, 1971), and (2) a redistribution of metals from root to shoot in the winter, as actively growing roots are capable of restricting movement of Pb to aerial parts (Jones *et al.*, 1973). Falling phosphate levels in grass leaves in the winter months would seem to correlate with increas-

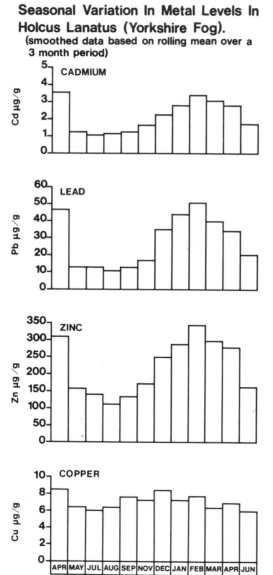

Seasonal Variation In Metal Levels In Holcus Lanatus (Yorkshire Fog).
(smoothed data based on rolling mean over a 3 month period)

Fig. 3-3   Seasonal variation in metal concentrations (dry weight basis) in *Holcus lanatus* sampled at Shipham. Smoothed data based on rolling mean over a 3-month period. (From Matthews and Thornton, 1982.)

SOIL/PLANT FACTORS THAT INFLUENCE ELEMENT AVAILABILITY/UPTAKE

ing metal burdens (Matthews and Thornton, 1982). It has been suggested (Jarvis *et al.*, 1976) that transport and mobility of Pb and Cd within the plant relates to the solubilities of their phosphates. Some recent investigations in Denmark using labeled isotopes of Pb and Cd have led to the suggestion that leaf-surface contamination from atmospheric deposition leads to foliar absorption and transportation and may account for as much as 20 to 60 percent of the total plant Cd (Hormand *et al.*, 1983) and 90 to 99 percent of the Pb (Tjell *et al.*, 1979). The latter authors account for the increase in plant Pb in autumn and winter as being due to a decrease in the dilution of atmospherically supplied Pb with new grown material after the midsummer peak in growth.

CONCLUSION

The chemical and physical processes controlling the forms and "species" of trace metals in the soil, and their availability to plants, are still far from being clearly understood. The role of organic ligands is particularly complex. However, the development of computer models, such as the thermodynamically based model GEOCHEM, continues to play a significant role in furthering the knowledge of the solid and solution phases of metals in the soil. There is still little that is known of the influence of soil bacteria and fungi on the mobility and availability of metals. This area requires intensive collaborative research between soil chemists and microbiologists.

It is clear that in planning any biogeochemical exploration, it is of paramount importance to consider sampling procedures that take into account potential differences in the metal content of plant species and plant parts as well as the effects of stages of maturity and seasonal cycles.

Finally, the importance of obtaining reliable background information on metal concentrations in both soils and plants is stressed. It has been shown in the sections above that variations in metal concentrations in soils and plants attributable to unmineralized bedrock geochemistry and to various pollution sources may be large; without good background data it is impossible to interpret the results of soil–plant investigations when they are used in the search for mineral deposits.

ACKNOWLEDGMENTS

Many of the examples cited in this paper are taken from the ongoing research program in environmental geochemistry of the Applied Geochemistry Research Group funded by the Agricultural Research Council and RTZ Services Ltd., Bristol. I am grateful to the many postgraduate students and colleagues who contributed to this research.

REFERENCES

Antonovics, J., Bradshaw, A., A., and Turner, R. G., 1971, Heavy metal tolerances in plants, *Adv. Ecol. Res.,* v. 7, p. 1–85.

Archer, F. C., 1971, Factors affecting the trace element content of pastures, *in Trace elements in soils and crops:* Min. Agric. Fish Food Tech. Bull. 21, p. 150–157.

Cannon, H. L., Connall, G. G., Epstein, J. B., Parker, J. G., Thornton, I., and Wixson, B. G., 1978, Rocks: the geologic sources of most trace elements, p. 17–31, *in Geochemistry and the Environment*, Vol. 3: National Academy of Science, Washington, D.C.

Davies, B. E., 1978, Plant-available lead and other metals in British garden soils: *Sci. Total Environ.*, v. 9, p. 243–262.

——, 1980, Trace element pollution p. 287–351 *in* Davies, B. E. (Ed.), *Applied Soil Trace Elements:* Wiley, Chichester, England.

Elgawharz, S. L., Lindsay, W. L., and Kemper, W. D., 1970, Effect of complexing agents and acids on the diffusion of zinc to a simulated root: *Proc. Soil Sci. Soc. Am.*, v. 34, p. 211–214.

Fleming, G. A., 1963, Distribution of major and trace elements in some common pasture species *J. Sci. Food Agric.*, v. 14, p. 203–208.

——, 1965, Trace element in plants with particular reference to pasture species: *Outlook Agric.*, v. 4, p. 270–285.

Hawkes, H. E., and Webb, J. S., 1962, *Geochemistry in Mineral Exploration:* Harper & Row, New York.

Hormand, M. F., Tjell, J. C. and Mosback, H., 1983, Plant uptake of airborne cadmium: *Environ. Pollut.*, v. 30, p. 27–38.

Jarvis, S. C., and Jones, L. H. P., 1978, Uptake and transport of cadmium by perennial ryegrass from flowing solution culture with a constant concentration of cadmium: *Plant Soil*, v. 49, p. 333–342.

——, Jones, L. H. P., and Hopper, M. J., 1976, Cadmium uptake from solution by plants and its transport from roots to shoots: *Plant Soil*, v. 44, p. 179–191.

——, Jones, L. H. P., and Clement, C. R., 1977, Uptake and transport of lead by perennial ryegrass from flowing solution culture with a controlled concentration of lead: *Plant Soil*, v. 46, p. 371–377.

Jones, L. H. P., Clement, C. R., and Hopper, M. J., 1973, Lead uptake from solution by perennial ryegrass and its transport from roots to shoots: *Plant Soil*, v. 38, p. 403–414.

Lagerwerff, J. V., 1972, Lead, mercury and cadmium as environmental contaminants, p. 593–636 *in* Mortredt, J. J., Gordiano, P. M., and Lindsay, W. L. (Eds.), *Micronutrients in Agriculture:* Soil Sci. Soc. Am., Madison, Wis.

Levinson, A. A., 1974, *Introduction to Exploration Geochemistry:* Applied Publishing, Calgary.

Loneragan, J. F., 1975, The availability and absorption of trace elements in soil–plant systems and their relation to movement and concentration of trace elements in plants, p. 109–134 *in* Nicholas, D. J. D., and Egan, A. R. (Eds.), *Trace Elements in Soil–Plant–Animal,* Academic Press, New York.

Matthews, H. M., and Thornton, I., 1982, Seasonal and species variation in the content of cadmium and associated metals in pasture plants at Shipham: *Plant Soil,* v. 66, p. 181–193.

Mattigod, S., and Page, A. L., 1983, Assessment of metal pollution in soils, *in* Thornton, I. (Ed.), *Applied Environmental Geochemistry:* Academic Press, London.

——, Sposito, G., and Page, A. L., 1981, Factors affecting the solubilities of trace elements in soils, p. 203–221 *in Chemistry in the Soil Environment:* Am. Soc. Agric., Madison, Wis.

Mercer, E. R., and Richmond, J. L., 1970, Fate of nutrients in soil: copper: *Letcombe Laboratory Annu. Rep.*, p. 9.

Mitchell, R. L., 1964, Trace elements in soils, p. 320–368 *in* Bear, F. E. (Ed.), *Chemistry of the Soil:* Reinhold, New York.

——, 1974, Trace element problems in Scottish soils: *Neth. J. Agric. Sci.,* v. 22, p. 295–304.

——, and Burridge, J. C., 1979, Trace elements in soils and crops: *Philos. Trans. R. Soc., Lond.,* v. B288, p. 15–24.

——, and Reith, J. W. S., 1966, The lead content of pasture herbage: *J. Sci. Food Agric.,* v. 17, p. 437–440.

——, Reith, J. W. S., and Johnston, I. M., 1957, Trace element uptake in relation to soil content: *J. Sci. Food Agric.* (Suppl. Issue), v. 8, p. 551–559.

Nye, S. M., and Peterson, P. J., 1975, The content and distribution of selenium in soils and plants from seleniferous areas in Eire and England, p. 113–121 *in* Hemphill, D. D. (Ed.), *Trace Substances in Environmental Health,* v. 9, University of Missouri, Columbia, Mo.

Nye, P. H., and Tinker, P. B., 1977, *Solute Movement in the Soil–Root System:* Blackwell Scientific, Oxford.

Olson, B. H., 1983, Microbial mediation of biogeochemical cycles, *in* Thornton, I. (Ed.), *Applied Environmental Geochemistry:* Academic Press, London.

Rains, D. W., 1971, Lead accumulation by wild oats (*Avena fatua*) in a contaminated area: *Nature (Lond.),* v. 233, p. 210.

Siegal, F. R., 1974, *Applied Geochemistry:* Wiley, New York.

Smith, C. A., 1983, The distribution of selenium in some soils developed on Silurian, Carboniferous and Cretaceous systems in England and Wales: Ph.D. thesis, University of London.

Sposito, G., 1983, The chemical forms of trace metals in soils *in* Thornton, I. (Ed.), *Applied Environmental Geochemistry:* Academic Press, London.

Swaine, D. J., and Mitchell, R. L., 1960, Trace element distribution in soil profiles: *J. Soil Sci.,* v. 11, p. 347–368.

Thomas, B., Rougham, J. A., and Walters, E. D., 1972, Lead and cadmium content of some vegetable foodstuffs: *J. Sci. Food Agric.,* v. 23, p. 1493–1496.

Thornton, I., 1980, Geochemical aspects of heavy metal pollution and agriculture in England *in Inorganic Pollution and Agriculture,* v. 2, Min. Agric. Fish Food Ref. Book 326: HMSO, London, p. 105–125.

——, and Kinniburgh, D., 1978, Intake of lead, copper and zinc by cattle from soil and pasture, p. 499 *in* Kirchgessner, M. (Ed.), *Trace Element Metabolism in Man and Animals,* v. 3, Friesing, West Germany.

Tjell, J. C., Hormand, M. F., and Mosback H., 1979, Atmospheric lead pollution of grass grown in a background area of Denmark: *Nature (Lond.),* v. 280, p. 425–426.

Tinker, P. B., 1978, Effects of vesicular-arbuscular mycorrhizas on plant nutrition and plant growth: *Physiol. Veg.,* v. 16, p. 743–751.

——, 1980, Root–soil interactions in crop plants. p. 1–34 *in* Tinker, P. B. (Ed.), *Soils and Agriculture (Critical Reports on Applied Chemistry),* Vol. 2, Blackwell Scientific, Oxford.

——, 1981, Levels, distribution and chemical forms of trace elements in food plants, *Philos. Trans. R. Soc. Lond.,* v. B294, p. 41–55.

Warren H. V., 1980, Biogeochemistry, trace elements and mineral exploration, p. 353–380 *in* Davies, B. E. (Ed.), *Applied Soil Trace Elements:* Wiley, Chichester, England.

West, T. S., 1981, Soil as the source of trace elements: *Philos. Trans. R. Soc. Lond.,* v. B294, p. 19–39.

Peter J. Peterson
Monitoring and Assessment Research Center
King's College London
University of London
London, England

4

TOLERANCE: A PLANT'S RESPONSE TO METAL STRESS

The evolutionary development of tolerance in plants to metalliferous soils has been recognized for many years as giving rise over geologic time to specialized floras containing endemic species. Metal-contaminated soils arising from mining and smelting industries lead to the development of metal-tolerant subspecies and physiotypes. Various studies have shown that tolerance is metal specific and is an inherited characteristic. With endemic species tolerance has been described as an absolute phenomenon, whereas with the recently evolved subspecies, a gradation exists between low and high degrees of tolerance.

The mechanisms involved or implicated in metal tolerance can range from general botanical strategies, through modified physiological processes, to specific biochemical mechanisms developed to sequester high concentrations of metals initially present in ionic forms. Low-molecular-weight organometallic compounds which are synthesized by plants and considered of significance include selenoamino acids, nickel–organic acids, chromium–organic acids, gold cyanide, copper proline, zinc-malate, and fluoroacetate. The formation of higher-molecular-weight compounds developed in response to metal stress include copper and cadmium metallothioneins. Differences in cell membrane permeability, immobilization of metals in root cell walls, enzyme adaptations, and differing cytoplasmic sensitivities to metals have also been developed by plants to avoid, restrict, or alleviate metal toxicity, making it unlikely that a unifying principle lies behind the development of tolerance.

INTRODUCTION

The evolutionary development of tolerance in plants to metalliferous soils has been recognized for many years as giving rise over geologic time to specialized adapted communities, containing many endemic species, such as the seleniferous, nickeliferous, uraniferous, calamine, and serpentine floras (Cannon, 1971; Wild and Bradshaw, 1977; Wild, 1978). More recently, metal-contaminated soils and waters arising especially from mining and smelting industries, from the widespread use of fungicides and pesticides, and from the use of galvanized fences and pylons have led to the rapid development of metal-tolerant plants (Antonovics et al., 1971; Bradshaw, 1976) variously described as tolerant subspecies, races, varieties, ecotypes, forms or physiotypes, but in general few have been subjected to experimental taxonomic investigation to determine their exact status (Peterson, 1983).

There is much evidence which suggests that plants growing on metalliferous soils accumulate the element, or elements (Fig. 4-1), which necessarily implies that they are tolerant of toxic concentrations of the metals (Peterson, 1971, 1979). Indeed, the tolerance is metal specific and is an inherited characteristic (Antonovics et al., 1971). With endemic species metal tolerance has been described as an absolute phenomenon (Peterson, 1983), whereas with the recently evolved metal-tolerant subspecies, varieties, and races, a gradation between low and high degrees of tolerance can be measured (Wu et al., 1975; Baker, 1978). Mechanisms involved, or implied, in metal tolerance are outlined in this chapter. They can range from general botanical strategies, including periodic

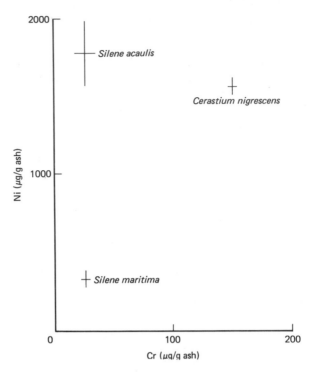

Fig. 4-1   Relationship between concentrations of chromium and nickel in the shoots of three species of plant collected from a serpentine site at Unst, United Kingdom. The arithmetic means are marked with crosses, the limits of which show the standard deviation. (From Shewry and Peterson, 1976.)

leaf fall, through modified physiological transport processes, to specific biochemical mechanisms developed to sequester high concentrations of metals initially present in ionic forms. Low-molecular-weight organometallic compounds which are synthesized by plants and considered of significance in metal tolerance include selenoamino acids, nickel–organic acids, fluoroacetate, and others. The formation of higher-molecular-weight compounds developed in response to metal stress, including the metallothioneins, may also constitute tolerance mechanisms.

Immobilization of metals in root cell walls and differences in cell membrane permeability provide additional examples of plant adaptations to toxic metals. Enzyme adaptations and differing cytoplasmic sensitivities to metals have also been noted between tolerant and nontolerant plants.

METAL EXCLUSION AND RESTRICTED TRANSPORT

The exclusion of metal ions from the whole plant, or from the shoots brought about by a restriction in long-distance transport, are possible physiological responses to metal-rich habitats. Surely, a plant's best strategy would be based on the avoidance of metal accumulation, and hence the need for the development of a tolerance mechanism would not arise. However, the concept of metal exclusion from plants is not well established because most studies have been based on an assessment of metal accumulation; the higher the concentration, the greater the interest. Such an imbalance of research effort requires correction because metal excluding crops could be of benefit, especially on areas polluted by sewage sludge and mine and smelter wastes.

There are many examples in the literature in which metal ions are apparently restricted in varying degrees to the roots. For instance, lead is normally considered to be an element associated with root accumulation with little transport to the leaves (Peterson, 1978). Indeed, the data of Nicholls *et al.* (1965) working with the grass *Triodia pungens* and more recently Barry and Clark (1978) working on *Agrostis tenuis, Festuca ovina,* and *Minuartia verna* have reported that there was little increase in foliar lead concentrations despite a wide range of soil lead concentrations extending over several orders of magnitude.

Cadmium, on the other hand, can be transported more readily to the shoots and cultivars of soybeans, and other crops are known where metal concentrations in plant tops may vary 100-fold or more between the extremes (Peterson and Alloway, 1979).

Additional information on species responses to restrictions in transport can be gained from an examination of the data of Reilly and Reilly (1973) on the copper concentrations in plants collected from some copper-rich Zambian soils. The data in Table 4-1 show that the two accumulators *Becium homblei* and *Combretum psidioides* accumulate much copper in their leaves, whereas the two grasses *Trachypogon spicatus* and *Stereochlaena cameronii* contain most of their copper within the roots. It is evident that further studies are required to establish the physiological basis for differential transport between root and shoot.

TABLE 4-1 Copper Leaf/Root
Ratios for Plants
from Zambia

Species	Ratio
Trachypogon spicatus	0.003
Stereochlaena cameronii	0.05
Becium homblei	1.9
Combretum psidioides	2.4

Source: Data from Reilly and Reilly (1973).

Metal tolerance in clones of *A. tenuis* resistant to either copper, zinc, and arsenic (Peterson, 1969; Porter and Peterson, 1977) and in *A. stolonifera* resistant to copper, zinc, or nickel (Peterson, 1969; Wu and Antonovics, 1975; Shewry and Peterson, 1976) has been associated with increased amounts of the metal within roots and an accompanying reduction of transport to the shoot with increasing metal stress (Fig. 4-2). In such cases the tolerant plants accumulated larger amounts of metal within the roots yet continued to grow, whereas growth of the nontolerant populations ceased at lower concentrations. Copper-tolerant clones *A. gigantea* also accumulate large amounts of copper in their roots (Hogan and Rauser, 1981), but more copper was translocated to the shoots

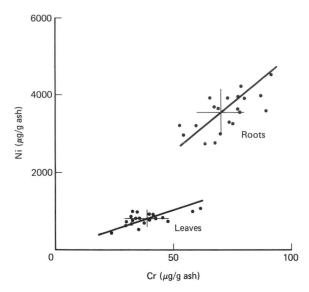

Fig. 4-2   Relationship between chromium and nickel concentrations in the roots and shoots of plants of *Agrostis stolonifera* collected from a serpentine site at Unst, United Kingdom. Regression lines are shown and the arithmetic means are marked with crosses, the limits of which show the standard deviations. (From Shewry and Peterson, 1976.)

of the tolerant than the nontolerant clones, a trend exactly opposite to that obtained with *A. tenuis* and *A. stolonifera*. Furthermore, the amount of copper in *A. gigantea* shoots as a proportion of the total absorbed by the plants was similar for both tolerant and nontolerant clones.

The data on metal tolerance in *Agrostis* species indicate that tolerance is not a mere capacity of tolerant plants to retain metal within their roots, nor the development of reduced translocation to the shoot. Tolerance must therefore imply different patterns of compartmentalization of intracellular metal complexation with low- or high-molecular-weight substances.

SPECIFIC METABOLIC ADAPTATIONS

Considerable information has been published documenting element concentrations in plants collected from a wide range of geochemically anomalous soils and polluted environments. Extreme accumulations of many elements, in addition to those considered essential for plant growth, have been reviewed, including aluminum, barium, chromium, cobalt, platinum, palladium, silver, thallium, tin, tungsten, uranium, vanadium (Peterson and Girling, 1981), nickel (Jaffré *et al.*, 1979), fluorine (Peterson, 1971), and rhenium (Myers and Hamilton, 1960), as well as the metalloids selenium, arsenic, and antimony (Peterson *et al.*, 1981). Irrespective of the element, concentration of an element must have placed the species under physiological, biochemical, and presumably genetic stress since the evolutionary development of flowering plants in the Cretaceous. Plants must, therefore, have evolved a tolerance to these conditions which allowed them to survive and compete under what are clearly adverse growth conditions. For convenience, low-molecular-weight complexes, which could be involved in metal tolerance, are discussed first, followed by high-molecular-weight biopolymers. Possible tolerance mechanisms, such as cell wall binding, differential membrane permeability, and enzymatic modifications are considered separately.

Low-Molecular-Weight Compounds

Selenoamino acids Certain species and floras characteristic of selenium-rich soils from various regions in the United States have been recognized for many years as being particularly toxic to livestock (Rosenfeld and Beath, 1964). Similar disorders in grazing animals have been reported from the Republic of Ireland, Australia, South Africa, and elsewhere; in South America, seeds of the accumulator *Lecythis ollaria* have been identified as being toxic to humans (Kerdel-Vegas *et al.*, 1965). Seleniferous flora in the Great Plains area of the United States contain many selenium accumulators, including several dozen species of the legume genus *Astragalus* (e.g., *A. pectinatus, A. bisulcatus*) as well as members of the Cruciferae (*Stanleya pinnata*) and Compositae (*Oonopsis condensata*), where the selenium concentration may exceed 1 percent of the dry weight, especially in mature seed. In all cases selenoamino acids comprise the majority of the selenium present in the plant. Because the selenium accumulators belong to a variety of unrelated genera, it seems likely that selenium accumulation arose through convergent evolution rather than from a single selenium-accumulating ancestor.

Selenocystathionine and Se-methylselenocysteine are the two principal amino acids (Fig. 4-3), and their occurrences are outlined in Table 4-2. These amino acids have

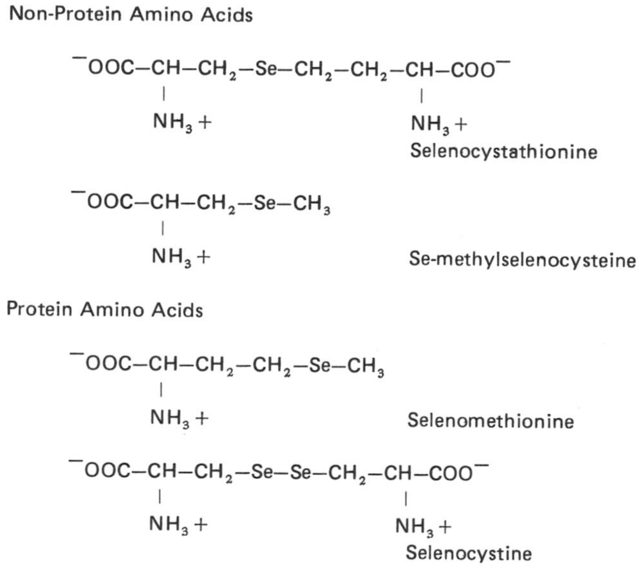

Non-Protein Amino Acids

$$^-OOC-CH-CH_2-Se-CH_2-CH_2-CH-COO^-$$

$$NH_3+ \qquad\qquad NH_3+$$
Selenocystathionine

$$^-OOC-CH-CH_2-Se-CH_3$$

$$NH_3+ \qquad \text{Se-methylselenocysteine}$$

Protein Amino Acids

$$^-OOC-CH-CH_2-CH_2-Se-CH_3$$

$$NH_3+ \qquad \text{Selenomethionine}$$

$$^-OOC-CH-CH_2-Se-Se-CH_2-CH-COO^-$$

$$NH_3+ \qquad\qquad NH_3+$$
Selenocystine

Fig. 4-3   Structures of various selenoamino acids isolated and characterized from the leaves of various plant species referred to in the text.

TABLE 4-2 Selenoamino Acids in Various Plant Species[a]

Compound	Species	Reference
Selenocystathionine	*Neptunia amplexicaulis*	Peterson and Robinson (1972)
	Morinda reticulata	Peterson and Butler (1971)
	Lecythis ollaria	Kerdel-Vegas *et al.* (1965)
	Astragalus pectinatus	Shrift and Virupaksha (1965)
	Stanleya pinnata	Shrift and Virupaksha (1965)
Se-methylselenocysteine	*Astragalus* spp.	Shrift and Virupaksha (1965)
Selenomethionine	Many species	Peterson and Butler (1962)
Selenocysteine	*Vigna radiata*	Brown and Shrift (1980)

[a] Also dimethylselenide and dimethyldiselenide as volatiles.

also been isolated from seeds as the γ-glutamyl derivative, with both isomeric γ-glutamyl derivatives of selenocystathionine being present in *A. pectinatus* (Nigam and McConnell, 1976). Nonaccumulator and selenium-sensitive plants, on the other hand, synthesize predominantly selenomethionine and to a lesser extent selenocysteine (Fig. 4-3), both amino acids being incorporated into proteins via aminoacyl-tRNA synthetase enzymes. Biochemical differences responsible for tolerance to selenium are of two kinds: first, accumulators synthesize large amounts of nonprotein selenoamino acids, whereas these compounds are scarcely detectable in nonaccumulators; second, the proteins of accumulators contain very little selenium and less than selenium-sensitive plants (Peterson and Butler, 1967). Confirmation of this latter exclusion mechanism comes from more recent work on selenium analysis of proteins of three *Astragalus* accumulators and three nonaccumulators (Brown and Shrift, 1981). Thus in *Neptunia amplexicaulis* and in at least some *Astragalus* species, selenium tolerance is associated with exclusion of selenium from enzymes and other proteins. A more detailed investigation of the mechanism of protein synthesis has revealed that the cysteinyl-tRNA synthetase from the accumulator *A. bisulcatus* did not link selenocysteine to its tRNA, but the equivalent preparation from a selenium-sensitive *Astragalus* was able to form cysteinyl-tRNA (Burnell and Shrift, 1979). Presumably, other mechanisms exist, because the cysteinyl-tRNA synthetase from *N. amplexicaulis* was able to use selenocysteine as a substrate, but the amino acid was not found in protein (Burnell, 1981).

Discrimination during protein synthesis therefore appears to be a likely mechanism for the exclusion of selenocysteine from proteins. Selenomethionine has not been reported in accumulators.

Nickel–organic acids The nickel concentration in plants varies around a mean of 0.05 μg per gram of dry weight tissue. Plants growing on ultrabasic soils, however, may exceed 1000 μg/g (Table 4-3). Many serpentine endemic species have an inordinate capacity to accumulate very large quantities of nickel. Nickel accumulators have now been described from various countries, such as Italy [e.g., *Alyssum bertolonii* (Minguzzi and Vergnano, 1948)], New Caledonia [e.g., *Hybanthus austro-caledonicus* (Brooks *et al.*, 1974)], Australia [e.g., *H. floribundus* (Severne and Brooks, 1972)], Zimbabwe [e.g., *Pearsonia metallifera* (Wild, 1974)], and more recently from Canada [e.g., *Arenaria humifusa* (Roberts, 1980)] and the United States [e.g., *Arenaria rubella* (Samiullah, Kruckeberg, and Peterson, unpublished data)].

TABLE 4-3   Nickel Concentrations in Various Plant Species [Percent (Ash Weight)]

Species	Plant	Soil	Ratio	Reference
Hybanthus floribundus	13	0.08	162.5	Severne and Brooks (1972)
H. austro-caledonicus	27	0.50	54.0	Brooks *et al.* (1974)
Pearsonia metallifera	15.3	0.55	27.8	Wild (1974)
Alyssum serpyllifolium	10.3	0.40	25.8	Menezes de Sequeira (1968)
Dicoma niccolifera	2.8	0.70	4.0	Wild (1970)
Pimelia suteri	0.59	0.33	1.8	Lyon *et al.* (1971)
Silene acaulis	0.18	0.33	0.5	Shewry and Peterson (1976)

An assessment of nickel accumulation in relation to evolutionary status (Sporne, 1969) indicates that accumulation occurs especially in "primitive" families. Aluminum accumulation is also correlated with "primitive" characters (Chenery and Sporne, 1976), whereas phylogenetic considerations suggest that manganese accumulation is a later evolutionary development (Jaffré, 1979). These relationships are discussed in greater detail in Peterson (1983).

Pelosi *et al.* (1976) and Pancaro *et al.* (1978) have shown that nickel is complexed with malic and malonic acids in the nickel accumulator *Alyssum bertolonii* and more detailed studies by Lee *et al.* (1978) have revealed that these organic acids are also important complexing agents in *A. serpyllifolium* ssp. *lusitanicum* and in *Pearsonia metallifera*. Pancaro *et al.* (1978) have confirmed the occurrence of high concentrations of malic acid in leaves of *A. serpyllifolium* ssp. *lusitanicum* collected from serpentine locations; however, it is present only in small amounts in *Alyssum* plants grown in garden soil. The presence of nickel-malate and malonate has been confirmed in *A. serpyllifolium* ssp. *lusitanicum* by Brooks *et al.* (1981) and in the subspecies *Malacitanum* from Spain but not in *A. serpyllifolium* spp. *serpyllifolium*, which grows throughout the Iberian Peninsula on normal soils.

In many other nickel accumulators, such as *Hybanthus austrocaledonicus, H. caledonicus, Sebertia acuminata, Psychotria douarrei, Geissois pruninosa, Homalium francii*, and other *Homalium* spp., the citratonickel(II) complex predominates (Lee *et al.*, 1977b, 1978). Indeed, a strong correlation was reported between the concentration of nickel and citric acid in the leaves of seventeen New Caledonian accumulators. High concentrations of hydrated Ni^{2+} were also recorded in some species (e.g., *S. acuminata*).

An anionic complex of nickel, with electrophoretic behavior similar to that of nickel-citrate, has also been reported by Tiffin (1971) in various nonaccumulator plants. Perhaps this element is transported largely as the citrate complex. Nevertheless, no clear evidence has been presented to establish whether this complex is formed metabolically or nonmetabolically. In any event, the occurrence of iron as the citrate complex in xylem sap of various plants (Tiffin, 1971) provides further indication of the important role of organic acids in the chelation of metals within plants.

The accumulation of organic acids in nickel-accumulating plants is now well established, but its adaptive significance is still uncertain. It is important to recall that nickel toxicity has been associated with an induced iron deficiency in many plants (Proctor and Woodell, 1971), yet iron chlorosis conditions are known to increase concentrations of organic acids, especially citrate and malate in a wide range of plants (Landsberg, 1981;

Wallace, 1981). The exact role of organic acids in plant adaptation to nickel-rich soils remains to be established.

Chromium trioxalate Few chromium accumulators have been discovered. In all cases the plant/soil ratio was less than 1 (Table 4-4), which contrasts strongly with the nickel accumulators, where values greater than 1 are relatively commonplace. An interesting chromium-accumulating epiphytic bryophyte, *Aerobryopsis longissima* from a New Caledonian serpentine area (Lee *et al.*, 1977a) contained 5000 μg/g chromium. This concentration was about 20 times higher than the chromium concentrations in moss, which represents another interesting example of tolerance, although no biochemical studies have yet been reported on this plant.

TABLE 4-4   Chromium Concentrations in Various Plant Species (μg/g Ash Weight)

Species	Plant	Soil	Ratio	Reference
Sutera fodina	48,000	125,000	0.38	Wild (1974)
Leptospermum scoparium	2,470	8,950	0.28	Lyon *et al.* (1971)
Dicoma niccolifera	30,000	115,000	0.24	Wild (1974)
Cerastium nigrescens	147	1,800	0.08	Shewry and Peterson (1976)
Silene maritima	22	1,800	0.01	Shewry and Peterson (1976)
Aerobryopsis longissima[a]	4,740	—	—	Lee *et al.* (1977a)
Homalium guillainii	112	8,550	0.01	Lee *et al.* (1977a)

[a]Epiphytic on *H. guillainii.*

Sutera fodina, Pearsonia metallifera, and *Dicoma niccolifera*, all from Zimbabwe, have been reported to contain high concentrations of chromium in their ash—up to 5 percent by weight (Wild, 1974)—while *Leptospermum scoparium* from an abandoned mine in New Zealand contained up to 1 percent chromium by weight (Lyon *et al.*, 1971). The biochemical form of much of the soluble chromium in leaf tissue of *L. scoparium* has been characterized as the trioxalatochromate(III) ion (Lyon *et al.*, 1969a,b). Chromium was, however, transported in the xylem sap as chromate, indicating that metabolism to the complex took place within the leaf tissues. The form of chromium in the other accumulators has not been described. Presumably, the function of the chromium–organic acid complex would be to reduce the cytoplasmic toxicity associated with Cr^{3+} and chromate ions.

Gold cyanide An examination of the gold concentrations in different plants has revealed that *Phacelia sericea* has the ability to accumulate considerably more gold than can many other species (Girling *et al.*, 1979). Chemical tests have shown that this species is a cyanogenic plant which can liberate free cyanide into the soil solution, with resultant dissolution of gold (Girling and Peterson, 1978). The gold–cyanide complex is readily absorbed by the plant and translocated to its leaves, where gold concentrations can increase to values higher than are found when gold is supplied as an inorganic salt. In the latter case, gold accumulates in the root and shoot cell wall with little soluble gold remaining. Gold in the form of the cyanide complex binds only slightly to soluble proteins and nucleic acids, but gold supplied as the chloride rapidly binds to these cell constituents. Presumably gold–cyanide formation is a mechanism that enables the plant to avoid the more toxic inorganic salts, thus permitting significant gold accumulation to take place.

Fluoroacetate Fluorine-accumulator plants reported from various geographical regions of the world belong to rather diverse families (Peters, 1960; Oelrichs and McEwan, 1961). *Gastrolobium grandiflorum* converts the absorbed inorganic fluoride to fluoroacetate, as much as 120 μg of fluorine per gram of dry weight (McEwan, 1964). Species that do not metabolize fluoride are apparently unable to accumulate it to high levels. The common tea plant *Camellia sinensis*, although containing elevated levels of fluorine compared with many agronomic plants, is not an effective accumulator and contains mainly the fluoride ion. It is of interest that *C. sinensis* has also been reported to be an aluminum accumulator (Sivasubramanian and Talibudeen, 1971). The significance of the production of fluoroacetate by accumulator plants remains obscure.

Zinc–organic acids Zinc-tolerant clonal populations of a range of plants, including *Agrostis tenuis, Deschampsia caespitosa, Silene cucubalus, Minuartia verna, Thlaspi alpestre, Anthoxanthum odoratum*, and *Plantago lanceolatum*, accumulate higher concentrations of malate, and to a lesser extent citrate, than do nontolerant clones. This finding has led to the view that zinc-malate may be involved in zinc transfer across the cytoplasm and perhaps in tolerance as well (Ernst, 1975; Ernst *et al.*, 1975; Mathys, 1977; Brookes *et al.*, 1981; Thurman and Rankin, 1982). Figure 4-4 shows the effect of increasing zinc concentration on the concentration of malate in zinc/cadmium-tolerant

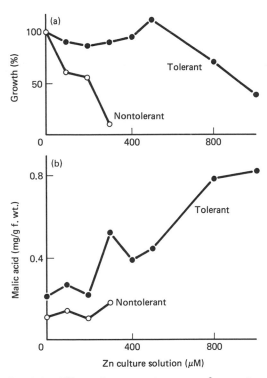

Fig. 4-4   Effect of exposure to a range of concentrations of zinc for 7 days on (a) growth and (b) malic acid contents of roots of zinc/cadmium-tolerant (●) and nontolerant (○) clones of *Deschampsia caespitosa.* (From Thurman and Rankin, 1982.)

and nontolerant clones of *Deschampsia caespitosa* (Thurman and Rankin, 1982). Increasing concentrations of zinc had no effect on malate concentrations in the nontolerant population up to 300 μM in culture, at which point growth was severely reduced, whereas the malate concentration steadily increased in the tolerant plants despite growth beyond 500 μM zinc in the culture solutions. Similarly, increasing the concentration of cadmium in culture also increased the malate content of the tolerant clones (but not the citrate content), compared with the control population.

The large malate pool is presumably genetically associated with increasing concentrations of zinc, because copper-, cobalt-, nickel-, or mercury-tolerant populations of many of these species do not stimulate malate production (Ernst, 1975; Thurman and Rankin, 1982). This suggests that some fundamental differences exist between zinc-tolerant populations and populations tolerant of copper and other metals. The lack of increased citrate concentration with increasing cadmium is, however, difficult to interpret if cadmium-citrate formation represents the tolerance mechanism for these plants.

The relationship between metal tolerance and accumulation of organic acids has been interpreted by Ernst (1975) and Ernst *et al.* (1975) as indicating that malate production is a mechanism for zinc tolerance. Thurman and Rankin (1982) have, however, cast serious doubts on this hypothesis. First, the amount of organic acids in zinc-tolerant *D. caespitosa* would account for only about one-third of the zinc, leaving the majority of the zinc available for complexation with unknown compounds other than organic acids. Second, the stability constants of zinc-citrate and zinc-malate are of a similar order of magnitude as the copper complexes; however, the plants are not copper-tolerant and indeed the mechanism of copper tolerance seems to be different from that for zinc. These authors concluded that increased zinc in the culture solution increased zinc uptake and stimulated the plants to produce organic acids to balance the increased cellular cations. Additional anion formation would thus be the result of the operation of a tolerance mechanism that permits the increase to occur, rather than the mechanism itself.

A possible direct correlation of high malate concentration with zinc tolerance in *Silene cucubalus* led Verkleij *et al.* (1981) to examine isoenzyme variation in three enzymes involved in malate metabolism: fumerase, the malic enzyme, and malic dehydrogenase. A clear difference was noted in isoenzyme variation of the malic enzyme but not in the other two enzymes. The authors also noted that the isoenzyme pattern of the malic enzyme changed during the development of tne plant in relation to increasing leaf age and hence zinc concentration. Neverthless, the relationship between isoenzyme variation and zinc tolerance remains obscure. Further adaptive enzyme differences between tolerant and nontolerant populations are discussed later in this section.

Copper–proline A specific copper–proline complex has been isolated from the roots of *Armeria maritima* (Farago and Mullen, 1979), but the relevance of the complex to copper tolerance is uncertain because high copper concentrations were found in the carbohydrate fraction of cell walls (Farago *et al.*, 1980). It is of interest that Farago and Mullen (1979) detected high concentrations of proline in the tissues of their tolerant plants. Because high concentrations of proline are produced by plants under arid conditions (Hegarty and Peterson, 1973) and because plants growing on tailings undoubtedly experience physiological water stress, the high proline values may not be a direct response to metal toxicity. Reilly (1972) has also described the occurrence of copper–amino acid complexes in both the copper accumulating *Becium homblei* and in a nontolerant ecotype. He did not, however, consider that the complexes were of importance in the development of copper tolerance.

High-Molecular-Weight Compounds

Metallothionein Metallothioneins have been isolated, characterized, and their possible functions closely examined, throughout the animal kingdom, during these past several decades. Typically, metallothioneins account for 70 to 95 percent of the total cadmium in liver, kidney, and shellfish tissues. However, it is only within the past few years that comparable studies on plants have been undertaken and metallothionein-like compounds identified in fungi, blue-green algae, and various higher plants.

Premakumar *et al.* (1975) reported that a copper-binding protein low in sulfur was produced in plants in response to copper exposure. The protein was termed a *copper chelatin*, although Bremner and Young (1976) suggested that it was an artifact of isolation. Nicholson *et al.* (1980), working with mung bean shoots and roots, concluded that the suspected copper chelatin was in reality a plastocyanin and they found no evidence of either a copper-binding chelatin or of a copper metallothionein.

Sulfur-rich metallothionein-type proteins have been identified in metal-stressed roots of corn, soybean, and beans (Rauser, 1981; Casterline and Barnett, 1982) and in tomato, cabbage, and tobacco (Bartolf *et al.*, 1980; Wagner and Trotter, 1982). Wagner and Trotter found evidence for a concentration-dependent induction of the cadmium metallothionein in cabbage and tobacco leaves (Fig. 4-5). The formation of this cadmium protein may represent a cadmium-tolerance mechanism. Many investigators, working with leaves from plants grown under metal stress for relatively short periods of time, have failed to detect other than trace amounts of metallothioneins. Wagner and Trotter (1982) have recently shown that although there is a low concentration of the metallothionein present in the leaves of control plants, the induced metallothionein is only

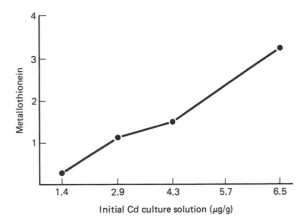

Fig. 4-5   Induction of a cadmium metallothionein–type complex measure after [109]Cd binding in the leaves of *Brassica capitata*. (Modified from Wagner and Trotter, 1982.)

synthesized in quantity after 2 weeks or more of cadmium stress (Fig. 4-6), thus suggesting an explanation for the variable results reported in some studies.

In the only study of its type, Curvetto and Rauser (1979), Rauser and Curvetto (1980) and Rauser (1981) have extracted a copper-binding protein from copper-tolerant clones of *Agrostis gigantea* and have shown that the protein belongs to the copper metallothionein class. Quantitative estimates and concentration-dependent induction studies were, however, not made. Hence the relevance of these unusual proteins to metal tolerance in tolerant varieties, ecotypes, and so on, remains to be established.

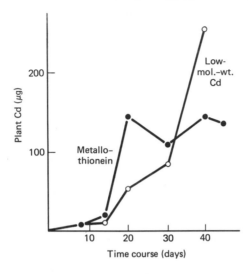

Fig. 4-6   Time course of occurrence of cadmium metallothionein–type complex (●) and <2000 molecular weight cadmium (○) in the leaves of *Brassica capitata* supplied with constant cadmium. (Modified from Wagner and Trotter, 1982.)

Enzyme Changes

Enzyme changes conferring plant resistance to an element would provide an excellent biochemical basis for the evolutionary development of metal tolerance in plants, but most experimental results to date fail to support such a hypothesis. For example, metal-resistant nitrate reductase, malate dehydrogenase, isocitrate dehydrogenase, or glucose-6-phosphate dehydrogenase enzymes could not be detected in in vitro experiments with enzymes from zinc-tolerant, copper-tolerant, and nontolerant populations of *Silene cucubalus* (Mathys, 1975). Similarly, there was no difference in cytoplasm sensitivity to zinc with malate and glutamate dehydrogenase extracted from the roots of zinc-tolerant and nontolerant clones of *Anthoxanthum odoratum* (Brookes *et al.*, 1981).

More recently, Thurman and Rankin (1982) have examined the zinc sensitivity of several enzymes involved in organic acid metabolism in crude extracts of roots of *Deschampsia caespitosa*, namely, aconitase malate dehydrogenase, pyruvate kinase, PEP carboxylase, and citrate synthase. No significant differences were detected in the zinc sensitivity of any of these enzymes in crude root extracts from either zinc/cadmium-tolerant or nontolerant clones. Perhaps the absence of any difference in sensitivity, with these or other enzymes from tolerant and nontolerant populations, indicates that zinc occurs in the cytoplasm not as ionic zinc but as a zinc–organic acid complex, or that zinc is accumulated in the vacuoles. Thus there are no absolute requirements for enzymes with altered substrate or metal sensitivities.

Wainwright and Woolhouse (1975), working with copper-tolerant and nontolerant *Agrostis tenuis*, have concluded that cell wall acid phosphatases, from plants growing on contaminated sites, formed less stable complexes than the enzymes from plants growing on uncontaminated locations. The enzyme studies of Cox and Thurman (1978), based on zinc-tolerant and nontolerant clones of *Anthoxanthum odoratum*, revealed that the tolerant clones possessed significantly higher K_i values for soluble and cell wall acid phosphatases than did the corresponding values for the nontolerant clones. The overall correlation of high specific activity for acid phosphatases with the index of tolerance was suggested but there was no correlation between one set of these enzymes and the index of tolerance. Such results support the view that enzyme adaptations may apply only to root cell wall enzymes that come into contact with high metal concentrations at the soil–root interfaces. Although such changes of specificity are undoubtedly of physiological and biochemical importance, they may not be directly relevant to the concept of whole plant tolerance.

CELLULAR ADAPTATIONS

Cytoplasmic Sensitivity

The plasmatic resistance of various populations of plants against graduated concentrations of metal-containing solutions has been tested by the method of comparative protoplasmatology. Leaf epidermal cells from nontolerant *Indigofera setiflora* retain their capacity for plasmolysis only up to 0.004 mM nickel, whereas cells from the nickel-tolerant plants required 0.4 mM nickel to induce the same response (Ernst, 1972). Similarly, a cytoplasmic resistance to either copper or zinc was noted in copper- or zinc-tolerant *I. setiflora* and *I. dyeri*. This effect was shown to be metal specific because neither the copper- or zinc-tolerant plants exhibited cytoplasmic sensitivity to nickel.

More recently, studies by Vergnano-Gambi (personal communication) have shown that leaf epidermal cells of the nickel-accumulating *Alyssum bertolonii* show greater resistance to that element than does the control *A. argenteum*.

Root Cell Wall Binding

The reduced transport of copper from roots to shoots of some grasses has already been mentioned (Table 4-1) but this restriction is not confined to this element. A number of studies have shown that zinc-tolerant as well as copper-tolerant grasses, and *Silene maritima*, also restrict the transport of these elements to leafy tissues (Antonovics *et al.*, 1971; Baker, 1974, 1978). Studies with *S. maritima* demonstrated that the correlation between the edaphic adaptation and zinc tolerance was significant because the root/shoot zinc concentration ratio was correlated with the total soil zinc (Fig. 4-7). Similar work by Matthews and Thornton (1982) on a cadmium-toxic site at Shipham, United Kingdom, has shown that cadmium tolerance, as measured by relative root extension, relates clearly to the amount of cadmium present in the soils from which the plant was collected (Fig. 4-8). These plants also exhibit a zinc tolerance and zinc and cadmium tolerance were directly correlated. However, the degree of zinc tolerance appeared to be lower than that

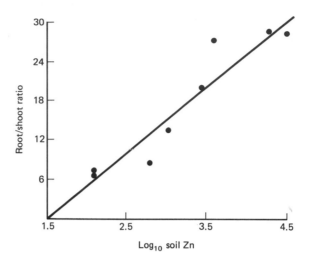

Fig. 4-7   Linear regression of root/shoot zinc concentration ratio onto log-transformed total soil zinc levels for population samples of *Silene maritima* grown in solution cultures. (From Baker, 1978.)

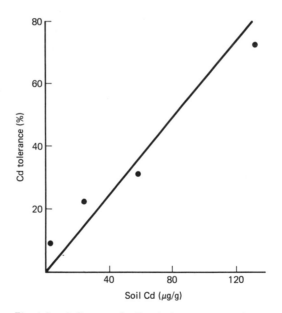

Fig. 4-8   Influence of soil cadmium concentration on the degree of cadmium tolerance in roots of *Holcus lanatus* from Shipham, United Kingdom pastures. (From Matthews and Thornton, 1982.)

for cadmium. Plants that exhibited a 70 percent cadmium tolerance based on root extension measurements were only 20 percent tolerant of zinc.

Immobilization of zinc and copper in the roots poses the question of the cellular location of these metals. Differential solvent extraction of root tissue homogenates has revealed the washed cell wall fraction as the major site of zinc or copper localization in tolerant ecotypes of *A. tenuis* and *A. stolonifera* (Peterson, 1969; Turner, 1970; Turner and Marshall, 1971, 1972). Furthermore, the extracted residues from tolerant plant roots contained a higher percentage of zinc than did those of the nontolerant plants (Peterson, 1969). Indeed, the greater the tolerance to zinc, the greater the amount of zinc in the cell wall (Fig. 4-9). The zinc distribution in the copper-tolerant plants was similar to that in the nontolerant plants, indicating that the tolerance mechanism for the two elements is different. Chemical fractionation revealed that much of the zinc (greater than 50 percent) was released into solution when the pectates were solubilized, but less than 20 percent was solubilized following digestion with the proteolytic enzyme Pronase. Wyn Jones *et al.* (1971), although finding that trypsin released significant quantities of zinc from cell walls, reported that 60 percent of the bound zinc was released by cellulase digestion and this fraction chromatographed on Sephadex G25 predominantly as a major band.

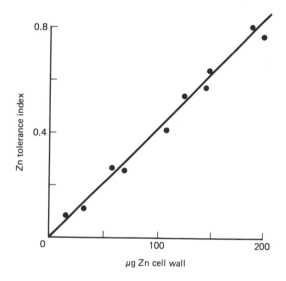

Fig. 4-9 Relationship between the zinc-tolerance index in roots and the concentration of zinc in the cell wall fraction of *Agrostis tenuis* clones collected from sites exhibiting different zinc concentrations in the soil. (From Turner and Marshall. 1972.)

Comparable solvent extraction experiments by Farago *et al.* (1980) have also shown that much of the copper in the tolerant *Armeria maritima* is associated with the pectates and other carbohydrates in the roots. Their results are in agreement with earlier work indicating that tolerance to copper is somehow linked with metal storage in the cell walls.

These observations undoubtedly lead to the hypothesis that in some copper- and zinc-tolerant plants, metal binding in root cell wall is an important process in the development of a tolerance mechanism, but the exact nature of the mechanism has not been elucidated. Further studies are required on the molecular structure of the cell walls which appear to confer the observed element specificity for metal binding. Moreover, accumulation by root cell walls is not a universal feature of metal tolerance, and additional mechanisms must also operate in plants.

Cell Membrane Permeability

In view of the number of membrane-bound enzymes and the importance of membrane integrity for cellular processes, it is surprising that so little work has been directed toward the study of the relationships between accumulation and membrane protection from such high metal concentrations. The most relevant study is a series of experiments performed by Wainwright and Woolhouse (1975) designed to measure the leakage of K^+ ions from roots of copper- and zinc-tolerant *Agrostis tenuis* plants. Increasing the concentration of copper caused increasing leakage of K^+ from nontolerant and zinc-tolerant roots but much less leakage from copper-tolerant roots. These results were interpreted as meaning that zinc ions did not damage the plasmalemma whereas copper ions did, suggesting that plasmalemma modifications are involved in copper tolerance but not in zinc tolerance. A similar study by Benson and Peterson (unpublished) on copper- and arsenic-tolerant races of *A. tenuis* revealed that increasing arsenic concentrations did not affect the leakage of K^+ ions but increasing copper had the same effect as that described earlier by Wainwright and Woolhouse (1975). These results again indicate that plants have responded differently toward copper, zinc, and other elements during the evolutionary development of their accumulation and tolerance mechanisms.

CONCLUSION

The mechanisms involved in plant tolerance of toxic soils have largely been shown to comprise differences of cellular and subcellular distributions of metals and organometals characteristic of particular species, ecotypes, and varieties. Yet further detailed biochemical research is required to equate the phenomenon of organometal complex formation with the adaptive response. In many studies the concentration of metals and organometals is considered from the plant chemistry viewpoint as being static parameters rather than active processes.

Brookes *et al.* (1981), in one of the few studies of its type, carried out compartmental flux analysis using zinc and reported that only the zinc-tolerant clone of *Deschampsia caespitosa* when exposed to concentrations of up to 1 mM zinc was capable of actively pumping zinc from the cytoplasm of root cells across the tonoplast into the vacuoles. This process was, however, inhibited in the nontolerant clone above 0.1 mM zinc. Metal tolerance or components of tolerance in many plants are undoubtedly associated with active processes; two of these are metallothionein formation and the development of cytoplasmic resistance in some accumulators. In vitro and in vivo enzymatic studies should also be a fruitful area for research, yet results to date are uniformly disappointing, with the exception of the selenoamino acid biosynthesis.

It can be concluded that the strategies adopted by plants to avoid, restrict, or alleviate potentially toxic conditions are many and varied, making it unlikely that a unifying adaptive principle lies behind the development of tolerance throughout the plant kingdom.

REFERENCES

Antonovics, J., Bradshaw, A. D., and Turner, R. G., 1971, Heavy metal tolerance in plants: *Adv. Ecol. Res.*, v. 7, p. 1–85.

Baker, A. J. M., 1974, Heavy metal tolerance and population differentiation in *Silene maritima* With.: Ph.D. thesis, University of London.

———, 1978, Ecophysiological aspects of zinc tolerance in *Silene maritima* With.: *New Phytol.*, v. 80, p. 635–642.

Barry, S. A. S., and Clark, S. C., 1978, Problems of interpreting the relationship between the amounts of lead and zinc in plants and soil on metalliferous wastes: *New Phytol.*, v. 81, p. 773–783.

Bartolf, M., Brennan, E., and Price, L., 1980, Partial characterization of a cadmium-binding protein from the roots of cadmium-treated tomato: *Plant Physiol.*, v. 66, p. 438–441.

Bradshaw, A. D., 1976, Pollution and evolution, p. 135–159 *in* Mansfield, T. A. (Ed.), *Effects of Air Pollutants on Plants:* Cambridge University Press, London.

Bremner, I., and Young, B. S., 1976, Isolation of (copper, zinc)-thioneins from the livers of copper-injected rats: *Biochem. J.*, v. 157, p. 517–520.

Brookes, A., Collins, J. C., and Thurman, D. A., 1981, The mechanism of zinc tolerance in grasses: *J. Plant Nutr.*, v. 3, p. 695–705.

Brooks, R. R., Lee, J., and Jaffré, T., 1974, Some New Zealand and New Caledonian plant accumulators of nickel: *J. Ecol.*, v. 62, p. 493–499.

———, Shaw, S., and Marfil, A. A., 1981, The chemical form and physiological function of nickel in some Iberian *Alyssum* species: *Physiol. Plant*, v. 51, p. 167–170.

Brown, T. A., and Shrift, A., 1980, Identification of selenocysteine in the proteins of selenate-grown *Vigna radiata: Plant Physiol.*, v. 66, p. 758–761.

———, and Shrift, A., 1981, Exclusion of selenium from proteins of selenium-tolerant *Astragalus* species: *Plant Physiol.*, v. 67, p. 1051–1053.

Burnell, J. N., 1981, Selenium metabolism in *Neptunia amplexicaulis: Plant Physiol.*, v. 67, p. 316–324.

———, and Shrift, A., 1979, Cysteinyl-tRNA synthetase from *Astragalus* species: *Plant Physiol.*, v. 63, p. 1045–1097.

Cannon, H. L., 1971, The use of plant indicators in ground water surveys, geologic mapping and mineral prospecting: *Taxon*, v. 20, p. 227–256.

Casterline, J. L., and Barnett, B. M., 1982, Cadmium-binding components in soybean plants: *Physiol. Plant.*, v. 69, p. 1004–1007.

Chenery, E. M., and Sporne, K. R., 1976, A note on the evolutionary status of aluminium-accumulators among dicotyledons: *New Phytol.*, v. 76, p. 551–554.

Cox, R. M., and Thurman, D. A., 1978, Inhibition by zinc of soluble and cell wall acid phosphatases of roots of zinc tolerant and non-tolerant clones of *Anthoxanthum odoratum: New Phytol.*, v. 80, p. 17–22.

Curvetto, N. R., and Rauser, W. E., 1979, Isolation and characterization of copper-binding proteins from roots of *Agrostis gigantea* tolerant to excess copper: *Plant Physiol.*, v. 63, p. 559.

Ernst, W., 1972, Ecophysiological studies on heavy metal plants in South Central Africa: *Kirkia*, v. 8, p. 125–145.

———, 1975, Physiology of heavy metal resistance in plants, p. 121–136 *in Heavy Metals in the Environment*, Vol. 2: University of Toronto, Toronto.

_____, Mathys, W., and Janiesch, P., 1975, Physiologische grundlagen der Schwermetallresistenz-enzymattivitäten und organische saüren: *Forschungsber. Landes Nordrhein-Westfalen*, v. 2496, p. 1–50.

Farago, M. E., and Mullen, W. A., 1979, Plants which accumulate metals. Pt. IV. A possible copper-proline complex from the roots of *Armeria maritima: Inorg. Chim. Acta. Lett.*, v. 32, p. L93–94.

_____, Mullen, W. A., Cole, M. M., and Smith, R. F., 1980, A study of *Armeria maritima* (Mill.) Willdenow growing in a copper-impregnated bog: *Environ. Pollut.*, v. A21, p. 225–244.

Girling, C. A., and Peterson, P. J., 1978, Uptake, transport and localization of gold in plants: *Trace Sub. Environ. Health*, v. 12, p. 105–118.

_____, Peterson, P. J., and Warren, H. V., 1979, Plants as indicators of gold mineralization at Watson Bar, British Columbia, Canada: *Econ. Geol.*, v. 74, p. 902–907.

Hegarty, M. P., and Peterson, P. J., 1973, Amino acids, p. 1–62 *in* Butler, G. W., and Bailey, R. W. (Eds.), *Chemistry and Biochemistry of Herbage:* Academic Press, New York.

Hogan, G. D., and Rauser, W. E., 1981, Role of copper binding, absorption, and translocation in copper tolerance of *Agrostis gigantea* Roth.: *J. Exp. Bot.*, v. 32, p. 27–36.

Jaffré, T., 1979, Accumulation du manganèse par les Protéacées de Nouvelle-Calédonie: *C. R. Acad. Sci. Paris D*, v. 289, p. 425–428.

_____, Kersten, W., Brooks, R. R., and Reeves, R. D., 1979, Nickel uptake by Flacourtiaceae of New Caledonia: *Proc. R. Soc. Lond. B*, v. 205, p. 385–394.

Kerdel-Vegas, F., Wagner, F., Russell, P. B., Grant, N. H., Alburn, H. E., Clark, D. E., and Miller, J. A., 1965, Structure of the pharmacologically active factor in the seeds of *Lecythis ollaria: Nature (Lond.)*, v. 205, p. 1186–1187.

Landsberg, E. C., 1981, Organic acid synthesis and release of hydrogen ions in response to Fe deficiency stress of mono- and dicotyledonous plant species: *J. Plant Nutr.*, v. 3, p. 579–591.

Lee, J., Brooks, R. R., Reeves, R. D., and Jaffre, T., 1977a, Chromium accumulating bryophyte from New Caledonia: *Bryologist*, v. 80, p. 203–205.

_____, Reeves, R. D., Brooks, R. R., and Jaffre, T., 1977b, Isolation and identification of a citrato-complex of nickel from nickel-accumulating plants: *Phytochemistry*, v. 16, p. 1503–1505.

_____, Reeves, R. D., Brooks, R. R., and Jaffre, T., 1978, The relationship between nickel and citric acid in some nickel-accumulating plants: *Phytochemistry*, v. 17, p. 1033–1035.

Lyon, G. L., Peterson, P. J., and Brooks, R. R., 1969a, Chromium-51 distribution in tissues and extracts of *Leptospermum scoparium: Planta (Berl.)*, v. 88, p. 282–287.

_____, Peterson, P. J., and Brooks, R. R., 1969b, Chromium-51 transport in the xylem sap of *Leptospermum scoparium* (Manuka): *N.Z. J. Sci.*, v. 12, p. 541–545.

_____, Peterson, P. J., Brooks, R. R., and Butler, G. W., 1971, Calcium, magnesium and trace elements in a New Zealand serpentine flora: *J. Ecol.*, v. 59, p. 421–429.

Mathys, W., 1975, Enzymes of heavy-metal-resistant and nonresistant populations of *Silene cucubalus* and their interaction with some heavy metals *in vitro* and *in vivo: Physiol. Plant.*, v. 33, p. 161–165.

_____, 1977, The role of malate, oxalate and mustard oil, glucosides in the evolution of zinc-resistance in herbage plants: *Physiol. Plant:* v. 40, p. 130–136.

Matthews, H., and Thornton, I., 1982, Seasonal and species variation in the content of cadmium and associated metals in pasture plants at Shipham: *Plant Soil*, v. 66, p. 181–193.

McEwan, T., 1964, Isolation and identification of the toxic principle *Gastrolobium grandiflorum: Nature (Lond.)*, v. 201, p. 827.

Menezes de Sequeira, E., 1968, Toxicity and movement of heavy metals in serpentine soils (north-eastern Portugal); *Agron. Lusit.*, v. 30, p. 115–154.

Minguzzi, C., and Vergnano, O., 1948, Il contenuto di nichel nelle ceneri di *Alyssum bertolonii* Desv.: *Mem. Soc. Tosc. Sci. Nat.*, v. A55, p. 49–74.

Myers, A. T., and Hamilton, J. C., 1960, Rhenium in plant samples from the Colorado Plateau: *Geol. Soc. Am. Bull.*, v. 71, p. 1934.

Nicholson, C., Stein, J., and Wilson, K. A., 1980, Identification of the low molecular weight copper protein from copper-intoxicated mung bean plants: *Plant Physiol.*, v. 66, p. 272–275.

Nicolls, O. W., Provan, D. M. G., Cole, M. M., and Tooms, J. S., 1965, Geobotany and geochemistry in mineral exploration in the Dugald River area, Cloncurry, Australia: *Trans. Inst. Min. Metall.*, v. 74, p. 695–799.

Nigam, S. N., and McConnell, W. B., 1976, Isolation and identification of two isomeric glutamylselenocystathionines from the seeds of *Astragalus pectinatus: Biochim. Biophys. Acta*, v. 437, p. 116–121.

Oelrichs, P. B., and McEwan, T., 1961, Isolation of the toxic principle in *Acacia georginae: Nature (London)*, v. 190, p. 808–809.

Pancaro, L., Pelosi, P., Vergnano-Gambi, O., and Galoppini, C., 1978, Ulteriori indagini sul rapporto tra nichel e acidi malico e malonico in *Alyssum: Giorn. Bot. Ital.*, v. 112, p. 141–146.

Pelosi, P., Fiorentini, R., and Galoppini, C., 1976, On the nature of nickel compounds in *Alyssum bertolonii* Desv., Pt. 2: *Agric. Biol. Chem.*, v. 40, p. 1641–1649.

Peters, R., 1960, Fluorine compound in African plants: *Biochem. J.*, v. 76, p. 32.

Peterson, P. J., 1969, The distribution of zinc-65 in *Agrostis tenuis* Sibth., and *A. stolonifera* L. tissues: *J. Exp. Bot.*, v. 20, p. 863–875.

——, 1971, Unusual accumulation of elements by plants and animals: *Sci. Prog. (Oxf.)*, v. 59, p. 505–526.

——, 1978, Lead and vegetation, p. 355–384 *in* Nriagu, J. O. (Ed.), *The Biogeochemistry of Lead in the Environment:* Elsevier/North-Holland, Amsterdam.

——, 1979, Geochemistry and ecology: *Philos. Trans. R. Soc. Lond. B*, v. 288, p. 169–177.

——, 1983, Unusual element accumulation as a taxonomic character, p. 167–173 *in* Metcalfe, C. R., and Chalk, L. (Eds.), *Anatomy of the Dicotyledon: leaves, stems and wood in relation to taxonomy*, 2nd ed., Vol. 2: Clarendon Press, Oxford.

——, and Alloway, B. J., 1979, Cadmium in soils and vegetation, p. 45–92 *in* Webb, M. (Ed.), *The Chemistry, Biochemistry, and Biology of Cadmium:* Elsevier/North-Holland, Amsterdam.

——, and Butler, G. W., 1962. The uptake and assimilation of selenite by higher plants: *Aust. J. Biol. Sci.*, v. 15, p. 126–146.

——, and Butler, G. W., 1967, Significance of selenocystathionine in an Australian selenium-accumulating plant, *Neptunia amplexicaulis: Nature (Lond.)*, v. 213, p. 599–600.

——, and Butler, G. W., 1971, The occurrence of selenocystathionine in *Morinda reticulata* Benth., a toxic seleniferous plant: *Aust. J. Biol. Sci.*, v. 24, p. 175–177.

_____ , and Girling, C. A., 1981, Other trace metals, p. 213–278 *in* Lepp, N. W. (Ed.), *Effect of Heavy Metal Pollution on Plants*, Vol. 1: Applied Science Publishers, Barking, Essex, England.

_____ , and Robinson, P. J., 1972, L-cystathionine and its selenium analogue in *Neptunia amplexicaulis: Phytochemistry*, v. 11, p. 1837–1839.

_____ , Benson, L. M., and Zieve, R., 1981, Metalloids, p. 279–342 *in* Lepp, N. W. (Ed.), *Effect of Heavy Metal Pollution on Plants*, Vol. 1: Applied Science Publishers, Barking, Essex, England.

Porter, E. K., and Peterson, P. J., 1977, Arsenic tolerance in grasses growing in mine waste: *Environ. Pollut.*, v. 14, p. 255–267.

Premakumar, R., Winge, D. R., Wiley, R. D., and Rajagopalan, K. V., 1975: Copper-chelatin: isolation from various eucaryotic sources: *Arch. Biochem. Biophys.*, v. 170, p. 278–288.

Proctor, J. and Woodell, S. R. J., 1971, The plant ecology of serpentine. I. Serpentine vegetation of England and Scotland: *J. Ecol.*, v. 59, p. 375–395.

Rauser, W. E., 1981, Occurrence of metal-binding proteins in plants, p. 281–284 *in* *International Conference on Heavy Metals in the Environment:* CEP Consultants Ltd., Edinburgh.

_____ , and Curvetto, N. R., 1980, Metallothionein occurs in roots of *Agrostis* tolerant to excess copper: *Nature (Lond.)*, v. 287, p. 563–564.

Reilly, C., 1972, Amino acids and amino acid copper complexes in water-soluble extracts of copper-tolerant and non-tolerant *Becium homblei: Z. Pflanzenphysiol.*, v. 66, p. 294–296.

Reilly, A., and Reilly, C., 1973, Zinc, lead and copper tolerance in the grass *Stereochlaena cameronii* (Stapf.) Clayton: *New Phytol.*, v. 72, p. 1041–1046.

Rosenfeld, I., and Beath, O. A., 1964, *Selenium:* Academic Press, New York.

Roberts, B. A., 1980, Concentration of micronutrients "trace elements" in native plants growing on serpentine soil from western Newfoundland: *Bot. Soc. Am., Misc. Ser. Publ. 158*, p. 95.

Severne, B. C., and Brooks, R. R., 1972, A nickel accumulating plant from Western Australia: *Planta (Berl.)* v. 103, p. 91–94.

Shewry, P. R., and Peterson, P. J., 1976, Distribution of chromium and nickel in plants and soil from serpentine and other sites: *J. Ecol.*, v. 64, p. 195–212.

Shrift, A., and Virupaksha, T. K., 1965, Seleno-amino acids in selenium accumulating plants: *Biochim. Biophys. Acta*, v. 100, p. 65–75.

Sivasubramanian, S., and Talibudeen, O., 1971, Effect of aluminium on growth of tea (*Camellia sinensis*) uptake of potassium and phosphorus: *J. Sci. Food Agric.*, v. 22, p. 325–329.

Sporne, K. R., 1969, The ovule as an indicator of evolutionary status in angiosperms: *New Phytol.*, p. 555–565.

Tiffin, L. O., 1971, Translocation of nickel in xylem exudate of plants: *Plant Physiol.*, v. 48, p. 273–277.

Thurman, D. A., and Rankin, J. L., 1982, The role of organic acids in zinc tolerance in *Deschampsia caespitosa: New Phytol.*, v. 91, p. 629–635.

Turner, R. G., 1970, The subcellular distribution of zinc and copper within the roots of metal-tolerant clones of *Agrostis tenuis* Sibth.: *New Phytol.*, v. 69, p. 725–731.

_____ , and Marshall, C., 1971, The accumulation of [65]-zinc by root homogenates of zinc-tolerant and non-tolerant clones of *Agrostis tenuis* Sibth.: *New Phytol.*, v. 70, p. 539–545.

——, and Marshall, C., 1972, The accumulation of zinc by subcellular fractions of roots of *Agrostis tenuis* Sibth., in relation to zinc tolerance: *New Phytol.*, v. 71, p. 671–676.

Verkleij, J. A. C., Marissen, A., and Lugtenborg, T. F., 1981, Genetic aspects of heavy metal stress on different metal-tolerant populations of *Silene cucubalus,* p. 296–299 *in International Conference on Heavy Metals in the Environment:* CEP Consultants Ltd., Edinburgh.

Wagner, G. J., and Trotter, M. M., 1982, Inducible cadmium binding complexes of cabbage and tobacco: *Plant Physiol.*, v. 69, p. 804–809.

Wainwright, S. J., and Woolhouse, H. W., 1975, Physiological mechanisms of heavy metal tolerance in plants, p. 231–257 *in* Chadwick, M. J., and Goodman, G. T. (Eds.), *The Ecology of Resource Degradation and Renewal:* Blackwell Scientific, Oxford.

Wallace, A., 1981, Some physiological aspects of iron deficiency in plants: *J. Plant Nutr.*, v. 3, p. 637–642.

Wild, H., 1970, Geochemical anomalies in Rhodesia. Pt. 3. The vegetation of nickel-bearing soils: *Kirkia*, v. 7, p. 1–62.

——, 1974, Indigenous plants and chromium in Rhodesia: *Kirkia*, v. 9, p. 233–242.

——, 1978, The vegetation of heavy metal and other toxic soils, p. 1301–1332 *in* Weger, M. J. A. (Ed.), *Biogeography and Ecology of Southern Africa:* Junk, The Hague.

——, and Bradshaw, A. D., 1977, The evolutionary effects of metalliferous and other anomalous soils in South Central Africa: *Evolution*, v. 31, p. 282–293.

Wu, L., and Antonovics, J., 1975, Zinc and copper uptake by *Agrostis stolonifera*, tolerant to both zinc and copper: *New Phytol.*, v. 75, p. 231–237.

——, Bradshaw, A. D., and Thurman, D. A., 1975, The potential for evolution of heavy metal tolerance in plants. III. The rapid evolution of copper tolerance in *Agrostis stolonifera: Heredity*, v. 34, p. 165–187.

Wyn Jones, R. G., Sutcliffe, M., and Marshall, C., 1971, Physiological and biochemical basis for heavy metal tolerance in clones of *Agrostis tenuis*, p. 575–581 *in* Samish, R. M. (Ed.), *Recent Advances in Plant Nutrition*, Vol. 2: Gordon and Breach, New York.

J. H. Richards
Department of Range Science and The Ecology Center
Utah State University
Logan, Utah

5

ROOT FORM AND DEPTH DISTRIBUTION IN SEVERAL BIOMES

Genetic and environmental factors, with the latter usually dominating, operate to produce great variability in the form of plant root systems. Some plant groups exhibit little plasticity in root development, however, resulting in deep-rooting (e.g., phreatophytes) or shallow-rooting (e.g., cacti) patterns. Foremost among the environmental factors controlling root development are water and macronutrient availability. Root depth distribution profiles from several biomes and plant community types illustrate the role of these factors and indicate that the great majority of plant roots occur in the upper 1 to 2 m of soil. Detailed studies of the seasonal timing and location of root growth emphasize the importance of moisture, adequate temperatures, and macronutrient availability in determining the location of root activity in the soil profile. Secondary environmental factors controlling root development include soil physical characteristics such as pore sizes, hardness, and aeration. Species capable of deep rooting may be limited in rooting depth by these soil features, causing significant habitat-dependent variability in root form and depth within a single species. Despite this variability, plant root systems explore very large soil volumes, sometimes at distinctive depths. These characteristics can be exploited to advantage for mineral exploration purposes.

INTRODUCTION

The development of root systems in the soil is a process in which expression of the potential, genetically determined root morphology is usually controlled more by the physical properties of the soil and the environmental conditions during the growth of the plant than by the genetic constitution of the plant. This process leads to within-species variability in root form and indicates the need for habitat-specific determinations of the rooting patterns of plants if their rooting depth and distribution must be known for planning or interpretation of biogeochemical surveys. Even within any one site, however, the variability of rooting system form and depth distribution between species is great. This results partially from the different reactions of individual species to environmental conditions, and also from the fact that some species have relatively nonplastic development of root systems. The latter group of species usually has more rigid habitat requirements than do species that show plasticity in their root development. Thus, in addition to variability within one species over a variety of sites, there is also a great deal of variability in root system form and depth distribution at any one site due to the particular makeup of the flora at that site and the root systems of those species.

Some generalizations can be made regarding how environmental factors affect the development and depth distribution of root systems, and in some cases generalizations can be made about the depth distribution of roots of the whole plant community in a particular habitat. Because of the variability of root development between species and habitat, it should be possible to select individual species that would be of particular use in biogeochemical surveys. Second, it should be possible through an understanding of the environmental factors affecting root development to predict which zones of the soil will not be sampled by plant roots.

I have avoided including a large number of references in the text of this chapter to provide easier reading. The ideas expressed are, however, supported by examples and authors cited in the figures and the table.

FORM OF ROOT SYSTEMS

The great variety of root system form is based on differential development of various parts of two basic types of root systems. One of the basic root system forms is a taproot system, which develops from a primary root and its branches or lateral roots (Fig. 5-1a). This type of system is found in gymnosperms and dicotyledonous plants. Variation in form of taproot systems is usually due to the relative development of the primary root and the branches, and to the angle and degree of secondary branching of lateral roots. In generalized taproot systems the primary root and laterals are developed to approximately equal extent, while in specialized taproot systems either the primary root or the

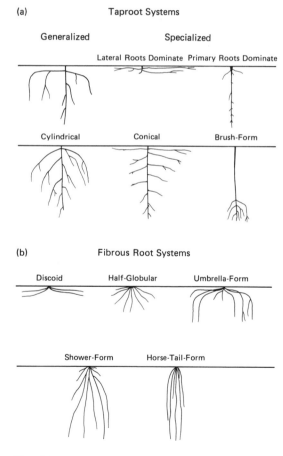

Fig. 5-1 Schematic diagrams of some root system forms. Modified after van Donselaar-ten Bokkel Huinink (1966) and Cannon (1911).

ROOT FORM AND DEPTH DISTRIBUTION IN SEVERAL BIOMES

laterals dominate. This differential development may be genetically or environmentally mediated.

The scale of taproot systems is also extremely variable. In annual plants size is usually fairly limited, but in perennial plants, which are the focus of this chapter, root systems can develop to great size. This fact alone is of significance to biogeochemical exploration. Large root systems explore and absorb ions from very large volumes of substrate. Additionally, perennial plant roots reexplore much of the same soil mass each growing season (Fernandez and Caldwell, 1975). Thus perennial plants sample a much larger volume of soil than could be analyzed in a soil survey, and concentrate and potentially accumulate ions of interest in a relatively small plant biomass.

Another factor contributing to variation in form of taproot systems is the degree of secondary growth exhibited by the various parts of the root system. The classification shown in Fig. 5-1 is quite subjective and certainly does not include all types of root systems. It is shown, however, to provide a framework for the examples and discussion that follow. Ludwig (1977) has suggested a way to make such root form classification more quantitative, and Fitter (1982) outlines an application of morphometric analysis to root systems.

The second basic type of root system is a fibrous root system, which is usually found in monocotyledons. Fibrous root systems are usually composed of a relatively large number of adventitious roots, all of which are approximately equal in importance and development. In perennial plants these adventitious roots can be truly perennial, although they seldom have any secondary thickening. Variation in form of fibrous root systems, some types of which are illustrated in Fig. 5-1b, depends mainly on the angle of penetration, the number, and the degree of branching of the adventitious roots. Fibrous root systems often have a large number of lateral branches on each adventitious root, leading to high root densities and intensive exploration of the soil volume. This contrasts with the often extensive nature of taproot systems.

Examples of the variation in form of both fibrous and taproot systems are given in scale drawings to be found in some of the classic works on root systems. For example, the works of Weaver (1919, 1920) and Clements *et al.* (1929) illustrate the great variety of root systems of plants found in the prairie regions of the United States. Cannon (1911) has described in detail the rooting habits of plants in the Sonoran Desert. More recently, van Donselaar-ten Bokkel Huinink (1966) has described a wide variety of root systems of species growing in savannas of northern Surinam.

These descriptive studies illustrate very well the great variation in root system form that can be achieved by an individual species growing in different soil conditions in adjacent habitats, and by different species growing in the same site. For example, comparing the root habits of plants in tall grass prairie and in drier, mixed and sandhills prairie associations, Weaver (1920) found that in the tallgrass prairie the roots did not tend to spread very widely in the surface soil, and that the depth at which the largest number of absorbing roots occurred was greater. This is illustrated by the lateral spread of roots of little bluestem, *Schizachyrium scoparium* [*Andropogon scoparius* in Weaver (1920)]. In the true prairie the little bluestem root system spread about 0.3 m and had a horsetail form, while in the sandhill mixed prairie it spread about 1 m and had a half-globular form. These differences in rooting habits appear to be correlated with the availability of moisture at greater depths in the soil in the tallgrass prairie.

Cannon (1911), studied the root systems of creosote bush, *Larrea tridentata* [*Covillea tridentata* in Cannon (1911)], in the Sonoran Desert. He found that creosote bush growing on the bajada, where there is a hard caliche layer at 35 to 50 cm, had root systems limited to the upper 50 cm of soil. Plants growing in the neighboring floodplain had a deep main root which would penetrate at least 2 m into the soil. Bur sage, *Ambrosia deltoidea* [*Franseria deltoidea* in Cannon (1911)], showed similar variability. The cacti, however, were found always to exhibit a specialized root system with highly developed, wide-spreading, shallow lateral roots. These differences illustrate the importance of examining the root systems of individual species growing in the habitats where they will be sampled if one is properly to interpret the zones of soil in which their roots are growing and absorbing ions. For the purposes of an orientation survey, sampling need not be done in nearly the detail obtained by these early workers. Observation and rapid mapping of several profile walls would usually suffice to determine the soil layers in which the majority of roots are active, and horizontal and vertical distribution of roots of the dominant species in the community, and to identify species with particularly deep roots. Rapid profile wall mapping techniques are described by Böhm (1979).

MAXIMUM DEPTH AND SPREAD OF ROOT SYSTEMS

One aspect of root system form that is of particular interest from a mineral exploration viewpoint is the maximum depth and spread that various species' root systems may attain. There are reports in the literature of roots being found at great depth (Table 5-1). Phillips (1963) reported finding 7.5 m of root material 6 to 12 mm in diameter in a gravel lens exposed at 53 m depth in the soil by a mining operation in Arizona. Anatomical observations of these roots suggest that they were probably from *Prosopis* spp. Cannon (1960) also reports having received oral communications indicating that *Juniperus monosperma* roots were found in mine shafts at a depth of 61 m. Douglas fir (*Pseudotsuga menziesii*) roots occur at 12 m depth in limestone caves in southern Oregon. Cannon (1960) also suggested that the effective depth of tree root penetration in various parts of the Colorado Plateau varied from 9 to 21 m. Even these smaller values seem quite large when one examines the distribution of maximum rooting depths shown in Table 5-1 and root distributions shown in Figure 5-3. Although these reports (and several others) indicate that some species are capable of rooting to great depth, they must be interpreted with caution. Extensive and detailed sampling of roots of these same species and many other deep-rooted species has failed to reveal consistent rooting depths much greater than 3 to 5 m for the majority of individual plants. The probability of sampling the exceptional, deep-rooted individuals in a biogeochemical survey is extremely small.

The maximum lateral length of roots shown in Table 5-1 indicates that the spread of root systems is almost always greater than the lateral spread of the aboveground canopy of the tree or shrub of interest. In some tree species [i.e., aspen (*Populus tremuloides*) and Gambel oak (*Quercus gambelii*)], growth is clonal, with many apparently individual trees actually being shoots from an interconnected network of roots. This makes the effective lateral spread of these species very great. Root grafting, which has been demonstrated in most forest types, also increases the effective lateral spread of many tree species. Mycorrhizal associates of plants increase the effective absorbing surface of the below-ground system, absorb ions that might otherwise not be absorbed by

TABLE 5-1   Reported Maximum Root Depth and Lateral Spread of Some Deeply or Broadly Rooting Species[a]

Species	Maximum Depth (m)	Maximum Lateral Length (m)	Location	Reference
Andira humilis	18.0	–	Brazil	Rawitscher (1948)
Zygophyllum dumosum	1.6	4.0	Israel	Kummerow (1980)
Banksia ornata	3.0	7.0	California	and authors cited therein
Adenostoma fasciculatum	0.6	3.0	California	
	8.0	4.0	California	
Arctostaphylos pungens	0.6	4.5	California	
A. glauca	2.8	10.0	California	
Ceanothus gregii	0.6	3.5	California	
	1.5	4.0	California	
Quercus dumosa	9.0	3.3	California	
Q. turbinella	6.0	6.9	Arizona	
Prosopis sp.	53.3	–	Arizona	Phillips (1963)
P. velutina	15.0	15.0	Arizona	Cannon (1911)
Larrea tridentata	2.1	4.0	Arizona	
Carnegiea gigantea	0.8	9.7	Arizona	
Quercus gambelii	4.0	–	Utah–Colorado	Cannon (1960)
Ephedra viridis	5.5	–	Utah–Colorado	and authors cited therein
Sarcobatus vermiculatus	5.8 (17.4)	–	Utah–Colorado	
Amelanchier utahensis	6.4	–	Utah–Colorado	
Juniperus monosperma	5.8 (61.0)	–	Utah–Colorado	
Pinus ponderosa	24.4	–	Utah–Colorado	
Artemisia tridentata	9.1	–	Utah–Colorado	
	2.0	1.5	–	Tisdale and Hironaka (1981)
Amorpha canescens	5.0	–	Nebraska	Weaver (1920)
Ceanothus ovatus	4.4	1.5	Nebraska	
Liatris punctata	4.8	0	Nebraska	
Rosa arkansana	5.5	–	Nebraska	
Yucca glauca	2.1	9.7	Nebraska	
Pinus sylvestris	2.3	–	–	Brooks (1972)
	3.5	–	–	and authors cited therein
Betula verrucosa	3.7	–	–	
Populus alba	2.7	–	–	

[a]These values should be interpreted as representing exceptional cases.

plant roots, and provide potential interplant connections that increase the effective lateral spread of root systems. In using these values for maximum depth and spread, a great deal of caution must be exercised since these values represent only individual plants in particular environmental conditions. In Table 5-1 the values given for maximum depth of *Adenostoma fasciculatum* range from 0.6 to 8.0 m, and this variation was correlated with the depth of the soil. Large variation in rooting depth and spread is also shown for different species of *Arctostaphylos, Quercus,* and *Artemisia tridentata* and *Pinus sylvestris.* Again, individual species can be highly variable in root depth and spread when growing in different habitats.

One widespread group of plants, termed *phreatophytes*, is notable for almost always having deep or wide-spreading roots. This rooting habit of phreatophytes allows them to obtain a permanent water supply throughout the growing season. The development of the deep taproots or highly developed, wide-spreading lateral roots of phreatophytes appears to be facultative and dependent on the depth of the water table or the distance to a nearby wash or arroyo (Fig. 5-2). It has been suggested that this facultative development of various parts of a generalized taproot system in phreatophytes gives these plants a great deal of flexibility and thus allows them to adapt to a wide range of environmental and soil conditions. The occurrence of some species of phreatophytes has been shown to be limited to areas where the depth to the water table is less than some threshold value. This is, of course, dependent on the soil type and the species of concern. Generally, phreatophytes are distributed along washes and paleovalleys, or associated with geologic features that bring the groundwater within reach of their root systems. These plants have been extremely useful as indicators of subsurface water in arid and

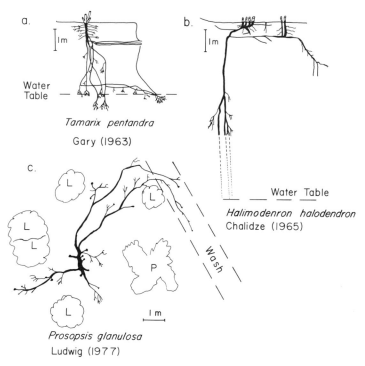

Fig. 5-2   Diagrams of root systems of three phreatophytes, illustrating extensive development of taproots and lateral roots. Note fine root proliferation both near the surface and in the capillary fringe above the water table. (a) *Tamarix pentandra* was excavated from the banks of the Salt River in central Arizona; (b) *Halimodendron halodendron* was excavated in the alluvial-delta valley of the Syr Dar'ya River in Kzyl-Ordinsk Province, USSR; (c) *Prosopis glandulosa* was excavated from near a small arroyo on the U.S./IBP Desert Biome site in southern New Mexico. In part (c) solid circles show the location of vertical roots and canopies of neighboring *Larrea tridentata* (L.) and *P. glandulosa* (P.) plants are shown. Redrawn from the references shown.

ROOT FORM AND DEPTH DISTRIBUTION IN SEVERAL BIOMES

semiarid regions, and may also be useful for mineral exploration because of their uptake of trace elements that move in subsurface waters. The site of origin of such trace elements may, however, be difficult to find, because of the wide dispersion of water-transported trace elements.

DEPTH DISTRIBUTION OF ROOTS

While the preceding examples have shown that some species and groups of plants have very deep roots, the majority of roots in nearly all plant communities occur in the upper meter or two of soil. In some areas this may be due to soil limitations, such as heavy clay soils, caliche and ironpan layers in the soil, or permafrost, all of which prevent or inhibit deep root penetration. In other areas, however, even where the soils are amenable to deeper root growth, roots are not found at great depth. The examples of depth distributions of roots in plant communities shown in Figure 5-3 bear this out for a wide range of plant communities and habitats. The important question, from both an ecological and a mineral exploration viewpoint, is to what extent the distribution of biomass represents the distribution of root activity, especially in regard to mineral uptake at the various depths. Usually, in layers of soil with a large biomass of roots there is also a large biomass of young absorbing roots. Certainly, we could expect that where there are more roots, most of the mineral nutrients (or water) would be taken up. Thus we can generalize that for most plant communities it is the upper meter, or at most 2 m, of soil which the plant root system is sampling. In nearly all of the examples given, if samples had been taken at depths deeper than that shown in the diagrams, a few roots may have been found, but their importance in mineral nutrient and trace element uptake is probably small. However, data that might verify this are lacking. Analysis of plants for trace elements known to occur at specific depths and proven to be absent in the layers above may be a technique that would be useful in answering this question. Initially, this could be done on artificially layered soil columns in glasshouses.

Other techniques that have proven useful in the study of depth and distribution of roots include placement of radioisotopes (usually ^{32}P) at specific locations around and beneath individual plants and analysis of the shoot system of those plants for the appearance of that radioisotope (Nye and Foster, 1961; McClure and Harvey, 1962). Alternatively, soil-immobile herbicides have been used in a similar approach (N. E. West, personal communication). Recently, Currie and Hammer (1979) compared the use of ^{32}P and excavation techniques to determine root depth and spread of four arid-land species occurring in the ponderosa pine–bunchgrass vegetation of the central and southern Rocky Mountains. They found that both techniques indicated similar rooting depth, ranging from 70 to 90 cm for all four species. Studies with ^{32}P identified two individual plants with roots to 120 cm depth, however. Much greater lateral spread of roots was always detected with the ^{32}P technique than with the excavation technique. This difference was attributed to loss of roots during washing and to individual plant variability, but might also be due to phosphorus transport by mycorrhizae.

Ecological factors that influence the depth at which large portions of the root biomass are found relate mainly to water and macronutrient (N, P, K, etc.) uptake. In Figure 5-3b, a gradient of moisture availability is shown for a tropical region. In the rain

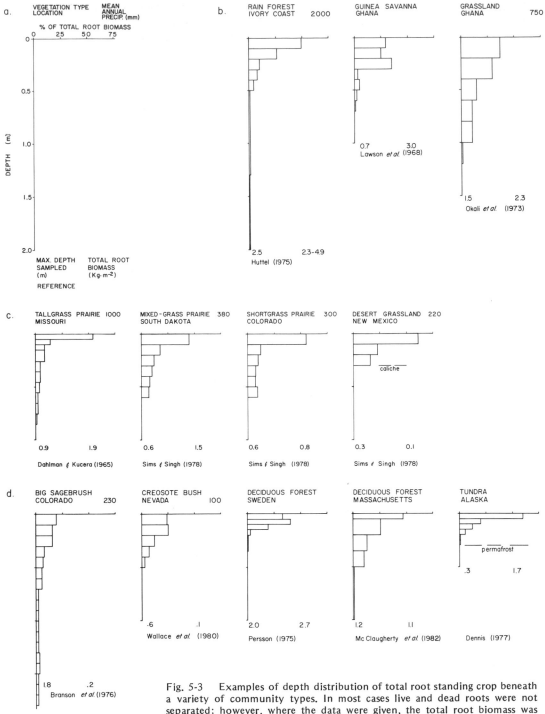

Fig. 5-3   Examples of depth distribution of total root standing crop beneath a variety of community types. In most cases live and dead roots were not separated; however, where the data were given, the total root biomass was correlated to live, fine root biomass. All figures are drawn to the scale given in part (a). Calculated from data given by the authors listed.

forest vegetation the percentage of roots found between 0.5 and 1 m is very small. This percentage increases as one moves toward the drier grassland regions. The tropical grassland in Ghana, for which root distribution is shown, occurs in a zone where water pentration is fairly deep and there is adequate precipitation to recharge the soil profile to a fairly great depth. Because of the drought that occurs at the end of the rainy period, it is advantageous for plants to be able to extract soil moisture from the deeper portion of the profile. However, in the rain forest, the high levels of precipitation coming throughout the whole year preclude any need for plant roots deep in the soil profile, since there will almost always be water available in the upper layers. In addition, the upper 10 cm of soil in the rain forest is much richer in macronutrients than that deeper in the soil profile. In the Guinea savanna diagram the large biomass of roots shown in the 20- to 30-cm layer is due to this zone being penetrated by spreading, woody roots of shrubs and trees. The zone above it is intensively penetrated with grass roots, only a few of which penetrate deeper into the soil. In all of these cases there is adequate precipitation for recharge of the soil moisture deep into the profile.

In Figure 5-3c the root distribution diagrams for tallgrass prairie, mixed-grass prairie, shortgrass prairie, and desert grassland all indicate a very large percentage of the root biomass occurring in the upper 10 cm of soil. In these areas the summer-growing-season precipitation is a large percentage of the total yearly precipitation, and thus it is of advantage to plants to have a large absorbing surface in the layers that are moistened by small intermittent rains. This would be particularly true in the shortgrass prairie and in the desert grassland. Of course, there are exceptional species that do not follow this pattern and root almost exclusively in the deeper portion of the profile. In the tallgrass prairie, however, there is adequate precipitation to recharge the soil profile to quite a great depth, so that deep-rooted species, which form the majority of the flora, are able to continue growth through dry periods at the end of the summer. In the desert grassland the caliche layer limits the amount of soil volume that can be sampled by roots of any species in this area. The few roots penetrating deeper do so by growing through cracks or fractures in the caliche.

Figure 5-3d contrasts the depth distribution of root biomass for temperate desert, temperate forest, and tundra plants. In the forest and tundra zones there is nearly always adequate moisture for plant growth in the upper 20 to 50 cm of soil. In the tundra, and in much of the boreal biome, permafrost prevents any deep penetration of roots even though some species (i.e., *Eriophorum angustifolium*) can grow roots at $0°C$ just above the thaw line (Billings *et al.*, 1976).

In the temperate forest regions adequate precipitation arrives during the summer period. Additionally, the decaying forest litter is a rich macronutrient source which is thoroughly penetrated by roots of forest trees and understory species (McClaugherty *et al.*, 1982; Persson, 1975, 1980; Vogt *et al.*, 1981, 1983). This leads to generally very shallow rooting depths (1 to 2 m) in temperate forests. Some species (box elder, honey locust, bur oak, black walnut, sycamore, red oak, shell bark hickory) growing on well-drained loess soils in Missouri have been reported to have roots penetrating to 3 to 5-m depths (Biswell, 1935). Similarly, studies on fruit trees have documented roots penetrating to 3+ m, although the majority of actively absorbing roots occur in the upper 50 cm of soil (Rogers and Head, 1969).

In the temperate desert regions a contrast can be seen between the summer precipitation zone, where moisture percolates only a small distance into the soil in Nevada, and the big sagebrush zone of Colorado, where because of the high percentage of winter and spring precipitation, the soil moisture percolates much deeper into the soil.

The following generalizations can be made from these root profile distributions as a means of predicting at what depth one would find the majority of roots of various plant communities. We would expect to find deeply rooted plants in areas where (1) the soil is recharged to great depth, either due to high levels of precipitation or because the precipitation comes during the winter and spring, allowing greater time for infiltration, and (2) there is a drought period during the growing season. This assumes that soil conditions allow deep root penetration. However, in areas where the growing season moisture is adequate, thus preventing severe plant water stress, the rooting systems will probably be quite shallow. In other areas, where there is only a small amount of precipitation, causing fairly shallow percolation, one would expect to find roots penetrating only to the depth to which moisture usually infiltrated. Phreatophytes provide an exception to this generalization.

SEASONAL DEVELOPMENT OF ROOT SYSTEMS

One approach to answering the question regarding which levels in the soil are most important for water and nutrient uptake has been to examine root growth at various depths in the profile through root observation chamber windows or through core sampling at different depths. Fernandez and Caldwell (1975) examined root growth rates of *Atriplex confertifolia* and other cool desert shrubs at different depths throughout the growing season of these species. They found that root growth in all regions of the rooting zone appeared to be important (Fig. 5-4a). Their study showed that the root growth rates of individual rootlets of *A. confertifolia* followed a wave pattern, such that as the upper soil layers dried out and thereby inhibited root growth, root growth in the lower layers increased. Thus there was nearly constant activity throughout the growing season. The activity, however, occurred in different depths of the soil profile. Persson (1980) studied root growth of *P. sylvestris* in central Sweden and found that there was no indication of overall periodicity in root growth. However, activity periods were dependent on moisture and temperature conditions during the growing season (Fig. 5-4b). The periods of reduced activity corresponded with several days of hot weather without any rainfall. In their study of *Quercus alba*, Reich *et al.* (1980) also found that environmental conditions, mainly soil temperature and soil moisture, were important controlling factors for root growth activity (Fig. 5-4c). However, they observed that root growth was highly suppressed during periods of rapid leaf growth even though environmental conditions were adequate for rapid root growth. In the oak, the pine, and the *Atriplex* studies, root growth activity occurred over a much longer period than the strictly defined growing season for the aboveground parts of the plant. In desert areas where plant growth depends on unpredictable rainfall, plants are able to rapidly initiate and grow fine roots to absorb the water and nutrients, which may be available for only a short period of time (Cannon, 1911; Jordan and Nobel, 1982). Root initiation and growth rates are positively correlated with high soil water and macronutrient availability when temperatures are adequate for root function.

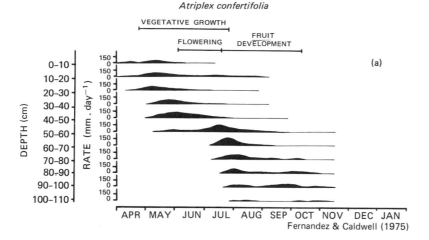

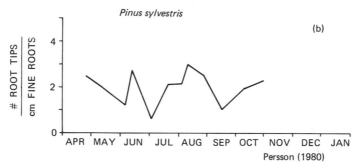

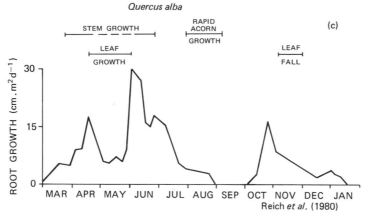

Fig. 5-4 Three types of seasonal root activity. (a) *Atriplex confertifolia* growing in a cool semidesert environment at Curlew Valley, Utah, shows increasing
root growth at depth as the upper layers of the soil dry out. (b) *Pinus sylvestris*
from central Sweden also has root growth activity throughout the summer, but
drought periods suppress new root initiation. (c) Root growth of *Quercus alba*
begins before the tree comes into leaf and continues after leaf fall. Low soil water
potential prevents root growth late in the summer, but root growth is also suppressed during rapid leaf expansion when environmental conditions are favorable.
Redrawn from the authors indicated.

FACTORS AFFECTING ROOT SYSTEM DEVELOPMENT

When soil conditions are ideal, the full genetic potential for root system development of an individual species can be expressed. Nye and Tinker (1977) have recently reviewed the factors affecting root system development, so only a few important factors will be considered here. In addition to toxic element or salt limitations, which are reviewed by other authors in this volume, there are two sets of environmental conditions that are very important in limiting the development of plant root systems in natural situations. Soil physical features, including hardness, mechanical resistance, aeration, water content, pore volume, and texture, have a strong influence on root system development. Poor aeration is an extremely important factor and is related to water content and soil texture. Reduced oxygen partial pressure around the root system has been shown to reduce root elongation rates rapidly in peas (Fig. 5-5a). In biomes and regions where large areas of land are wet or waterlogged, such as in the boreal forest regions, lowland tundra, and lowland tropical areas, root system development is essentially prevented in gleyed or waterlogged horizons. In addition, some areas with heavy clay soils, which reduce aeration because of the low pore volume, prevent the growth of all but the most tolerant species.

Soil mechanical resistance, which is also correlated to some extent with soil texture and water content, is a second major factor that can prevent root growth. Experimental studies have shown rapid decreases in root elongation rates of maize as the mechanical

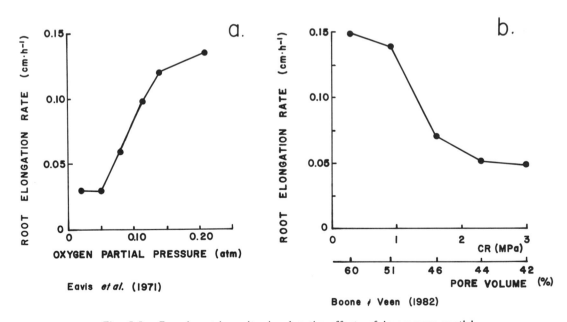

Eavis *et al.* (1971)

Boone / Veen (1982)

Fig. 5-5 Experimental results showing the effects of low-oxygen partial pressures and mechanical resistance (CR) of the soil on root elongation rate: (a) root elongation rate of pea seedlings when the roots were at the oxygen partial pressure shown but the cotyledons and shoots were at 0.21 atm oxygen partial pressure; (b) effects of the mechanical resistance (CR) of the soil on root elongation rate of maize seedlings. During growth the soil was well aerated and well watered. Redrawn after the authors indicated.

resistance of the soil increased (Fig. 5-5b). In vast areas of desert and semiarid landscape, soils have a caliche hardpan, and over large areas in tropical savannas and tropical deciduous or semievergreen forests, ironpans are present in the soil. These hard layers inhibit the penetration of roots, thus preventing plants from exploring large soil volumes except along fractures or cracks in these layers. Bedrock, of course, prevents root development except along fractures. Permafrost causes both a reduction in soil temperature and soil aeration, and presents a mechanical resistance to any root growth. Even in areas where permafrost is not present but soils are very cold, the low soil temperatures deep in the profile inhibit root growth.

CONCLUSION

Roots are stimulated to grow into regions with high resource availability, including water and/or macronutrients. Thus knowledge of the time and distribution of water and macronutrient availability is of critical importance in predicting rooting depths, but must be considered in light of the factors that inhibit root growth. These include poor aeration, soil mechanical resistance, and chemical characteristics of the soil. Application of these principles should allow biogeochemists to predict with fair accuracy the depths at which actively absorbing roots are present and thus determine from which soil layers metallic ions would be absorbed.

Plants exhibit great variability in the way they obtain the resources they require, and this is reflected in variable root system form and activity. Species can be selected for use based on rooting depth, to preferentially sample different soil layers, or based on total rooting volume to sample extensive volumes of soil. This variability presents some problems, but also promise for mineral exploration uses of plants.

ACKNOWLEDGMENTS

Support from the National Science Foundation (DEB-7907323 and DEB-8207171) and the Utah Agricultural Experiment Station is gratefully acknowledged. I thank Chuck Warner for drafting the figures.

REFERENCES

Billings, W. D., Shaver, G. R., and Trent, A. W., 1976, Measurement of root growth in simulated and natural temperature gradients over permafrost: *Arct. Alp. Res.*, v. 8, p. 247–250.

Biswell, H. H., 1935, Effects of environment upon the root habits of certain deciduous forest trees: *Bot. Gaz.*, v. 96, p. 676–708.

Böhm, W., 1979, *Methods of Studying Root Systems:* Springer-Verlag, Berlin.

Boone, F. R., and Veen, B. W., 1982, The influence of mechanical resistance and phosphate supply on morphology and function of maize roots: *Neth. J. Agric. Sci.*, v. 30, p. 179–192.

Branson, F. A., Miller, R. F., and McQueen, I. S., 1976, Moisture relationships in twelve northern desert shrub communities near Grand Junction, Colorado: *Ecology*, v. 57, p. 1104–1124.

Brooks, R. R., 1972, *Geobotany and Biogeochemistry in Mineral Exploration:* Harper & Row, New York.

Cannon, W. A., 1911, The root habits of desert plants. *Carnegie Inst. Wash. Publ. 131.*

Cannon, H. L., 1960, The development of botanical methods of prospecting for uranium on the Colorado Plateau: *U.S. Geol. Surv. Bull. 1085-A.*

Chalidze, F. N., 1965, Ecological characteristics and the root system structure of some hydrologic indicator species in the alluvial-delta valley of the Syr-Dar'Ya, p. 44–47 *in* Chikishev, A. G. (Ed.), *Plant Indicators of Soils, Rocks, and Subsurface Waters:* Consultants Bureau, New York.

Clements, F. E., Weaver, J. E., and Hanson, H. C., 1929, Plant competition: an analysis of community function: *Carnegie Inst. Wash. Publ. 398.*

Currie, P. O., and Hammer, F. L., 1979, Detecting depth and lateral spread of roots of native range plants using radioactive phosphorus: *J. Range Manage.*, v. 32, p. 101–103.

Dahlman, R. G., and Kucera, C. L., 1965, Root productivity and turnover in native prairie: *Ecology*, v. 46, p. 84–89.

Dennis, J. G., 1977, Distribution patterns of belowground standing crop in arctic tundra at Barrow, Alaska, p. 53–62 *in* Marshall, J. K. (Ed.), *The Belowground Ecosystem: A Synthesis of Plant-Associated Processes:* Range Science Department, Colorado State University, Fort Collins, Colo.

Eavis, B. W., Taylor, H. M., and Huck, M. G., 1971, Radicle elongation of pea seedlings as affected by oxygen concentration and gradients between shoot and root: *Agron J.*, v. 63, p. 770–772.

Fernandez, O. V. and Caldwell, M. M., 1975, Phenology and dynamics of root growth of three cool semi-desert shrubs under field conditions: *J. Ecol.*, v. 63, p. 703–714.

Fitter, A. H., 1982, Morphometric analysis of root systems: application of the technique and influence of soil fertility on root system development in two herbaceous species: *Plant Cell Environ.*, v. 5, p. 313–322.

Gary, H. L., 1963, Root distribution of five-stamen tamarisk, seep willow, and arrow-weed: *For. Sci.*, v. 9, p. 311–314.

Huttel, C., 1975, Root distribution and biomass in three Ivory Coast rain forest plots, p. 123–130 *in* Golley, F. B., and Medina, E. (Eds.), *Tropical Ecological Systems: Trends in Terrestrial and Aquatic Research:* Springer-Verlag, New York.

Jordan, P. W., and Nobel, P. S., 1982, Root responses by *Agave deserti* to drought and rainfall (Abstract): *Bull. Ecol. Soc. Am.*, v. 63, p. 91.

Kummerow, J., 1980, Adaptation of roots in water-stressed native vegetation p. 57–73 *in* Turner, N. C., and Kramer, P. J. (Eds.), *Adaptation of Plants to Water and High Temperature Stress:* Wiley, New York.

Lawson, G. W., Jenik, J., and Armstrong-Mensah, K. O., 1968, A study of a vegetation catena in Guinea savanna at Mole Game Reserve (Ghana): *J. Ecol.*, v. 56, p. 505–522.

Ludwig, J. A., 1977, Distributional adaptations of root systems in desert environments, p. 85–91 *in* Marshall, J. K. (Ed.), *The Belowground Ecosystem: A Synthesis of Plant-Associated Processes:* Range Science Department, Colorado State University, Fort Collins, Colo.

McClaugherty, C. A., Aber, J. D., and Melillo, J. M., 1982, The role of fine roots in the organic matter and nitrogen budgets of two forested ecosystems: *Ecology*, v. 63, p. 1481–1490.

McClure, J. W., and Harvey, C., 1962, Use of radio phosphorus in measuring root growth of sorghums. *Agron. J.*, v. 54, p. 457–459.

Nye, P. H., and Foster. W. N. M., 1961, The relative uptake of phosphorus by crops and natural fallow from different parts of their root zone: *J. Agric. Sci.*, v. 56, p. 299–306.

Nye, P. H., and Tinker, P. B., 1977, *Solute Movement in the Soil-Root System:* University of California Press, Berkeley, p. 196–210.

Okali, D. U. U., Hall, J. B., and Lawson, G. W., 1973, Root distribution under a thicket clump on the Accra plains, Ghana: its relevance to clump localization and water relations: *J. Ecol.*, v. 61, p. 439–454.

Persson, H., 1975, Deciduous woodland at Andersby, eastern Sweden: field layer and below-ground production: *Acta Phytogeogr. Suec.*, v. 62, p. 1–66.

——, 1980, Spatial distribution of fine-root growth, mortality and decomposition in a young Scots pine stand in central Sweden: *Oikos*, v. 34, p. 77–87.

Phillips, W. S., 1963, Depth of roots in soil: *Ecology*, v. 44, p. 424.

Rawitscher, F., 1948, The water economy of the vegetation of the Campos Cerrados in southern Brazil: *J. Ecol.*, v. 36, p. 237–268.

Reich, P. B., Teskey, R. O., Johnson, P. S., and Hinckley, T. M., 1980, Periodic root and shoot growth in oak: *For. Sci.*, v. 26, p. 590–598.

Rogers, W. S., and Head, G. C., 1969, Factors affecting the distribution and growth of roots of perennial woody species, p. 280–295 *in* Whittington, W. J. (Ed.), *Root Growth* (Proc. 15th Easter School Agric. Sci., Univ. Nottingham, 1968): Plenum, New York.

Sims, P. L., and Singh, J. S., 1978, The structure and function of ten western North American grasslands: *J. Ecol.*, v. 66, p. 547–572.

Tisdale, E. W., and Hironaka, M., 1981, *The Sagebrush-Grass Region: A Review of the Ecological Literature:* Bull. 33, Forest, Wildlife, and Range Experiment Station, University of Idaho, Moscow, Idaho.

Van Donselaar-ten Bokkel Huinink, W. A. E., 1966, Structure, root systems and periodicity of savanna plants and vegetations in northern Surinam: *Wentia*, v. 17, p. 1–162.

Vogt, K. A., Edmonds, R. L., and Grier, C. C., 1981, Seasonal changes in biomass and vertical distribution of mycorrhizal and fibrous textured conifer fine roots in 23- and 180-year old subalpine *Abies amabilis* stands: *Can. J. For. Res.* v. 11, p. 223–229.

——, Moore, E. E., Vogt, D. J., Redlin, M. J., and Edmonds, R. L., 1983, Conifer fine root and mycorrhizal root biomass within the forest floors of Douglas-fir stands of different ages and site productivities: *Can. J. For. Res.*, v. 13, p. 429–437.

Wallace, A., Romney, E. M., and Cha, J. W., 1980, Depth distribution of roots of some perennial plants in the Nevada test site area of the northern Mojave Desert: *Great Basin Nat. Mem. 4*, p. 201–207.

Weaver, J. E., 1919, The Ecological Relations of Roots: *Carnegie Inst. Wash. Publ. 286.*

——, 1920, Root Development in the Grassland Formation: A Correlation of the Root Systems of Native Vegetation and Crop Plants: *Carnegie Inst. Wash. Publ. 292.*

Robert R. Brooks
Department of Chemistry and Biochemistry
Massey University
Palmerston North, New Zealand

6

PRESENT STATUS
OF BOTANICAL EXPLORATION
USING HIGHER PLANTS

ABSTRACT

This chapter emphasizes current trends and new developments in botanical methods of exploration. In the section on geobotany, a substantial number of indicator plants is discussed and listed. Three systems of classifying indicator plants are reviewed. Geobotanical methods have shown significant progress in recent years in the use of computer-assisted techniques to evaluate field data.

Aerial photography and remote sensing of vegetation are rapidly developing fields. Studies have involved determining the effect of mineralization on the spectral reflectivity of vegetation with a view to applying this knowledge to aerial photography and satellite imagery. A number of case histories of these techniques are discussed.

Recent developments in biogeochemical prospecting involve the use of alternative sample types, such as roots, bark, and sap. These sample types have been studied by Kovalevsky, who has classified them as barrier free, high barrier, medium barrier, and low barrier, depending on the degree to which elements in the substrate are accumulated by the plant organs.

Perhaps one of the most significant advances in biogeochemical prospecting in recent years has been the development of the AIRTRACE and SURTRACE systems, whereby biological atmospheric particulates or mineralized foliar dust are sampled by helicopter and analyzed on-board by means of an automated analytical system (ICP). Case histories of these techniques are given and are discussed.

The use of herbarium material is also discussed. The technique involves analyzing very small leaf fragments to identify hyperaccumulators (>1000 μg/g in dried material) or indicator plants, and to use the details of collection localities, carried on the herbarium sheets, to pinpoint areas worthy of in situ investigations by more conventional prospecting techniques.

This review includes a number of case histories to illustrate each topic discussed and assesses trends in botanical prospecting as illustrated by other chapters in this volume. It is forecast that future developments will be centered around new methods of chemical analysis concomitant with improved data handling systems, as well as an increased emphasis on remote sensing of the environment, including satellite imagery of ever-improving resolution.

INTRODUCTION

Although geobotanical methods of exploration for minerals have been used for centuries, biogeochemical methods date back only to 1938. The difference in time of development is due to the very different nature of the two fields. Since geobotany originally depended on human vision alone, it had always been capable of execution ever since human exploitation of minerals began. By contrast, biogeochemical methods have been linked to the progress of analytical chemistry, and particularly to the development of speedy methods of analysis.

The boost to biogeochemistry caused by these new analytical methods has had its counterpart in geobotany, where the field has been revolutionized by development of methods of remote sensing of the environment. The rise of the computer has also been of great assistance for both botanical methods of exploration.

In this chapter it is proposed to outline some of the new and exciting developments in botanical methods of exploration which have appeared in the last decade, including new methods of remote sensing, as well as aerial sampling and analysis of vegetation. The technique of barrier-free biogeochemical prospecting developed by A. L. Kovalevsky of the Soviet Union is also described. In addition, some mention is given of the use of the computer in botanical methods of exploration. A new field, the use of herbarium material as an aid to prospecting, is also covered. There is little mention of classical techniques of botanical prospecting since these are already discussed in standard works, such as those of Brooks (1983) and Kovalevsky (1979). However, other papers in the present Rubey volume will be mentioned whether they concern classical methods or not.

Although a significant proportion of the papers in the 1983 workshop/colloquium at UCLA was concerned with microbiology in mineral exploration, this subject will not be covered in this present review because I do not feel that I have enough expertise to give it full justice. The botanical techniques described below are not a complete overview, but it is hoped that current developments and directions of research will have been adequately covered in this rapidly developing field.

GEOBOTANY IN MINERAL EXPLORATION

Geobotanical methods of prospecting involve the study of vegetation communities indicative of mineralization, of indicator plants, or of morphological changes in plants caused by the toxic effects of various ore elements. These techniques are the oldest of all botanical methods of prospecting and date back to the time of Agricola in the sixteenth century or even to Roman times, when the nature of the vegetation cover was studied to indicate the presence of subterranean water.

Until the 1960s, geobotanical studies were largely empirical in nature and were based mainly on field notes and observations. More recently, however, the field has been stimulated by two important developments. The first of these is the use of the computer to assist in carrying out various statistical procedures whereby subtle changes in vegetation communities undetectable by field notes alone can be determined by procedures that were impossible in the precomputer age.

The second boost to geobotany was the development of satellite imagery and aerial photography to detect vegetation patterns from the air or from outer space. A good example of this is the LANDSAT series of satellites, which affords good coverage of almost every part of the globe with a resolution of about 30 m.

Although a geobotanist should have some knowledge of a wide range of subjects, such as biochemistry, biogeography, biogeochemistry, botany, chemistry, ecology, geology, and plant physiology, it is highly unlikely that any one person will be found to have knowledge of all or even most of these fields. In spite of this, quite good results can be achieved with persons possessing only common sense and an ability to recognize and rerecognize a given species without necessarily even knowing its name.

Geobotany has the advantage of being inexpensive to perform and is well worthwhile carrying out provided that suitable personnel can be found to undertake it. Although there have been cases of the discovery of mineralization by geobotany alone (e.g., the use of the "copper flower," *Becium homblei*, in central Africa), the technique should preferably be part of a general scheme of exploration, including other techniques, such as geochemistry or geophysics.

Computer-Assisted Statistical Procedures in Geobotany

Figure 6-1 shows the results of a geobotanical survey in Western Australia in which the relationship between lithology and vegetation is well illustrated. However, in some cases the relationship is much less obvious and recourse has to be made to some sort of statistical procedure to enhance the data. Discrimant analysis by means of a computer is one of the best ways of doing this and in some cases depends on derivation of a suitable multiple regression equation of the form

$$y = a_1 x_1 + a_2 x_2 + a_3 x_3 + \cdots + a_n x_n + c$$

where a_1 through a_n are coefficients chosen so as to maximize differences in the value of y for separate populations of plants growing over different lithologies. The values x_1 through x_n are variables such as elemental concentrations in plants or numbers of different species growing over different lithologies, and c is a constant.

An example of the use of discriminant analysis in geobotanical exploration is given by the work of Nielsen *et al.* (1973) in a study carried out at Spargoville, Western Australia. In this work, 34 different species were recorded over 43 quadrats (30 m X 15 m) representing 26 amphibolites, 11 ultramafics, and 7 transitional zones. A numerical score was assigned to each quadrat based on the equation above and with values of x being the number of a given species present in a given quadrat. This numerical value of y was used to compute the Mahalanobis D^2 statistic (Mahalanobis, 1936). In general, the higher the numerical value of D^2, the greater the degree of discrimination between the three types of quadrats. The statistical data are shown in Table 6-1, from which it will be noted that it was possible to identify from the vegetation distributions all 26 amphibolite quadrats, 10 of 11 ultramafics, and 6 of 7 transitional zones. It is clear that a study of this nature was much more effective than a purely visual interpretation of the data, which showed no obvious trends.

Plant Communities and Metal Stress

As stated by Gough (Chapter 2, this volume), caution should be exercised in the interpretation of geobotanical data. Although well over 100 plants have the reputation of being indicators of mineralization, most species are not *obligate* (i.e., confined to mineralization because of a physiological requirement for the edaphic conditions existing over that mineralization). They are usually *facultative* (tolerating a wide variety of stress conditions which might not even be associated with mineralization). This principle has been well illustrated in an excellent quantitative study on the flora of a mineralized area (copper and nickel) at Ngai in Zimbabwe, where Thomas *et al.* (1977) used statistical procedures to show that the dominant factor controlling the vegetation was not the

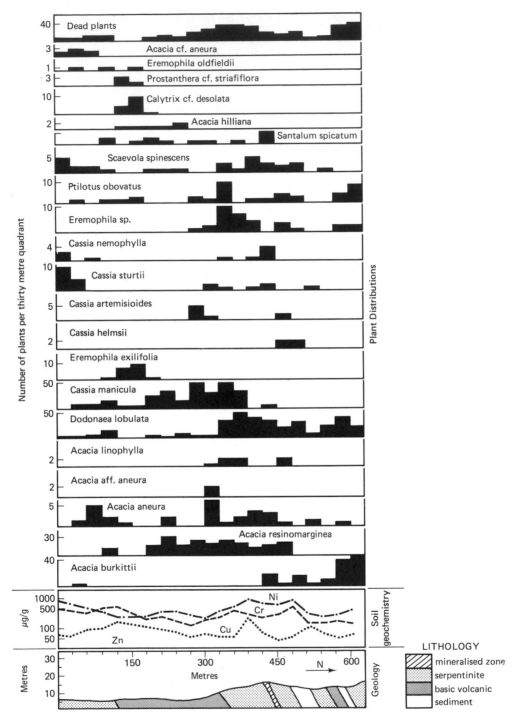

Fig. 6-1   Belt transect across Marriott Prospect, Mt. Clifford, Leonora, Western Australia. (From Severne, 1972.)

TABLE 6-1 Characterization of Substrates by Discriminant Analysis (D^2 Statistic) of Geobotanical Data from Spargoville, Western Australia

Species[a]	D	Number of Correct Predictions[b]		
		A (26 quad.)	UB (11 quad.)	A/UB (7 quad.)
21	0.5	11	6	0
23	5.7	2	11	1
21, 23	6.2	10	6	1
21–23	9.4	15	3	1
6, 21, 23	7.4	16	7	1
2, 21, 23	8.3	21	7	1
2, 3, 21, 23	9.1	16	8	1
2, 4, 21, 23	12.1	17	8	1
2, 4, 14, 21, 23	21.1	19	9	1
2, 4, 14, 15, 21, 23	29.1	22	10	1
2, 4, 14–16, 21, 23	29.2	22	11	2
2, 4, 14–16, 21, 23, 25	29.4	20	10	2
2, 4, 14–16, 21, 23, 27	33.0	21	10	2
2, 4, 12, 14–16, 21, 23, 27	41.3	22	10	3
2, 4, 12, 14–16, 21, 23, 24, 27	42.5	22	10	4
1, 2, 4, 12, 14–16, 21, 23, 24, 27	48.2	23	10	3
1, 2, 4, 5, 12, 14–16, 21, 23, 24, 27	50.4	24	10	3
1, 2, 4–6, 12, 14–16, 21, 23, 24, 27	51.8	24	10	3
1, 2, 4, 5, 8, 12, 14–16, 21, 23, 24, 27	56.6	24	10	4
1, 2, 4, 5, 7, 8, 12, 14–16, 21, 23, 24, 27	58.7	24	10	4
1, 2, 4–8, 12, 14–17, 21, 23, 24, 27	67.8	24	10	4
1, 2, 4–8, 12, 14–17, 19, 21, 23, 24, 27	76.2	24	10	5
1, 2, 4–8, 12, 14–17, 19, 21, 23, 24, 27, 28	89.8	24	9	4
1, 2, 4–8, 12, 14–17, 19, 21, 23, 24, 27, 29	81.3	23	9	5
1, 2, 4–8, 12, 14–17, 19, 21, 23, 24, 27, 29	100.3	24	10	6
1, 2, 4–8, 12, 14–17, 19, 21, 23, 24, 27, 31	119.4	25	9	6
1–8, 12, 14–17, 19, 21, 23, 24, 27, 31, 32	137.5	25	10	6
1–9, 12, 14–17, 19, 21, 23, 24, 27, 31, 32	138.1	25	10	6
1–8, 12–17, 19, 21, 23, 24, 27, 31, 32	147.2	25	10	6
1–8, 12, 14–19, 21, 23, 24, 27, 31, 32	144.3	24	10	6
1–8, 12, 14–19, 21, 23, 24, 26, 27, 31, 32	142.8	24	10	6
1–8, 12, 14–19, 21, 23, 24, 27, 30–32	173.9	26	10	5

[a]1, *Acacia acuminata*; 2, *A. colletoides*; 3, *Westringia cephalantha*; 4, *Acacia erinacea*; 5, *A. graffiana*; 6, *Alyxia buxifolia*; 7, *Cratystylis microphylla*; 8, *C. subspinescens*; 9, *Dodonaea filifolia*; 10, *D. lobulata*; 11, *D. microzyga*; 12, *D. stenozyga*; 13, *Eremophila caerulea*; 14, *E. dempsteri*; 15, *E. ionantha*; 16, *E. oppositifolia*; 17, *E. pachyphylla*; 18, *E.* sp. I; 19, *E.* sp. II; 20, *Eucalyptus calycogona*; 21, *E. lesouefii*; 22, *E. longicornis*; 23, *E. salubris*; 24, *E. torquata*; 25, *Exocarpus aphyllus*; 26, *Kochia pyramidata*; 27, *Melaleuca sheathiana*; 28, *Olearia muelleri*; 29, *Pittosporum phillyraeoides*; 30, *Rhagodia* sp.; 31, *Santalum spicatum*; 32, *Scaevola spinescens*; 33, *Trymalium ledifolium*.
[b]A, Amphibolites; UB, ultramafic rocks.
Source: Nielsen *et al.* (1973). By permission of Blackwell Scientific Publications Ltd.

mineralization but rather a combination of other factors, including soil development, man-made disturbance of the soil profile, and the nutrient status of the substrate.

Indicator Plants

Indicator plants are species that are found over mineralized ground and are of two main types: *local* and *universal*. Local indicators are found to have only local indicating properties, whereas at other sites they may well grow over nonmineralized ground. By contrast, universal indicators are said to indicate minerals at all sites where they occur. Despite the widely held view that universal indicators are found only over mineralization, this is not always the case and indeed in the case of the famous "copper flower," *Becium homblei*, of central Africa, an isolated population was found growing over noncupriferous granite near Harare in Zimbabwe. Another example is the discovery of an isolated population of the "serpentine endemic," *Dicoma niccolifera*, growing over "normal" soils near Lusaka in Zambia. Despite these examples, universal indicators are to all intents and purposes confined to mineralized ground and are therefore of great use in mineral population.

Indicator plants have been known for many years and indeed the seventeenth-century copper miners of Scandinavia were guided to their targets by the local indicator *Viscaria alpina*, which was known as the "kisplante" (pyrite plant).

The study of indicator plants has been hindered by the existence of persistent folklore on the alleged indicating ability of some species. A myth once established is often perpetuated by a string of review articles that merely quote previous reviewers. A good example of this is furnished by the work of Nemec *et al.* (1936), who alleged that the horsetail (*Equisetum palustre*) in Czechoslovakia contained 610 μg/g gold in its ash. Subsequent work by Cannon *et al.* (1968) indicated that the ash never contained more than 1 μg/g. It is not suggested that Nemec *et al.* (1936) falsified their data, but they appear to have been mistaken in that they precipitated sulfides from solutions of plant material, weighed the precipitate, and assumed that it was all gold. In fact, as we now know, copper, arsenic, and lead, which all form insoluble sulfides, are much more common in horsetails than is gold. In a later paper, Brussell (1978) claimed to have discovered gold in the emission spectrum of the ash of *E. hyemale*. The spectral lines used were weak lines of gold subject to massive interference from other elements in the plant material.

The myth of gold accumulation by *Equisetum* species was finally laid to rest by the work of Brooks *et al.* (1981b), who found massive quantities of arsenic (781 μg/g in dried material) in horsetails growing over gold mines in Nova Scotia. In no case was gold detectable in the samples.

Specific Indicator Plants

A list of indicator plants is given in Table 6-2 and is an update of a list given by Brooks (1983). The table is conservative in that it does not include some of the "classical" indicators unless modern work has confirmed their status as such. A feature of the table is the very large number of taxa that indicate copper. This is because of the importance and extent of the copper/cobalt deposits of central Africa, which support a diverse copper/cobalt-tolerant flora described in more detail by Brooks and Malaisse (1985).

TABLE 6-2 Plant Indicators of Mineral Deposits

Species[a]	Family	Locality	References[b]
	Boron		
Eurotia ceratoides (L)	Chenopodiaceae	USSR	1
Limonium suffruticosum (L)	Plumbaginaceae	USSR	1
Salsola nitraria	Chenopodiaceae	USSR	1
	Cobalt		
Crassula alba (L)	Crassulaceae	Zaire	2
Crotalaria cobalticola (U)	Leguminosea	Zaire	3, 4
Haumaniastrum robertii (U)	Labiatae	Zaire	5
Silene cobalticola (U)	Caryophyllaceae	Zaire	4
	Copper		
Acalypha cupricola (U)	Euphorbiaceae	Zaire	38
A. dikuluwensis (U)	Euphorbiaceae	Zaire	6
Aeolanthus biformifolius (U)	Labiatae	Zaire	7
Anisopappus hoffmanianus (U)	Compositae	Zaire	6
Armeria maritima (L)	Plumbaginaceae	Wales	8–10
Ascolepis metallorum (U)	Cyperaceae	Zaire	6
Becium empetroides (U)	Labiatae	Zambia	38
B. homblei (L)	Labiatae	Zaire/Zambia	11
B. peschianum (U)	Labiatae	Zaire	6
Bulbostylis barbata (U)	Cyperaceae	Australia	12
B. burchelli (L)	Cyperaceae	Australia	13
Commelina zigzag (U)	Commelinaceae	Zaire	6
Crotalaria cobalticola (U)	Leguminosae	Zaire	4
C. francoisiana (U)	Leguminosae	Zaire	6
Cyanotis cupricola (U)	Commelinaceae	Zaire	6
Ecbolium lugardae (L)	Acanthaceae	S.W. Africa	13, 14
Elsholtzia haichowensis (L)	Labiatae	China	15
Eschscholzia mexicana (L)	Papaveraceae	United States	16
Gladiolus actinomorphanthus (U)	Iridaceae	Zaire	6
G. klattianus ssp. *angustifolius* (U)	Iridaceae	Zaire	6
G. peschianus (U)	Iridaceae	Zaire	6
G. tshombeanus ssp. *parviflorus* (U)	Iridaceae	Zaire	6
Gutenbergia cuprophila (U)	Compositae	Zaire	6
Gypsophila patrinii (L)	Caryophyllaceae	USSR	17
Haumaniastrum katangense (U)	Labiatae	Zaire	6
H. robertii (U)	Labiatae	Zaire	5, 6
Helichrysum leptolepis (L)	Compositae	S.W. Africa	13, 14
Impatiens balsamina (L)	Balsaminaceae	India	18
Lindernia damblonii (U)	Scrophulariaceae	Zaire	6
L. perennis (U)	Scrophulariaceae	Zaire	6
Lychnis alpina (L)	Caryophyllaceae	Fennoscandia	19, 20
Minuartia verna (L)	Caryophyllaceae	United Kingdom	8
Pandiaka metallorum (U)	Amaranthaceae	Zaire	6
Penstemon cyanocaulis (L)	Scrophulariacae	Texas	39
Polycarpaea corymbosa (L)	Caryophyllaceae	India	21
P. spirostylis (L)	Caryophyllaceae	Australia	22, 23
Rendlia cupricola (U)	Gramineae	Zaire	6
Sopubia metallorum (U)	Scrophulariaceae	Zaire	6

TABLE 6-2 Plant Indicators of Mineral Deposits (*Continued*)

Species[a]	Family	Locality	References[b]
	Copper		
S. neptunii (U)	Scrophulariaceae	Zaire	6
Sporobolus stelliger (U)	Gramineae	Zaire	6
S. deschampsioides (U)	Gramineae	Zaire	6
Tephrosia s. nov. (L)	Gramineae	Queensland	12
Vernonia cinerea (L)	Compositae	India	21
V. ledocteanus (U)	Compositae	Zaire	6
	Iron		
Acacia patens (L)	Leguminosae	Australia	24
Burtonia polyzyga (L)	Leguminosae	Australia	24
Calythrix longiflora (L)	Myrtaceae	Australia	24
Chenopodium rhadinostachyum (L)	Chenopodiaceae	Australia	24
Eriachne dominii (L)	Gramineae	Australia	24
Goodenia scaevolina (L)	Goodeniaceae	Australia	24
	Lead		
Alyssum wulfenianum (U)	Cruciferae	Australia/Italy	25
Thlaspi rotundifolium			
ssp. *cepaeifolium* (&)	Cruciferae	Australia/Italy	25
	Manganese		
Crotalaria florida			
var. *congolensis* (L)	Leguminosae	Zaire	4
Maytenus bureauvianus (L)	Celastraceae	New Caledonia	26
	Nickel		
Alyssum bertolonii (U)	Cruciferae	Italy	27
A. pintodasilvae (U)	Cruciferae	Portugal	28
A. spp. of Section *Odontarrhena* (U)	Cruciferae	S. Europe	29
Hybanthus austrocaledonicus (U)	Violaceae	New Caledonia	30
H. floribundus (L)	Violaceae	Australia	31, 32
Lychnis alpina			
var. *serpentinicola* (L)	Caryophyllaceae	Fennoscandia	33
	Selenium and Uranium		
Aster venusta (L)	Compositae	United States	34
Astragalus albulus (L)	Leguminosae	United States	34
A. argillosus (L)	Leguminosae	United States	34
A. confertiflorus (L)	Leguminosae	United States	34
A. pattersoni (U)	Leguminosae	United States	34
A. preussi (U)	Leguminosae	United States	34
A. thompsonae	Leguminosae	United States	34
	Zinc		
Armeria halleri (L)	Plumbaginaceae	Pyrenees	35
Hutchinsia alpina (L)	Cruciferae	Pyrenees	35
Minuartia verna (L)	Caryophyllaceae	W. Europe	36

TABLE 6-2 Plant Indicators of Mineral Deposits (*Continued*)

Species[a]	Family	Locality	References[b]
	Zinc		
Thlaspi calaminare (U)	Cruciferae	W. Europe	36
Thlaspi ssp. (U)	Cruciferae	S. Europe	37
Viola calaminaria (U)	Violaceae	W. Europe	36

[a]L, Local indicator; U, universal indicator.
[b]1, Buyalov and Shvyryayeva (1961); 2, Malaisse *et al.* (1979); 3, Brooks *et al.* (1977b); 4, Duvigneaud (1959); 5, Brooks (1977); 6, Duvigneaud and Denaeyer-de Smet (1963); 7, Malaisse *et al.* (1978); 8, Ernst (1969); 9, Farago *et al.* (1980); 10, Henwood (1857); 11, Howard-Williams (1970); 12, Nicolls *et al.* (1965); 13, Cole (1971); 14, Cole and Le Roex (1978); 15, Se Sjue-Tzsin and Sjuj Ban-Lian (1953); 16, Chaffee and Gale (1976); 17, Nesvetaylova (1961); 18, Aery (1977); 19, Brooks and Crooks (1980); 20, Brooks *et al.* (1979b); 21, Ventakesh (1964); 22, Brooks and Radford (1978); 23, Skertchly (1897); 24, Cole (1965); 25, Reeves and Brooks (1983a); 26, Jaffre (1977); 27, Minguzzi and Vergnano (1948); 28, Menezes de Sequeira (1969); 29, Brooks *et al.* (1979a); 30, Brooks *et al.* (1974); 31, Cole (1973); 32, Severne and Brooks (1972); 33, Rune (1953); 34, Cannon (1957); 35, Pulou *et al.* (1965); 36, Ernst (1968); 37, Reeves and Brooks (1983b); 38, Brooks and Malaisse (1984); 39, Olwell (Abstract C in Chapter 12, this volume).
Source: After Brooks (1983). With permission from John Wiley & Sons, Inc.

Classification of Indicator Plants

There has already been some mention of a simple classification of indicator plants as local or universal. This is a simplified approach that can be improved by consideration of two other systems of classification. The first of these is due to Duvigneaud (1959) and Denaeyer-de Smet (1963) and refers specifically to the copper/cobalt flora of Shaba Province in Zaïre. The principles enumerated can, however, be applied to any heavy-metal flora. Details of this system are given in Table 6-3, which in essence examines the question of copper tolerance of the flora. The second system is that of Jacobsen (1967, 1968, 1969), who used the term *specific indicator value* (s.i.v.) to describe indicator plants. The term was used originally for copper-tolerant plants but may be used equally well for other heavy metals. The s.i.v. is a function of the range of copper values found in the soils supporting the indicator plant and is inversely related to the mean content of the element in the soil as follows:

$$\text{s.i.v.} = \frac{\text{highest copper level in soil} - \text{lowest copper level in soil}}{\text{mean copper level in soil}}$$

Jacobsen suggested that good indicators should have s.i.v.'s of 4 or less.

Rhythmic Disturbances in Plants Caused by Mineralization

The effect of geochemical stress on plants is often shown as a disturbance of the flowering or foliation patterns. Sometimes there is a second flowering in the fall, as was found by Buyalov and Shvyryayeva (1961) in the case of plants exposed to high boron levels in the

TABLE 6-3 Classification of Vegetation in Zaire in Relation to Copper Tolerance

Class	Description	Subclass	Description	Typical Species
I: Cuprophytes	Plants growing on soils with highest copper values and often restricted to them (eucuprophytes)	1	Polycuprophytes found on soils with 0.5–1.0% copper	Annual Labiates, (*Haumaniastrum*), geophytes (*Icomum*), grasses (*Eragrostis*)
		2	Oligocuprophytes found on soils with 0.08–0.20% copper	Perennial Labiates (*Becium, Triumfetta, Thunbergia*); *Xerophyta, Ipomoea*
		3	Eurycuprophytes found on wide range of copper levels	*Becium homblei*
		4	Local cuprophytes restricted locally to copper soils	*Becium homblei, Uapaca robynsii, Olax obtusifolia*
II: Cuprophiles	Found on lower copper levels (0.05–0.10%) and not confined to cupriferous soils			
III: Cuproresistant species	Ubiquitous species resistant to weak, medium, or strong copper concentrations	1	Eucupro-resistant; tolerant to all copper ranges up 2%	*Monocymbidium ceresiforme, Eragrostis boehmii, Crotalaria cornetii*
		2	Oligocupro-resistant species of cupricolous steppe and rocky outcrops; also adventives on mine waste	*Andropogon filifolius, Gladiolus robiliartianus, Aeolanthus* spp., *Pennisetum* spp.
IV: Cuprifuge	Species that never occur over copper			*Hyparrhena* spp.

Source: Adapted from Wild (1978) and based on data from Duvigneaud and Denaeyer-de Smet (1963). By permission of Dr. W. Junk, Publishers.

soil. Canney *et al.* (1979) have advanced the interesting thesis that under geochemical stress, deciduous species may show the autumn colors at an earlier date and that such vegetation might also be more susceptible to insect attack.

Canney *et al.* (1979) suggested that the reason for early autumn coloration may be related to depression of chlorophyll production (a common consequence of metal stress) and a consequent early predominance of red, orange, and yellow anthocynanin pigments in the fall.

Some Case Histories of Geobotanical Exploration

Table 6-4 lists a number of case histories of successful use of geobotany in different parts of the world. Most of this work has been carried out in arid or semiarid regions such as in Australia and Africa and has usually been of a retrospective nature (i.e., studies were performed over known mineralization). However, Cole and Le Roex (1978) were able to detect previously unknown copper anomalies in Botswana beneath a cover of Kalahari sands, which are a great barrier to mineral exploration in this part of the world.

Perhaps the best known of all geobotanical studies of a nonretrospective nature is the extensive work carried out in central Africa in the 1950s on the use of the "copper flower," *Becium homblei*, for discovering copper (Anon., 1959). It is in central Africa that indicator plants have their greatest multiplicity and diversity, and as already men-

TABLE 6-4   Examples of the Use of Plant Communities in Geobotanical
Exploration in Various Countries

Country	Elements Sought	Major Community Species	References
Australia	Copper. lead, zinc	*Polycarpaea glabra, Eriachne mucronata*	Nicolls *et al.* (1965)
	Lead, zinc	*Polycarpaea synandra* var. *gracilis, Tephrosia* sp.	Cole *et al.* (1968)
	Iron	*Acacia patens*	Cole (1965)
	Nickel	*Hybanthus floribundus, Acacia burkittii*	Cole (1973), Severne (1964), Severne and Brooks (1972)
Botswana	Copper	*Ecbolium lugardae*	Cole and Le Roex (1978)
New Caledonia	Nickel	*Phyllanthus serpentinus, Homalium kanaliense, Hybanthus austrocaledonicus*	Jaffré (1980)
Papua–New Guinea	Copper	*Albizzia* sp.	Cole (1980)
South-West Africa	Copper	*Helichrysum leptolepis*	Cole and Le Roex (1978)
United States	Selenium, uranium	*Astragalus preussi*	Cannon (1957)
Zaire	Cobalt	*Crotalaria cobalticola, Silene cobalticola*	Duvigneaud (1959)
	Copper, cobalt	*Haumaniastrum robertii, H. katangense*	Malaisse *et al.* (1979)
Zimbabwe	Copper	*Celosia trigyna, Becium homblei*	Wild (1968)
	Nickel	*Dicoma niccolifera*	Wild (1970)

Source: Brooks (1983). By permission of John Wiley & Sons, Inc.

tioned previously, over 50 of them are indicators of copper/cobalt mineralization (Brooks and Malaisse, 1984).

REMOTE SENSING OF VEGETATION

It is probably the field of remote sensing of vegetation that has the greatest potential for geobotany in mineral exploration. Until the development of artificial satellites, aerial photography in the visible and infrared part of the spectrum was the only effective method of aerial survey. Today, however, the field has widened tremendously with the enormous potential afforded by artificial satellites capable of remote sensing of the environment with multichannel sensors giving a resolution down to 30 m. When this work is combined with studies on reflectance patterns from vegetation subjected to geochemical stress, we have a new exploration tool of great potential.

Reflectance Patterns of Vegetation

Typical reflectance patterns for vegetation are shown in Figure 6-2. They show a small maximum at 550 nm in the green part of the spectrum and a much larger one in the red end. It has been established by several workers (e.g., Horler *et al.*, 1981; Howard, 1971; Press, 1974; Yost and Wenderoth, 1971) that metal-induced stress results in reduced absorption of radiation in the green part of the spectrum and an increased reflectance at this wavelength. An excellent review by Horler *et al.* (1981) deals with this question

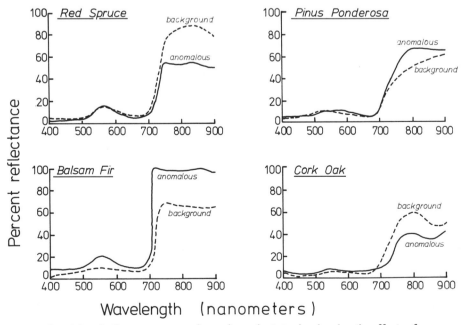

Fig. 6-2   Reflectance spectra for various plant species showing the effects of geochemical stress. (From Press, 1974. By permission of Environmental Research Institute of Michigan.)

PRESENT STATUS OF BOTANICAL EXPLORATION USING HIGHER PLANTS

and its relationship to geochemical stress. This subject is a very complex one since reflectance depends on a large number of factors, such as the area, structure, pigment content, and water content of stressed leaves, as well as the extent of ground coverage of the vegetation, topography, and solar illumination. These variables are so numerous and of such magnitude that often, stress induced by mineralization can be obscured completely. It is therefore clear that extensive study of reflectance patterns is necessary before confident judgments can be made about the presence or absence of metal stress in vegetation.

Horler *et al.* (1981) have suggested use of the first derivative reflectance spectra ($dR/d\lambda$) using a spectrophotometer. In plants there is usually a change of reflectance (R) at about 700 nm at the limit of chlorophyll absorption. This is known as the red edge (λ_D). The wavelength at the red edge is a function of the chlorophyll content of the leaf and is illustrated in Figure 6-3.

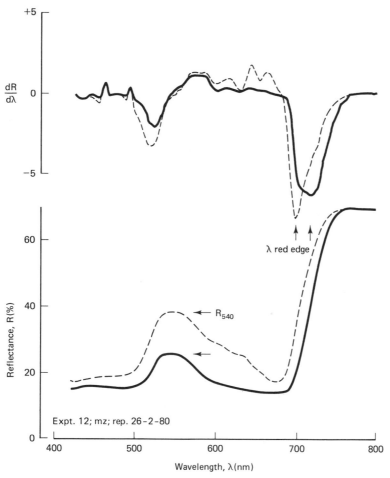

Fig. 6-3 Reflectance and first derived reflectance spectra of stressed (dashed lines) and unstressed (solid lines) maize leaves. (From Horler *et al.*, 1981. By permission of Remote Sensing Society of London.)

Spectral analysis of the red end of the spectrum is the only sure way of isolating the vegetational component of a viewing area and has been studied by workers such as Horler *et al.* (1981) and Collins and Chang (Abstract E in Chapter 12, this volume). The latter carried out an airborne survey of part of the United States using specially developed sensors and detected known anomalies beneath the forest canopy. In a "blind" project at Pilot Mountain, North Carolina, the same system detected anomalies which later showed good correlation with ground truth. Collins and Chang (Abstracts E and F in Chapter 12, this volume) described instruments for airborne biogeophysical measurement of metal stress in plants. The instruments had a high spectral resolution (1.4- to 8-μm bandwidth) in the visible and near-infrared part of the spectrum. These workers based their observations mainly on the red-edge part of the spectrum.

From Figure 6-2 it will be noted that reflectance of all types of vegetation is far greater in the near infrared (800 nm) than in the green (550 nm). The reflectance ratio 550/800 nm is usually diagnostic of vegetational differences, which in some cases may be due to metal stress. The principle is of particular significance for satellite imagery and is discussed further below.

Aerial Photography in the Ultraviolet, Visible, and Near-Infrared Part of the Spectrum

Aerial photography in the visible part of spectrum has been established for at least 100 years. One of the earliest applications of the technique in mineral exploration was the systematic aerial mapping of central Africa with panchromatic film just before and just after World War II. This work was used to identify copper clearings (*dambos*), which are regions of barren ground caused by the phytotoxic effect of copper. The work resulted in the discovery of three major ore fields (Mendelsohn, 1961).

More recently, Birnie and Francica (1979) used the ratio 565/465 nm for aerial sensing of a part of Washington State and found that a ratio exceeding 1.7 in a forest dominated by *Pseudotsuga menziesii* (Douglas fir) indicated a pyrite halo from a porphyry copper deposit. They had an 87 percent success rate in this work. Although little work has been done in the ultraviolet end of the spectrum because of poor reflectance of vegetation in this region, Horler *et al.* (1981) have explored the feasibility of using solar-induced fluorescence using a Fraunhofer line discriminator (FLD) operated from aircraft or from the ground. The same authors found a lower chlorophyll fluorescence intensity in plants growing over copper–molybdenum mineralization in the United States. Solar-induced fluorescence is known as passive fluorescence, whereas fluorescence induced artifically by excitation sources such as laser beams (active fluorescence) is also a theoretical possibility, although still in its infancy.

Aerial photography in the near infrared has been used for over 40 years since Ives (1939) showed that different vegetation types could be differentiated by black-and-white infrared photography. Photography with false-color infrared film (Brooks, 1972b) has been used regularly for studying vegetation during the past 20 years or so. The technique suffers from the disadvantage that most of the infrared reflectance is absorbed by the atmosphere at increasing altitudes (i.e., 90 percent at 12,000 m). This attentuation was studied by Pease and Bowden (1969), who have recommended the use of auxiliary filters in addition to the obligatory Wratten 12 filter. Excellent infrared work has been carried out in Finland by various workers (Talvitie, 1979; Talvitie and Paarma, 1973) using multiband scanning of the terrain. One of the classical studies carried out by this procedure was the accurate delineation of the Sokli carbonatite formation in the northwest of the country (Talvitie, 1979).

Thermographic and Radar Imagery

The techniques of thermographic and radar imagery involve the scanning of terrain at much higher wavelengths than those discussed above. They do not involve a photographic process. Thermography operates at wavelengths > 1200 nm and depends on detection of the thermal emission from the ground by means of a scanner (Colwell, 1968) which converts thermal energy into an electrical signal that can appear on a video screen or can be transferred to tape for computer processing at a later date. Thermography has not been used to any great extent for vegetation studies, although Talvitie et al. (1981) were able to identify peat which at night is often colder than it surroundings and during the day is warmer. Although thermographic work on vegetation is still in its infancy, Horler et al. (1981) have shown that metal stress can indeed affect thermal emission from plants because of the effect of heavy metals in inhibiting stomatal opening in the leaves. In greenhouse studies, the above authors showed increases of canopy radiation of about $1°C$ in the case of plants stressed with copper and cadmium.

Radar imagery is only indirectly of use in vegetation studies. A high-frequency signal generated from an aircraft is reflected from the surface of the terrain and can be analyzed by some suitable form of electronic detector. Wavelengths of about 1 to 10 cm are commonly used in this type of work. The nature of the vegetation cover can sometimes be deduced by changes in frequency and phase of polarization of the reflected signal.

Multiband Satellite Imagery

Multiband satellite imagery is potentially one of the most effective tools for inexpensive monitoring of the vegetation cover. The LANDSAT series of earth-monitoring satellites is equipped with four-channel sensors recording reflected radiation in the ranges 500 to 600 nm (MSS4), 600 to 700 nm (MSS5), 700 to 800 nm (MSS6), and 800 to 1100 nm (MSS7). The sensors scan the terrain and record the reflectance of individual areas of about 0.6 ha. They assign numbers on a scale of 0 to 255 to indicate the degree of reflectance of each picture element (pixel). The pixels are assembled into scenes that cover 31,000 km^2 and comprise 7,636,760 individual pixels. The information is carried on computer-compatible tape (CCT) which gives unlimited latitude for computer processing of the data. Maps produced from the data can be uniband or a combination of any pair of the four channels. Ratioing of these bands can accentuate vegetational differences that would not otherwise be noticeable to the human eye. Examples of studies of this nature are shown in work by Ballew (1975), Bélanger (1980a,b), and Lyon (1975).

The ratio MSS7/MSS5 is particularly favorable for vegetational studies since the denser the vegetation, the more it absorbs MSS5, whereas the healthier it is, the more it reflects MSS7. This ratio therefore gives a good indication of the density and health of the vegetation (Bélanger, 1980b). A further advantage of band ratioing is that variation in solar illumination of the terrain can be eliminated.

Satellite imagery is also useful to indicate areas with a sparse vegetation cover, such as ultramafic rocks. For example, Brooks and McDonnell (1983) used computer processing of LANDSAT data to delineate ultramafic rocks in New Zealand. Each subscene was enhanced by histogram equalization of bands MSS 4, 5, and 6, and the data were written in blue, red, and green, respectively, onto positive color film. Better delineation of ultramafic rocks was obtained with this procedure than with incorporation of the MSS7 infrared band or by use of the more usual technique of band ratioing.

Case History of Satellite Imagery from Canada

Bélanger (1980a,b) used ratioed LANDSAT data to interpret the vegetation cover of an ultramafic region near Thetford in Quebec Province. Figure 6-4 shows histograms of pixel signatures for overflights by LANDSAT II in two different seasons (June and November, 1978). The figure shows great differences of reflectance for the two seasons due to

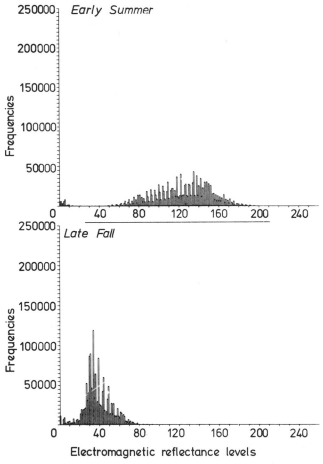

Fig. 6-4   Histograms of band MSS7 from a LANDSAT scene taken over the Thretford Mines area, Quebec, during two different seasons. (From Bélanger, 1980a. By permission of Geological Survey of Canada.)

senescence of leaves of trees and shrubs in the fall. The MSS7/MSS5 ratio (*Biomass Index*) was used for interpretation of the Thetford scene. The Biomass Index was plotted in false color using an Applicon color plotter (Bélanger, 1980b). The plotter ratios the data from the computer-compatible tapes and sprays ink jets at a computer-controlled density for the primary colors yellow, cyan, and red. The system provides a total of 25 shades of nine main colors. Water and terrain with sparse vegetation cover appear in various shades of red, whereas blue shades indicate dense healthy vegetation. When the scene was correlated with ground truth, it was possible to relate different shades with specific vegetation communities, some of which were related to ultramafic rocks. Table 6-5 shows the correlation between spectral signatures and vegetation communities.

TABLE 6-5   Reflectance Levels (Pixel Signatures) on LANDSAT Bands 4 to 7 for Five Vegetation Associations in the Thetford Mines Area, Quebec

Community Type	Reflectance Level			
	Band 4	Band 5	Band 6	Band 7
Scrub community	35	25	83	96
Coniferous forest	35–39	25–29	82–83	87–92
Mixed deciduous forest				
Poplar	42	26–31	125–131	145–158
Birch	36–39	25–27	106–147	125–172
Deciduous forest	40	28	139	154
Agricultural land	42	33	39	120

Source: Bélanger *et al.* (1979). By permission of Société Québecoise de Télédétection.

Other Case Histories of the Use of Satellite Imagery in Geobotany

Cole (1977) has used LANDSAT imagery in combination with thermography and multiband aerial photography for investigation of an orebody in the Mount Isa region of Queensland and used the data to identify plant communities typical of mineralization in the substrate.

Despite the relatively poor resolution of LANDSAT imagery compared with aerial photography, Bølviken *et al.* (1977) were able to use the MSS7/MSS5 ratio to delineate a previously known barren area of copper mineralization in Norway. Table 6-6 summarizes work by Lyon (1975) on the use of LANDSAT imagery for identifying specific ore occurrences by means of their anomalous vegetation cover.

Ballew (1975) carried out an excellent piece of quantitative work on LANDSAT data in which he correlated multispectral digital data from individual pixels, with geochemical data for six elements in the soils corresponding to each pixel investigated.

Another potential of satellite imagery has been explored by Cole (1980), who has shown that kimberlite pipes, host for diamonds, usually carry a more luxuriant vegetation than their surroundings because of higher levels of plant nutrients. Unfortunately, many kimberlites have a surface expression of only a few hectares, although Cole (1980) has cited the Orapa pipe in Botswana, which has a diameter of 1.5 km. It is likely that the largest kimberlite pipes would be detectable by existing satellite imagery, although this may change with the launching of new satellites, such as the French SPOT (Système Probatoire d'Observation de la Terre), which will have a resolution of only 20 m.

TABLE 6-6 Localities and Results of Some Case Histories of Geobotanical Anomalies Detected by LANDSAT Imagery

Nature of Cover	Location	Results of Investigations
Very low to zero	Yerington, Nevada	Oxidized and nonoxidized ore (sulfide) may be differentiated in the pit and sulfide rock discovered in the tailings pond
	Goldfield, Nevada	Alteration zones and gossans surrounding gold–alunite mineralization identified
Moderate	Pine Nut Mts., Nevada	A molybdenum-bearing skarn with a biogeochemical anomaly in the pinon pine and juniper, independently located by color-ratio images
Heavy	Karasjok, Norway	A known copper biogeochemical anomaly (Bølviken et al., 1977) was relocated by LANDSAT data
	Tifalmin, New Guinea	Vegetation over known copper-bearing intrusives was studied, although results inconclusive

Source: Lyon (1975).

BIOGEOCHEMISTRY IN MINERAL EXPLORATION

Biogeochemical methods of prospecting are heavily dependent on the existence of rapid and sensitive analytical techniques suitable for the analysis of plant material. In these days of speedy multielement analytical systems such as inductively coupled plasma emission spectrometry (ICP), where 20 elements can be determined in samples fed through the system at a rate of one every 2 minutes, it is difficult to appreciate the problems encountered by some of the earlier workers who relied on semiquantitative emission spectrography or on colorimetry to produce the data they needed.

Biogeochemical prospecting began just before World War II with the pioneering work of Tkalich (1938), Brundin (1939), and Vogt and his coworkers (e.g., Vogt, 1939). After the war the work was continued by Warren and his coworkers (e.g., Warren and Delavault, 1950; Warren, Prologue to this volume), who published over 30 papers on the subject during the period 1947–1970.

Since World War II most of the biogeochemical work has been carried out in the Soviet Union (Malyuga, 1964; Kovalevsky, 1979). The leading biogeochemist in the Soviet Union today is undoubtedly A. L. Kovalevsky, who has contributed many new ideas to the field, not the least of which is the concept of barrier-free sample types which will be discussed in more detail below.

Alternative Sample Types in Biogeochemical Prospecting: Barrier-Free Sample Materials

It has long been known that some elements are "difficult" for biogeochemical prospecting because the concentration of such elements in plant material remains relatively constant irrespective of their concentration in the substrate. This problem was investigated by

Timperley *et al.* (1970), who proposed that elements essential in plant nutrition, such as copper and zinc, had relatively constant levels in plant material because of internal control mechanisms operated by the species concerned. In the case of nonessential elements such as nickel or uranium, external mechanisms affect passive uptake by plants so that the concentrations in plant material reflect the concentrations of the same elements in the substrate. This problem was further investigated by Kovalevsky (1974), who proposed the use of appropriate sample types in biogeochemical prospecting. He investigated the use of such varied samples as roots, bark, needles, leaves, or even the entire aerial parts of plants. On plotting elemental concentrations in sample types as a function of the levels of the same elements in the substrate, he proposed that plots were of four main types, as illustrated in Figure 6-5. Plot (a) is a linear function with a very highly significant relationship between the two variables. Such a system is said to be barrier free. In plot (b),

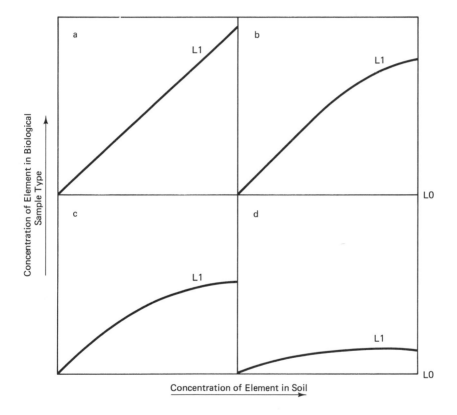

Fig. 6-5   Idealized plots of elemental concentrations in hypothetical biological sample types and in the supporting soil: (a) barrier-free system ($L_1/L_0 > 300$); (b) high-barrier system ($L_1/L_0 = 30$–300); (c) medium-barrier system ($L_1/L_0 = 3$–30); (d) low-barrier system ($L_1/L_0 = 2$–5). (L_0, Concentration in plant from background; L_1, concentration in specific plant sample. (From Brooks, 1983. By permission of John Wiley & Sons, Inc.)

there is partial exclusion of the element by the plant at higher concentrations of this element in the soil. This is said to be a high-barrier system. This term is somewhat confusing because in English the inference is that there is a high barrier to uptake, whereas in the Russian, the meaning is that exclusion occurs only for high values on the y axis. Plot (c) shows a medium barrier system, and plot (d) shows a low barrier where once again a certain amount of confusion could be caused. Low barrier means a high resistance to uptake by the plant, where "low" refers to the position on the y axis where exclusion occurs. To quantify still further the terms "high" and "low," the degree of exclusion can be indicated by the term L_1/L_0, where L_1 is the limiting value for the concentration of an element in the sample type and L_0 is the same value for samples growing in a non-mineralized "background" soil. Table 6-7 classifies some of the common ore elements in order of effectiveness for biogeochemical prospecting and is based on data in Kovalevsky (1979).

TABLE 6-7   Biological Sample Types Listed for Four Categories of Uptake of Ore Elements

Element	Number of Samples Tested	Percentage of Biological Sample Types within the Category[a]			
		Non-barrier	High Barrier	Medium Barrier	Low Barrier
Molybdenum	64	16	65	19	0
Gold	67	10	26	44	20
Lead	70	10	22	35	33
Tungsten	31	3	20	32	45
Silver	90	2	7	49	42
Beryllium	53	0	4	45	51
Uranium	65	0	3	17	80
Fluorine	60	0	3	7	90
Total	500	Aver. 5	Aver. 19	Aver. 31	Aver. 45

[a]Nonbarrier, no limit to BAC (biological absorption coefficient = plant/soil concentration ratio); high barrier, BAC 0.50–5.0; medium barrier, BAC 0.15–0.49; low barrier, BAC < 0.15. BAC values based on dry weight.
Source: Kovalevsky (1979).

Other Alternative Sample Types for Biogeochemical Prospecting

The interest in alternative sample types in biogeochemical prospecting is reflected in the number of chapters in this volume which are concerned with this subject. These various types are summarized in Table 6-8 and the reader is referred to Brooks (1983) for a further discussion of this important subject.

TABLE 6-8 Alternative Sample Types (excluding Leaves and Twigs)
in Biogeochemical Prospecting

Sample Types	Species	Elements Sought	Location	References
Algae	Various	Cu, Zn, As	California	Filipek (Abstract D in Chapter 12, this volume)
Bark	*Pinus* sp.	Pb	USSR	Kovalevsky (1979)
Fallen leaves	Various	Pb	USSR	Kovalevsky (1979)
Flowers	*Elsholtzia*	Cu	USSR	Ginzburg (1957)
Grain	Oats	Cu	USSR	Ostrovskaya (1961)
Herbarium sp.	Various	Various	Various	This chapter
Lichens	Various	Various	Various	Richardson *et al.* (1980)
Mosses				
Aquatic	Various	U	New Zealand	Whitehead and Brooks (1969)
Terrestrial	Various	Various	Various	Shacklette (1967)
Mull/humus	Various	Au, Ag, Cu, Pb, Zn, Mo	Colorado/Idaho	Curtin and King (Chapter 22, this volume)
Pollen	Various	Au	British Columbia	Warren (Prologue, this volume)
Roots	*Mimulus*	Ag, Pb, Zn, Cu	British Columbia	Fortescue (1983)
Tree sap	*Betula*	Au	USSR	Krendelev *et al.* (1978)
Wood	Douglas fir	Au	Idaho	Erdman and Leonard (Abstract A in Chapter 12, this volume)

AERIAL SAMPLING AND ANALYSIS OF VEGETATION

Aerial sampling of vegetation is an obvious method of taking advantage of the superior position of the vegetation canopy in global terrain. Vegetation poses a barrier for aerial sampling of soils, but if it could be sampled itself instead of the soils, the advantages are very obvious. Early attempts to sample vegetation from the air in New Zealand were unsuccessful because of safety problems associated with a foliage reaper suspended from a helicopter. Perhaps the earliest record of aerial sampling of vegetation was afforded by Weiss (1967), who employed a filter system for collecting vegetation particulates in air sucked into a device which then analyzed them by x-ray fluorescence spectrometry.

The original work of Weiss has been refined by A. R. Barringer and his associates in techniques known as AIRTRACE and SURTRACE, respectively. Basic details of these systems are given below.

It has been known for about 15 years that plants produce metal-rich exudates (Nemeryuk, 1970). Curtin *et al.* (1974) were able to show that forest trees such as *Pinus contorta, Picea engelmannii, and Pseudotsuga menziesii* produce exudates at their needle surfaces which contain elevated concentrations of metals present in the substrate below. Their experimental findings are summarized in Figure 6-6. Later work by Beauford *et al.* (1975, 1977) and Horler *et al.* (1980) showed that the "exudates" were, in fact, microscopic foliar dust particles and that they bore some relationship to the concentrations of elements in the substrate.

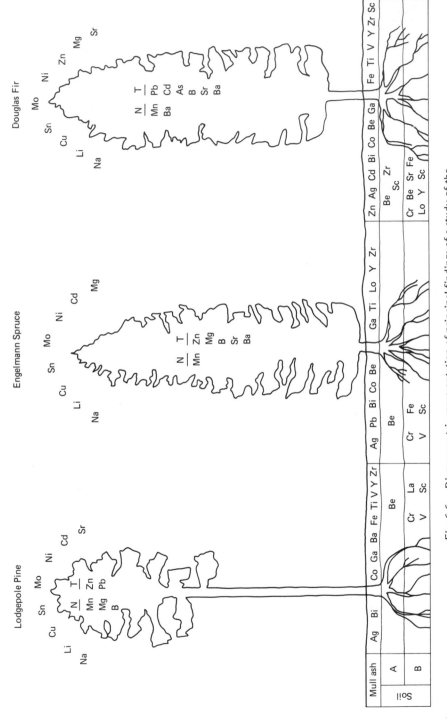

Fig. 6-6 Diagrammetric representation of principal findings of a study of the highest elemental concentrations in soils, humus (mull), vegetation, and the ash of elemental halos above trees. N, Ash of needles; T, ash of twigs. (From Curtin et al, 1974. By permission of Elsevier Science Publishers.)

Meteorological Factors
Affecting Atmospheric Particulates

Barringer (1977) has shown that atmospheric particulates are of two main types. The first of these comprises small particulates (<10 μm) which are of global origin and may have traveled hundreds of kilometers in the upper atmosphere before returning to the lower atmosphere. Larger particulates (>10 μm) form the second type and are usually of local origin. These are often of a biological nature derived from the foliar dust of vegetation.

Sampling of atmospheric particulates depends largely on the existence of suitable convection plumes since the surrounding cooler air does not usually contain the larger particulates, which are the target of the technique of aerial sampling and analysis. In addition to problems associated with the presence or absence of convection plumes, there is also the fact that there is also a marked diurnal cycle whereby particulates settle low at night and rise during the day due to solar radiation. A third factor to be reckoned with is seasonal. In winter there is less mixing of particulates in the lower atmosphere, although Barringer (1977) has shown that satisfactory results can be obtained even in winter with snow on the ground and with an air temperature of $-10°$C.

AIRTRACE and SURTRACE

Equipment designed by Barringer and his coworkers (Barringer, 1977) for sampling and analysis of atmospheric particulates is known as AIRTRACE. Surveys are carried out about 10 m above treetop height and air is sucked into a helicopter instrument at a predetermined rate. Thermal detectors eliminate material not sampled from convection plumes. The particulates collected are forced into a special adhesive tape which is fed into the source of an ICP instrument (see above). The material is vaporized by a laser beam before entering the source and some 14 elements are determined in situ. To make the technique quantitative (weighing of the particulates is obviously impossible), the data are normalized to the titanium content of the samples (titanium has a relatively constant abundance in inorganic material) or to the major constituents (SiO_2, Al_2O_3, etc.), from which an estimate of the mass can be made assuming that these major elements form 90 percent of the mass of silicate material. In the case of plant material, the problem is more complicated and in such cases it is customary to normalize to the copper content of the vegetation since this being an essential element (see above) it has a relatively constant abundance in plant material.

SURTRACE is a modification of AIRTRACE and was devised by Barringer et al. (1978). This technique depends on collecting and analyzing surface dust by means of a flexible tubular suction device trailed from a helicopter. The process of analysis is similar to that of AIRTRACE. To identify sampling sites, marks on the collecting tape (see above) are correlated with fiducial marks on 35mm film shot by the camera as the helicopter passes over the terrain. SURTRACE can also be operated from a backpack with analysis carried out later in the laboratory.

Case Histories of the Use
of AIRTRACE and SURTRACE

A case history of the use of AIRTRACE in Quebec is illustrated in Figure 6-7. It was possible to delineate with some precision a large base metal deposit in this province. The correlation between AIRTRACE and airborne EM conductimetric measurements was particularly good.

A case history of the use of SURTRACE is shown in Figure 6-8. In this instance

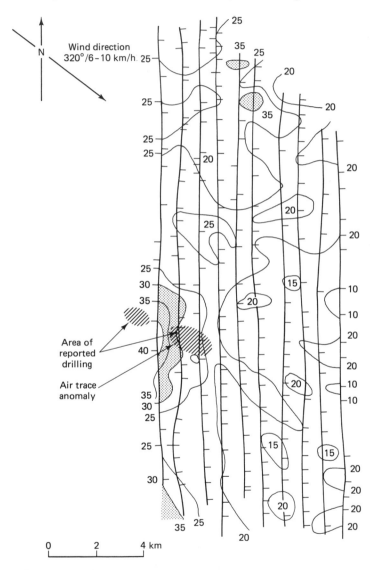

Fig. 6-7 AIRTRACE activity contours computed from multiple regression analysis of copper + zinc concentrations in aerial particulates. The overflight was over an area of base metal mineralization in Quebec. (From Barringer, 1977.)

PRESENT STATUS OF BOTANICAL EXPLORATION USING HIGHER PLANTS

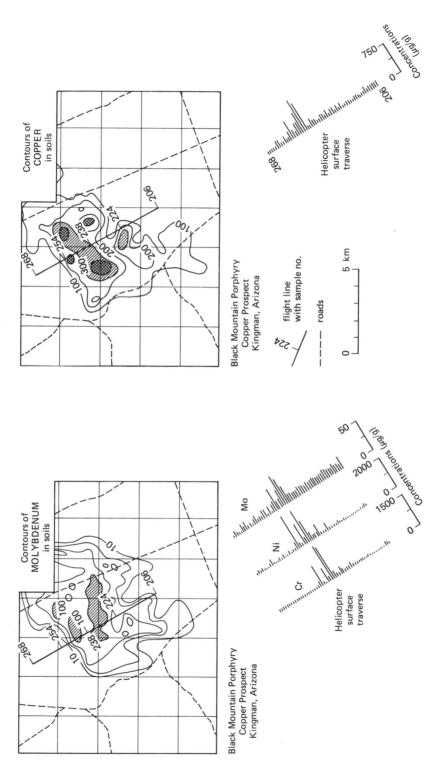

Fig. 6-8 SURTRACE measurements over the Black Mt. porphyry deposit, Kingman, Arizona. (From Barringer *et al.*, 1978.)

123

the survey was carried out at Kingman, Arizona, over a porphyry copper deposit. The area was sufficiently well vegetated to ensure that the material collected by SURTRACE was biological rather than inorganic. As will be seen from the figure, the copper and molybdenum mineralization was very well delineated and there was a good correlation between SURTRACE and soil geochemistry.

The methods described above have the advantage of very great speed of execution compared with more conventional methods such as soil geochemistry. It must, however, be emphasized that AIRTRACE depends on the existence of suitable atmospheric conditions, whereas this is not so to the same extent for SURTRACE, which probably has the better promise for the future.

BIOGEOCHEMICAL PROSPECTING IN THE HERBARIUM

Some 200 million specimens of dried plants are located in the world's herbaria (plant museums), where they are held as reference material for botanical and other studies. The oldest herbarium (Paris) was founded in 1635 and has about 5,000,000 specimens. The largest herbarium (6,000,000 specimens) is at Kew near London.

Herbarium material was not originally available for chemical analysis because of the large sample needed for older analytical techniques. However, some 35 years ago, Chenery (1948, 1951) carried out simple spot tests for aluminum on leaf fragments and tested all of the 259 recognized families of dicotyledones. Herbarium material was also used for geobotanical work when Persson (1956) studied the collection localities of "copper mosses" in Scandinavia and "discovered" three copper anomalies, of which one was an existing copper mine.

The development in recent years of analytical techniques requiring only a few milligrams of plant material have now rendered chemical analysis a feasible possibility.

Some Case Histories of Use of Herbarium Material for Biogeochemical Prospecting

Some years ago, Brooks *et al.* (1977a) determined nickel in some 2000 species of the genera *Homalium* and *Hybanthus* using leaf fragments with a mean mass of about 20 mg. These families were chosen because they contained previously known hyperaccumulators (>1000 $\mu g/g$ in dried material) of nickel. This herbarium survey resulted in the identification of most of the world's major serpentine occurrences in the tropical and warm-temperate latitudes from the collection localities recorded on the herbarium sheets. In addition, a further 10 hyperaccumulators of nickel were identified. Such species are almost always indicator plants with significance for geobotanical prospecting. Details of this and other case histories are summarized in Table 6-9.

Only two further case histories will be discussed in any detail. The first of these involved the discovery that *Rinorea bengalensis* was a hyperaccumulator of nickel. This taxon extends in a wide arc from Sri Lanka through southeast Asia down to Queensland in Australia. Brooks and Wither (1977) determined the nickel and cobalt content of over 100 specimens of this species and related high nickel contents to the type of substrate in

TABLE 6-9 Use of Herbarium Material in Mineral Exploration

Element	Locality	Object and Achievement	References
Cobalt	Zaire	Discovery of hyperaccumulators and indicator plants (15 taxa identified)	Brooks (1977), Brooks *et al.* (1980)
Copper	Zaire	Discovery of hyperaccumulators and indicator plants (13 taxa identified)	Brooks *et al.* (1980)
	Scandinavia	Relationship of *Lychnis alpina* and *Silene dioica* to minerals was investigated (numerous mineralized sites identified)	Brooks *et al.* (1979b)
	Australia	Numerous specimens of *Polycarpaea* were analyzed (sites worthy of follow-up were identified)	Brooks and Radford (1978)
	Salajar I.	Several sites on island with possibly anomalous values were pinpointed by analysis of plants	Brooks *et al.* (1978)
Lead	Scandinavia	See *Lychnis*, etc. above	Brooks *et al.* (1979b)
	Austria/Italy	Identification of hyperaccumulators of lead and zinc confined to mineralization (two taxa identified)	Reeves and Brooks (1983a)
Manganese	New Caledonia	Identification of hyperaccumulators of manganese (numerous taxa discovered)	Brooks *et al.* (1981a), Jaffré (1977)
Nickel	Worldwide	Identification of hyperaccumulators of nickel which indicate ultramafic areas in world (most extensive regions pinpointed)	Brooks *et al.* (1977a)
	Mediterranean	Identification of indicators and hyperaccumulators of nickel (about 50 species of *Alyssum* with this property identified)	Brooks *et al.* (1979a)
	S.E. Asia	Discovery of unknown ultramafic areas in S.E. Asia by analysis of *Rinorea bengalensis* and two other species (two areas identified)	Brooks and Wither (1977), Wither and Brooks (1977)
	New Caledonia	Numerous hyperaccumulators in eight genera sought and identified	Jaffré *et al.* (1979a,b), Jaffré (1980), Kersten *et al.* (1979)
Zinc	Austria/Italy	See under *Lead* above	Reeves and Brooks (1983a)

which the plant was known to be growing. One specimen from Nabire in Irian Jaya (Indonesian New Guinea) was clearly growing over an ultramafic substrate because of its very high nickel content. A form of crude geological mapping over inaccessible terrain was thereby effected without even visiting the country. It was further observed that the nickel/cobalt ratio was diagnostic of other types of substrate beside ultramafics. Figure 6-9 shows the distribution of *R. bengalensis* and the results of chemical analysis of this taxon.

An unusual example of biogeochemical prospecting in the herbarium is afforded by the work of Brooks *et al.* (1978), who analyzed the whole of a collection of plants made by Docters van Leeuwen (1937) in the island of Salajar just south of Sulawesi (Celebes) in Indonesia. A number of specimens showed very high copper levels for collection localities centered around the south of the island. Values as high as 600 μg/g were obtained for some specimens. When identical species from outside of Salajar were analyzed, these were found to have negligible concentrations of this element. To check

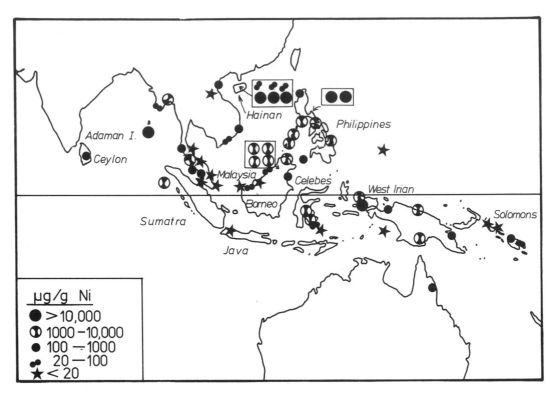

Fig. 6-9　Map of southeast Asia and Australasia, showing collection localities of specimens of *Rinorea bengalensis* and indicating nickel concentrations (μg/g dry weight) in leaf material. (From Brooks and Wither, 1977. By permission of Elsevier Science Publishers.)

that the high copper levels from Salajar were not due to fungicides or insecticides used by the herbarium, the paper of the original herbarium sheets was also analyzed. It would appear that the south of Salajar Island would be an appropriate place to carry out follow-up mineral exploration.

CURRENT AND FUTURE DEVELOPMENTS
IN BOTANICAL METHODS OF EXPLORATION

Although it is usually difficult and indeed unwise to predict the future, I would like, finally, to add my own opinion on the direction of future trends in botanical methods of prospecting. In an earlier work (Brooks, 1972a) I predicted developments in aerial sampling and analysis of vegetation. This forecast proved to be true with the later development of AIRTRACE and SURTRACE (see above). At present the future of these two systems seems to be in some doubt in the current economic climate, and it is earnestly hoped that there will be further developments in this direction in the immediate future once the exploration industry is back on its feet. Aerial and satellite sensing will also continue to flourish, particularly with the advent of better resolution, approaching that of aerial photography.

It is likely that biogeochemists will continue to profit from new and sophisticated analytical systems such as ICP (see above) which render possible the obtaining of vast amounts of data. Large amounts of data do, of course, carry with them the concomitant problems of data handling and processing, but I look to the future to provide us with newer and speedier computer-assisted statistical procedures which will enable the biogeochemist to detect subtle anomalies concealed in otherwise apparently random data and which will lead him to fresh discoveries of minerals beneath the earth's surface.

REFERENCES

Aery, N. C., 1977, Studies on the geobotany of Zawar mines: *Geobios* (Jodhpur), v. 4, p. 225–228.

Anon., 1959, A flower that led to a copper discovery: *Horizon* (RST Group, Ndola, Zambia), v. 1, p. 35–39.

Ballew, G. I., 1975, Correlation of LANDSAT-1 multispectral data with surface geochemistry: *Proc. 10th Int. Symp. Rem. Sens. Environ.*, Ann Arbor, Mich., v. II, p. 1045–1055.

Barringer, A. R., 1977, AIRTRACE—an airborne geochemical exploration technique: *U.S. Geol. Surv. Prof. Pap. 1015*, p. 231–251.

———, Davies, J. H., and Daubner, L., 1978, SURTRACE—an airborne geochemical system: *Proc. 12th Int. Symp. Rem. Sens. Environ.*, Manila, p. 975–990.

Beauford, W., Barber, J., and Barringer, A. R., 1975, Heavy metal release from plants into the atmosphere: *Nature*, v. 256, p. 35–36.

———, Barber, J., and Barringer, A. R., 1977, Release of particles containing metals from vegetation into the atmosphere: *Science*, v. 195, p. 571–573.

Bélanger, J. R., 1980a, Studies on Quaternary geology; an approach using remote sensing information: *Geol. Surv. Can. Pap. 80-1B*, p. 287–291.

———, 1980b, LANDSAT et les gisements miniers: *Geos* (Summer), p. 10–12.

———, Rencz, A. N., and Shilts, W. W., 1979, Patterns of glacial dispersion of heavy metals as reflected by satellite imagery, p. 59–73 *in* Bonn, F. (Ed.), *Télédétection et géstion des ressources:* Assoc. Québ. de Télédétection, Ste Foy, Québec.

Birnie, R. W., and Francica, J. R., Jr., 1979, Remote detection of geobotanical anomalies: a method for mineral exploration: *Abstr. Prog. Geol. Soc. Am.*, v. 11, p. 389.

Bølviken, B., Honey, F., Levine, S. R., Lyon, R. P. J., and Prelat, A., 1977, Detection of naturally heavy-metal-poisoned areas by LANDSAT-1 digital data: *J. Geochem. Explor.*, v. 8, p. 457–471.

Brooks, R. R., 1972a, *Geobotany and Biogeochemistry in Mineral Exploration:* Harper & Row, New York, 290 p.

———, 1972b, Aerial infrared photography as a guide to geological conditions: *Proc. Australas. Inst. Min. Metall.*, no. 242, p. 25–36.

———, 1977, Copper and cobalt uptake by *Haumaniastrum* species: *Plant Soil*, v. 48, p. 541–545.

———, 1983, *Biological Methods of Prospecting for Minerals:* Wiley, New York, 350 p.

———, and Crooks, H. M., 1980, Studies on uptake of heavy metals by the Scandinavian "kisplanten" *Lychnis alpina* and *Silene dioica: Plant Soil*, v. 54, p. 491–495.

———, and Malaisse, F., 1985, *The Heavy Metal Tolerant Flora of South Central Africa— An Interdisciplinary Approach:* Balkema, Rotterdam, 199 p.

———, and McDonnell, M. J., 1983, Delineation of New Zealand ultramafic rocks by computer processing of digital data from satellite imagery: *N.Z. J. Sci.*, v. 26, p. 65–71.

———, and Radford, C. C., 1978, An evaluation of background and anomalous copper and zinc concentrations in the "copper plant" *Polycarpaea spirostylis* and other Australian species of the genus: *Proc. Australas. Inst. Min. Metall.*, no. 268, p. 33–37.

———, and Wither, E. D., 1977, Nickel accumulation by *Rinorea bengalensis* (Wall.) O.K.: *J. Geochem. Explor.*, v. 7, p. 295–300.

———, Lee, J., and Jaffré, T., 1974, Some New Zealand and New Caledonian plant accumulators of nickel: *J. Ecol.*, v. 62, p. 523–529.

———, Lee, J., Reeves, R. D., and Jaffré, T., 1977a, Detection of nickeliferous rocks by analysis of herbarium specimens of indicator plants: *J. Geochem. Explor.*, v. 7, p. 49–57.

———, McCleave, J. A., and Malaisse, F., 1977b, Copper and cobalt in African species of *Crotalaria* L.: *Proc. R. Soc. Lond. B*, v. 197, p. 231–236.

———, Wither, E. D., and Westra, L. Y., 1978, Biogeochemical copper anomalies in Salajar Island: *J. Geochem. Explor.*, v. 10, p. 181–188.

———, Morrison, R. S., Reeves, R. D., Dudley, T. R., and Akman, Y., 1979a, Hyperaccumulation of *Alyssum* Linnaeus (Cruciferae): *Proc. R. Soc. Lond. B*, v. 203, p. 387–403.

———, Trow, J. M., and Bølviken, B., 1979b, Biogeochemical anomalies in Fennoscandia: a study of copper, lead and nickel in *Melandrium dioicum* and *Viscaria alpina: J. Geochem. Explor.*, v. 11, p. 73–87.

———, Reeves, R. D., Morrison, R. S., and Malaisse, F., 1980, Hyperaccumulation of copper and cobalt—review: *Bull. Soc. R. Bot. Belg.*, v. 113, p. 166–172.

———, Trow, J. M., Veillon, J.-M., and Jaffré, T., 1981a, Studies on manganese-accumulating *Alyxia* from New Caledonia: *Taxon*, v. 30, p. 420–423.

———, Holzbecher, J., and Ryan, D. E., 1981b, Horsetails as indicators of gold mineralization: *J. Geochem. Explor.*, v. 16, p. 21–26.

Brundin, N., 1939, Method of locating metals and minerals in the ground: U.S. Pat. 2,158,980.

Brussell, D., 1978, *Equisetum* stores gold: *Phytologia*, v. 38, p. 469–473.

Buyalov, N. I., and Shvyryayeva, A. M., 1961, Geobotanical methods in prospecting for salts of boron: *Int. Geol. Rev.*, v. 3, p. 619–625.

Canney, F. C., Cannon, H. L., Cathrall, J. B., and Robinson, K. 1979, Autumn colors, insects, plant disease and prospecting: *Econ. Geol.*, v. 74, p. 1673–1692.

Cannon, H. L., 1957, Description of indicator plants and methods of botanical prospecting for uranium deposits on the Colorado Plateau: *U.S. Geol. Surv. Bull. 1030-M*, p. 399–516.

———, Shacklette, H. T., and Bastron, H., 1968, Metal absorption by *Equisetum* (horsetail): *U.S. Geol. Surv. Bull. 1278-A*, 21 p.

Chaffee, M. A., and Gale, C. W., III, 1976, The California poppy *Eschscholzia mexicana*) as a copper indicator plant—a new example: *J. Geochem. Explor.*, v. 5, p. 59–63.

Chenery, E. M., 1948, Aluminium in the plant world: *Kew Bull.*, p. 173–183.

———, 1951, Some aspects of the aluminium cycle: *J. Soil Sci.,* v. 2, p. 97–103.

Cole, M. M., 1965, Biogeography in the service of man: Inaugural Lecture, Bedford College, London, 59 p.

———, 1971, The importance of environment in biogeographical/geobotanical and bio-geochemical investigations: *Can. Inst. Min. Metall. Spec. Vol. 11,* p. 414–425.

———, 1973, Geobotanical and biogeochemical investigations in the sclerophyllous wood-land and shrub associations of the Eastern Goldfields area of Western Australia with particular reference to the role of *Hybanthus floribundus* (Lindl.) F. Muell. as a nickel indicator and accumulator plant: *J. Appl. Ecol.,* v. 10, p. 269–320.

———, 1977, LANDSAT and airborne multispectral and thermal imagery used for geo-logical mapping and identification of ore horizons in Lady Annie–Lady Loretta and Dugald River areas, Queensland, Australia: *Trans. Inst. Min. Metall. Sec. B,* v. 86, p. 195–215.

———, 1980, Geobotanical expression of orebodies: *Trans. Inst. Min. Metall. Sec. B,* v. 89, p. 73–91.

———, and Le Roex, H. D., 1978, The role of geobotany, biogeochemistry, and geochemis-try in mineral exploration in South West Africa and Botswana: a case history: *Trans. Geol. Soc. S. Africa,* v. 81, p. 277–317.

———. Provan, D. M. J., and Tooms, J. S., 1968, Geobotany, biogeochemistry and geo-chemistry in the Bulman–Waimuna Springs area, Northern Territory, Australia: *Trans. Inst. Min. Metall. Sec. B,* v. 77, p. 81–104.

Colwell, R. N., 1968, Remote sensing of natural resources: *Sci. Am.,* v. 218, p. 54–69.

Curtin, G. C., King, H. D., and Mosier, E. L., 1974, Movement of elements into the atmosphere from coniferous trees in subalpine forests of Colorado and Idaho: *J. Geochem. Explor.,* v. 3, p. 245–263.

Docters van Leeuwen, W. M., 1937, Botanical results of a trip to the Salajar Islands: *Blumea,* v. 2, p. 239–383.

Duvigneaud, P., 1959, Plantes cobaltophytes dans le Haut Katanga: *Bull. Soc. R. Bot. Belg.,* v. 91, p. 111–134.

———, and Denaeyer-de Smet, 1963, Cuivre et la végétation au Katanga: *Bull. Soc. R. Bot. Belg.,* v. 96, p. 93–231.

Ernst, W., 1968, Das Violetum calaminariae westfalicum eine Schwermetallpflanzengesell-schaft bei Blankenrode in Westfalen: *Mitt. Florist.-Soziol. Arbeitsgem.,* v. 13, p. 263–268.

———, 1969, Pollenanalytischer Nachweis eines Schwermetallrasens in Wales: *Vegetatio,* v. 18, p. 393–400.

Farago, M. E., Mullen, W. A., Cole, M. M., and Smith, R. F., 1980, A study of *Armeria maritima* (Mill.) Willdenow growing in a copper-impregnated bog: *Environ. Pollut.,* v. 21, p. 225–244.

Fortescue, J. A. C., 1983, The use of *Mimulus guttatus* DC as an aid to prospecting in northern British Columbia: Program and abstracts for colloquium Organic Matter, Biological Systems and Mineral Exploration, Dept. Earth and Space Sciences, UCLA, Feb., 1983.

Ginzburg, I. I., 1957, *Experiments to Provide a Theoretical Basis for Geochemical Prospecting Methods* (in Russian): Gosgeoltekhizdat, Moscow.

Henwood, W. J., 1857, Notice of the copper turf of Merioneth: *Edinb. New Philos. J.,* v. 5, p. 61–63.

Horler, D. N. H., Barber, J., and Barringer, A. R., 1980, A multielement study of plant

Skertchly, S. B. J., 1897, The copper plant (*Polycarpaea spirostylis*, F. von Mueller): *Queensland Geol. Surv. Publ. 119*, p. 51–53.

Talvitie, J., 1979, Remote sensing and geobotanical prospecting in Finland: *Bull. Geol. Soc. Finl.*, v. 51, p. 63–73.

——, and Paarma, H., 1973, Recent prospecting in photogeology in Finland, p. 73–81 *in* Jones M. J. (Ed.), *Prospecting in Areas of Glacial Terrain:* Institution of Mining and Metallurgy, London.

——, Lehmuspelto, P., and Vuotesi, T., 1981, Airborne thermal surveying of the ground in Sokli Finland: *Geol. Surv. Finl. Rep. 50*, 13 p.

Thomas, P. I., Walker, B. H., and Wild, H., 1977, Relationship between vegetation and environment on an amphibolite outcrop near Nkai, Rhodesia: *Kirkia*, v. 10, p. 503–541.

Timperley, M. H., Brooks, R. R., and Peterson, P. J., 1970, The significance of essential and non-essential trace elements in plants in relation to biogeochemical prospecting: *J. Appl. Ecol.*, v. 7, p. 429–439.

Tkalich, S. M., 1938, Experience in the investigation of plants as indicators in geological exploration and prospecting (in Russian): *Vest. Dal'nevost. Fil. Akad. Nauk SSSR*, v. 32, p. 3–25.

Ventakesh, V., 1964, Geobotanical methods of mineral exploration in India: *Indian Miner.*, v. 18, p. 101.

Vogt, T., 1939, Chemical and botanical prospecting at Roros (in Norwegian): *Forh. Kong. Norsk. Videns. Selsk.*, v. 12, p. 82–83.

Warren, H. V., and Delavault, R. E., 1950, Gold and silver content of some trees and horsetails in British Columbia: *Geol. Soc. Am. Bull.*, v. 61, p. 123–128.

Weiss, O., 1967, Method of aerial prospecting which includes the step of analyzing each sample for element content, number and size of particles: U.S. Pat. 3,309,518.

Whitehead, N. E., and Brooks, R. R., 1969, Aquatic bryophytes as indicators of uranium mineralization: *Bryologist*, v. 72, p. 501–507.

Wild, H., 1968, Geobotanical anomalies in Rhodesia. 1. The vegetation of copper-bearing soils: *Kirkia*, v. 8, p. 1–71.

——, 1970, Geobotanical anomalies in Rhodesia. 3. The vegetation of nickel-bearing soils: *Kirkia*, v. 7 (Suppl.), p. 1–62.

——, 1978, The vegetation of heavy metal and other toxic soils, p. 1301–1332 *in* Werger, M. J. A. (Ed.), *Biogeography and Ecology of Southern Africa:* Junk, The Hague.

Wither, E. D., and Brooks, R. R., 1977, Hyperaccumulation of nickel by some plants of southeast Asia: *J. Geochem. Explor.*, v. 8, p. 579–583.

Yost, E., and Wenderoth, S., 1971, The reflectance spectra of mineralized trees: *Proc. 7th Int. Symp. Rem. Sens. Environ.*, Ann Arbor, Mich., p. 269–284.

Colin E. Dunn*
Saskatchewan Geological Survey
Regina, Saskatchewan, Canada

7

APPLICATION OF BIOGEOCHEMICAL METHODS TO MINERAL EXPLORATION IN THE BOREAL FORESTS OF CENTRAL CANADA

*Current address: Geological Survey of Canada, Ottawa, Canada.

ABSTRACT

The northern forests of central Canada are dominated by black spruce (*Picea mariana*) and jack pine (*Pinus banksiana*), with birch, poplar, alder, willow, and tamarack. The dominant shrubs are labrador tea (*Ledum groenlandicum*) and leather leaf (*Chamaedaphne calyculata*). Investigation of the element uptake of these and other plants is helping to identify which organ of a given plant is the most sensitive indicator of a particular element.

In an area of low-grade gold mineralization, the forest litter appears the most effective biogeochemical medium for outlining auriferous zones. Samples collected from a reported tungsten occurrence provided negative results, but data indicate that conifer twigs concentrate tungsten more than the 15 other sample media examined.

Of particular interest in Saskatchewan are the remarkable concentrations of uranium in black spruce twigs. In the ash of the latest 10 years of growth, concentrations range from 1 to 2270 ppm U with up to 1360 ppm U in areas of virgin forest. There is a 10,000-km^2 area across the eastern edge of the Athabasca Sandstone, within which all spruce have over 10 ppm U in their ashed twigs. This area encompasses several deeply buried uranium deposits, above which there are intense biogeochemical anomalies.

Plant parts that concentrate other elements include the trunks of black spruce and jack pine (Ag, Zn); jack pine trunk (Cd); jack pine cones and labrador tea roots (Ni, Cu); alder twigs and leaves, and blueberry leaves (Mo); black spruce twigs (rare earths); and Pb in the twigs and stems of most species.

INTRODUCTION

Only a few studies (e.g., Walker, 1979; Dunn, 1981) have been published on biogeochemical investigations for minerals in the boreal forests of central Canada (i.e., Alberta, Saskatchewan, and Manitoba). Data are available in provincial government assessment files, but they are often sketchy with regard to the exact nature of the sample medium, sample preparation, and analysis.

Currently, orientation studies are being conducted in Saskatchewan by the author to establish which parts of the common plant species are the most sensitive indicators of the presence of a particular element or a particular type of mineralization. Specific studies have been carried out over zones of uranium, gold, and reputed tungsten mineralization.

PHYSIOGRAPHY AND GEOLOGY

In central Canada the boreal forest occupies a zone extending from about 54° to 62° N. The ground is covered with glacial till of variable thickness and composition, upon which immature podzols, regosols, and peat bogs are developed. The geology is mainly Precambrian Shield composed of metamorphic rocks with some igneous intrusions. The large Athabasca Basin, containing unmetamorphosed clastic sediments of Helikian age, stretches across northern Saskatchewan into northeastern Alberta.

The severe continental climate has temperatures ranging from −50 to +35°C, with

40 to 60 cm of precipitation that occurs mainly during the short summer. These conditions sustain a boreal forest, the northern part of which is in an area of discontinuous permafrost. The forest comprises mainly black spruce [*Picea mariana* (Mill.) B.S.P.] and jackpine (*Pinus banksiana* Lamb.), with lesser numbers of white spruce [*Picea glauca* (Moench) Voss], balsam fir [*Abies balsamea* (L.) Mill.], tamarack [*Larix laricina* (Du Roi) K. Koch], birch (*Betula* sp.), poplar (*Populus* sp.), willow (*Salix* sp.), and alder (*Alnus* sp.). The latter two occur as tall shrubs rather than well-formed trees. The main low shrubs are labrador tea (*Ledum groenlandicum*), leather leaf (*Chamaedaphne calyculata*), and blueberry (*Vaccinium myrtilloides*). Numerous small shrubs, forbs, graminoids, lichens, and mosses cover the ground but have low practical significance for conducting a biogeochemical survey because of their sporadic development.

Of the above, the most ubiquitous species are black spruce, jack pine, birch, poplar, alder, and labrador tea. These comprise, therefore, the species most likely to be of use to the mineral explorationist.

METHODS

General

The main thrust of the biogeochemical investigations has been directed toward the exploration for uranium, since the Athabasca Basin contains some of the world's richest uranium deposits. However, many of the 3000 samples collected by the author for U analysis were analyzed also for a wide range of trace and major elements by AAS (atomic absorption spectrophotometry) and ICAP (inductively coupled argon plasma spectrometry), thereby building a broad data base of element distributions within the plants (Dunn, 1981). In addition, the biogeochemical response to gold and tungsten mineralization has been investigated (Fig. 7-1).

Studies have maintained a strong practical approach in order to provide the explorationist with a rapid sampling technique and an understanding of the problems and limitations that must be taken into account when interpreting results.

Sample Collection and Preparation

Over mineralized zones samples are collected at 15- or 30-m intervals, usually along cut lines spaced 30 m or 60 m apart. To obtain about 1 g of ash to permit multiple analyses of a given sample, from 100 to 400 g of fresh material is collected to allow for the weight loss due to moisture and organic components. This weight loss varies from one plant organ to the next.

Figure 7-2 is a flowchart of the sample preparation and analysis procedure adopted for multielement studies. The use of a microwave oven (preferably in the field) saves considerable time in drying the samples, and the pottery kiln permits the ashing of 50 to 100 samples each of about 100 g, in aluminum trays, in a single overnight firing.

For some elements (e.g., Au, As and Sb) the samples are not ashed in case they volatilize: instead, the dried samples are milled, then pressed into 8-g disks for neutron activation analysis.

LOCATION MAP

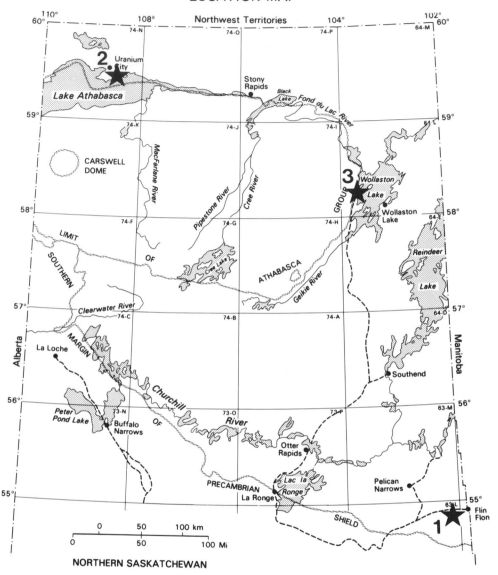

Fig. 7-1   Location map of northern Saskatchewan showing areas of study.
1, Au survey; 2, W survey; 3, U survey.

Each batch of 20 samples contains at least one standard sample and one duplicate, in order to quantify accuracy and precision, respectively. Batches not meeting the required standards (generally ±10 percent, depending on the element and its concentration) are discarded or reanalyzed. In general, the neutron activation data are considerably better than ±10 percent, and the aqua regia digest of ashed samples for AAS or ICAP

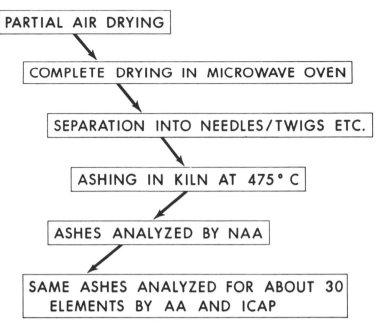

```
                  ┌──────────────────────────┐
                  │   PARTIAL  AIR  DRYING   │
                  └──────────────────────────┘
                              ↘
          ┌──────────────────────────────────────┐
          │ COMPLETE  DRYING  IN  MICROWAVE  OVEN │
          └──────────────────────────────────────┘
                              ↘
          ┌──────────────────────────────────────┐
          │ SEPARATION  INTO  NEEDLES/TWIGS  ETC. │
          └──────────────────────────────────────┘
                              ↙
          ┌──────────────────────────────────────┐
          │   ASHING  IN  KILN  AT  475° C        │
          └──────────────────────────────────────┘
                              ↙
          ┌──────────────────────────────┐
          │   ASHES  ANALYZED  BY  NAA   │
          └──────────────────────────────┘
                              ↙
      ┌──────────────────────────────────────────┐
      │  SAME  ASHES  ANALYZED  FOR  ABOUT  30    │
      │    ELEMENTS  BY  AA  AND  ICAP            │
      └──────────────────────────────────────────┘
```

Fig. 7-2   Flowchart of sample preparation and analysis.

analysis has proved to be a total digestion for most trace elements. Consistent results are obtained except for occasional fluctuations in Cu and Pb data.

GOLD

In August 1980 an orientation study was conducted over six low-grade gold showings (< 0.2 oz/ton) in the west channel of Amisk Lake, near Flin Flon (Fig. 7-3). The following account summarizes an earlier report on the study (Dunn, 1980).

The rocks in this area are Aphebian volcanics and sediments (Byers and Dahlstrom, 1954; Pearson, 1980), that commonly contain traces of gold. Many of the gold occurrences have been trenched, and one quartz-hosted native gold deposit was once mined (Monarch mine, Fig. 7-3). Most of the gold is finely disseminated in graywacke (locally carbonatized), feldspar porphyry, or quartz veins, and is commonly associated with pyrite or arsenopyrite (Pearson, 1980).

Sampling and Analytical Program

Vegetation sampling was restricted to vicinities of known mineralization, and short traverses were made normal to mineralized shear zones. The exception was Waverley Island (Fig. 7-3), which was sampled in more detail.

Eighty-seven sites were visited during the first few days of August, a time of the

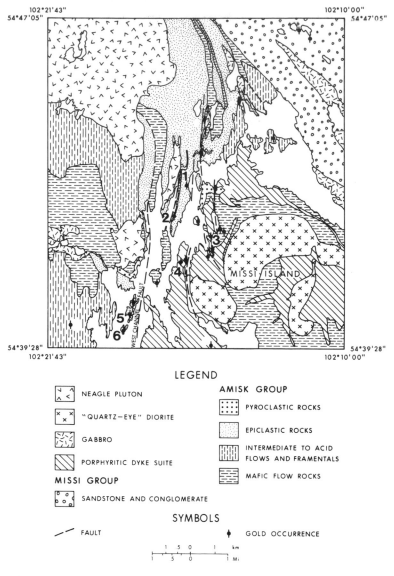

Fig. 7-3 Geological map of the west channel area of Amisk Lake, showing locations of gold occurrences studied. 1, Monarch mine; 2, Amisk gold syndicate; 3, Laurel; 4, Beaver; 5, Waverley Island; 6, Sonora Island. (Geology after Pearson, 1980.)

first appearance of fall colors. Seasonal variation in the uptake of gold by plants is considerable, with spring being the time when concentrations are greatest (Schiller *et al.*, 1973). However, sampling can be satisfactorily conducted at any time of the year provided that it is done within a short time frame, preferably not in the height of summer when growth is most active. These precautions do not, of course, apply to a forest litter sampling program.

Vegetative cover is mixed forest of white spruce, balsam fir, occasional black spruce, and jack pine; poplar and birch dominate the deciduous trees, with local shrub alder and willow. Low shrubs comprise rose, labrador tea, cranberry, and juniper. Forest litter is the only ubiquitous sample type.

Four or five types of vegetation each weighing about 50 g were collected at every site: typically, forest litter, birch bark, white spruce twigs and needles, and where present, rose leaves (recent growth), cranberry leaves, and juniper. In addition, a few samples of balsam fir, willow, alder, horsetail, and moss were collected, and isolated samples of a few other species. It proved impractical to sample the poplars because the branches were out of reach and the bark could not readily be skinned.

In recent years development of neutron activation techniques has lowered the detection limit for gold to 1 ppb with a precision of ±1 or 2 ppb (Minski *et al.*, 1977; Hoffman *et al.*, 1979) permitting direct analysis of dried vegetation. Previously, preconcentration by ashing was usually necessary to detect gold. Disadvantages of ashing are that quite large volumes of vegetation need to be collected to provide sufficient ash for analysis; and some gold may be volatilized during ashing, particularly if the plant species are cyanogenic (Girling *et al.*, 1979).

After drying and separation, samples were milled in a coffee grinder (for leaves) or a Wiley mill (for twigs). About 400 samples were analyzed for gold, arsenic, and antimony by neutron activation. The choice of As and Sb is guided by their common association with Au and their demonstrated uptake by plants (e.g., Warren *et al.*, 1968; Talipov *et al.*, 1976; Girling *et al.*, 1978).

Results

Gold was detectable in most samples (Table 7-1), but was commonly near the detection limit of 1 ppb. Highest concentrations (except for moss) occurred in the forest litter, and highest values for each sample medium were from mineralized trenches (Table 7-2). Three mosses collected from old trenches returned analyses of up to 1400 ppb Au, 2100 ppm As, and 20 ppm Sb. A few samples of birch bark, white spruce needles, and horsetails contained above-background levels of Au, but they are not good biogeochemical indicators of Au mineralization, unless Au in the bedrock is greater than about 1 ppm.

Horsetails contained above background levels of gold only where appreciable concentrations of arsenic were present (e.g., Sonora Island, Fig. 7-3). It seems that the horsetail is capable of absorbing moderate quantities of arsenic (cf. Brooks *et al.*, 1981), and it may therefore be an indirect indicator of gold and/or base metal arsenides.

All of the remaining 20 different types of vegetation samples contained less than 3 ppb Au, and relatively low levels of As and Sb. It is noteworthy, however, that samples of labrador tea stems (Dunn, 1981) and alder leaves collected from nonauriferous localities elsewhere in northern Saskatchewan have yielded much higher concentrations (up to 200 ppb Au) in *ashed* samples than in other plants and have, therefore, potential use in biogeochemical surveys for gold.

TABLE 7-1   Gold, Arsenic, and Antimony in Dried Vegetation from Low-Grade Gold Occurrences in the West Channel of Amisk Lake

Sample Medium	n	Au (ppb)		As (ppm)		Sb (ppm)	
		Range	Median	Range	Median	Range	Median
Forest litter	85	1–140	2.4	3–2100	5.5	0.3–8.5	0.6
Birch bark	63	<1–8	1	<1–420	1	<0.1–1.6	0.1
Birch leaves	5	<1–2	1	1–2	1	<0.1–0.1	0.1
Birch twigs	3	1–2	–	1	–	0.1	0.1
Spruce needles	78	<1–8	1	<1–16	1	<0.1–0.5	0.1
Spruce twigs	8	1–2	1	<1–3	1	<0.1–0.2	0.1
Horsetail	9	<1–8	1	<1–400	1.5	<0.1–0.9	0.1
Balsam fir needles	5	<1–1	1	<1–1	1	0.1–0.2	0.1
Balsam fir twigs	2	2	–	1	–	0.1	–
Willow leaves[a]	5	<1–3	1.25	1–9	5	0.1	–
Jack pine needles	1	1	–	1	–	0.2	–
Alder leaves	18	<1–2	1	<1–3	1	<0.1–0.1	0.1
Alder twigs	3	<1–1	–	<1–1	–	<0.1–0.1	–
Moss[b]	3	33–1400	–	35–2700	–	1.3–20.0	–
Rose	19	<1–3	1	<1–27	1	<0.1–0.2	0.1
Cranberry	22	<1–2	1	<1–26	1	<0.1–0.3	0.1
Juniper needles	21	<1–2	1	<1–3	1	<0.1–0.2	–
Juniper twigs	2	2–3	–	1–8	–	0.1–0.3	–
Grass[c]	3	<1–5	–	1–2	–	<0.1–0.1	–
Bearberry	1	1	–	<1	–	<0.1	–
Labrador tea leaves	6	1–3	1	<1–2	1	<0.1–0.1	<0.1
Labrador tea stems	3	1–2	–	<1–1	–	0.1	–
Fireweed[d]	1	3	–	8	–	0.1	–
Fungus	1	<1	–	16	–	0.2	–
Red bilberry	1	<1	–	2	–	0.1	–

[a]$n = 3$ for As and Sb.
[b]All sampled from mineralized pits.
[c]5 ppb near tailings, Monarch mine.
[d]Tailings pile, Monarch mine.

Each of the six gold occurrences examined (Fig. 7-3) had above the background levels of 1 ppb gold in the forest litter. Near the Waverley Island occurrence samples were collected from 33 sites at 15-m intervals along eight cut lines spaced 30 m apart (Fig. 7-4). Two graywacke samples from a trench in the southwestern part of the island yielded 160 ppb and 28 ppb Au.

Figure 7-4a shows data on gold in the dry forest litter, contoured at concentrations of 3 and 5 ppb. There is a good correspondence of the contours with the strike of the bedrock. Gold values are, however, only a few ppb above the analytical detection limit; therefore, the As data are used to help sort out real from spurious anomalies. Figure 7-4b is a map of Au values (ppb) multiplied by As values (ppm), with the data contoured arbitrarily at 20-"unit" intervals. The general picture is the same as that shown in Figure 7-4a, except that the anomaly is enhanced. A similar outline is produced by further multiplying the data by the Sb content, although the exercise adds nothing of value to the picture: the Sb data are merely confirmatory.

In northern Saskatchewan roses commonly grow over rocks containing carbonates

TABLE 7-2 Gold (ppb) in Rocks and Vegetation from Old Trenches into Gold Showings

Location (Fig. 7-1)	Bedrock[a]	Forest Litter	Birch Bark	White Spruce Needles	Other
Waverly Island	28 and 160	5	4	3	3 (juniper twigs)
Beaver	62	140	1	1	1 (rose)
Sonora	2,500[b]	52	8	1	8 (horsetail)
Amisk syndicate[c]	15,500	(705)[d]	2	2	<1 (rose)
Laurel	170	2	2	1	120 (moss)
Monarch	[e]	13	<1	2	1400 (moss), 1 (rose), 3 (grass), (31)[f] (horsetail)

[a]Rock analyses are from single specimens. Values cited may not be representative of the entire trench because of the wide local variation of gold concentrations that usually occurs.
[b]Arsenopyrite.
[c]Gossan.
[d]In ash (estimated to be 60 ppb in dry forest litter).
[e]Native gold in quartz veins, locally up to 90 ppm.
[f]In ash (estimated to be 8 ppb in dry horsetail).

and calc-silicates. Parslow and Watters (1980) note that carbonate-rich rocks in the area exhibit gold enrichments. The rose may therefore be an indirect geobotanical indicator of potential gold-enriched zones in auriferous areas such as this.

Discussion

Initially, it seems anomalous that the live species should all (except the moss) have much lower contents of Au, As, and Sb than the forest litter which they eventually constitute. Explanations for this situation are that (1) most Au in trees is present in the crowns (Mineyev, 1976), and a substantial portion of the forest litter is made up of leaves from the crowns of deciduous trees; and (2) gold can be dissolved by heterotrophic bacteria, transported in an amino acid and protein-rich solution, and accumulated by molds, which, in turn, are saprophytic on decaying broad-leaved plants (Mineyev, 1976). Thus the forest litter contains not only the plant tissues, but also bacteria and molds which can be contributing gold. Furthermore, mosses comprise part of the forest litter and they are able to concentrate gold by converting it from the ionic to colloidal form (see Zhukov, 1966).

In the study area sulfides are abundant. Oxidation of the deposits can take place by inorganic reactions and by microorganisms (e.g., thion bacteria), thereby dissolving gold, which can then be transported in solution at alkaline pH (Mineyev, 1976). It may then (1) be absorbed by the lower plants (e.g., molds, mosses), (2) be taken up in plants through fine root hairs, or (3) precipitate in an acid pH regime.

Hence the biogeochemical survey may simply reflect zones of oxidation (shear zones, lithological contacts, porous zones). Fortunately, such zones are host to mineral deposits, and therefore any means of delineating such zones is of use in defining drill targets.

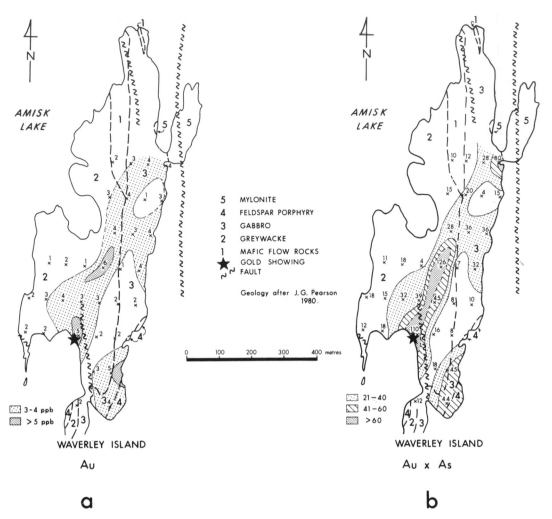

Fig. 7-4   Waverly Island: (a) contours of gold concentrations in dry forest litter; (b) contours of gold (ppb) times As (ppm) in dry forest litter. (From Dunn, 1980.)

TUNGSTEN

Christie (1953, p. 115) reported "a considerable amount of scheelite ($CaWO_4$)" in material collected from a radioactive occurrence 10 km east of Uranium City (Fig. 7-1). No further information has been published on the occurrence, so a soil, rock, and vegetation survey was conducted over the area in July 1982. Examination of the pit (mined recently for uranium) surrounding outcrop and drill core failed to locate any scheelite. A night survey of the area with an ultraviolet lamp showed spectacular green fluorescence of uranium minerals on joint planes but no sign of the bluish-white fluorescence characteristic of scheelite.

Thus the survey was restricted to a 33-site two-day orientation sampling of an area 200 m X 250 m. A forest fire is estimated to have burned part of the area 12 years ago, so the forest is dominated by young jack pine, birch, and alder, with some black spruce, labrador tea, and juniper. Immature podzols are present between outcrops of mildly radio-active felsic and biotite gneiss.

Sixteen sample media were prepared, providing over 300 samples for analysis. Samples were dried and ashed, according to the procedure outlined in Figure 7-2, then analyzed by neutron activation for several elements. Tungsten concentrations were close to normal background levels on all 130 samples selected for analysis (Table 7-3), and the mean W content of the rocks was 2 ppm (equal to the clarke for W).

Although no W mineralization was discovered, the biogeochemical survey has indicated that the most suitable sample media for a tungsten survey in this environment would be twigs of black spruce, jack pine, or labrador tea.

TABLE 7-3 Tungsten (ppm) in Rocks and Vegetation from the Vicinity of Reported Scheelite ($CaWO_4$) Showing near Uranium City: 130 Samples Comprising 16 Sample Media

Sample Medium	n	Tungsten (ppm)	
		Maximum	Mean
Rocks	21	6	2.0
Twigs (black spruce)	13	3	2.0
Twigs (labrador tea)	4	3	1.7
Twigs (jack pine)	5	2	0.7
Needles and leaves of above, and all leaves and twigs of alder, birch, willow, juniper	78	1	<0.7
B_F-horizon soils (−80 mesh)	9	1	<0.7

URANIUM

Extensive sampling of plant material since 1979 has outlined regional uranium biogeo-chemical patterns and the biogeochemical response to both shallow and deeply buried mineralization. These data are summarized in several papers (e.g., Dunn, 1981, 1982, 1983a,b), hence only a brief outline is given here.

Analysis of over 3000 samples for U following the procedure of Fig. 7-2 has indicated clearly that the twigs of black spruce are the most effective concentrators of U in the boreal forests. In general, the order of U concentration is twigs > leaves > roots > trunk (Dunn, 1982).

It is of the utmost importance that a similar number of years of twig growth is collected from each site. Analysis of dry twigs shows that the 2- to 5-year-old growth contains the highest concentrations of uranium, and there is progressively less in the older growth (Dunn, 1981). A practical amount to collect is the latest 10 years of growth, since it takes only about 1 minute to accumulate the desired 200 g. Data on U quoted in this section refer solely to concentrations in the latest 10 years' growth of ashed black spruce twigs.

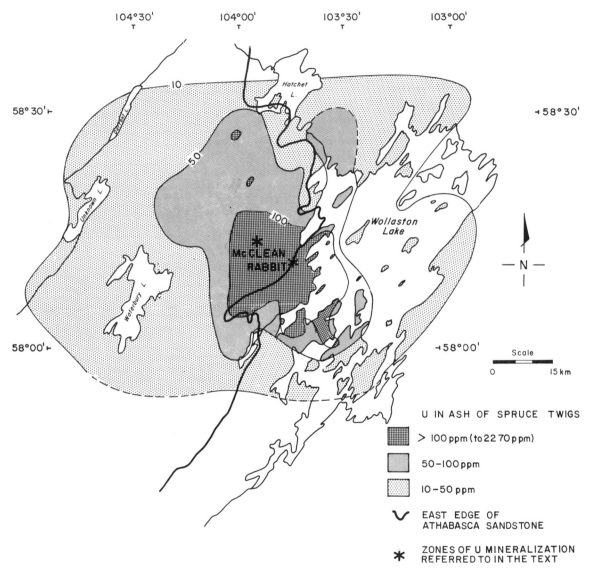

Fig. 7-5 Wollaston Uranium Biogeochemical Anomaly: concentrations of uranium in the ashed twigs (latest 10 years' growth) of black spruce (*Picea mariana*)—about 1000 sample sites. (Modified after Dunn, 1983a.)

APPLICATION OF BIOGEOCHEMICAL METHODS

The sensitivity of black spruce to uranium has rendered this species very useful in determining the effects of natural influences, such as local variation between neighboring trees; tree height, age, and health; the annual and seasonal variations in uranium accumulation; variations within a tree; and the nature of the terrain. For the spruce twigs, these natural variations account for about ±15 percent of data variability (Dunn, 1982); therefore, successive contour intervals for a biogeochemical map of U data should be incremented by at least this amount in order to reflect significant variation in the data.

Concentrations greater than 1 ppm U in plant ash are considered to be anomalous in the carnotite districts of the Colorado Plateau (Cannon, 1952). Throughout much of northern Saskatchewan the black spruce twigs usually contain over 2 ppm U. However, near Wollaston Lake (Fig. 7-1) there is an area of 10,000 km² within which all spruce trees analyzed have over 10 ppm U in the ash of their twigs (Fig. 7-5). Within this Wollaston Uranium Biogeochemical Anomaly the 50-ppm contour encompasses an area of 3000 km², which in turn contains a core of 1000 km² in which all values are over 100 ppm, ranging up to a maximum of 2270 ppm U. There is a uranium mine at Rabbit Lake, but it cannot possibly account for contamination over an area of this size, and air monitoring stations confirm that exceedingly small concentrations of U are present in the atmospheric dust. There is no other U mine within 300 km of the area, and most of the known U deposits occur at depths of 60 to 250 m. There is no discernible correlation between the U content of the twigs and U in the bedrock, lodgement tills, peats, or soils—all of these media have normal background concentrations of U. Formation waters near the orebodies do contain elevated concentrations of U (up to about 50 ppb); hence it is concluded that the anomaly is the result of predominantly upward movement of slightly uraniferous formation waters from uraniferous sources at the base of the Athabasca Group. In addition, there is some lateral movement of waters along relatively permeable strata of the Athabasca Group, thereby dispersing uranium and providing an exceptionally high background of uranium in plants throughout the area. Above Aphebian metasediments (i.e., east of the Athabasca Group) the uranium source must be local, and the high concentrations in the plants are attributed to the presence of labile uranium in fractures and joints that was emplaced prior to the last glaciation when an Athabasca cover existed over the area (Dunn, 1983a).

Within the regional Wollaston Anomaly, intense biogeochemical anomalies occur above zones of deeply buried uranium mineralization (e.g. Fig. 7-6) and in areas of virgin forest as yet undrilled (Dunn, 1982), where concentrations up to 1360 ppm U in twig ash are recorded.

The McClean North Zone of uranium mineralization (Fig. 7-6) is beneath 150 m of Athabasca sandstones, yet biogeochemical anomalies with up to 500 ppm U occur above, but laterally displaced from the mineralized zone.

It is concluded that this and other local biogeochemical anomalies over sediments of the Athabasca Group result from the upward migration of U ions in solution along steeply inclined fractures through the Athabasca sandstones—the result is a "rabbit's ear"-type of U anomaly in the trees that are above but laterally displaced from the ore bodies. It would appear that detailed application of this biogeochemical technique may help delineate more U ore bodies in this highly uraniferous region.

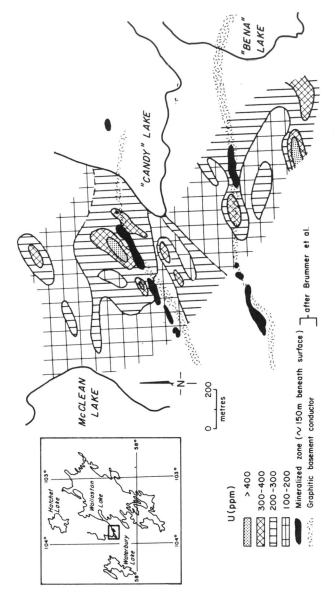

Fig. 7-6 Relationship between the McClean Zones of uranium mineralization and the uranium content of ashed twigs (latest 10 years' growth) of black spruce (*Picea mariana*). Approximately 200 sample sites. (Modified after Dunn, 1982.)

From the hundreds of multielement analyses performed on the easily accessible parts of the common species in Saskatchewan's boreal forests, a picture is emerging of the sample media which appear to be the most sensitive indicators to the presence of particular elements and which commonly exhibit enrichments of an element or suite of elements. Multielement analyses from two areas of northern Saskatchewan are published in Walker (1979; Key Lake area) and Dunn (1981; McClean Lake area). Table 7-4 summarizes the information in these papers and is supplemented by some of the author's unpublished data. It should be noted, however, that a simple increase in the concentration of an element in a given species does not necessarily indicate the approach of mineralization, because the stress induced by metal poisoning of vegetation (e.g., Cu and Pb) may actually reduce the metal content of the plants (Labovitz *et al.*, 1982). Thus a negative anomaly may occur which, coupled with other factors, such as early senescence of leaves, may provide an exploration target. A phenomenon of this sort underscores the importance of considering the full spectrum of botanical, chemical, biological, and geological parameters in the interpretation of biogeochemical data—simply plotting the numbers is not enough.

TABLE 7-4 Plant Organs that Concentrate Certain Elements[a]

Element	Plant Organs
U	Black spruce twigs
Au	Forest litter; mosses; alder leaves
W	Black spruce twigs; labrador tea stems
Mo	Alder twigs (and leaves), blueberry leaves
Ag, Zn	Black spruce and jack pine trunkwood and bark
Cd	Jack pine trunkwood
REE	Black spruce twigs; labrador tea stems
As, Sb	Forest litter; mosses; horsetails; birch bark; (black spruce twigs)
Ni, Cu	Jack pine cones; (labrador tea roots)
Pb	Twigs and stems of most species

[a]Parentheses indicate less efficient concentrators.

CONCLUSION

The pristine boreal forest of central Canada provides an excellent field laboratory for biogeochemical investigations. Studies conducted over gold mineralization have provided encouraging results for delineating auriferous zones by mapping Au and As concentrations in forest litter. Examination of an area reported to contain tungsten disclosed no tungsten mineralization, but the study indicated that black spruce twigs may be a suitable biogeochemical sample medium for tungsten.

Extensive sampling of northern Saskatchewan for uranium has shown that black spruce twigs are highly enriched in uranium over an area of 10,000 km^2. This Wollaston

Uranium Biogeochemical Anomaly contains several intense anomalies which occur both above zones of deeply buried uranium mineralization and in virgin forest where no mineralization is known. The upward migration of uranium, dissolved in formation waters, along steeply inclined fractures is invoked as the process most likely to have caused the anomalies.

It is concluded that the careful use of biogeochemical methods has exciting potential for providing a "window" through the extensive glacial deposits which blanket much of the ground beneath the boreal forest of central Canada.

ACKNOWLEDGMENTS

I thank the many people who have contributed information and ideas to this study; my field and laboratory assistants; and the Saskatchewan Geological Survey for support of this project.

REFERENCES

Brooks, R. R., Holzbecher, J., and Ryan, D. E., 1981, Horsetails (*Equisetum*) as direct indicators of gold mineralization: *J. Geochem. Explor.*, v. 16, p. 21–26.

Brummer, J. J., Saracoglu, N., Wallis, R. H., and Golightly, J. P., 1981, McClean uranium deposits, Saskatchewan. Part 2. Geology (amended by R. H. Wallis): CIM Geology Div. Field Excursion Guidebook (Sept. 8-13, 1981, Sask.): *Can. Inst. Min. Metall.*, Montreal, p. 51–64.

Byers, A. R., and Dahlstrom, C. D. A., 1954, Geology and mineral deposits of the Amisk-Wildnest Lakes area, Saskatchewan: *Sask. Dept. Min. Res. Rep. 14,* p. 177.

Cannon, H. L., 1952, The effect of uranium-vanadium deposits on the vegetation of the Colorado Plateau: *Am. J. Sci.*, v. 250, no. 10, p. 735–770.

Christie, A. M., 1953, Goldfields-Martin Lake map-area, Saskatchewan: *Geol. Surv. Can. Mem. 269,* p. 126.

Dunn, C. E., 1980, Gold biogeochemistry investigations, *in* Christopher, J. E., and Macdonald, R. (Eds.), *Summary of Investigations 1980, Saskatchewan Geological Survey:* Sask. Dept. Min. Res. Misc. Rep. 80-4, p. 81–85.

—— , 1981, The biogeochemical expression of deeply buried uranium mineralization in Saskatchewan, Canada: *J. Geochem. Explor.*, v. 15, p. 437–452.

—— , 1982, The massive Wollaston Uranium Biogeochemical Anomaly in the boreal forest of northern Saskatchewan, Canada: p. 477–491 *in Uranium Exploration Methods: Review of the NEA/IAEA Research and Development Programme*, Paris, June 1-4, 1982: OECD Nuclear Energy Agency.

—— , 1983a, Uranium biogeochemistry of the NEA/IAEA Athabasca test area: p. 127–132 *in* Cameron, E. M. (Ed.), *Uranium Exploration in Athabasca Basin, Saskatchewan Canada:* Geol. Surv. Can. Pap. 82-11.

—— , 1983b, Detailed biogeochemistry of the NEA/IAEA Athabasca test area: p. 259–272 *in* Cameron, E. M. (Ed.), *Uranium Exploration in Athabasca Basin, Saskatchewan, Canada:* Geol. Surv. Can. Pap. 82-11.

Girling, C. A., Peterson, P. J., and Minski, M. J., 1978, Gold and arsenic concentrations in plants as an indication of gold mineralization: *Sci. Total Environ.*, v. 10, p. 79–85.

_____ , Peterson, P. J. and Warren, H. V., 1979, Plants as indicators of gold mineralization at Watson Bar, British Columbia, Canada: *Econ. Geol.* v. 74, p. 902–907.

Hoffman, E. L., Brooker, E. J., and White, M. V., 1979, The determination of gold by neutron activation analysis: *Yellowknife Gold Symp.*, Dec. 4, 1979, p. 26.

Labovitz, M. L., Masuoka, E. J., Bell, R., Siegrist, A. W., and Nelson, R. F., 1982, The application of remote sensing in geobotanical exploration for metal sulfides—results from 1980 field season: *NASA Tech. Memo. 83935*, p. 44.

Mineyev, G. G., 1976, Organisms in the gold migration–accumulation cycle: *Geochem. Int.*, v. 13, no. 2, p. 164–168.

Minski, M. J., Girling, C. A., and Peterson, P. J., 1977, Determination of gold and arsenic in plant material by neutron activation analysis: *Radiochem. Radioanal. Lett.*, v. 30, p. 179–186.

Parslow, G. R., and Watters, B. R., 1980, Flin Flon base metals project: p. 64–69 *in* Christopher, J. E., and Macdonald, R. (Eds.), *Summary of Investigations 1980, Saskatchewan Geological Survey:* Sask, Dept. Min. Res. Misc. Rep. 80-4.

Pearson, J., 1980, Flin Flon gold project: p. 70–80 *in* Christopher, J. E., and Macdonald, R. (Eds.), *Summary of Investigations 1980, Saskatchewan Geological Survey:* Sask. Dept. Min. Res. Misc. Rep. 80-4.

Schiller, P., Cook, G. B., Kitzinger-Skálová, A., and Wolfl, E., 1973, The influence of the season variation for gold determination in plants by neutron activation analysis: *Radiochem. Radioanal. Lett., 13*, nos. 5-6, p. 283–286.

Talipov, R. M., Glushchenko, V. M., Tverskaya, K. L., and Nishanov, P., 1976, Some characteristics of gold and antimony distribution in plants in ore-deposits of the Chatkal-Kurama region: *Uzb. Geol. Zh.*, v. 3, p. 65–69.

Walker, N. C., 1979, Trace elements in vegetation and soils over the Key Lake uranium-nickel orebody, northern Saskatchewan, Canada: p. 361–369 *in* Watterson, J. R., and Theobold, P. K. (Eds.), *Geochemical Exploration, 1978:* Proc. 7th Int. Geochem. Explor. Symp., Assoc. Explor. Geochemists, Rexdale, Ontario.

Warren, H. V., Delavault, R. E., and Barakso, J., 1968, The arsenic content of Douglas fir as a guide to some gold, silver and base metal deposits: *Can. Min. Metall. Bull.*, v. 61, p. 860–867.

Zhukov, A. N., 1966, *Tsvetn. Met.*, v. 10.

W. E. Baker
Department of Mines
Rosny Park
Tasmania, Australia

8

GOLD IN VEGETATION
AS A PROSPECTING METHOD
IN TASMANIA, AUSTRALIA

ABSTRACT

A moderately rapid method has been developed for the determination of gold in shredded plant material. This involves wet oxidation with nitric acid, extraction with heptan-2-one and gold determination by carbon furnace atomic absorption.

   The method has been applied successfully to orientation and prospecting studies over the Lisle Valley alluvial gold deposits in northeastern Tasmania, where gold-bearing areas were found to be coincident with biogeochemical anomalies exhibiting high contrasts. In the vicinity of the Jane River alluvial gold deposits in western Tasmania, plants were found to contain less gold than at Lisle and the biogeochemical contrast was low.

INTRODUCTION

Reports on the gold content of plants are few in number and conflicting data have appeared in the results of various studies. A brief extract of some of the published results is given in Table 8-1 and a more extensive compilation of the gold content of vascular plants which could be of use in biogeochemistry has been presented by Brooks (1982). Studies of one plant group in particular, the horsetail rush (*Equisetum* spp.), have produced highly conflicting data. Babicka (1943) recorded an incredible maximum of 610,000 ng/g (ppb) in the ash of this genus, while Warren and Delavault (1950) found only 340 ng/g. The extreme value quoted was most likely the result of analytical error, although Boyle (1979) has returned some respect to the genus, as an accumulator of gold, by recording a value of 8500 ng/g.

   In view of this literature record the Tasmania Department of Mines decided to examine the gold content of local plant species and to attempt to develop a rapid method of gold analysis suitable for biogeochemical prospecting. The Lisle Valley (147°19'E, 41°13'S) in northeastern Tasmania, the site of extensive alluvial workings which produced about 7800 kg of gold around the turn of the century, was selected as a test area. This valley is a wide basin-like structure measuring about 4 km across. The floor comprises soil and alluvium up to 25 m in depth overlying deeply weathered Devonian granodiorite, while the valley sides are of steeply dipping Silurian metasediments. The origin of the alluvial gold is unknown and a covering of deep scree over the granodiorite–metasediment contact makes prospecting for reef deposits very difficult. It is likely that the gold in the Lisle deposits is the result of residual concentration of the products of weathering of numerous small quartz stringers in the metamorphic aureole surrounding the granite.

   A second area, the Jane River in western Tasmania (146°02'E, 42°23'S) which produced minor gold in the 1920s, was also studied briefly. This goldfield is relatively small, with the main prospect of Reward Creek occupying an area less than 2 km². A highly variable distribution of gold occurs in an alluvial cover known to reach 3 m in thickness, which is spread over a bedrock consisting of relatively unmetamorphosed Precambrian dolomite, mudstone, and sandstone.

ANALYTICAL TECHNIQUES

In the preliminary stage of the analytical studies a standard approach of dry ashing, aqua regia leach, 2,4-methylpentanone (methyl isobutyl ketone, MIBK) extraction, and carbon furnace atomic absorption (CFAA) determination of gold was followed. The results ob-

TABLE 8-1 Some Literature Values of Gold Content of Plants

Plant		Au Content (ppb in ash)	Location	Reference
Polytrichum hyperboreum	(moss)	300	Yakut, USSR	Razin and Rozhkov (1966)
Camptothecium nitens	(moss)	10,700	Yakut, USSR	Razin and Rozhkov (1966)
Carex pediformis	(grass)	2,400	Yakut, USSR	Razin and Rozhkov (1966)
Agrostis alba	(grass)	5	Yakut, USSR	Razin and Rozhkov (1966)
Equisetum palustre	(horsetail rush)	610,000	Oslany, Czechoslovakia	Babicka (1943)
Equisetum sp.	(horsetail rush)	340	British Columbia, Canada	Warren and Delavault (1950)
Equisetum sp.	(horsetail rush)	400	Kentucky, United States	Cannon et al. (1968)
Equisetum sp.	(horsetail rush)	8,500	Ontario, Canada	Boyle (1979)
Salsola rigida	(thistle)	1,400	Kyzyl-Kum, USSR	Khotamov et al. (166)
Corylus avellana	(hazelnut)	20,000	Eule, Czechoslovakia	Babicka (1943)
Salix caprea	(willow)	30	Yakut, USSR	Razin and Rozhkov (1966)
Salix sp.	(willow)	1,020	British Columbia, Canada	Warren and Delavault (1950)
Pinus sibirica	(pine)	1,200	Yakut, USSR	Razin and Rozhkov (1966)
Pinus contorta	(pine)	<30	British Columbia, Canada	Warren and Delavault (1950)
Pseudotsuga taxifolia	(Douglas fir)	650	British Columbia, Canada	Warren and Delavault (1950)
Populus tremula	(aspen)	2,100	Yakut, USSR	Razin and Rozhkov (1966)
Populus tremuloides	(aspen)	890	British Columbia, Canada	Warren and Delavault (1950)

tained by this approach yielded values for gold that rarely exceeded 20 ng/g in the ash. It was noticed during ashing that many crucibles developed a mauve bloom around their upper surfaces, a color reminiscent of colloidal gold. Analytical tests of this feature by procedures described below confirmed that the color was due to gold and that loss of this metal during dry ashing of *Pinus radiata* was 65 percent at 300°C, 80 percent at 500°C, and 95 percent at 800°C.

Prior to analysis plants were dried at 40°C under infrared lamps and finely shredded by use of a cross-beater mill. Because of the gold loss noted above, wet ashing techniques were used and it was found that concentrated nitric acid was capable of destroying the finely shredded plant matter. The use of perchloric acid was rarely necessary. Samples weighing 1 g were placed in 100-ml Erlenmeyer flasks with 10 ml of concentrated nitric acid, allowed to react without heating for 10 minutes, after which the temperature was raised to 140°C and the acid volume reduced to 5 ml by fuming. If organic matter remained at this stage, it was oxidized by the addition of 1 ml of perchloric acid. Any gold present was converted to the aurichloride complex by the addition of 10 ml of concentrated hydrochloric acid and fuming to near dryness. The gold complex was leached with 10 ml of water, filtered into extraction vessels (Fig. 8-1), and washings of the flasks added.

Extraction of gold from the leachates with the commonly used MIBK presented a problem in that the solubility of this solvent in an aqueous phase was such that where use of small volumes (1 ml) was planned very little separated for use in analysis. This problem was solved by substituting heptan-2-one (amyl methyl ketone, AMK) for MIBK. This ketone was found to be almost completely recoverable at 1 ml volume and was insensitive to variation of acid concentration in the range 1 to 6 M. One-milliliter volumes of AMK were added to the extraction vessels and these were shaken for 5 minutes. After the AMK had separated it was brought into the neck of the extraction vessel by the addition

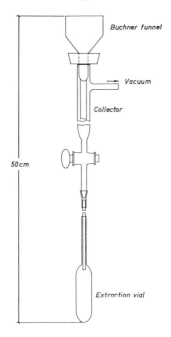

Buchner funnel

Vacuum

Collector

50 cm

Extraction vial

Fig. 8-1 Arrangement for rapid filtering and collection of wet oxidation products of plant matter.

of 2 M hydrochloric acid. The gold content was determined by CFAA on 5-μl aliquots. The difficulties associated with loading organics in CFAA were reduced by use of a hot loading technique. Where the recently developed platform furnaces and autosamples are available, the gold determination should be more easily achieved. For prepared gold standards, the relative standard deviation (RSD) of replicate analyses at an absolute gold content of 0.2 ng was 8 percent. Under operating conditions, the addition of variability within the bulk sample and processing factors increased the RSD to 26 percent. The detection limit was 1 ng/g dry plant material and this was subsequently found adequate for exploration applications. Lower limits were available for orientation studies by reducing the volume of AMK used to 0.5 ml and repeated loading of the furnace before atomization. Provided that laboratory equipment and services are adequate, the method allows the processing of 40 to 50 samples per day.

GOLD IN TASMANIAN PLANTS

The analytical procedure was applied to an orientation survey in the region of the Lisle Valley. Three samples each of a wide variety of species were collected from the margin of the valley and also from sites within the valley that were known to have yielded gold in the past. Where possible, sampling was of second-year twigs. These were cut to 2-cm

TABLE 8-2   Gold Content of Lisle Valley Plants

| Species | Average Ash Content (%) | Gold Content[a] (ng/g) | | | |
| | | Average Background | | Maximum | |
		Dry Wt.	Ash	Dry Wt.	Ash
Acacia dealbata (common wattle)	1.6	1.1	70	1,000	62,000
Acacia melanoxylon (blackwood)	8.5	6.6	80	850	10,000
Blechnum wattsii (hardwater fern)	12.6	7.7	60	7,800	62,000
Dicksonia antartica (man fern)	17.1	10.8	60	1,800	10,500
Eucaluptus obliqua (stringy bark)	1.9	0.9	50	180	9,500
Gahnia sp. (rush)	8.1	3.4	40	1,300	16,000
Melaleuca ericifolia (tea tree)	5.2	2.0	40	750	14,500
Pinus radiata (radiata pine)	1.1	0.5	50	1,900	180,000
Pteridium aquilinum (bracken fern)	10.0	8.0	80	3,500	35,000
Prostanthera lasianthos (wild lilac)	1.1	0.4	40	85	7,500
Pomaderris apetala (dogwood)	1.1	0.2	20	130	11,000

[a]In samples of second-year twigs except for ferns and rush.

lengths in the field and approximately 50 g taken to the laboratory for shredding. The results (Table 8-2) show that many species are well endowed with gold. One pine, carrying 1900 ng/g dry weight (180,000 ng/g on an ash basis) contains about 125 g of gold (4 oz troy). Unfortunately, the likely costs of extraction preclude such trees from being economic sources of the metal! Various parts of selected samples of *Acacia dealbata, Eucalyptus obliqua,* and *Pinus radiata* were also analyzed for gold and the results are given in Table 8-3. These show that there is considerable variability in the gold content of different plant parts and also in the ash content of these. For *Pinus radiata*, as an example, the cones contain about 2.5 times the amount of gold found in the needles, but due to the lower ash content of the cones, this factor increases to 9.5 on an ash basis. While the presentation of biogeochemical data is continually under discussion, this writer prefers ash basis figures which remove the effect of the highly variable organic content of the plant matter. Although second-year growth was sampled for the results reported in this study the values in Table 8-3 suggest that other parts (such as trunkwood) may have been a better choice. All plants studied appear to respond to the presence of gold, and this may be a climatic feature. The results differ from those for Siberian plants (Kovalevskii, 1978) which revealed that only 10 percent of the species studied yielded information on gold distribution in the substrate.

TABLE 8-3   Gold Content of Various Parts of Some Lisle Valley Plants

Species and Part	Ash Content (%)	Gold Content (ng/g)	
		Dry Wt.	Ash
Acacia dealbata			
(common wattle)			
Fronds	3.01	160	5,500
2-year twigs	1.34	100	7,500
Bark	1.49	140	9,500
Trunkwood	0.54	80	15,000
Eucalyptus obliqua			
(stringy bark)			
Leaves	1.97	120	6,000
2-year twigs	1.49	150	10,000
3–5-year branch	0.89	60	6,500
Bark	1.47	70	4,500
Trunkwood	0.70	60	8,500
Pinus radiata			
(radiata pine)			
Needles	1.88	220	12,000
2-year twigs	0.90	760	84,000
3–5-year branch	0.71	340	48,000
Bark	0.72	420	58,000
Trunkwood	0.45	620	138,000
Cones	0.49	570	116,000

The encouraging results of the orientation survey suggested that more extensive sampling might prove interesting. The numerous roads and tracks throughout the Lisle Valley were sampled at 100-m intervals. The results of part of this study are given in Figure 8-2. The gold values obtained indicate several localities of interest, such as that

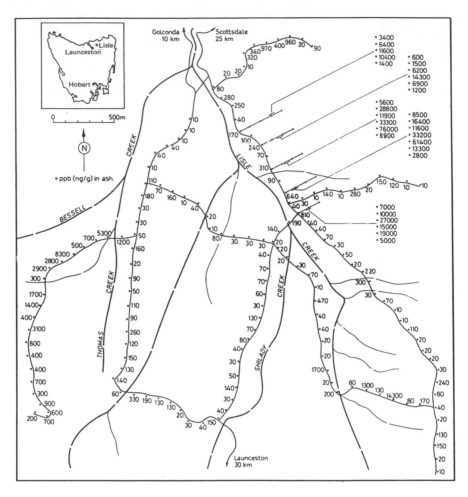

Fig. 8-2 Gold distribution in plants (ng/g ash) in the Lisle Valley. Road sampling at 100-m intervals. Traverse sampling (upper right) at 20-m intervals with highly anomalous values shown in offsets.

east of Lisle Creek near the upper center of the sampled area (310 to 970 ng/g ash) and that along the southwestern margin (300 to 5300 ng/g ash). A point anomaly (14,300 ng/g ash) in the southeast is also of interest, as it stands some 150 m above the valley floor and possibly indicates the presence of minor reef gold in the contact aureole of the granodiorite that underlies the Lisle Valley. The contrast (geochemical relief) evident in the results is substantial and ranges from a minimum of 10 to a maximum exceeding 200. The eastern locality, being partly covered by a pine plantation, was reasonably accessible and was chosen for detailed sampling. Five lines, 200 m apart, were sampled at 20-m intervals and the results for anomalous sections of these lines are given at the top right of Figure 8-2. These results were found to define an auriferous streak in alluvial material which was 15 m in width by 800 m in length and covered by at least 2 m of soil. Only one bulk sample (0.25 m^3) of this alluvial material has been tested to date, giving a gold recovery by tabling of 10 g/m^3 ($\frac{1}{3}$ oz troy/yd^3).

Similar procedures were applied to sampling of the Jane River area. The results (Table 8-4) show the anomalous values to be far lower than those found in the Lisle Valley, with contrasts rarely exceeding 10. Thus the Jane River appears to be a less attractive prospect on the basis of biogeochemical studies.

Only a few plants and associated soils have been analyzed to date. The plant/soil ratios have been found to be highly variable and range from 0.02 to 0.90 (1.57 to 100.43 on an ash basis).

TABLE 8-4   Gold Content of Jane River Plants

		Gold Content[a] (ng/g)			
		Average Background		Maximum	
Species	Average Ash Content (%)	Dry Wt.	Ash	Dry Wt.	Ash
Acacia sp. (wattle)	1.0	4	400	14	1460
Cenarrhenes nitida (native plum)	0.7	8	1270	28	2680
Eucalyptus sp. (peppermint)	1.4	4	290	18	1130
Pteridium aquilinum (bracken fern)	2.4	2	80	28	760
Gahnia sp. (reeds)	3.6	8	220	16	900
Gahnia grandis (cutting grass)	2.1	3	140	4	240
Gymnoschoenus sphaerocephalus (button grass)	1.9	4	210	36	2160

[a]In samples of second-year twigs except for ferns and rush.

CONCLUSION

The Department of Mines, Tasmania, studies of biogeochemistry are only at an early stage of development, although the results to date are sufficiently encouraging to warrant further work. The sample of pine giving the extremely high gold value will be studied more thoroughly, including roots and substrate, in an attempt to find the reason for this gold content. The rapid plant processing system will be adapted for use in the determination of other metals in the near future.

As far as the biogeochemical studies of gold are concerned, it appears that the close sampling required by the capricious nature of this metal will limit the application to detailed investigations. The marked response of plants to gold in the Lisle Valley indicates that in an alluvial target area, biogeochemistry could considerably reduce the amount of mechanical testing necessary to evaluate a deposit.

REFERENCES

Babicka, J., 1943, Gold in Lebewesen: *Microchim. Acta*, v. 31 (in German).

Boyle, R. W., 1979, The geochemistry of gold and its deposits: *Geol. Surv. Can. Bull., 280*, p. 74–83.

Brooks, R. R., 1982, Biological methods of prospecting for gold: *J. Geochem. Explor.*, v. 17, p. 109–122.

Cannon, H. L., Shacklette, H. T., and Bastron, H., 1968, Metal absorption by *Equisetum* (horsetail): *U.S. Geol. Surv. Bull., 1278-A*, p. 1–21.

Khotamov, Sh., Lobanov, E. M., and Kist, A. A., 1966, The problem of the concentration of gold in organs of plants within ore fields: *Dokl. Akad. Nauk. Tadzh. SSR*, v. 9, p. 27–30.

Kovalevskii, A. L., 1978, Mineral forms of gold in plants: *Dokl. Akad. Nauk. SSR*, v. 242, p. 340–433.

Razin, L. V., and Rozhkov, I. S., 1966, *Geochemistry of Gold in the Crust of Weathering and in the Biosphere in Gold-Ore Deposits of the Kuranakh Type:* Nauka, Moscow, 254 p.

Warren H. V., and Delavault, R. E., 1950, Gold and silver content of some trees and horsetails in British Columbia: *Bull. Geol. Soc. Am.*, v. 61, p. 123–128.

E. L. Hoffman
Nuclear Activation Services Limited
McMaster Nuclear Reactor
Hamilton, Ontario, Canada

E. J. Booker
X-ray Assay Laboratories Limited
Don Mills, Ontario, Canada

9

BIOGEOCHEMICAL PROSPECTING FOR GOLD USING INSTRUMENTAL NEUTRON ACTIVATION ANALYSIS

With Reference To Some Canadian Gold Deposits

ABSTRACT

Two major reasons why the mineral exploration community has not taken advantage of biogeochemical exploration techniques in the past include high analytical cost and a poor understanding of biogeochemical cycles.

With recent technological advances in the field of analytical chemistry some of the problems associated with analyzing biogeochemical samples have been solved. Instrumental neutron activation analysis (INAA) provides an accurate, fast, and inexpensive method for large-scale biogeochemical surveys. In addition to gold, several of the so-called pathfinder elements, such as arsenic, antimony, and tungsten, can be determined simultaneously.

Some sampling techniques and the analytical methods chosen for the analysis will be discussed as well as some of the problems that may be encountered. Case histories will be described from several Canadian gold deposits.

INTRODUCTION

Biogeochemical prospecting for many metals has been an academic technique used infrequently by the mining industry in the Western world. Reasons for this have been the lack of information available on biogeochemical prospecting, the inherent high cost of analysis, the generally unacceptable detection limits of the many analytical methods used for this purpose, as well as the possible poor analytical data put forth by commercial laboratories unaware of some of the problems involved in analyzing biological materials (e.g., possible volatilization of gold on ashing). Recently, modern instrumentation has advanced to the stage of solving several of the analytical problems. Some of the most notable advances have been in the field of instrumental neutron activation analysis.

Biogeochemical prospecting techniques essentially have involved sampling either of plant species or of humic material (mull). Sampling of plant species has been a more academic subject in Canada perhaps because most of the people involved in routine sampling on a large scale are not trained botanists, and feel that they would have difficulty distinguishing between species of vegetation. The cost of collecting specific parts of certain species may also be substantially higher than sampling humic material.

Sampling of humic material has been used successfully. Curtin *et al.* (1968) have described the use of humic material from the Empire District, Colorado, in locating gold mineralization beneath colluvium and glacial drift. Gleeson and Boyle (1979) and Boyle (1979) have shown that biogeochemical sampling for gold exploration has been successful in the Abitibi greenstone belt of Canada and elsewhere if it is undertaken properly. Curtin and King (Chapter 22, this volume) have presented impressive results from two properties in Colorado and Idaho using humic material in providing superior anomaly definition as opposed to sampling soil horizons. Dunn (1985) has studied the use of live vegetation and organic litter (humus) for mineral exploration in the boreal forests of Canada and concluded that the organic litter, as opposed to live vegetation, provided the best sampling material from the area studied.

This chapter deals mainly with the humic or mull sampling technique and with analytical techniques that have been used for gold analysis of organic materials. It concentrates on instrumental neutron activation, a technique used almost exclusively in Canada for the analysis of biogeochemical samples from gold exploration programs. Information is also presented on sampling, and some examples from the Abitibi and Timmins–Porcupine greenstone belts of northern Quebec and Ontario are discussed.

ANALYSIS OF GOLD IN BIOGEOCHEMICAL SAMPLES

Analytical Techniques Available

A variety of techniques have been cited in the literature for the analyses of biogeochemical samples. These techniques include: (1) ashing followed by fire assaying, (2) ashing followed by flameless atomic absorption, and (3) instrumental neutron activation.

The ashing-fire assaying technique is the oldest of the methods and was used by Lungwitz (1900) and Warren and Delavault (1950). The obvious disadvantage of the technique is the ashing procedure to obtain 10 g of ash. Assuming an ash content of 5 percent, 200 g of organic material would have to be ashed to supply the material for fire assay. Our own experiments and those of Girling et al. (1979) indicate that dry ashing would most certainly result in the loss of gold if the gold were tied up in the plant material as gold cyanogenic complexes. Wet ashing of a sample of 22 g would be very expensive as well as impractical. Fire assaying in itself would add gold at trace levels from the fire assay reagents, such as litharge. Recent advances in analytical technology—fire assaying with a dc plasma, flameless atomic absorption, or INAA finish, for example—have eliminated the need for gravimetric weighing of the fire assay bead. These are not the best techniques available, however, because the problems of possible contamination and losses due to ashing still remain.

The dry ashing–flameless atomic absorption technique also suffers from the dry ashing problems just discussed. Baker (Chapter 8, this volume) described a method whereby 2-g samples were wet ashed to prevent loss of gold and the gold was extracted from the solution by an organic solvent. The gold content of the solvent was then determined by graphite furnace atomic absorption. This method was particularly slow and only 100 samples per week could be processed. The detection limit as routinely determined was 4 ng/g on the original plant material. Ward and Brooks (1978) determined gold on vegetation by dissolving plant ash in acidic solutions and then determining the gold by flameless atomic absorption. The detection limit was 100 ng/g on the ash or about 5 ng/g on a dry weight basis. In an improvement of the technique, Brooks (1982), reported that detection limits better than 5 ng/g on a dry weight basis could be obtained by extracting the gold from a large volume of acid into an organic solvent and then running the organic solvent by flameless atomic absorption for gold.

The third method of analysis, that of instrumental neutron activation analysis (INAA), has been used extensively in biogeochemical exploration for gold deposits. Girling et al. (1979) have described a technique where 250 mg of dried, homogenized plant material was compressed into disks and irradiated in a nuclear reactor. Following a 3-day period, the samples were counted for gold and arsenic. Erdman and Leonard (Abstract A in Chapter 12, this volume) also used INAA on plant ash to determine gold in vegetation. Hoffman et al. (1979) and Hoffman and Brooker (1982) have described a somewhat similar technique which is described in some detail in the next section.

Instrumental Neutron Activation Analysis

The INAA technique is applicable to any biogeochemical sample including humic forest cover. Samples of approximately 20 to 50 g of material are collected, dried, macerated, and screened (−30 mesh). Eight grams of this material is briquetted in a press at 30,000 psi to form a 40-mm briquette about 6 mm thick (Fig. 9-1). Briquettes are then batch irradiated under thermal or epithermal neutron fluxes, depending on the elements besides gold that are to be determined. The irradiated samples are allowed to decay from 4 days to 1 week, after which they are counted singly using a hyperpure germanium detector linked to a multichannel analyzer–computer system (Fig. 9-2). Detection limits may vary with the type of material being analyzed but will usually be in the range 0.1 to 1 ng/g. Briquettes are quite simply made and have been prepared in the field by one exploration company.

The advantages of the technique are many, including the ability to analyze many common gold pathfinder elements (As, Sb, W, and Cr) simultaneously. Since the technique avoids dry or wet ashing of the sample, it reduces the possible ashing losses and also contamination of the sample. The method is very cost-effective, rapid, and the sensitivity for gold is unrivaled by any of the previously mentioned analytical techniques.

Every method has its drawbacks and this technique is no exception. Some materials may have large quantities of certain elements, which can cause an effective increase in the detection limits. An example of this is the bromine content of some pine needles. In general, though, this has not proved to be a problem since 99.9 percent of the samples submitted for analysis do not contain interfering elements. In humic biogeochemical surveys the amount of inorganic materials in the sample must be kept to a minimum to maintain a low sample background. Inclusion of large amounts of inorganics, however, will tend to impede forming of a briquette without a binding agent, and this provides a self-regulating

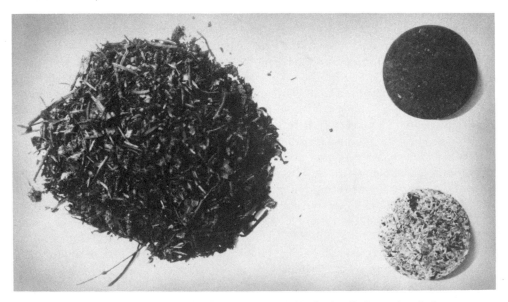

Fig. 9-1   Humus material and briquettes suitable for irradiation and analysis by INAA. Briquettes are 40 mm in diameter.

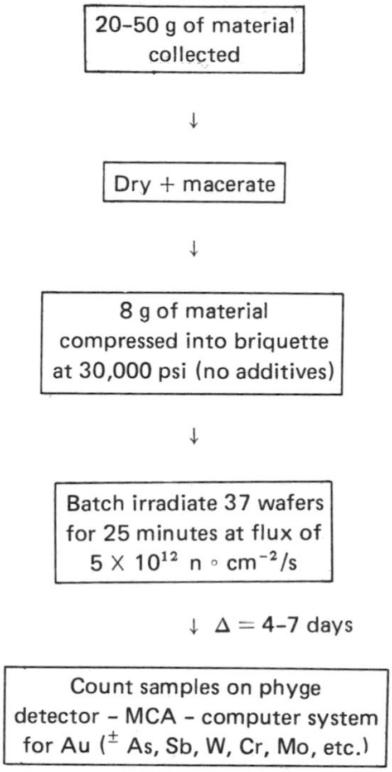

```
┌─────────────────────────┐
│   20–50 g of material   │
│        collected        │
└─────────────────────────┘
             ↓
      ┌────────────────┐
      │ Dry + macerate │
      └────────────────┘
             ↓
┌─────────────────────────────────┐
│        8 g of material          │
│   compressed into briquette     │
│  at 30,000 psi (no additives)   │
└─────────────────────────────────┘
             ↓
┌─────────────────────────────────┐
│  Batch irradiate 37 wafers      │
│  for 25 minutes at flux of      │
│  $5 \times 10^{12}$ n ∘ cm$^{-2}$/s │
└─────────────────────────────────┘
             ↓  $\Delta = 4\text{–}7$ days
┌─────────────────────────────────────────┐
│      Count samples on phyge             │
│ detector – MCA – computer system        │
│ for Au ($^{\pm}$ As, Sb, W, Cr, Mo, etc.) │
└─────────────────────────────────────────┘
```

Fig. 9-2   Flowchart for biogeochemical analysis by INAA.

deterrent to the use of material which would not be classed as humus but which is commonly submitted as such by field personnel.

Elimination of ashing losses is one of the main benefits of analyzing vegetation by INAA. The data in Table 9-1 indicate that dry ashing losses may vary from extreme to

TABLE 9-1   Determination of Gold Losses During Dry Ashing[a]

Briquette	Ashed Sample
380	12
6	3
12	10
3	3
3	2
100	100
74	76
830	120
47	17
16	14

[a]The gold content (ppb) of samples was determined on an organic briquette by INAA. The briquettes were then dry ashed, the ash fire assayed, and the gold determined by INAA on the fire assay bead.

negligible. Recent experiments we have conducted in selecting samples submitted from random batches of samples from different customers indicate that the losses will occur in from 20 to 30 percent of the batches. Information is not available to show whether other gold organometallic compounds in addition to gold cyanogenic complexes are volatile, but indications are that carbonyl, chloride, iodide, and bromide complexes are also volatile. In fact, gold in humic materials may be tied up in a great many organic complexes which have not yet been recognized (E. M. Perdue, personal communication, 1983) and which may be volatilized during dry ashing.

SAMPLING CONSIDERATIONS PECULIAR TO GOLD

One of the aims in undertaking biogeochemical rather than soil sampling programs is to improve the detection of erratic particulate gold in soils. Some experiments we have undertaken on soil samples from Latin America illustrate the problem very well. The soil samples were analyzed by fire assay on 20-g aliquots, with the fire assay beads being run by INAA (detection limit 1 ppb). Every fifth sample was analyzed in duplicate by the same technique. It was not uncommon on the rerun of samples to obtain very anomalous values where the initial run was at background levels, or vice versa. If one were to examine the soil data, the pattern presented would be very difficult to interpret. Ideally, a much larger soil sample could be run, perhaps as much as 500 g according to sampling theory and gold particle size (Clifton *et al.*, 1969), to make the results statistically valid. Cost would not, however, allow this to be a practical solution.

Biogeochemical surveys may tend to minimize, although not necessarily eliminate, the problem of sampling particulate gold in two ways. Girling *et al.*, (1979) and Peterson, (Chapter 4, this volume), for example, have discussed the ability of roots of cyanogenic plants to produce a cyanide compound which will dissolve gold and incorporate it into the plant structure. Baker (Chapter 23, this volume) and Schnitzer (Chapter 24, this volume) have also described the ability of humic and fulvic acid to dissolve metals, including native gold and gold compounds, so that the metal is bioavailable and may be taken into the plants and trees (see also Baker, Chapter 8, this volume). In addition, the root systems of plants are complex and may extend tens of meters in the search for water or nutrients (Richards, Chapter 5, this volume), enabling the geochemist to obtain a much larger coverage than would be possible with a single soil sample. Experience has shown that for many types of biogeochemical samples replication of results can be very good (Table 9-2). In other cases agreement may be much poorer. Usually, however, anomalous values can be repeated, although not necessarily at the same absolute level. This can be confirmed with resampling on a yearly or seasonal basis.

In some areas contamination of the environment is a definite concern in sampling. Certain areas of the Abitibi belt show substantially elevated (20 to 30 times) background levels of both gold and arsenic compared to other greenstone belts, possibly due to the smelting activity in the region, although there are numerous gold deposits in the belt which can contribute to the background levels as well.

The next section deals with some examples of gold deposits in the Noranda-Timmins areas of the Precambrian Shield of Canada.

TABLE 9-2 Reproducibility of Biogeochemical Samples
for Au by INAA, with Subsequent Analysis
by Atomic Absorption[a]

Sample	INAA (first cut)	INAA (second cut)	AA (second cut)
119	79	86	100
133	88	84	100
172	210	230	220
174	280	220	220
176	240	210	220
257	120	130	120
259	220	200	190
260	190	140	190
261	140	170	180
279	62	53	60

[a]Biogeochemical wafers were analyzed by INAA, then ashed, fire assayed, and analyzed by AA.

BIOGEOCHEMICAL RESULTS AROUND SOME KNOWN GOLD DEPOSITS

Biogeochemical sampling was undertaken by White *et al.* (1979) in the vicinity of 25 gold deposits from the Abitibi and Timmins-Porcupine greenstone belts of Canada. Sample materials collected consisted of live vegetation (leaves) as well as the upper portion of the decaying humus material. Twenty-three of the 25 deposits sampled showed anomalous biogeochemical zones associated with the gold deposit.

The decaying humus layer demonstrated a pronounced variation in gold content versus that for live vegetation, as is shown for background levels in the upper left-hand insert of Figure 9-3. Live foliage tended to have background gold levels ranging from 0 to 4 ppb, while the organic humus layer showed background levels of from 8 to 23 ppb. Anomalous values ranged as high as 230 ppb in the humus. A similar relationship is reported by Dunn (Chapter 7, this volume). Some explanations for this enrichment of humic material suggested by Dunn include the findings of Mineyev (1976), that (1) gold is enriched in the leaves from the crown of deciduous trees which make up much of the forest litter, and (2) gold may be dissolved by heterotrophic bacteria, transported in amino acid and protein-rich solutions, and accumulated by molds which thrive on decaying broad-leaved plants.

The Powell-Rouyn mine is located 1 to 2 miles west northwest of the Noranda smelter. The orebody is located in a distal facies of an exhalative chert horizon containing copper and gold lying near the contact of rhyolite and andesite flows and tuffs (R. H. Ridler, personal communication, 1983). About $1\frac{1}{2}$ million tons of ore grading 0.2 oz/ton were mined between 1935 and 1957. Reserves as of 1978 were 266,000 tons

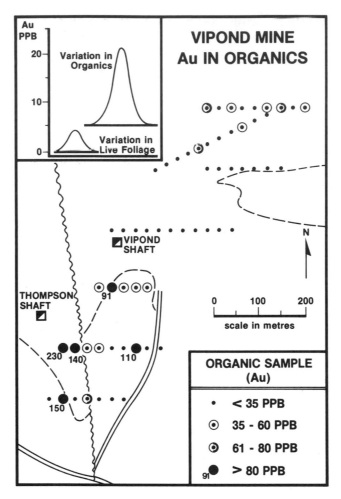

Fig. 9-3 Biogeochemical gold (ppb) survey from the Vipond mine, Timmins–Procupine area of northwestern Quebec. The variation of gold in live vegetation and gold content of humus material is illustrated. (From White *et al.*, 1979.)

grading 0.132 oz/ton (Annis *et al.*, 1978). The left portion of Figure 9-4 illustrates an alteration zone determined by a whole rock geochemical survey undertaken by White *et al.* (1979). The right side of the figure illustrates the anomalous gold in humus correlating with the rock alteration zone. The highest value obtained in humus was 210 ppb, with general background levels between 10 and 30 ppb. Background levels in this area are probably substantially elevated because of the proximity to the Noranda smelter. Care was taken not to sample in the vicinity of the shaft or mine dump areas, to avoid contamination.

A final example is that of the Arntfield gold mine in northwestern Quebec, shown in Figure 9-5. The rocks in this area are basic to acidic volcanics and intercalated pyroclastics. These are cut by the Arntfield Shear and by alteration zones striking approxi-

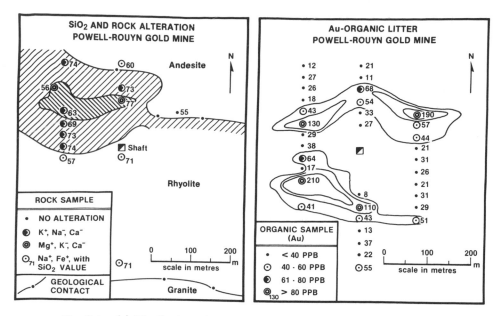

Fig. 9-4   (a) Distribution of SiO₂ and rock alteration in the vicinity of the Powell–Rouyn mine, northwestern Quebec (alteration zones are indicated by crosshatching); (b) gold in humus, same area. (From White *et al.*, 1979.)

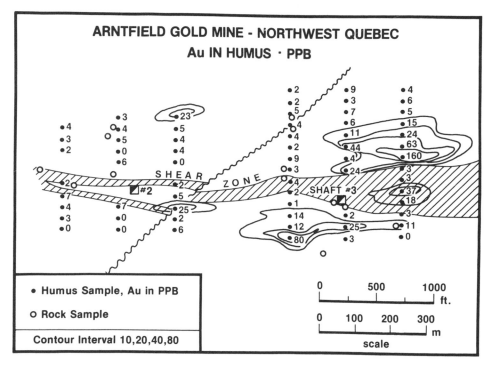

Fig. 9-5   Gold (ppb) in humus overlying the Arntfield gold mine, northwestern Quebec. (From White *et al.*, 1979.)

mately east-west and dipping 45 to 60° to the north. The ore zones are associated with shear zones. The seven located orebodies have an average length of 175 ft and an average thickness of from 6 to 30 ft (Dresser and Dennis, 1949). General background levels in humus range from about 0 to 5 ppb. Anomalous values as high as 160 ppb are located to the north of the Arntfield shear zone.

CONCLUSION

The field of biogeochemistry is still in its infancy, but the technique of humus sampling in the search for gold deposits can be very useful if undertaken properly. Recent advances in analytical technology have eliminated the need for ashing samples together with the problems inherent in ashing. The sampling of humus material is a rapid and inexpensive method for large-scale geochemical surveys for gold when undertaken with the proper orientation surveys. In fact, biogeochemical sampling of humus is probably a less expensive exploration technique than soil sampling and perhaps more effective. Under certain conditions the humus sampling may provide an enhancement of the anomaly relative to the soil horizons and may provide a simpler and more easily understood geochemical pattern. This is particularly true because of the gold "nugget" effect inherent in soil sampling.

As for live plant matter, although the levels of gold and other trace elements are extremely low, the INAA technique now allows detection of background levels without ashing. Sampling of certain parts and species of plants is becoming more prevalent, particularly in the United States, where there are large arid areas that do not generate humic material. The excellent sensitivity for gold and other metals obtained by INAA will make plant sampling a very cost-effective exploration technique. Published biogeochemical data on gold exploration in boreal forested areas are very sparse, and virtually nonexistent for tropical climates, where there is usually an abundance of vegetation and deeply weathered soil profiles. The latter environment should be very suitable for this kind of exploration.

REFERENCES

Annis, R. C., Cranstone, D. A., and Vallée, M., 1978, A survey of known mineral deposits in Canada that are not being mined: *EMR Bull. MR 181.*

Boyle, R. W., 1979, The geochemistry of gold and its deposits: *Geol. Surv. Can. Bull.,* 280, 584 p.

Brooks, R. R., 1982, Biological methods of prospecting for gold: *J. Geochem. Explor.,* v. 17, p. 109–122.

Clifton, H. E., Hunter, R. E., Swanson, F. J., and Phillips, R. L., 1969, Sample size and meaningful gold analysis: *U.S. Geol. Surv. Prof Pap. 625-C,* 17 p.

Curtin G. C., Lakin, H. W., Neuerburg, G. J., and Hubert, A. L., 1968, Utilization of humus-rich forest soil (mull) in geochemical exploration for gold: *U.S. Geol. Surv. Circ. 562,* p. 1–11.

Dresser, J. A., and Dennis, T. C., 1949, Geology of Quebec: Vol. III: Economic geology: *Quebec Dep. Mines, Rep. 20,* p. 283.

Girling, C. A., Peterson, P. J., and Warren, H. V., 1979, Plants as indicators of gold mineralization at Watson Bar, British Columbia, Canada: *Econ. Geol.,* v. 74, p. 902–907.

Gleeson, C. F., and Boyle, R. W., 1979, Consider geochemistry when seeking gold: *North. Miner,* v. 64, no. 52, March 8, 1979, p. C1.

Hoffman, E. L., Brooker, E. J., and White, M. V., 1979, The determination of gold by neutron activation, *in* Morton, R. D. (Ed.), *Proc. Gold Workshop,* Yellowknife, N.W.T., Canada.

———, and Brooker, E. J., 1982, The determination of gold by neutron activation analysis, p. 69–78 *in* Levinson, A. A. (Ed.), *Precious Metals in the Northern Cordillera:* Assoc. Explor. Geochemiste, Rexdale, Ontario.

Lungwitz, E. E., 1900, The lixiviation of gold deposits by vegetation and its geological importance: *Min. J. (London),* p. 318–319.

Mineyev, G. G., 1976, Organisms in the gold migration–accumulation cycle: *Geochem. Int.,* v. 13, no. 2, p. 164–168.

Ward, N. I., and Brooks, R. R., 1978, Gold in some New Zealand plants: *N.Z.J. Bot.,* v. 16, p. 175–177.

Warren, H. V., and Delavault, R. E., 1950, Gold and silver content of some trees and horsetails in British Columbia: *Bull. Geol. Soc. Am.,* v. 61, p. 123–128.

White, M. V., Brooker, E. J., and Eagles, T. E., 1979, Some geological and geochemical criteria for the selection of exploration targets for base metals and gold: Unpublished manuscript.

W. C. Riese*
Anaconda Minerals Company
Denver, Colorado

G. K. Arp
ARCO Oil and Gas Location
Plano, Texas

10

BIOGEOCHEMICAL EXPLORATION FOR PLATINUM DEPOSITS IN THE STILLWATER COMPLEX, MONTANA

*Current Address: ARCO Exploration and Technology Company
Houston, Texas

Platinum mineralization in the Stillwater Complex of southern Montana is hosted in an Archean age layered, tholeiitic intrusion. The zone of platinum mineralization, the Howland Reef, hosts mineralization only in sulfide phases: pyrrhotite, pentlandite, chalcopyrite, and platinum-group minerals.

Geochemical work performed by Anaconda Minerals Company in 1982 consisted of an orientation survey over known mineralization to select optimum sample media, spacing, and chemical elements for evaluation of this area. Stream sediment samples, B-horizon soil samples, and biogeochemical samples (*Pseudotsuga menziesii*) (Douglas fir) were collected. In the case of the soil and biogeochemical samples, both were collected at each sample point. The elements selected for analysis were Pt, Pd, Ni, and Cr. These were chosen on the basis of mineralogical studies of occurrences in this area.

Our study demonstrated that both Pt and Pd are more mobile in the surface geochemical environment than is generally thought: both were found in the few stream sediment samples collected as well as in the soils developing on tills beneath which the platinum-bearing horizon subcrops. Interpretation of these data and implementation of the results of sampling soils developed on glacial materials must proceed cautiously, however, for factors other than those immediately apparent may contribute to forming a geochemical anomaly in this medium.

Our study further demonstrated that twigs of *P. menziesii* are an effective accumulator of Pt and that this element in this medium is a useful exploration tool. Neither Ni nor Cr are selectively enriched by this species and are not useful indicators of mineralization.

INTRODUCTION

The Stillwater mining district is located in south-central Montana, immediately north and east of Yellowstone National Park (Fig. 10-1). Its general geologic setting and the character of the platinum/palladium-bearing horizon were most recently described by Bow *et al.* (1982); the following generalized description is taken from that work.

The Stillwater Complex is a large, stratiform, tholeiitic intrusion that was intruded into a stable Archean craton approximately 2700 m.y. ago. Platinum-group elements are concentrated in the lower part of the banded zone of the complex in a unit referred to as the Howland Reef. Significant concentrations of platinum-group elements within the reef are restricted to sulfide minerals; the dominant minerals belong to the braggite-vysotskite (Pt, Pd, Ni)S solid solution series.

The Howland Reef is continuous over the exposed length of the Stillwater Complex. It strikes approximately N70°W and dips steeply to the north at between 65 and 70° in the area covered by this study. High-angle reverse faults cause this zone of platinum-group-bearing sulfides to be repeated in places.

The geochemical orientation survey (Fig. 10-2) performed over the Howland Reef portion of the Stillwater Complex in 1982 was designed to test the utility of lithogeochemical, pedogeochemical, and biogeochemical samples in mapping the outcrops/

Fig. 10-1   Location map for the Stillwater mining district, Montana.

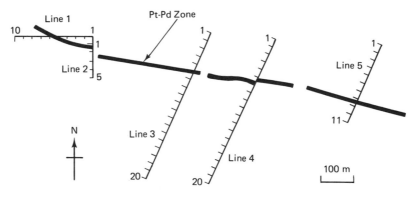

Fig. 10-2   Detailed location map showing line and station locations relative to the Pt–Pd zone in the Howland Reef.

subcrop of this zone. We also hoped that the more platinum-rich zones could be mapped or projected using any, or some combination of, these media.

Lithogeochemical samples were quickly ruled out as a feasible sampling medium because of the paucity of outcrop. Rock chip samples from C-horizon float also could not be used effectively because this region has been the site of alpine glaciation and much of the area is covered by kame terrace deposits which are smeared downslope by solifluction. In short, there is no way to relate the geochemistry of these samples to the underlying lithologies.

For the same reasons, pedogeochemical samples were not anticipated to be useful. Our hypothesis, however, was that because the platinum- and palladium-bearing phases are sulfides, these elements should be available to oxidizing groundwater solutions and might migrate sufficient distances to produce break-in-slope anomalies downslope from the more platinum-rich zones. This process of anomaly formation would be further augmented if organic acids were available to the groundwater system (Huang and Keller, 1970a,b, 1971; Riese *et al.*, 1978). Healthy plants exude very strong organic acids (pH 4 to 5) at their actively growing root tips. These are efficient leaching agents and the organic functional groups that result from their oxidation are fairly effective transporting molecules (Schmidt-Collerus, 1969). The abundance of healthy vegetation in the area suggests that such acids are available and pedogeochemical samples were therefore collected. Each sample was collected from the B soil horizon and represents a three-pit composite sample.

Biogeochemical sample media, which have not been widely used by Anaconda, were added to this orientation survey for two reasons. First, we were attempting to establish a series of case history files that would document the use of this medium in areas with good surface and subsurface geologic control—the Stillwater deposit is an area with such control. Second, the glacial tills that overlie the subcropping ore-bearing horizons are post-ore in age and as such constitute a real obstacle to the utility of standard surface geochemical sampling techniques. We felt that if biogeochemical samples could be shown to reflect the presence of the known mineral deposits in the area, they might contribute significantly to further evaluation of the Stillwater Complex.

Rationale for using biogeochemical samples in geochemical exploration are discussed elsewhere in this volume. Very briefly, the assumption is made that an element in the substrate will be accumulated by the plant and that anomalous amounts in the vegetation will indicate anomalies in the substrate. The proper and more useful medium is determined by an orientation survey. One very important point to remember, which is often overlooked, is that only the actively growing portions of the root system will accumulate elements and consequently, soil samples taken around a tree or line of trees may not reflect similar anomaly distributions. This is because actively growing portions of a root system are commonly deep and/or distal. In our experience in areas of known mineralization, samples collected from 40-ft-tall pine trees, for example, reflect elements collected about 40 ft below the surface if the substrate is soil. However, samples collected from a 40-ft-tall pine tree rooted in thin soil and bedrock may reflect element abundances 50, 60, or more feet below the surface. We believe that this is a manifestation of root depth under the particular geological and climatic conditions of our studies. If any sort of postmineral cover is present in the first 40 to 60 ft, soil samples may not reflect mineralization. Because glacial tills cover most of the area discussed in this chapter, it was felt that vegetation might be the most suitable medium.

The biogeochemical samples were all three-tree composites of *Pseudotsuga menziesii* (Douglas fir) twigs and represented 5 to 7 years of growth in a logged and reseeded area. Needles were also collected, but initial analyses found them not to be within the analytical working range, so they were not studied further. All samples were collected at heights between 5 and 6 ft and from all sides of each tree. All trees sampled were the same size and age.

With the exception of one paper by Fuchs and Rose (1974) and the observation of Rankama and Sahama (1950) that Pt is enriched in the ash of certain marine flora, no studies have been done to examine the utility of Pt in biogeochemical sampling media. Indeed, the work by Fuchs and Rose (1974) concluded that Pt in biogeochemical samples was not useful to exploration. In light of the fact that only four samples formed the basis for this conclusion, this can hardly be considered a defensible position.

One reason why so little work has been done with these elements is the difficulty in analyzing for Pt and Pd in biological materials (J. Arp and L. Walters, personal communication, 1983). Accordingly, a substantial amount of time was devoted to analytical procedure.

The pedogeochemical samples were analyzed for Cu, Zn, Ni, Co, Cr, Pt, and Pd. The biogeochemical samples were analyzed for Ni, Cr, Pt, and Pd. They were not analyzed for Cu or Zn under the assumption that the uptake of these essential nutrients is closely regulated by plants (Epstein, 1972; Trudinger and Swaine, 1979). [See also Berry, Chapter 1, this volume. Ed.] Nor was cobalt obtained on the biogeochemical samples because no acceptable analytical techniques have yet been developed. It was only possible, therefore, to compare soil geochemistry with the biogeochemical results for Ni, Cr, Pt, and Pd. However, Cu, Zn, and Co are important pathfinders in soils in this geologic setting and did provide additional control in that generally accepted sampling medium.

The survey was conducted in early August. Sixty-five samples of each medium were collected at 25-m intervals along lines which were approximately 300 m apart (Fig. 10-2).

SAMPLE PREPARATION AND ANALYSIS

The soil samples were prepared at the Anaconda Minerals Company sample preparation facility in Monte Vista, Colorado. They were first dried, sieved to −80 mesh, and then shipped to Bondar-Clegg, Inc., in Lakewood, Colorado, for analysis. A split to be analyzed for Cu, Zn, Ni, and Co was taken up in aqua regia and then analyzed by atomic absorption. A second split to be analyzed for Cr, Pt, and Pd was fused, taken up in hydrofluoric acid, and analyzed by atomic absorption. Quoted detection limits on the reported results were given as follows: Cu, 1 ppm; Zn, 1 ppm; Ni, 2 ppm; Co, 1 ppm; Cr, 2 ppm; Pt, 50 ppb; and Pd, 5 ppb.

The biogeochemical samples were both prepared and analyzed by the ARCO Research Laboratory in Plano, Texas. The samples were first dried at 110°C. This is not only a requisite step in the preparation procedure but facilitates the separation of needles from twigs as well. This accomplished, the twigs were then ashed at 550°C. The ash was then digested in aqua regia. Pd, Ni, Cr, and Pt were then analazyed by ICP. Quoted detection limits are as follows: Ni, 0.6 ppm; Cr, 0.6 ppm; Pd, 0.4 ppm; and Pt, 0.2 ppb. All raw data resulting from this work are summarized in Tables 10-1 and 10-2 and Figures 10-3 through 10-17.

TABLE 10-1   Stillwater Biogeochemical Data: 1982 Orientation Survey

Line	Stations	Sample Number	Ni (ppm)	Cr (ppm)	Pd (ppm)	Pt (ppb)
1	1	20556	49	25	<0.4	<0.2
	2	20712	33	21	<0.4	241
	3	20638	30	19	8	173
	4	20667	33	20	15	337
	5	20562	25	12	1	711
	6	20593	24	15	<0.4	1318
	7	20624	18	10	5	<0.2
	8	20742	27	18	6	652
	9	20519	8	12	<0.4	<0.2
	10	20617	18	10	3	<0.2
2	2	20563	52	42	<0.4	3000
	3	20517	41	37	0.6	<0.2
	4	20745	40	25	<0.4	<0.2
	5	20522	38/122[a]	30/43[a]	<0.4/13	<0.2/<0.2
3	1	20599	54	13	<0.4	<0.2
	2	10633	56	11	7	<0.2
	3	20581	68	23	<0.4	1027
	4	20607	129	44	<0.4	441
	5	20666	72	15	6	403
	6	20527	171	44	8	<0.2
	7	20506	114	30	<0.4	195
	8	20652	118	29	2	617
	9	20702	60	32	<0.4	<0.2
	10	20577	57	23	1	984
	11	20616	88	29	<0.4	192
	12	20571	47	23	<0.4	552
	13	20725	69	31	4	<0.2
	14	20689	58	58	4	702
	15	20630	108	24	3	<0.2
	16	20692	133	39	<0.4	<0.2
	17	20595	155	33	1	437
	18	20637	30	15	9	140
	19	20750	74	36	<0.4	104
	20	20584/20647[a]	44/61[a]	16/55[a]	7/0	2000/1028
4	1	20713	94	20	<0.4	<0.2
	2	20699	94	19	15	<0.2
	3	20588	72	13	<0.4	1861
	4	20540	185	14	<0.4	823
	5	20654	38	8	<0.4	1240
	6	20587	43	21	5	443
	7	20704	53	27	1	<0.2
	8	20569	62	25	<0.4	1243
	9	20579	111	17	3	811
	10	20734	131	18	<0.4	<0.2
	11	20615	110	9	4	<0.2
	12	20511	273	8	<0.4	<0.2
	13	20691	116	33	5	15
	14	20614	147	27	<0.4	1255
	15	20596	89	27	7	94
	16	20690	701	36	2	316
	17	20636	162	16	3	<0.2

TABLE 10-1   Stillwater Biogeochemical Data: 1982 Orientation Survey *(Continued)*

Line	Stations	Sample Number	Ni (ppm)	Cr (ppm)	Pd (ppm)	Pt (ppb)
	18	20737	157	20	13	<0.2
	19	20749	83	14	4	<0.2
	20	20514	39	22	4	32
5	1	10645	78	13	<0.4	578
	2	20674	230	15	<0.4	854
	3	20610	188	14	2	<0.2
	4	20678	122	29	<0.4	1482
	5	20550	132	14	5	234
	6	20719	189	6	<0.4	<0.2
	7	20729	123	21	14	<0.2
	8	20684	137	29	<0.4	647
	9	20606	100	28	1	454
	10	20721	57	41	<0.4	<0.2
	11	20686	78	23	3	24

[a]Repeat analyses.

TABLE 10-2   Still Water Pedogeochemical Data: 1982 Orientation Survey

Line	Station	Sample Number	Pt (ppb)	Pd (ppb)	Cu (ppm)	Zn (ppm)	Ni (ppm)	Co (ppm)	Cr (ppm)
1	1	20662	<50	65	76	34	132	11	15
	2	20747	<50	<5	71	37	135	23	15
	3	20650	<50	35	51	20	144	19	15
	4	20618	<185[a]	15	55	35	115	19	15
	5	20694	<250[a]	<25[a]	55	30	111	18	15
	6	20741	85	<5	50	33	121	19	15
	7	20518	<50	15	50	27	101	19	15
	8	20700	<50	15	54	28	109	17	15
	9	20592	<50	<5	43	30	112	15	20
	10	20668	<185[a]	<20[a]	42	25	100	18	20
2	1	20662	<50	65	76	34	132	22	15
	2	20740	<250[a]	<25[a]	62	37	108	20	10
	3	20594	110	105	48	26	110	20	10
	4	20558	<250[a]	<250[a]	42	30	121	17	15
	5	2054	<50	15	65	32	244	39	35
3	1	20735	<185[a]	<20[a]	83	34	163	29	20
	2	20512	<50	20	94	35	175	28	20
	3	20644	485	50	64	27	153	26	15
	4	20572	<185[a]	65	67	32	182	29	20
	5	20547	<185[a]	25	59	32	204	29	20
	6	20730	<185[a]	<20[a]	56	27	195	31	25

Line	Station	Sample Number	Pt (ppb)	Pd (ppb)	Cu (ppm)	Zn (ppm)	Ni (ppm)	Co (ppm)	Cr (ppm)
	7	20687	<50	<15[a]	58	26	183	30	25
	8	20628	<70	15	59	28	172	27	25
	9	20551	<50	<5	63	29	171	28	25
	10	20640	<50	<5	69	25	166	26	30
	11	20669	<50	<5	63	26	187	30	25
	12	20675	<50	<5	61	24	154	26	25
	13	20609	50	<5	66	26	156	25	20
	14	20732	65	<5	70	24	151	28	25
	15	20509	<250[a]	<25[a]	65	31	153	22	20
	16	20564	<50	5	68	27	157	27	20
	17	20739	<50	<5	69	29	150	28	15
	18	20516	<50	<5	92	25	168	27	15
	19	20659	<50	5	76	21	162	27	25
	20	20538	<50	5	117	37	167	28	20
		20790	115	<5	117	36	171	30	
4	1	20557	50	20	62	27	251	38	20
	2	20523	<50	15	64	36	208	35	30
	3	20746	140	170	65	25	153	31	20
	4	20676	<185[a]	65	73	30	182	28	25
	5	20707	<185[a]	50	66	31	181	29	20
	6	20552	<185[a]	215	70	33	205	31	20
	7	20524	<185[a]	125	73	31	205	34	20
	8	20642	575	725	76	32	223	35	20
	9	20611	<50	15	76	33	220	36	25
	10	20670	<50	<5	71	36	192	36	30
	11	20600	<185[a]	35	79	42	230	35	20
	12	20565	<185[a]	20	79	37	237	37	25
	13	20632	1010	25	72	34	211	32	30
	14	20738	<50	5	74	29	225	39	30
	15	20566	<185[a]	<20[a]	98	32	280	43	30
	16	20515	<50	55	124	32	292	41	35
	17	20543	<250[a]	<25[a]	151	36	405	45	30
	18	20660	<50	10	79	27	237	36	30
	19	20597	<50	<15[a]	93	35	410	60	25
	20	20710	<50	10	68	23	245	40	35
5	1	20582	<185[a]	60	69	33	247	39	30
	2	20570	<185[a]	<20[a]	69	36	245	39	30
	3	20545	<50	15	65	32	244	39	35
	4	20549	<185[a]	<20[a]	59	31	272	40	30
	5	20574	<250[a]	35	69	37	256	40	30
	6	20604	70	20	64	35	263	40	30
	7	20530	<185[a]	<25[a]	61	35	265	39	35
	8	20504	<50	25	63	30	285	40	40
	9	20573	<50	10	59	32	307	42	40
	10	20528	<250[a]	110	61	34	280	36	30
	11	20507	<250[a]	60	61	35	270	36	35

[a]Lower detection limit due to smaller sample size.

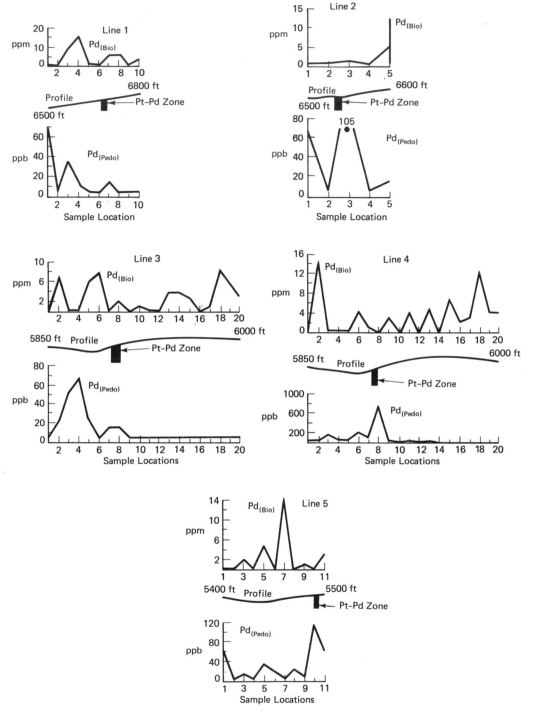

Figs. 10-3 to 10-7   Line plots showing Pd in biogeochemical versus pedogeo-chemical samples, Stillwater orientation survey, lines 1 to 5.

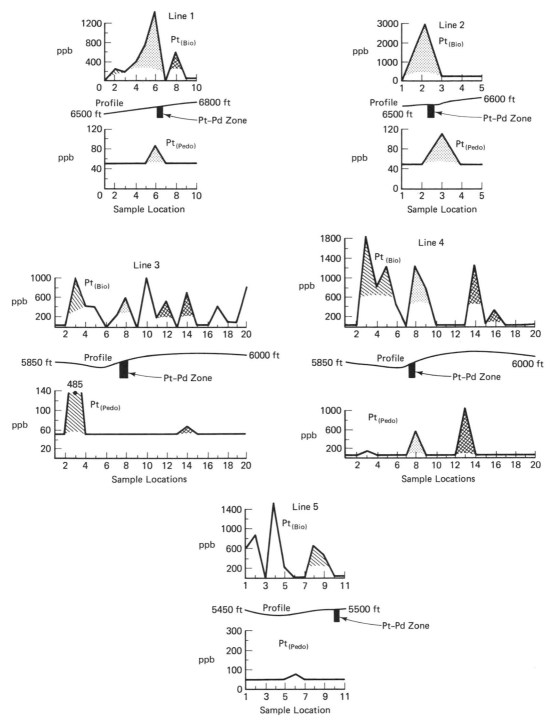

Figs. 10-8 to 10-12   Line plots showing Pt in biogeochemical versus pedogeo-chemical samples, Stillwater orientation survey, lines 1 to 5.

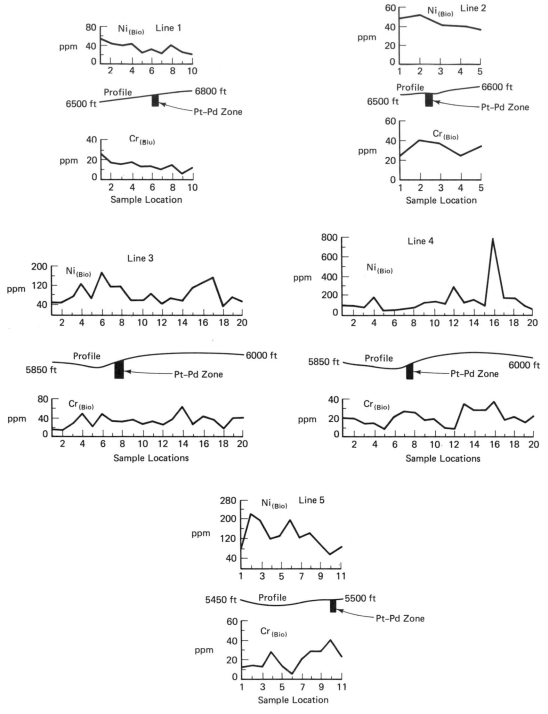

Figs. 10-13 to 10-17   Line plots showing Ni and Cr in biogeochemical samples, Stillwater orientation survey, lines 1 to 5.

Platinum and Palladium Biogeochemistry and Lithogeochemistry

Palladium analyses of tree and soil samples are plotted in Figures 10-3 through 10-7. The palladium in biogeochemical samples was very difficult to analyze, so the quality of these data may not be as good as it should be. Lines 1, 3, 4, and 5 all show anomalous Pd concentrations in biogeochemical samples collected downslope of the main Howland Reef zone. The interpretation given to this distribution is that the Pd, being more readily oxidized and complexed in solution (Pourbaix, 1966), is more mobile and as such is migrating down hydrologic gradient to be picked up by the plants from groundwater and not directly from the rock. This, of course, represents a displaced anomaly, so care should be taken in following up these anomalies.

The platinum in biogeochemical samples (Figs. 10-8 through 10-12) consistently mapped the occurrence of the main Howland Reef platinum–palladium zone. The anomalies of this element in lines 3 and 4 (Figs. 10-10 and 10-11) suggest that the zone of interest may be wider here than mapped: this has yet to be confirmed.

The soil Pt analyses revealed the location of the main platinum–palladium zone in all lines except line 5. On this line a weak Pt anomaly several hundred feet downslope from the projected subcrop of the main zone may mark the development of a hydromorphically displaced, break-in-slope anomaly.

Nickel and Chromium Biogeochemistry

In general, neither Ni nor Cr were found to be either acceptable or reliable pathfinders for the main platinum–palladium zone of the Howland Reef (Figs. 10-13 through 10-17). The two exceptions to this are the Ni profiles in lines 3 and 4 (Figs. 10-15 and 10-16). In line 3, the Ni in the biogeochemical samples does appear to be anomalous over the main zone. In line 4, the Ni maps the subcrop of a fault sliver of this zone quite well (compare Ni in Fig. 10-16 with the biogeochemical Pt in Fig. 10-11).

CONCLUSIONS

This orientation survey was designed to test the utility of pedogeochemical and biogeochemical samples to map the Howland Reef platinum–palladium zone of the Stillwater Complex. Palladium in soils and platinum in both soils and Douglas fir appear to be useful for this purpose.

In selecting an element and sampling medium for a geochemical survey, low noise-to-signal ratio, high anomaly-to-background contrast, and ranges of concentration at analytically reliable levels are all of primary importance. These criteria were best met by the Pt analyses in Douglas fir obtained in this survey. In general, the noise-to-signal ratio is extremely low in these data, certainly lower than is found in the pedogeochemical data. The concentration range is approximately one order of magnitude higher in the biogeochemical data than in the pedogeochemical data and is at levels sufficiently high to eliminate any question about analytical reliability. All these features point to Pt in biogeochemical samples as the preferred system for further work in this area.

REFERENCES

Bow, C., Wolfgram, D., Turner, A., Barnes, S., Evans, J., Zdepski, M., and Boudreau, A., 1982, Investigations of the Howland Reef of the Stillwater Complex, Minneapolis adit area: stratigraphy, structure, and mineralization: *Econ. Geol.,* v. 77, p. 1481–1492.

Epstein, E., 1972, *Mineral Nutrition of Plants: Principles and Perspectives:* Wiley, New York, 412 p.

Fuchs, W. A., and Rose, A. W., 1974, The geochemical behavior of platinum and palladium in the weathering cycle in the Stillwater Complex, Montana: *Econ. Geol.,* v. 69, p. 332–346.

Huang, W. H., and Keller, W. D., 1970a, Dissolution of rock-forming silicate minerals in organic acids: *Am. Min.,* v. 55, p. 2076–2084.

——, and Keller, W. D., 1970b, Dissolution of rock-forming silicate minerals in organic acids, simulated first stage weathering of fresh mineral surfaces: *Am. Min.,* v. 55, p. 2084–2096.

——, and Keller, W. D., 1971, Dissolution of clay minerals in dilute organic acids at room temperature: *Am. Min.,* v. 56, p. 1082–1095.

Pourbaix, M., 1966, *Atlas of Electrochemical Equilibria in Aqueous Solutions:* Pergamon Press, Elmsford, N.Y., 683 p.

Rankama, K. K., and Sahama, T. G., 1950, *Geochemistry:* University of Chicago Press, Chicago, 912 p.

Riese, W. C., Brookins, D. G., and Della Valle, R., 1978, The effectiveness of organic acids in providing uranium mineralization in the Grants Mineral Belt, New Mexico: an experimental study (abstract): *Bull. N.M. Acad. Sci.,* v. 18, no. 1, p. 18.

Schmidt-Collerus, J., 1969, Research in uranium geochemistry—investigation of the relationship between organic matter and uranium deposits, Part 2. Experimental investigations: *Open File Rep., U.S. Dept. Energy GJO-933-2,* 192 p.

Trudinger, P. A., and Swaine, D. J. (Eds.), 1979, *Biogeochemical Cycling of Mineral-Forming Elements:* Elsevier, Amsterdam.

S. Clark Smith
Minerals Exploration Geochemistry
Reno, Nevada

Rene E. Fournier
Condor Minerals Management, Inc.
Arvada, Colorado

11

BIOGEOCHEMICAL AND SOIL GEOCHEMICAL METHODS APPLIED TO MASSIVE SULFIDE EXPLORATION AT PARMACHENEE, MAINE

ABSTRACT

Biogeochemical exploration methods were used in conjunction with soil sampling in an area of known base metal mineralization near Parmachenee, Maine. Approximately 1500 white spruce and balsam fir twigs and 1800 B2-horizon soils were taken on 60-m (200-ft) intervals in an area of about 3 km² (1 square mile). Data from each tree species were statistically treated to determine background, threshold, and anomalous concentrations. Composite biogeochemical maps were then compared to soil geochemical maps and other data to determine the location, orientation, and attitude of the stratiform mineral body.

The study area is in a region underlain by Cambrian subaqueous metavolcanics and metasediments which host several Cu-Zn-Ag-bearing massive sulfide deposits. At Parmachenee, cupriferous pyritic massive sulfide outcrops are present at the metavolcanic-metasediment contact along an estimated strike length of 4.5 km (3 miles). Outcrop density in the area is about 5 percent. Overburden thickness is variable but is not known to exceed 12 m (40 ft). Forest cover is a mixture of coniferous and deciduous trees where density and maturity are variable.

The Parmachenee survey was preceded by a small orientation survey at the Milan mine, West Milan, New Hampshire. The results of that survey indicated an optimum sample density, expected trace element concentrations in balsam, spruce, and soils, and the feasibility of biogeochemistry for typical base metal deposits. The data and conclusions are detailed in the body of the chapter.

The Parmachenee survey indicates that shallow-rooted conifers such as white spruce and balsam fir are useful for base metal exploration. In areas where one tree species is unevenly distributed, two species can be sampled if the analytical data from each are treated separately. After statistical treatment, the normalized data can be merged into consistent biogeochemical maps representing the entire survey area. Comparison of these maps to the soil geochemical data demonstrate downslope displacement of soil anomalies, significantly less displacement of plant anomalies, and areas of near-surface and deep mineralization.

INTRODUCTION

A detailed soil and biogeochemical survey of the Parmachenee area in western Maine was begun in 1977 to further exploration for volcanogenic massive sulfide base metal mineralization. The survey continued for 4 years and was terminated in 1981 after 1500 biogeochemical and 1800 soil samples had been collected. The results of the survey offer an opportunity to compare the sensitivity of biogeochemical methods directly to soil geochemical methods in an area where the mineralized body has been defined by drilling. Biogeochemical sampling of spruce and balsam twigs concurrent with soil sampling significantly added to the authors' ability to predict the location, orientation, and attitude of the stratiform mineral body.

This chapter reports on the sampling, analytical, and statistical methods and on the interpretation of results at Parmachenee, Maine. Prefacing the study is a review of a small orientation study at the Milan mine, West Milan, New Hampshire. It is this orientation study which suggested the usefulness of sampling spruce and balsam twigs to the exploration for massive sulfide mineralization.

LOCATION AND REGIONAL SETTING

The Parmachenee area (Fig. 11-1) is located in western Maine near the New Hampshire and the Quebec, Canada, boundaries. Terrain in the area is mountainous with an average relief of 150 m (500 ft) at an average elevation of 600 m (2000 ft). The project area itself is located on the flanks of Thrasher Peaks in northeast Parmachenee Township and comprises an area of 3.5 km^2 (1.3 square miles).

Climate is characterized by severe winters and moderate summers. Temperatures range from −10°C in January to 16°C in July. Average precipitation is 130 cm (50 in.) per year, with most falling as snow from November through April. The area is heavily forested, with variably mature stands of coniferous and deciduous trees as well as thick undergrowth. Deciduous species include beech (*Fagus grandifolia*), sugar maple (*Acer saccharum*), yellow birch (*Betula alleghaniensis*), white birch (*Fraxinus americana*), basswood (*Tilia americana*), red maple (*Acer rubrum*), red oak (*Quercus rubra*), and white elm (*Ulmus americana*). Coniferous species include white spruce (*Picea glauca*), balsam fir (*Abies balsamea*), and white pine (*Pinus strobus*) (Brockman, 1968). Soils are developed on glacial overburden which may have thicknesses up to 12 m (40 ft). Soils are podzolic in character and have well developed profiles to an average depth of 30 cm (12 in.).

Figure 11-2 shows the location of Parmachenee with respect to the location of several massive sulfide deposits in the northern Appalachians. The Parmachenee project is in a belt of pre-Silurian eugeocynclinal island arc volcanics which are the host for the Elizabeth, Milan, Bald Mountain, and Bathurst Camp massive sulfide deposits. Another belt, thought to be Silurian, extends northeastward along the Maine coast and hosts the Harborside and Blue Hill massive sulfide deposits. Parmachenee is within 3 km (2 miles) of the Ledge Ridge deposit announced by Superior Mining Company, and within 15 km (10 miles) of Dome's Clinton River Mine near Woburn in southewast Quebec, Canada.

ORIENTATION STUDY AT MILAN MINE

The old Milan mine area, West Milan, New Hampshire, served as a geochemical orientation site precursory to the work at Parmachenee. The Milan mine operated briefly in the late nineteenth century and was mapped by Emmons (1910). The surface projection of the 115 ft level is shown in Figure 11-3. At this level the orebody dips to the west 40 to 70 degrees, steepening to near vertical at depth. Emmons reported the grade to be 2.25 percent Cu, 7.26 percent Zn, 1.57 percent Pb, 32.9 percent Fe, 52 g/metric ton (1.5 oz/ton) Ag, and 2.5 g/metric ton (0.07 oz/ton) Au.

The Milan mine survey area (Fig. 11-3) includes a bog that is bounded on the south by a hill of approximately 60 m (200 ft) relief, and topographic highs to the north and

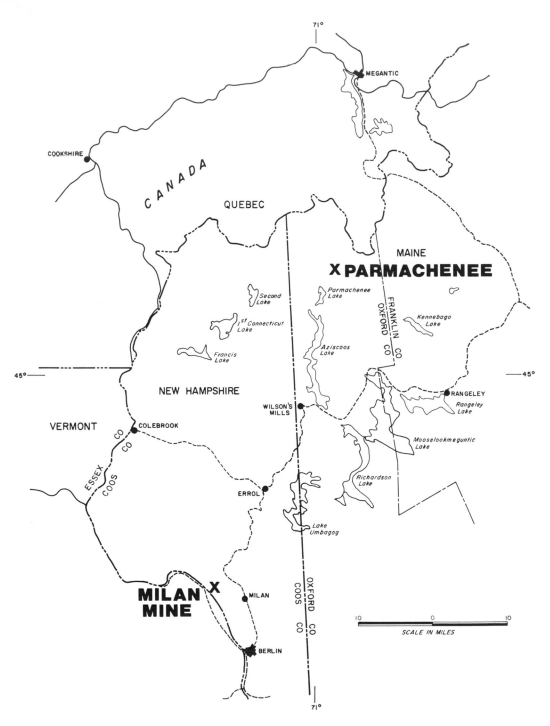

Fig. 11-1   Location of the Milan mine orientation study and the Parmachenee survey area.

BIOGEOCHEMICAL AND SOIL GEOCHEMICAL METHODS

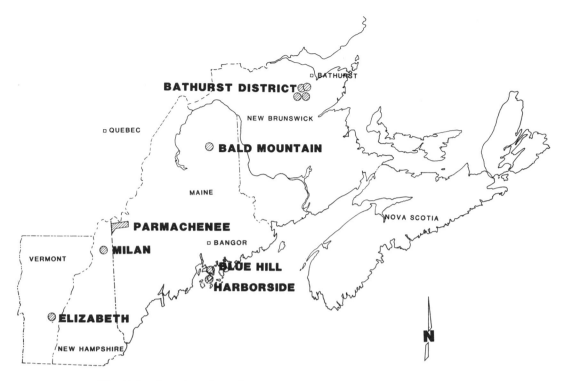

Fig. 11-2   Location of the Parmachenee area compared to location of other known massive sulfide deposits and mines in the northern Appalachians.

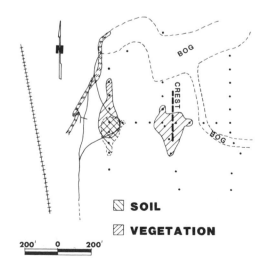

Fig. 11-3   Milan mine orientation study, showing the general topography and cultural features, the surface projection of the 115-ft level, the sample site distribution, and the soil and vegetation anomalies as defined by mean + 2s.

SOIL

VEGETATION

200'   0   200'

S. CLARK SMITH, RENE E. FOURNIER

west. The bog drains in two directions: into the mine area to the west and through the central survey area.

Thirty-eight soil samples were carefully taken on 30-m (100 ft) intervals from the B2 horizon. Vegetation samples were twigs from balsam fir (*Abies balsumea*) and some white spruce (*Picea glauca*), 30 cm (12 in.) long, less needles, cones, pollen sacs, and the most recent year of growth. Ten twigs from one tree were taken from branches 62 to 215 cm (2 to 7 ft) from the ground and equally distributed around the trunk.

Soils were dry sieved and the −80 mesh fraction was digested with hot nitric-perchloric acid and analyzed by atomic absorption spectrophotometry for Cu, Pb, and Zn. Vegetation parts were ashed and similarly analyzed.

The geochemical results (Table 11-1) indicate concentrations representative of a typical massive sulfide ore deposit. Mean concentrations are normal background values, mean plus two standard deviations are representative of threshold concentrations, and maxima are the result of contamination due to previous mining activity. Traditional statistical intervals were not used to define rigid target selection criteria. Rather, these levels were used at Parmachenee to determine mineral potential on a loosely comparative basis.

TABLE 11-1   Statistical Results from the Milan Mine Orientation Study[a,b,c]

	Soil			Balsam (Spruce)[d]		
	Cu	Pb	Zn	Cu	Pb	Zn
Mean	15	25	60	100	160	2190
+1s	50	130	140	130	250	3010
+2s	170	640	290	170	390	4130
+3s	600	3220	630	220	−	−
Max.	1000	6500	700	210	350	4000
n	38	38	38	38	38	38

[a]Mean and standard deviation have been calculated from $\log_{10}$ transformed data; mean is the Tukey estimator of central tendency (Andrews *et al.*, 1972).
[b]Dash (−) indicates computed values that exceed the observed maximum.
[c]All values are in parts per million.
[d]Despite poor biogeochemical practice, balsam and spruce data are mixed. This practice was not continued at Parmachenee since it was obvious that each species obtained different mean values. At the Milan mine, however, it was not apparent that trace element accumulation was species dependent.

The survey results were particularly encouraging with respect to the areal distribution of the vegetation anomalies relative to the soil anomalies, topographic features, and orebody orientation. Soil copper, lead, and zinc anomalies are coincident near and within the topographic low around the mine site. Anomalous soil and vegetation concentrations (+2s) are quickly attenuated within 60 to 90 m (200 to 300 ft) of the mine site. A second feasible target is suggested by balsam alone, 120 to 170 m (400 to 600 ft) east of the mine site. The northwest dip of the orebody suggests that the biogeochemical anomaly can be explained as a limb of the same body or as a parallel limb that pinches out below the soil/bedrock surface, therefore creating a biogeochemical anomaly, but no soil anomaly.

The Milan mine orientation survey suggested the following guidelines for the Parmachenee area:

1. Balsam and spruce reinforce soil anomalies in some areas and in others suggest additional legitimate and feasible targets.

2. Balsam and spruce are ubiquitous and sufficiently distributed to be used in a gridded survey.

3. Soil sample spacing should not exceed 60 m (200 ft). This is especially important since vertical dips would mean restricted surface cross sections. Also, glacial overburden may restrict metal dispersion.

4. Dense tree stands do not yield live twigs within easy reach. Therefore, both live and dead twigs from living trees would have to be taken to obtain even coverage. The sampling priority would be (1) live balsam twigs, (2) live spruce twigs, (3) dead balsam twigs, and (4) dead spruce twigs. Separately, each would have to be treated statistically.

THE BIOGEOCHEMICAL AND SOIL GEOCHEMICAL SURVEY AT PARMACHENEE

Parmachenee Geology and Topography

The geology and mountainous topography at Parmachenee is illustrated on Figure 11-4. The heavy dashed line is the Thrasher Peaks ridge crest. The watershed is to the northwest and southeast of this line into stream drainages represented by thinner lines. Relief is 150 m (500 ft) at an average elevation of 600 m (2000 ft). Contour lines indicate steep slopes on the south side of the ridge crest and more gentle slopes to the northwest.

Diamond drill hole locations are indicated by solid dots south of Thrasher Peaks. Figures 11-5, 11-6, and 11-7 show the same drill locations, but with dots that indicate where mineralization, minor mineralization, and no mineralization were encountered. Specific details concerning grade and tonnage will not be discussed.

For brevity, the geology of the Parmachenee project area (Fig. 11-4) is simplified into three provinces based on lithology and structural deformation style (Fournier, 1981). Stratified rocks in the project area are Cambro-Ordivician in age and the intrusives are Early Silurian (Eisenberg, 1982). All units are in the lower greenschist metamorphic facies. The provinces are described below.

1. *Northern Province:* The Northern Province consists of slate, pillowed metabasalts, pyroclastics, agglomerate, siliceous iron formation, and sulfide-bearing siliceous chemical sediments. These rocks are intruded by an Early Silurian age microgranite. The stratified units dip steeply to the northwest. The volcanics are altered to epidote-rich greenstones.

2. *Central Province:* The Central Province is composed of a mixture of pillowed andesite, volcanoclastics, and graywacke, which dominate to the southeast. This package of rocks is strongly sheared parallel to strike of the various units. The majority of the rocks are now quartz-sericite-chlorite schists. All units appear to dip steeply to the northwest. The province is now thought to be part of a major northeast-trending cataclasite zone. This zone is also the site of cupiferous pyritic massive sulfide mineralization. Several massive sulfide outcrops are known in this zone and diamond drilling has intersected mineralization consisting of copper and minor zinc and silver sulfides.

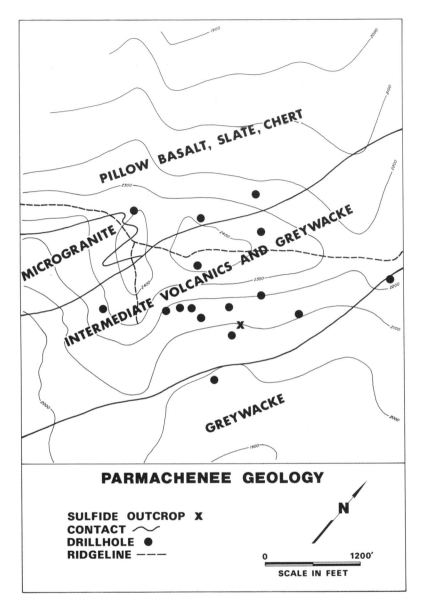

Fig. 11-4 Generalized map of the Parmachenee survey area, showing the topography, major lithologic and structural provinces, drill hole locations, and massive sulfide outcrop.

3. *Southern Province:* The Southern Province is predominantly graywacke with minor slate and conglomerate units. These rocks are isoclinally folded.

Although not shown on the geological map (Fig. 11-4), north-trending vertical faults are suspected in the project area from air photographic studies.

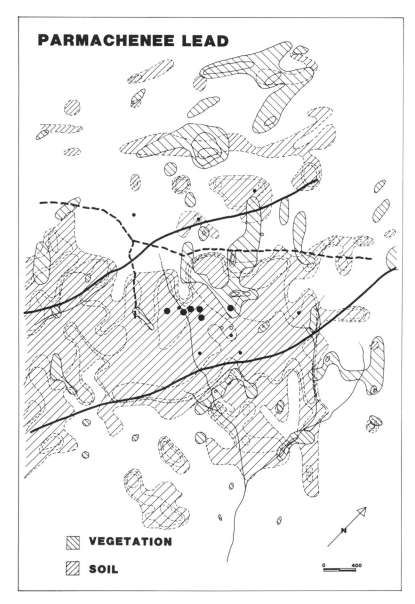

Fig. 11-5   Areal distribution of soil and vegetation lead anomalies at Parmachenee. Hachured areas represent lead concentrations above 1s. Contours within these areas represent lead concentrations above 2s and 3s. Also shown are drill hole locations. Large dots represent mineralized core. Small dots represent minor mineralization in core, and hollow dots represent no mineralization.

Sampling Methods

Over a 4-year period from 1977 through 1980, 1800 soil samples and 1500 balsam and spruce samples were collected in the Parmachenee area. The bulk of the sampling occurred in 1977 and 1978 in the area overlying the Central and Southern Provinces. The area

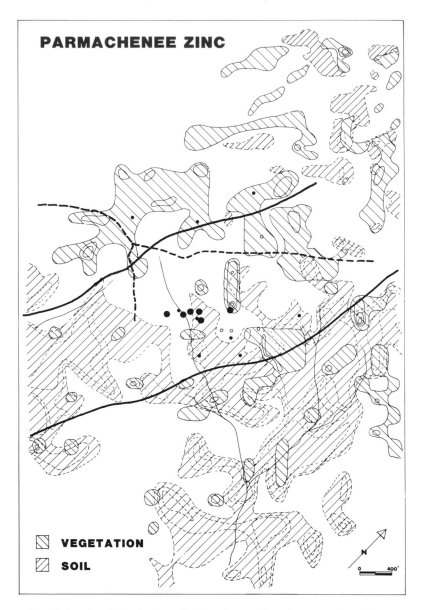

Fig. 11-6   Areal distribution of soil and vegetation zinc anomalies at Parmachenee. Hachured areas represent lead concentrations above 1s. Contours within these areas represent lead concentrations above 2s and 3s. Also shown are drill hole locations. Large dots represent mineralized core. Small dots represent minor mineralization in core, and hollow dots represent no mineralization.

overlying the microgranite intrusive was sampled in 1979, and the area overlying the Northern Province was sampled in 1980. Therefore, differences in statistical values for each year (Tables 11-2, 11-3, and 11-4) are partly due to the geology. As well, different laboratories were used through the 4-year survey.

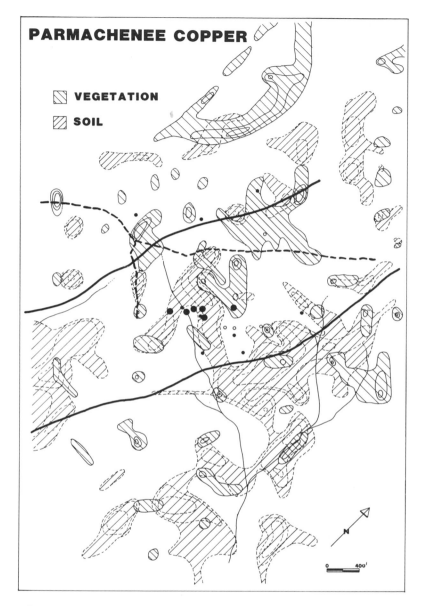

Fig. 11-7 Areal distribution of soil and vegetation copper anomalies at Parmachenee. Hachured areas represent lead concentrations above 1s. Contours within these areas represent lead concentrations above 2s and 3s. Also shown are drill hole locations. Large dots represent mineralized core. Small dots represent minor mineralization in core, and hollow dots represent no mineralization.

In each year, vegetation analytical data were statistically treated as four distinct media. Only one species type was sampled at each location using the priority: live balsam, live spruce, dead balsam, and finally dead spruce. Each was found to have a significantly

TABLE 11-2 Statistical Results for Soil and Vegetation Zinc Concentrations from the Parmachenee Survey, Compared to Means and Maxima from the Milan Mine Study[a,b,c]

a. Parmachenee Zinc

| | Vegetation[d] | | | | | | | | | | | | Soil | | |
| | 1977 + 1978 | | | | 1979 | | | | 1980 | | | | | | |
	LB	LS	DB	DS	LB	LS	DB	DS	LB	LS	DB	DS	'78	'79	'80
Mean	2,300	1,640	5,330	5,130	2,850	3,430	4,130	3,920	3,290	4,480	4,460	4,220	70	30	50
+1s	3,290	2,660	7,260	7,180	3,670	4,830	5,350	5,100	4,560	6,380	6,750	6,660	140	60	80
+2s	4,700	4,310	9,190	9,240	–	6,230	–	6,630	5,830	8,280	9,040	–	280	100	120
+3s	6,710	6,980	–	–	–	–	–	–	–	–	11,330	–	570	–	190
Max.	9,350	7,400	10,500	11,200	4,200	7,000	6,000	6,800	6,500	8,790	12,530	8,470	750	130	200
n	120	73	68	98	23	44	28	14	134	35	104	45	764	114	270

b. Milan Mine Zinc

| | Live Balsam | | Soil | |
	Mean	Max.	Mean	Max.
	2,190	4,000	60	700

[a]Mean and standard deviation have been calculated from $\log_{10}$-transformed data; mean is the Tukey estimator of central tendency (Andrews et al., 1972).
[b]Dash (–) indicates computed values that exceed the observed maximum.
[c]All values are in parts per million.
[d]LB, live balsam; LS, live spruce; DB, dead balsam; DS, dead spruce.

different mean concentration. Therefore, to avoid spurious anomalies due to species and the differences between live and dead twig trace element concentrations, means and standard deviations were calculated separately, and only later were the results synthesized into consistent regional biogeochemical maps.

Zinc Statistics

The zinc statistical summary (Table 11-2) demonstrates how widely the geochemical results for species and live and dead twigs differ with respect to mean, standard deviation, and maxima in one given year. Between years, the differences are due to the differences in geology (Fig. 11-4) and laboratory procedure. Note the high mean and maximum zinc values for vegetation in 1980, reflecting the higher zinc clarke in metabasalts. Conversely, zinc soil concentrations in 1980 are lower than zinc soil concentrations in 1977 + 1978. This difference is probably due to thick glacial deposits on the north-facing slope of Thrasher Peaks, which masks anomalous soil concentrations but does not affect positive biogeochemical response.

High variation due to outcropping mineralization, like that found in the area surveyed in 1977 + 1978, is a function of residual soil chemistry. Using live balsam for comparison, results from the Milan mine orientation survey indicate that the volcanoclastics at Parmachenee are a favorable ore host.

Copper Statistics

Copper concentrations (Table 11-3) in dead twigs are almost constantly two times the concentrations in the live counterparts. It is generally true that after ashing, dead plant material has higher trace element concentrations than live plant material.

Comparing results between years demonstrates the effect of variable felsic and mafic bedrock on vegetation trace element concentrations. Mean copper values are higher from areas overlying metabasalts than over volcanoclastics. The relationships are not the same for soils that seem rather to be affected by accumulation in drainages and bogs. For instance, the 1979 survey was in an area of relatively high topographic relief. Compare the mean and maximum for 1979 to 1977 + 1978, where relief is much more gentle and where soils are poorly drained.

One other noteworthy feature of the copper data is the somewhat lower soil concentrations compared to those at Milan mine, suggesting that Parmachenee is not favorably mineralized. Yet, vegetation concentrations indicate that Parmachenee is mineralized. This is further evidence that vegetation, better than soils, characterizes the massive sulfide potential.

Lead Statistics

Similar to copper, lead concentrations in dead vegetation often are two or more times higher than in live matter (see Table 11-4). Also, overall Parmachenee vegetation lead concentrations are higher than at Milan mine. Contrary to the vegetation data, however, soil lead concentrations are lower. One would deduce that the vegetation better characterizes

TABLE 11-3 Statistical Results for Soil and Vegetation Copper Concentrations from the Parmachenee Survey, Compared to Means and Maxima from the Milan Mine Study[a,b,c]

a. Parmachenee Copper

| | Vegetation[d] | | | | | | | | | | | | Soil | | |
| | 1977 + 1978 | | | | 1979 | | | | 1980 | | | | | | |
	LB	LS	DB	DS	LB	LS	DB	DS	LB	LS	DB	DS	'78	'79	'80
Mean	180	230	450	470	230	200	420	340	370	400	530	450	15	10	10
+1s	250	380	610	750	320	280	560	450	810	940	1,060	890	30	20	20
+2s	350	620	760	1,020	430	400	710	600	1,780	2,220	2,130	—	50	40	30
+3s	490	1,010	—	1,290	590	580	860	—	3,910	5,250	4,260	—	110	—	50
+4s	680	—	—	1,560	—	830	—	—	8,580	—	—	—	220	—	80
+5s	—	—	—	—	—	1,200	—	—	18,830	—	—	—	440	—	120
Max.	580	850	760	1,700	610	1,440	760	540	20,000	7,770	7,200	1,680	470	70	170
n	120	73	68	98	23	44	28	14	134	35	104	45	764	114	270

b. Milan Mine Copper

| Live Balsam | | Soil | |
Mean	Max.	Mean	Max.
100	210	15	1,000

[a]Mean and standard deviation have been calculated from $\log_{10}$-transformed data; mean is the Tukey estimator of central tendency (Andrews, et al., 1972).
[b]Dash (—) indicates computed values that exceed the observed maximum.
[c]All values are in parts per million.
[d]LB, live balsam; LS, live spruce; DB, dead balsam; DS, dead spruce.

TABLE 11-4 Statistical Results for Soil and Vegetation Lead Concentrations from the Parmachenee Survey, Compared to Means and Maxima from the Milan Mine Study[a,b,c]

a. Parmachenee lead

| | Vegetation[d] | | | | | | | | | | | | Soil | | |
| | 1977 + 1978 | | | | 1979 | | | | 1980 | | | | | | |
	LB	LS	DB	DS	LB	LS	DB	DS	LB	LS	DB	DS	'78	'79	'80
Mean	430	640	1,800	1,880	220	100	290	110	530	370	860	920	20	15	15
+1s	820	1,240	3,990	6,210	370	260	500	200	970	820	1,520	1,750	40	30	20
+2s	1,540	2,420	–	–	520	–	720	380	1,770	1,830	2,700	–	80	50	30
+3s	2,910	4,710	–	–	–	–	–	–	–	–	4,810	–	180	80	40
+4s	–	–	–	–	–	–	–	–	–	–	–	–	370	–	60
+5s	–	–	–	–	–	–	–	–	–	–	–	–	–	–	80
Max.	3,400	5,900	7,650	15,200	540	440	900	280	2,050	3,620	8,540	3,170	560	120	90
n	120	73	68	98	23	44	28	14	134	35	104	45	764	114	270

b. Milan Mine Lead

| | Live Balsam | | Soil | |
	Mean	Max.	Mean	Max.
	160	350	60	700

[a]Mean and standard deviation have been calculated from $\log_{10}$-transformed data; mean is the Tukey estimator of central tendency (Andrews, et al., 1972).
[b]Dash (–) indicates computed values that exceed the observed maximum.
[c]All values are in parts per million.
[d]LB, live balsam; LS, live spruce; DB, dead balsam; DS, dead spruce.

the lead potential of the Parmachenee massive sulfide mineralization. Yet, as far as is known, Parmachenee mineralization does not carry significant amounts of lead. A study of lead accumulation at the Hubbard Brook Experimental Forest in central New Hampshire by Siccama and Smith (1978) indicates that lead concentrations in pristine forests are in part due to aerial contributions. They calculated that the forest was retaining 97 percent of the total lead input and that most of that accumulation was in the upper portion of the watershed and possibly due to the higher aerial exposure of vegetation to the windstream.

Siccama and Smith's study suggests that, in part, the higher vegetation lead concentration on the southeast side of Thrasher Peaks is due to the retention of lead from an aerial source. The lead that is captured by the surfaces of living and dead plant material and that eventually becomes part of the forest mull is not in turn being leached and reconstituted in the iron-rich B2 soil horizon. Therefore, vegetation at Parmachenee inaccurately suggests that mineralization is lead-rich.

Areal Distribution Composites

Based on the statistics presented in Tables 11-2, 11-3, and 11-4, vegetation composite maps were made. Soil composites were similarly produced; however, they are contoured only on 1980 results because using earlier results would not have been sufficiently low to capture 1980 details. This has exaggerated the soil anomalies in areas of outcropping and subcropping mineralization.

The topographic and geographic features in Figures 11-5, 11-6, and 11-7 are the same as those in Figure 11-4. The heavy dashed line is the ridge crest. The watershed is to the northwest and southeast of this line into stream drainages represented by thinner lines. The heavy solid lines separate the three geological provinces described earlier. Large, small, and hollow dots represent drill intersections with significant mineralization, minor mineralization, and no mineralization, respectively.

Lead Distribution

The areal distribution of lead concentrations for both soil and vegetation are represented on Figure 11-5. The hachured areas represent lead concentrations greater than mean plus one standard deviation based on data that have been $\log_{10}$ transformed. Within each hachured area, successive contours represent $+2s$ and $+3s$ standard deviations. The ppm equivalents of those levels are represented in Table 11-4. The most obvious feature in Figure 11-5 is the downslope displacement of the soil lead anomaly from the mineralized area. The vegetation lead anomalies are farther uphill and more accurately define the location of mineralization.

A general feature of Figure 11-5, and of Figures 11-6 and 11-7, are sinuous and predominantly downslope soil anomaly patterns. These are far more conspicuous in the mean to mean plus one standard deviation range (not shown on these figures). They appear to be due to trace element entrainment in the areas of high surface and ground-water flow along northwest and southeast joints and fractures. Measured soil pH, using Hydrion pH paper with sensitivity to 0.5 pH unit, is higher (5.0 to 5.5) in areas of high runoff than in background areas (3.5 to 4.5). The pH of hydrolysis can be invoked to explain a correlation between high pH and high trace element concentrations, but this is untenable with respect to the highly organic nature of western Maine's podzolic soils. A more reasonable explanation is offered by Skogerboe and Wilson (1981), who suggest that fulvic acid's reduction capacity increases with increasing pH. This results from increased ionization of acidic functional groups within the complex fulvic acid molecule. Fulvic acid is a stronger reducing agent than humic acid and therefore primarily responsible for these sinuous patterns.

Sinuous patterns are similarly produced by vegetation contours, but these have a negative correlation with areas of high soil pH. Therefore, vegetation patterns probably result from groundwater flow along joints and fractures, and at some depth.

As mentioned in the earlier discussion on lead statistics, the authors believe that much of the vegetation lead may be aerially contributed. Because of this factor, it is suggested that mineral surveys may be adversely affected when either humus soils or vegetation are used if the aerial contribution of lead is not considered.

Zinc Distribution

Figure 11-6 shows the relationship of soil zinc and vegetation zinc anomalies. The hachured areas represent the mean plus one standard deviation concentration level for both media, and the inner contours go up to the mean plus three standard deviations. All statistics are based on $\log_{10}$ transformed data. The ppm equivalents of those levels are represented in Table 11-2.

Zinc is known to have high downslope mobility, which is apparent when compared to the lead distribution shown in Figure 11-5. Far more than vegetation zinc anomalies, soil anomalies are much farther downslope from the source of known mineralization.

Within the volcanoclastics, zinc more than lead suggests a northeast trend to the mineralization. To the southwest, soils have high zinc concentration, but vegetation does not. This suggests that the mineralization is plunging to the northeast or at least is at a deeper level. In part, this feature has been verified by drilling and by the presence of massive sulfide outcrops beyond the area of Figure 11-5 to the southwest.

Within the area underlain by metabasalts, vegetation zinc anomalies are probably due to the higher zinc background of mafic rocks. The bedrock is buried beneath a thick glacial overburden, which may explain the lack of soil anomalies.

Copper Distribution

Parmachenee mineralization is copper- and zinc-rich, as is evident in Figures 11-6 and 11-7. Like soil lead and soil zinc, soil copper anomalies are generally confined to topographic lows.

Vegetation copper anomalies very clearly locate the mineralized area. Soil anomalies up to mean plus three standard deviations are also present and similarly locate the mineralization. The two anomalies together better define the target than either one alone.

Like zinc, the copper anomaly pattern suggests that the mineralization is plunging to the northeast. Soil copper anomalies to the southwest and predominant vegetation copper anomalies to the northeast suggest that mineralization is at greater depth to the northeast.

Within the volcanoclastics, the mineralization strike length appears to be on the order of 300 to 600 m (1000 to 2000 ft). This is an extrapolation of the geochemical pattern based on the hypothesis that the mineralization is plunging and that soil and vegetation are variably sensitive because of structural deformation and glacial overburden.

West of the volcanoclastic/basalt contact is a large enigmatic arcuate vegetation copper anomaly. Soil anomalies are few in this area, possibly because of a thicker glacial overburden. Certainly there is very limited outcrop on this north-facing slope. The arcuate pattern may be the result of an intrusive body, where ring fractures act as a conduit for mineral-rich groundwater percolating toward the bedrock–glacial overburden interface. The balsam and spruce have accumulated copper from these waters, but the soils remain unaffected. On the other hand, displacement along northwest-trending faults may also explain the arcuate shape.

DISCUSSION

It is commonly thought that biogeochemical exploration methods are effective because of the proximity of taproots to buried ore. Balsam and spruce do not have taproots and are known to be highly susceptible to windfall, implying that they have quite shallow root morphology. At Parmachenee, they prove to be good, if not excellent, indicators of subsurface mineralization. Deep taproots are not, therefore, requisite to successful biogeochemical surveys. Evapotranspiration creates a moisture and chemical gradient that carries trace elements from mineralized rock through tens of vertical feet of barren overburden. Root systems, whether deep or shallow, in contact with large volumes of soil and broken rock, accumulate trace metals as they migrate from depth to the near-surface soil strata. Tissue throughout the plant, but especially in the twigs, accumulates proportional amounts of the trace constituents. Responsiveness to subsurface mineralization appears to be better from biogeochemical samples than from non-residual soils or soils in areas where mineralization is not at the surface.

It has been recently reported by Crisci, et al., (1982) that certain pines native to Italy are not generally sensitive to massive sulfide mineralization. By contrast, the Milan Mine orientation study and the Parmachenee survey indicate that certain other conifers can be effective biogeochemical samples.

At Parmachenee, trace element anomalies in balsam and spruce appear to be related to groundwater in fracture and joint systems. Soils on the other hand are anomalous in areas of outcropping mineralization, high soil pH, drainage and catchment basins. Consequently, soils are adversely affected by solifluction in areas of moderate to high relief which tends to displace trace element concentrations downslope of the mineral source. Vegetation sampling contemporaneous with soil sampling often indicates to what degree displacement is a factor in geochemical interpretation.

The distribution of vegetation anomalies and soil anomalies better defines the location and attitude of the mineralization than either medium alone. At Parmachenee, soil anomalies exist where mineralization is near surface. Vegetation anomalies with few if any soil anomalies exist where mineralization is buried. Anomaly patterns of this sort were obtained in the Milan mine area, and northwest and northeast of Thrasher Peaks where it is apparent that the ore horizon is plunging.

Statistical analysis applied to biogeochemical data involves accurate plant species identification and standard normalization of the raw data on a species by species basis. As applied at Parmachenee, the authors were able to create composite maps of the entire survey area even though neither balsam nor spruce are uniformly distributed. In addition, different plant parts, such as live and dead twigs, can be sampled in the same survey as long as they receive separate statistical treatment. This technique permits far greater use of biogeochemical methods where one species is not uniformly available.

Finally, biogeochemistry is not without complications. Geologists and geochemists should be aware that the trace element concentrations in plants may not be related to bedrock and soil mineralization. Plants interact with the atmosphere as well as with the soil and, therefore, certain aerial trace element accumulations may be interfering and misleading.

ACKNOWLEDGMENTS

The authors would like to thank Houston International Minerals Corporation, a subsidiary of Tenneco, Inc., and the Boise Cascade Corporation for permission to use the geochemical data presented in this chapter. We would also like to express our appreciation for the support Houston Oil and Minerals Company management gave to the study.

REFERENCES

Andrews, D. F., Bickel, P. J., Hampel, F. R., Hubel, P. J., Rogers, W. H., and Tukey, J. W., 1972, *Robust Estimates of Location: Survey and Advances:* Princeton University Press, Princeton, N.J., 376 p.

Brockman, F. C., 1968, *Trees of North America:* Western Publishing Co., Golden Press, New York, 280 p.

Crisci, G. M., De Vivo, B., LaFratta, R., La Valva, V., and Lima, A., 1982, Metal response in plants to sulfide mineralization in the Longobucco area (Calabria, southern Italy): *J. Geochem. Explor.*, v. 17, p. 187–204.

Eisenberg, R., 1982, Chronostratigraphy and lithogeochemistry of Lower Paleozoic rocks from the Boundary Mountains, Maine: Unpublished dissertation, University of California, Berkeley, 180 p.

Emmons, W. H., 1910, Some ore deposits in Maine and the Milan mine: *U.S. Geol. Surv. Bull.*, 432, 62 p.

Fournier, R. E., 1981, Maine Exploration Project: Houston International Minerals Corporation (unpublished).

Siccama, T. G., and Smith, W. H., 1978, Lead accumulation in a northern hardwood forest: *Environ. Sci. Technol.*, v. 12, p. 593–594.

Skogerboe, S. K., and Wilson, S. A., 1981, Reduction of ionic species by fulvic acid: *Anal. Chem.*, v. 53, p. 228–232.

12

INDEPENDENT ABSTRACTS FOR LARGER PLANTS IN GEOCHEMICAL EXPLORATION

A. GOLD IN DOUGLAS-FIR: A CLUE TO HIDDEN MINERALIZATION AT THE RED MOUNTAIN STOCKWORK, YELLOW PINE DISTRICT, IDAHO

James A. Erdman and B. F. Leonard
U.S. Geological Survey
Denver Federal Center
Denver, Colorado

Evidence from this study indicates the presence of exploration targets beyond the stockwork not only for concealed deposits of Au and Mo, but also for W and Sn. The targets are new and, to a considerable degree, unexpected from other surface evidence. The biogeochemical anomalies that help define the targets are extensive, and the deposits that might be sought are presumably of low grade.

Red Mountain, which lies on the western edge of the ring-fracture zone of the Eocene Quartz Creek cauldron, has been prospected for Au and Ag for at least 50 years. A biogeochemical study was conducted in 1980–1981 in an attempt to better assess the mineral potential of the stockwork area. Bedrock contacts are concealed by colluvium, glacial deposits, and forest cover. Soil and plant samples were collected on 200-yd centers over an area of 3600 ft × 8400 ft. The wood of douglas-fir (*Pseudotsuga menziesii*) and the leaves of beargrass (*Xerophyllum tenax*) were used because they concentrate Au and Mo, respectively. Results of the soil-metal contents were trivial, although they did indicate a W anomaly south of the stockwork. Analysis of ashed wood by instrumental neutron activation yielded Au values of 0.07 to 14.2 ppm and revealed two distinct Au populations. More important, the highly anomalous samples (>4 ppm) are concentrated in the southern quarter of the grid in an area that has no anomalous Au in the soils, has not been prospected, and lies within inclusion-bearing granodiorite, not stockwork. Beargrass samples, which typically contain 20 ppm Mo, contained <5 to >500 ppm. A belt of above-median values of Mo transects some part of every map unit except the quartz body at the summit. The great extent and the continuity of this belt require some comparably extensive bedrock source of the Mo. The location, shape, and Mo content of the bedrock source remain conjectural, but the source must be large. Subsequent geomagnetic traverses confirmed the belt configuration.

Footnote: Additional information on this case study is contained in...

Erdman, J. A., Leonard, B. F., and McKown, D. M. (in press), A case for plants in exploration: gold in Douglas-fir at the Red Mountain stockwork, Yellow Pine district, Idaho, *in* McIntyre, D. H. (Ed.), *Symposium on the Geology and Mineral Deposits of the Challis 1° × 2° Quadrangle*, Idaho: U.S. Geol. Surv. Bull. 1658.

Leonard, B. F., and Erdman, J. A., 1983, Preliminary report on geology, geochemical exploration, and biogeochemical exploration of the Red Mountain stockwork, Yellow Pine district, Valley County, Idaho: *U.S. Geol. Surv. Open-File Rep. 83–151*, 49 p.

B. BIOGEOCHEMICAL PROSPECTING IN THE GEORGETOWN (Au) DISTRICT, MONTANA

W. C. Riese
Anaconda Minerals Company
Denver, Colorado

D. W. Charlton
Charlton International
Missoula, Montana

Mineralization in the Georgetown district of western Montana is hosted by Precambrian (Beltian) and Paleozoic sedimentary rocks intruded and mineralized by shallow outliers of the Laramide Boulder batholith. The orebodies are veins that have been mined principally for Au, although credits are reported for Ag, Cu, and W.

A geochemical orientation survey of the district demonstrates that Zn, Ag, Cu, As, and Au in *Pinus contorta* (lodgepole pine) are the most useful elements and media for geochemical evaluation of the area. These provide threshold enhancement sufficient to avoid analytical problems and the ability to sample through the soil mantle to bedrock. This is crucial in areas of steep terrain, such as the Georgetown district, where solifluction can displace soil anomalies several hundred feet downslope.

C. GEOBOTANICAL STUDY OF *PENSTEMON CYANOCAULIS* PAYSON IN LISBON VALLEY, UTAH

Margaret Olwell

U.S. Fish and Wildlife Service
Office of Endangered Species
P.O. Box 1306
Albuquerque, New Mexico

Penstemon cyanocaulis Payson was observed on copper mine tailings throughout Lisbon Valley, Utah. Because of the distribution of *P. cyanocaulis* and earlier reports suggesting that there may be some edaphic influence on the genus, it was suspected *P. cyanocaulis* might be a geobotanical indicator species for copper mineralization. This hypothesis was tested by mapping the distribution of *P. cyanocaulis* over areas of known copper mineralization and on areas with no copper mineralization. Using the random quadrat method, plant and soil samples were collected from four research sites located in Lisbon Valley. The samples were analyzed for copper, molybdenum, uranium, and vanadium concentrations. The results of the distribution mapping of *P. cyanocaulis* and the random quadrat study in Lisbon Valley strongly suggest that *P. cyanocaulis* can be successfully utilized as a local copper indicator species. However, the density of the population must be considered together with the presence of the plant in evaluating the potential of the area for copper mineralization. The increasing density of *P. cyanocaulis* with the increasing copper mineralization of the soil presents a distinct trend in the data which is not found with any of the other elements (i.e., molybdenum, uranium, or vanadium). The biogeochemical data also substantiate the potential of *P. cyanocaulis* as a target species for copper. Because the plant exhibits a very definite tolerance for copper-mineralized soils and accumulates high concentrations of copper in both the roots and the aboveground portion, *P. cyanocaulis* may be a valuable biogeochemical target species for copper mineralization.

D. ORGANIC-METAL INTERACTION IN A STREAM CONTAMINATED BY ACID MINE DRAINAGE

Lorraine H. Filipek

U.S. Geological Survey
Denver Federal Center
Denver, Colorado

Organic matter tends to act as a metal scavenger in aquatic environments. However, this effect is expected to be pH dependent. In the present study, the effect of pH on the scavenging ability of algae was examined by sampling water, sediments, and algae along Squaw Creek, a stream in the West Shasta mining district, California, that receives large inputs of acid mine drainage. Dissolved metal concentrations in the stream were found to increase with decreasing pH, not only because of metal input from the mine drainage, but also because of the corrosive effect of the acid waters on the underlying sediments of Squaw Creek itself. The abundance of algae in the stream also tended to increase with decreasing pH. At low pH, the algae were mainly green filamentous. They grew in intimate association with iron oxides and/or jarosite $[KFe(SO_4)_2(OH)_6]$ and appeared to aid in the precipitation of these iron minerals. High concentrations of heavy metals were associated with most samples of algae (range: copper, 2100 to 8500 ppm; zinc, 1500 to 4200 ppm; arsenic, 400 to 5300 ppm—all dry unashed weight). However, the ratio of adsorbed to dissolved metal decreased rapidly with decrease in pH for those metals that exist as cationic species. In contrast, the anionic arsenic was almost completely scavenged from solution within a short distance from its mine source. These results indicate that hydrochemical variables such as pH must be considered when using metal–organic ratios in mineral exploration.

E. FIELD APPLICATIONS OF AIRBORNE BIOGEOPHYSICAL TECHNIQUES

William Collins and Sheng-Huei Chang

Aldridge Laboratory of Applied Geophysics
Columbia University
New York, New York

The specially developed airborne sensor systems for detecting metal-induced plant stress have been flown over several test sites of heavy forest cover in the northwest and southeast United States. In each test site the instruments were successful in mapping geochemically stressed forest canopies overlying the known mineral bodies.

A copper-zinc-lead vein system in Cotter Basin, Montana, has a known geochemical zone in the soil. In a section of the system the very high metal content has caused severe stunting of the trees and a natural clearing where trees cannot grow. The airborne surveys show that the metal effects extend along other parts of the geochemical zone where the forest cover appears normal accept in the infrared properties observed with the survey instrument.

The same survey techniques were also successful in detecting mineral-induced stress in the forest canopies over several zones near Spirit Lake, Washington. The airborne methods are especially applicable in this terrain, where the thick forest is tedious to penetrate and where overburden and volcanic sediment layers can cover deeper soil anomalies near the bedrock.

In Pilot Mountain, North Carolina, the airborne surveys were flown in a "blind" test situation where the data analysts did not know the locations or patterns of the geochemical anomalies. The aircraft survey results were later compared with U.S. Geological Survey ground maps, with which there was a very good correlation.

F. BIOGEOPHYSICAL METHODS OF DETECTING HIDDEN OREBODIES UNDER HEAVY FOREST CANOPIES

William Collins and Sheng-Huei Chang
Aldridge Laboratory of Applied Geophysics
Columbia University
New York, New York

Biogeophysical methods have been developed to detect metal-induced stress in forest canopies through the use of specially designed airborne survey instruments and computer processing techniques.

Airborne spectroradiometer instruments with high spectral resolution (1.4- to 8-μm bandwidth) in the visible and near-infrared spectral region are able to detect subtle changes in the chlorophyll properties of plants affected by metals. Laboratory studies show that copper, zinc, nickel, and selenium are especially toxic to plants. The metals in concentrations of less than 100 ppm appear to inhibit the normal development of the chlorophyll system in living plants.

In both laboratory plants and trees growing in known geochemical anomaly soils, the metal effects are seen as a change in the chlorophyll absorption bands. The chlorophylls and pigments absorb sunlight strongly in the visible region of the spectrum. The absorption ends abruptly in the near infrared at about 750 nm. This absorption limit, which is well defined for normal plants, shifts spectrally in plants growing in soils with higher metal contents.

The metal-induced spectral changes have been observed over several known metallic mineral bodies using the airborne sensors.

John R. Watterson
U.S. Geological Survey
Denver Federal Center
Denver, Colorado

II

AN INTRODUCTION TO MICROORGANISMS IN GEOCHEMICAL EXPLORATION — A NEW APPROACH

The idea for including a session on microbiology in the 1983 Rubey Colloquium began early in 1982 following discussions between myself and Betty Olson, and Don Carlisle, Isaac Kaplan, and Wade Berry after seminars in the Department of Earth and Space Sciences at UCLA in which Betty Olson and I presented preliminary data.

The belief that microorganisms can be of use in mineral exploration is based on ideas and observations outlined here as an introduction to the nine chapters in this section. Many of these chapters deal with the toxicity of elements and the apparently predictable development of bacterial resistance to metals and/or antibiotics in metalliferous environments.

First, we should consider the origin of metal toxicity in biologic evolution. Before the most primitive prebiotic structures or the components of their assembly were defined, "toxicity" had no meaning. There were only relative probabilities of encounter between various ionic and molecular species, and possible chemical interactions. The origin and evolution of complex, polymeric, prebiotic structures and self-replicating biomolecules must have depended on their probability of encounter with various ionic or molecular species. It may not be a coincidence that cobalt, the rarest of the elements for which at least one atom is present in a cubic micrometer of seawater (a reasonable size for a protoorganism), should be the rarest trace element widely employed in biochemistry, or, by the same token, that most of the exclusively toxic elements should occur at lower concentrations. If a new tentative protein, say an enzyme, were quickly inactivated by ions with a high probability of encounter, such as Ca^{2+}, that enzyme, and any DNA coding for it, could not have survived for long. However, an enzyme that was inactivated only by ions with a low probability of encounter, such as Hg^{2+}, could have been an enzyme with a future.

In other words, essential biologic structures and processes must, from the very first, have been constrained by the probability of encountering the different elements; and the chemical *forms* adopted by living structures under these constraints, must have defined chemical toxicity as it is observed today. We can observe today that the toxicity of ions to most bacteria is generally inversely proportional to the probability of encounter with that ion. The probability of encounter with an ion of any element is related, as a first-order approximation, to the crustal abundance of that element (Schroeder, 1973). Indeed, the relative toxicity of the ions of mercury, cadmium, copper, nickel, and zinc to bacteria inhabiting an unmineralized wilderness soil (Duxbury, 1981) follows in inverse order the relative abundance of these elements in crustal rocks (Rose *et al.*, 1979).

Now consider ore deposits and geochemical exploration. H. W. Lakin (1979) pointed out that in the vicinity of mineral deposits, much greater enrichment of the very rare elements can occur. Table 1 shows the impressive enrichment factors obtained by Gott and McCarthy (1966) in the vicinity of the porphyry copper deposit near Ely, Nevada. Significant enrichment, rather than absolute concentration, is the central concept of geochemical exploration. A few parts per million of tellurium is a most significant indication of nearby mineralization (Watterson *et al.*, 1977).

The parallel relationship between the order of element toxicities and their potential for enrichment near ore deposits is the theoretical basis for the use of microbiology in mineral exploration. The biologically available elements that are the rarest are generally the most toxic to bacteria and other forms of life. The toxicity of elements to bacteria is thus roughly proportional to the potential enrichment of these elements in the vicinity of ore deposits of many types.

TABLE 1   Enrichment of Some Metals at Ely, Nevada

Element	Abundance in Igneous Rocks (ppm)	Ely District Maximum Enrichment Factor
Tellurium	0.001	10,000,000
Silver	0.02	100,000
Antimony	0.3	33,000
Arsenic	2.0	5,000
Copper (ore)	70.0	100

Source: Gott and McCarthy (1966).

Most ordinary soil bacteria cannot live in the presence of even small enrichments of rare/toxic indicator elements such as mercury, cadmium, and arsenic without developing resistance to those metals. Such metal resistances are quite easy to test in bacterial populations. It is much easier and cheaper to test the resistance of soil bacteria in biologically available toxic elements than it is to do a chemical analysis of the soil for those elements. Mercury, which is bedeviled by sampling as well as analytical problems, is a good candidate for such an approach. Populations of bacteria living in an environment containing harmful levels of available mercury can be expected to have statistically integrated their response to the presence of mercury over long periods of time. Bacterial resistance to soil mercury may therefore be far more meaningful and reliable from an exploration standpoint than the snapshot information represented by analytical data.

These are some reasons why exploration geochemists and exploration managers should consider the development and use of microbiological methods. We hope that the invited papers in this section will contribute to the opening of a new field of investigation that will expand the horizons of both geochemistry and microbiology.

This part of Rubey Volume V begins with a historical perspective written by D. M. Updegraff, a geomicrobiologist who participated in some of the pioneering research on soil and aquatic bacteria and their role in geological processes. Updegraff reviews the early use of microbiology in petroleum exploration in the Soviet Union and the United States and looks at the prospect of using microorganisms in mineral exploration either by means of possible indicator species or by means of indirect soil metal assays.

All but one of the remaining chapters are by microbiologists who, for a variety of reasons other than their potential application in geochemical exploration, have studied interactions between heavy metals and microorganisms in soil or water. In Chapter 14, H. L. Ehrlich reviews the physiology and biochemistry of metal-microbe interactions and points out some of the complexities that may attend metal-resistance studies. Bio-oxidation and bioreduction of metals, including magnetite deposition in bacteria, growth inhibition by metals, and metal tolerance, are discussed. Chapter 15, by G. Stotzky and H. Babich, is a careful examination of the possible pitfalls of an exclusively microbiological approach. The authors point out that physicochemical environmental factors, such as pH, Eh, and clay content of soils, can influence metal–microbe interactions and thus could complicate interpretations of the metal content of an environment if studies were based solely on metal–microbe interactions. Some practical techniques based on their observations are suggested. Chapter 16, by A. O. Summers, gives an excellent discussion of the genetic machinery used by bacteria in defense against specific toxic metals and also against antibiotics. Readers needing to become familiar with microbiological terminology may wish to begin with this chapter. In Chapter 17, R. R. Colwell and her colleagues, drawing upon extensive research conducted in her laboratories, describe the exceedingly interesting association between multiple metal resistance and multiple antibiotic resistance observed in Chesapeake Bay bacterial populations growing in metal-polluted environments. A similar relationship between heavy metal and antibiotic resistance is described in Chapter 18 by G. G. Michaels, J. J. Hill, and D. J. Schneck in a high-altitude mineralized watershed in Colorado remote from large human populations. The association of heavy metal and antibiotic resistance was first recognized in bacteria in hospital environments in the late 1960s. Plasmids, small circular units of extrachromosomal DNA carrying heavy metal- and antibiotic-resistance genes, were later identified in these bacteria. R. R. Colwell and her colleagues followed up on this discovery by documenting the presence of such plasmids in resistant bacteria isolated from sites polluted by heavy metals. B. H. Olson and T. Barkay present data in Chapter 19 which show unmistakable links between the heavy metal content of soils and the metal-resistance characteristics of the indigenous microbiota. Other evidence of this interesting association is given in the abstract by S. Tripp, T. Barkay, and B. H. Olson at the end of this section. Finally, J. R. Watterson and two colleagues, L. A. Nagy and D. M. Updegraff, present evidence in Chapter 20 that penicillin resistance in soil bacteria can be used as an index of soil metal content over two mineral deposits and suggest that the distribution of the penicillin-resistant end member *Bacillus, B. cereus*, should be further investigated.

REFERENCES

Duxbury, T., 1981, Toxicity of heavy metals to soil bacteria: FEMS Microbiol. Lett., v. 11, p. 217-220.

Gott, G. B., and McCarthy, J. H., 1966, Distribution of gold, silver, tellurium, and mercury in the Ely mining district, Nevada: *U.S. Geol. Surv. Circ. 535,* 5 p.

Lakin, H. W., 1979, Keynote address—Sharpen your tools, *in* Watterson, J. R., and Theobald, P. K., (Eds.), *Geochemical Exploration 1978:* Assoc. Explor. Geochemists, Rexdale, Ontario, 504 p.

Rose, A. W., Hawkes, H. E., and Webb, J. S., 1979, *Geochemistry in Mineral Exploration:* Academic Press, New York, 657 p.

Schroeder, H. A., 1973, Recondite toxicity of trace elements: *Essays in Toxicity*, v. 4, p. 109-199.

Watterson, J. R., Gott, G. B., Neuerburg, G. J., Lakin, H. W., and Cathrall, J. B., 1977, Tellurium, a guide to mineral deposits: *Journal of Geochemical Exploration*, v. 8, p. 31-48.

Some useful standard references are:

Ehrlich, Henry Lutz, 1981, *Geomicrobiology*; Marcel Dekker, Inc., New York, 393 p.

Gould, G.W., and Hurst, A., 1969, *The Bacterial Spore*; Academic Press, London, 724 p.

Singleton, Paul, and Sainsbury, Diana, 1978, *Dictionary of Microbiology*; A Wiley-Interscience Publication, John Wiley and Sons, 481 p.

Stanier, Roger Y., Adelberg, Edward A., and Ingraham, John L., 1976, *The Microbial World*, Fourth Edition; Prentice-Hall, Inc., Englewood Cliffs, New Jersey, 871 p.

David M. Updegraff
Department of Chemistry and Geochemistry
Colorado School of Mines
Golden, Colorado

13

GEOMICROBIOLOGICAL PROSPECTING: PAST AND FUTURE

ABSTRACT

Prospecting with bacteria is not a blue-sky idea. Research carried out in the Soviet Union and the United States beginning in the 1930s clearly showed that oil and gas deposits deep within the earth could be detected by enumerating bacteria in near-surface soil which oxidize gaseous hydrocarbons as a sole source of energy. Many oil and gas fields have been discovered with the aid of such methods.

Since bacteria are by far the most versatile of living organisms and are found at greater depths in the earth's crust than other living things, it might be expected that both the kinds and the numbers of bacteria will reflect the geochemical environment in which they live. Thus it is possible that the detection of certain kinds of bacteria or of anomalous numbers of bacteria, or even the absence of certain bacteria may be used as indicators of other economically important deposits such as toxic heavy metals, toxic nonmetals, or radioactive isotopes. Furthermore, the environment affects the biochemistry and development of bacteria and fungi, and we may expect genetic changes in the presence of biologically active chemicals in the environment, as well as changes in the rate of cell division, sporulation, antibiotic synthesis, and other vital processes. Research on these processes may well lead to new geochemical prospecting methods.

PROSPECTING FOR PETROLEUM

The idea of using microorganisms as a tool in exploring for economically useful deposits in the earth's crust was first put forward by the Soviet geologist, Mogilevskii, who, between 1937 and 1940 had been carrying out geochemical surveys of oil and gas fields in order to detect and contour hydrocarbon gas emanations from subsurface petroleum deposits (Mogilevskii, 1940).

Mogilevskii obtained the assistance of the Soviet microbiologist, Butkevich, and more than 3000 soil samples were collected and analyzed for bacteria that oxidize gaseous hydrocarbons as a source of energy. In addition, they analyzed for the gaseous hydrocarbons themselves. The rather surprising conclusion was that although only negligible concentrations of gaseous hydrocarbons were found in the soil, over known gas deposits, methane-oxidizing bacteria were commonly detected over gas deposits.

Meanwhile, scientists in the U.S. oil industry became aware of these developments, and patents began to appear on the use of gaseous hydrocarbon-oxidizing bacteria (Taggart, 1941; Sanderson, 1942) and of products of their metabolism (Blau, 1942, 1943) in petroleum prospecting. These methods were evidently never fully evaluated, and have not been used in petroleum prospecting. It appears that little intensive research was done in the United States until the end of World War II. Between 1945 and 1948 two U.S. oil companies, Texaco and Mobil, instituted major research projects, including both laboratory and field work, to evaluate the usefulness of the method in exploration. This resulted in conflicting patent applications, filed at almost exactly the same time in 1953 by Texaco and Mobil (Strawinsky, 1954; Updegraff and Chase, 1958). The methods described in both patents were nearly identical. Both measured, by manometric or volumetric means, the ability of hydrocarbon-oxidizing bacteria in subsurface soil samples to consume ethane, propane, or butane gas admixed with air. An interference suit resulting from these conflicting claims was settled by cross-licensing in 1958.

The early Soviet work employed an extremely simple qualitative method for detecting anomalies of methane-oxidizing bacteria in subsoils. A medium comprising a solution of all the essential mineral nutrients was placed in culture tubes, sterilized, inoculated with soil, placed in a jar containing a mixture of methane and oxygen, and incubated for 12 to 14 days. When methane-oxidizing bacteria were present in the soil, they formed a pellicle of growth on the surface of the mineral medium. The method employed was first described by the German microbiologist Söhngen in 1906 in his classical paper on the isolation of bacteria utilizing methane as a source of carbon and energy.

Butkevich was well aware of the fact that methane was commonly formed in sediments by the anaerobic digestion of organic matter such as cellulose. Accordingly, he tested the soil samples for cellulolytic bacteria by incubating strips of filter paper in mineral salts medium inoculated with the soil samples and counted as positive results for methane-oxidizing bacteria in the absence of cellulolytic bacteria. This is obviously not a very logical procedure, since the presence of cellulolytic bacteria does not indicate methanogenesis. Methanogenesis can come about only through the agency of three or four different kinds of bacteria acting together under strictly anaerobic conditions, the final step of methanogenesis being carried out by methanogenic bacteria.

Many other Soviet workers have followed up the leads provided by Mogilevskii and coworkers. Bokova *et al.* (1947) described laboratory experiments as well as field surveys on the oxidation of gaseous hydrocarbons by bacteria as a basis of microbiological prospecting for petroleum. They reported the isolation of bacteria which oxidized ethane and some higher hydrocarbons, but not methane. Davis *et al.* (1956) also isolated ethane-oxidizing bacteria, described as the new species *Mycobacterium paraffinicum*, and observed that these bacteria oxidize ethane, propane, and higher hydrocarbons, but not methane. Surveys carried out by the Mobil Oil Company over known oil and gas fields invariably showed significant anomalies of ethane-oxidizing bacteria over such fields, but usually failed to show anomalous numbers of methane-oxidizing bacteria. This finding was expected because of the ubiquitous production of methane by methanogenic bacteria in anaerobic sediments and soils. Subbota (1947) and Kartsev *et al.* (1959) found 5 to 15 ppm of methane in soils at depths of a few feet in areas not associated with petroleum and only about a twofold increase over petroliferous areas. Concentrations of ethane and propane are much lower, but these higher-molecular-weight hydrocarbons are relatively specific to petroleum gas emanations. Davis and Squires (1954) found that no more than 7.0 ppm ethane and only traces of ethylene, propane, propylene, and acetylene were in gases produced by the anaerobic digestion of cellulose and glucose. Testing for the presence of bacteria which utilize these gases thus eliminated much of the problem of false positives encountered with methane oxidizers. Natural gas, on the other hand, often contains from 1 to 10 percent of ethane and propane, and up to 95 percent methane.

Several different methods can be found in both the patent literature and in scientific journals for enumerating bacteria that oxidize gaseous hydrocarbons as a means of prospecting for petroleum. The simplest method, used by the earliest Soviet workers, of observing a pellicle of growth on a tube of mineral salts medium inoculated with subsurface soil and incubated under an atmosphere of the hydrocarbon gas and oxygen, has already been mentioned. A more sophisticated method was employed by the Soviet workers, as well as by Strawinski (1954), Updegraff and Chase (1958), and Brown (1962) in the United States. This involved measuring gas consumption as a function of time in either a volumetric or manometric apparatus. Approximately 10 g of soil is a convenient inoculum for manometric systems.

A direct plating method, in which mineral salts agar plates were sprinkled with 0.1 g of prospect soil and incubated for 20 to 30 days in an atmosphere of ethane and air gave typical yellow, waxy wrinkled colonies of *M. paraffinicum* when soils were collected over petroliferous areas. Several other kinds of bacteria found in soils can adapt to grow on ethane, but also grow on other kinds of soil organic matter, and hence are of no diagnostic value in prospecting for gas. The plating method is quicker and simpler than the gas uptake method and appears to be just as accurate (Davis, 1967). *Mycobacterium* has a waxy coat of lipid material surrounding the cell, which appears to give it a great deal of resistance to adverse environmental factors, such as drying and lack of nutrients. Thus these bacteria act as "integrating agents" for petroleum emanations in the soil. They multiply, using gaseous hydrocarbons as an energy source, when these gases are present. Then, if the gaseous emanations are interrupted because of earth movements or flooding by rainfall, they may remain dormant in the soil for months, years, or perhaps centuries until gas emanations again reach them, when they will resume multiplication. Thus these bacteria could provide clear evidence of subsurface petroleum gases even in the absence of a geochemical hydrocarbon anomaly. This "integrating effect" may, by analogy, be useful in sensing emanations of H_2S, CO_2, COS, and other gases characteristic of certain mineral deposits.

The oxidation of ^{14}C-labeled ethane in soil samples provides an exceedingly sensitive method of measuring the rate of oxidation of ethane by bacteria in soil, but the method does not appear to be specific to petroleum-gas emanations, probably because of the presence of bacteria in the soil other than *M. paraffinicum*, which will oxidize ethane and grow on organic matter other than gaseous hydrocarbons. Davis (1967) showed that the $^{14}CO_2$ resulting from the oxidation of ethane could be converted to barium carbonate and measured in a Geiger-Müller counting tube, or colonies of ethane-oxidizing bacteria grown on labeled ethane could be readily identified by radioautography.

Where Davis and coworkers emphasized the utilization of a highly specific prospecting parameter, namely a single species almost entirely specific to gaseous petroleum hydrocarbons, Sealey (1974) has taken a completely different approach. He has collected soil samples from a depth of about 2 ft over and off 173 oil fields in Texas and to a limited extent in New Mexico, Utah, and Louisiana, and has analyzed for many different kinds of bacteria in order to develop a composite scoring system indicative of subsurface petroleum. He has reported that an evaluation of his method shows that it correlated with subsurface oil fields in 86 percent of the fields tested. Unfortunately, he gives no details of his method, since he is offering his services, using his own proprietary method, to oil companies for prospecting purposes.

In spite of the fact that the research described above, carried out in the USSR and by several oil companies in the United States, has clearly shown that microbes can detect very slight emanations of petroleum hydrocarbon gases from underlying oil and gas deposits, the method is apparently not used in exploration for petroleum in the United States. By contrast it has been used routinely in the Soviet Union for more than 40 years to detect promising areas for detailed exploration (Kuznetsov, 1979). Once a general area has been shown to yield geomicrobiological anomalies of hydrocarbon-oxidizing bacteria, detailed geological and geophysical exploration is undertaken to pinpoint exact locations for drilling exploratory wells. Kuznetsov (1979) reports that a number of Soviet oil fields have been located by this procedure. It is puzzling to me why such a proven oil-finding tool has been neglected by U.S. companies. I fervently hope that this simple and relatively cheap prospecting method will not be forgotten in future years by companies exploring for oil in remote areas outside the United States where geological conditions are less well known.

PROSPECTING FOR METALLIC AND NONMETALLIC ELEMENTS

Although the use of microorganisms in prospecting for petroleum is well established, there are no well-documented examples of the use of microbiological methods in prospecting for other valuable minerals in the earth's crust. It is quite clear, however, that there is a good theoretical basis for believing that such methods can be developed for a number of metals and nonmetals in the near future.

Geobotanical methods involving the use of higher plants to detect anomalous concentrations of selenium in the soil were shown by the elegant work of Helen Cannon (1964) to provide an excellent method of prospecting for uranium since selenium is enriched in the uranium ore. All that was necessary was to look for two species of selenium indicator plants, *Astragalus pattersoni* and *A. preussi*, which grow *only* on soil heavily enriched in selenium. Thus these lowly leguminous loco weeds are specific indicator plants for the direct detection of selenium and the indirect detection of uranium just as *M. paraffinicum* is a direct detector of petroleum gas.

There are many good reasons to believe that many other such indicator organisms may exist, particularly among the bacteria. These most primitive of living things are also the most environmentally versatile of all organisms. They grow and multiply everywhere in the biosphere, from the frozen deserts of Antarctica to the geysers of Yellowstone National Park at temperatures of nearly 100°C and the black smoker vents along the East Pacific rise at 300°C (Baross and Deming, 1983), from the surface of the soil to depths of thousands of meters beneath the earth, and from the surface of the waters to the greatest depths of the sea. They obtain their energy by every known mechanism used by other living things—photosynthesis, heterotrophic oxidative respiration, and anaerobic fermentation—but also by other mechanisms unique to the bacteria, for example anaerobic photosynthesis using H_2S as a hydrogen donor instead of H_2O, chemosynthetic autotrophy wherein the oxidation of inorganic materials provides their energy, and anaerobic respiration where nitrate or sulfate serves as terminal electron acceptors instead of oxygen as in higher organisms. Miller (1983) has found several indicator species for sulfur deposits among both autotrophic and heterotrophic and both aerobic and anaerobic bacteria.

Letunova and Koval'sky (1978) have published a remarkable book summing up their studies from 1954 to 1975 on the geochemical ecology of microorganisms in soils and sediments. Their conclusions, which might have been expected from our present knowledge of microbial ecology, show that bacteria reflect the mineral composition of their environment both in their requirements for minerals as nutrients and in their resistance to toxic mineral elements. Many methods were employed in their studies, including direct microscopic examination of soil, water, and sediments by scanning electron microscopy and fluorescence microscopy, the elegant capillary methods of Perfil'ev, the use of laboratory soil columns, the isolation of pure cultures on agar plates and liquid media enriched with the minerals of interest, and the determination of the minimum optimum and maximum tolerated levels of the elements of interest by bacteria isolated from environments containing low amounts of these minerals and from others containing very high concentrations of these minerals. The elements studied were cobalt, molybdenum, copper, vanadium, selenium, boron, and uranium. All of these elements except uranium have been shown to be essential in the nutrition of some living organism (Mertz, 1981), and many bacteria require cobalt, molybdenum, and copper, since these elements are essential cofactors for many different enzymes. Vanadium and selenium are required by mammals, but their roles in metabolism are not clearly understood, and they may not be required by all bacteria. Boron is required by plants but not by animals, and probably not by many bacteria as well.

Regardless of whether or not the foregoing elements are essential nutrients for bacteria, all are toxic in high concentration. An exceedingly important contribution of Letunova and Koval'sky is the finding that the toxic level of all these elements to bacteria depends on the level of the element in the natural environment from which the bacteria were isolated. Thus pure cultures of *Bacillus megaterium* isolated from soils with high concentrations of molybdenum, copper, and vanadium showed a much greater tolerance to the toxic effects of high concentrations of compounds of these elements than did *B. megaterium* cultures isolated from soils with low concentrations of these elements. Similar effects were noted with the same and other species of bacteria in relation to cobalt, selenium, uranium, and boron.

The work of Tuovinen and Kelly (1974) sheds light on a possible biochemical mechanism to account for such effects. These researchers isolated a wild-type strain of *Thiobacillus ferrooxidans* which was completely inhibited by uranyl ion at 0.2 to 0.9 mM. By successive subculturing in progressively higher concentrations of uranium the organism was induced to produce mutant strains which grew at 5 mM uranyl ion.

Once a few members of the bacterial population in a given environment have achieved a high degree of resistance to a toxic element, they can then pass on this resistance to other members of their species or even to different species or genera by the usual processes of prokaryotic sexual reproduction, including conjugation, transformation, plasmid transfer, or transduction by bacterial viruses. Thus, thanks to recent developments in molecular biology, we now have a sound theoretical basis for believing that the resistance of bacteria to toxic mineral elements will reflect the concentrations of these elements in their environments.

Timoney *et al.* (1978) have clearly shown that mercury-resistant *Bacillus* species occur in large numbers on the seafloor off New York City where sewage sludge containing large amounts of heavy metals has been dumped. These bacteria also have a high frequency of resistance to ampicillin. Sizemore and Colwell (1977) have described plasmid-mediated heavy-metal and antibiotic resistance in marine *Pseudomonas* and *Vibrio* species. Similarly, Troyer *et al.* (1980) have described greatly increased heavy-metal resistance in soil bacterial populations in highly metalliferous soil.

The application of this knowledge to prospecting is now obvious. It should be possible to prospect for any highly toxic element by preparing culture media with different levels of a soluble compound of this element and testing the growth of isolated bacterial cultures on these media. Large populations of bacteria resistant to high concentrations of the toxic elements are indicative of mineralization.

Alternatively, it would be possible to employ one of the many agar plate diffusion assays now used for antibiotics, such as the paper disk diffusion assay. Such methods have been very widely employed by clinical bacteriologists to determine the antibiotic resistance of clinical isolates causing human disease. They have been perfected and automated to the point where hundreds of tests per day can be carried out rapidly and cheaply by a single laboratory technician. A simple research approach would be to inoculate a melted agar gel containing suitable nutrients with from 10^4 to 10^5 cells of bacteria selected from the prospect area, allow the gel to solidify in a petri dish, and place small paper disks containing appropriate amounts of the test compounds on the surface of the agar gel. The plates would then be incubated for 24 to 48 hours and observed for the presence of clear zones of no bacterial growth surrounding the disks containing toxic elements. The diameter of the clear zone would indicate the degree of sensitivity of the bacteria to the toxic element.

The development of simple, rapid prospecting methods based on these principles appears feasible with a high level of probability. A U.S. Geological Survey project headed by Watterson is already deeply involved in the development and testing of such prospecting methods. This project is also testing the use of antibiotic resistance in bacteria as an exploration parameter, since it has been shown that antibiotic resistances and heavy-metal resistances, which may be carried on common plasmids in several species, can be characters jointly selected for by certain metalliferous environments (Watterson *et al.*, Chapter 20, this volume).

But resistance to toxic elements is not the only potentially useful prospecting parameter. We have already alluded to indicator organisms which are specific to a particular organic substrate, for example *M. paraffinicum* for ethane and other higher gaseous hydrocarbons. In a similar way, several species of chemolithotrophic *Thiobacillus* utilize specific compounds of sulfur as their sole energy source and are an obvious prospecting tool for sulfur deposits. A detailed search for other indicator organisms may turn up microbial prospecting parameters which we cannot yet imagine. The prospects for such exploration methods seem limited only by the imagination of the researchers involved and, of course, by the willingness of exploration managers to try new methods.

REFERENCES

Baross, J. A., and Deming, J. W., 1983, Growth of "black smoker" bacteria at temperatures of at least 250°C: *Nature,* v. 303, p. 423–426.

Blau, L. W., 1942, Process for locating valuable subterranean deposits: U.S. Pat. 2,269,889.

———, 1943, Process for locating valuable subterranean desposits: U.S. Pat. 2,337,443.

Bokova, E. N., Kuznetsova, V. A., and Kuznetsov, S. I., 1947, Oxidation of gaseous hydrocarbons by bacteria as a basis of microbiological prospecting for petroleum: *Dokl. Akad. Nauk. SSSR,* v. 56, p. 755–757.

Brown, L. R., 1962, Hydrocarbon prospecting: U.S. Pat. 3,033,761.

Cannon, H. L., 1964, Geochemistry of rocks and related soil and vegetation in the Yellow Cat area, Grand County, Utah: *U.S. Geol. Surv. Bull. 1176,* p. 127.

Davis, J. B., 1967, *Petroleum Microbiology:* Elsevier, New York, p. 604.

———, and Squires, K. M., 1954, Detection of microbially produced gaseous hydrocarbons other than methane: *Science,* v. 119, p. 381–382.

———, Chasc, II. H., and Raymond, R. L., 1956, *Mycobacterium, paraffinicum* n. sp., a bacterium isolated from soil: *Appl. Microbiol.,* v. 4, p. 310–315.

Kartsev, A. A., Tabasaranskii, Z. A., Subbota, M. I., and Mogilevskii, G. A., 1959, *Geochemical Methods of Prospecting and Exploration for Petroleum and Natural Gases:* University of California Press, Berkeley, p. 341.

Kuznetsov, S. I., 1979, Personal communication at the Institute of Microbiology in Moscow.

Letunova, S. V., and Koval'sky, V. V., 1978, *Geochemical Ecology of Microorganisms:* Academy of Sciences, USSR, Moscow (in Russian).

Mertz, W., 1981, The essential trace elements: *Science,* v. 213, p. 1332–1338.

Miller, C. L., 1983, The development of a geomicrobiological technique for sulfur exploration: M. S. thesis, Colorado School of Mines, Golden, Colo.

Mogilevskii, G. A., 1940, The bacterial method of prospecting for oil and natural gas: *Razved. Nedr.,* v. 12, p. 32–43.

Sanderson, R. T., 1942, Geomicrobiological prospecting: U.S. Pat. 2,294,425.

Sealey, J. Q., 1974, A geomicrobiological method of prospecting for petroleum: *Oil Gas J.,* v. 72, Apr. 8, 1974, p. 142–146.; Apr. 15, 1974, p. 98–102.

Sizemore, R. K., and Colwell, R. R., 1977, Plasmids carried by antibiotic-resistant marine bacteria: *Antimicrob.-Agents Chemother.,* v. 12, p. 373–382.

Söhngen, N. L., 1906, Uber bakterien welche Methan als Kohlenstoffnahurng Energiequelle gebrauchen: *Zentralbl. Bakteriol. Parasitenkd. Abt. II,* v. 15, p. 513–517.

Strawinski, R. J., 1954, Prospecting: U.S. Pat. 2,665,237.

Subbota, M. I., 1947, A complex study of the causes of seasonal variation of data in soil gas surveys: *Neft. Khoz.,* v. 25, p. 13–17.

Taggart, M. S., 1941, Oil prospecting method: U.S. Pat. 2,234,637.

Timoney, J. F., Port, J., Giles, J., and Spanier, J., 1978, Heavy metal and antibiotic resistance in the bacterial flora of sediments of New York Bight: *Appl. Environ. Microbiol.,* v. 36, p. 465–472.

Troyer, L. S., Olson, B. H., Hill, D. C., Thornton, I., and Matthews, H., 1980, Assessment of metal availability in soil through the evaluation of metal resistance: p. 129–141, *in* Hemphill, D. D. (Ed.), *Trace Substances in Environmental Health,* Vol. 14: University of Missouri, Columbia, Mo.

Tuovinen, O. H., and Kelly, D. P., 1974, Studies on the growth of *Thiobacillus ferrooxidans: Arch. Microbiol.,* v. 95, p. 153–164.

Updegraff, D. M., and Chase, H. H., 1958, Microbiological petroleum prospecting method: U.S. Pat. 2,861,921.

Henry L. Ehrlich
Department of Biology
Rensselaer Polytechnic Institute
Troy, New York

14

INTERACTIONS
OF HEAVY METALS
AND MICROORGANISMS

ABSTRACT

Interactions of microorganisms with heavy metals may be growth enhancing, growth suppressing, or the interactions may have no detectable effect on their growth. Growth-enhancing interactions may, on the one hand, involve growth stimulation by very small amounts of metal compound. The metal in that instance may be needed for enzymatic function. Examples of such metals are Co, Cu, Fe, Mo, Mn, Ni, and Zn. On the other hand, growth-enhancing interactions may involve large-scale use of a metal compound as energy sources or terminal electron acceptors in respiration is restricted to prokaryotes pounds or ionic species that can serve as energy sources are Cu_2S, Fe^{2+}, Mn^{2+}, and U^{4+}. Examples of metal compounds or ionic species that can serve as electron acceptors are chromate, Fe^{3+}, ferric oxides, Mn (IV) oxides, and molybdate. Use of heavy metals as energy sources of terminal electron acceptors in respiration is restricted to prokaryotes (archaebacteria and eubacteria).

A special group of taxonomically unrelated bacteria can accumulate iron intracellularly in the form of membrane-bound magnetite (Fe_3O_4) crystals. These crystals act as tiny compasses to the bacteria, helping to guide them to their preferred habitats, namely reducing environments in sediments and such.

Growth-suppressing interactions of heavy metals with microorganisms involve the poisoning of cells by a metal. The poisoning reaction may involve specific enzymes or the cell genetic substance (DNA), or it may be nonspecific. Organisms differ in their susceptibility to specific toxic metals. Some microorganisms are able to convert a toxic metal compound into a less toxic form. Metals that are especially toxic to certain microorganisms at low concentration include Ag, Cd, Cu, Cr, Hg, Mo, Pb, Sn, and W. Some microorganisms may undergo genetic modification and acquire resistance to a toxic metal.

Many prokaryotic microorganisms can bind a variety of metals in ionic (especially cationic) form to cell structures such as peptidoglycan, teichoic and teichuronic acids, or glycocalyx, and, similarly, some eukaryotic microorganisms can bind metal ions to their cell envelope. This binding is relatively nonspecific and involves small quantities of metal per cell. Some interactions of microorganisms with heavy metals can be exploited commercially.

INTRODUCTION

Microorganisms are always exposed to various heavy metals in their natural environment, albeit at very low concentrations in most instances. The definition of heavy metals is somewhat arbitrary. For this discussion they will include all elements in the periodic table with metallic properties that have an atomic number greater than 20. Such metals may exist in the metallic state, as salts, or as oxides. They may be dissolved or in solid amorphous or crystalline forms. Typical concentrations of some heavy metals in fresh water are listed in Table 14-1. Typical concentrations of heavy metals in some rock types are listed in Table 14-2.

TABLE 14-1   Heavy-Metal Concentrations
in Fresh and Marine Waters

Metal	Average Concentrations (µg/liter)	
	Fresh water[a]	Sea water[b]
Ag	0.3	0.3
As	0.5	2.6
Au	0.002	0.01
Cd	0.1	0.1
Co	0.2	0.4
Cr	1	0.5
Cu	3	3
Fe	500	3[c]
Hg	0.1	0.2
Mn	8	2[d,e]
Mo	0.5	10
Ni	0.5	7
Pb	3	0.03
Sn	0.009	0.8
Ti	5	1
U	0.4	3
V	0.5	2
W	0.03	0.1
Zn	15	10

[a]Data based on compilation published by Bowen (1979)

[b]Data based on compilation by National Academy of Sciences (1971).

[c]Gordon et al. (1982) reported 0.15 to 0.3 nmol dissolved Fe per kilogram in surface waters of the northeast Pacific.

[d]Landing and Bruland (1980) found in the range of 0.2 to 6 nmol of dissolved Mn per kilogram in north Pacific waters, including coastal, open ocean surface, and bottom waters.

[e]Klinkhammer and Bender (1980) reported a range of 0.3 to 3 nmol of dissolved Mn per kilogram in Pacific surface waters.

TABLE 14-2   Heavy-Metal Concentrations in Rocks

Metal	Average Concentration (µg/kg)		
	Granite	Shale	Sandstone
Ag	40	70	250
As	1,500	13,000	1,000
Au	1.8	0.5	3
Cd	90	220	50
Co	1,000	19,000	300

TABLE 14-2   Heavy-Metal Concentrations in Rocks
(Continued)

Metal	Average Concentration (μg/kg)		
	Granite	*Shale*	*Sandstone*
Cr	4,000	90,000	35,000
Cu	13,000	39,000	30,000
Fe	27,000	40,000	29,000
Hg	80	180	290
Mn	400,000	850,000	460,000
Mo	2,000	2,600	200
Ni	?	68,000	9,000
Pb	24,000	23,000	10,000
Sn	3,500	6,000	500
Ti	2,300,000	5,000,000	3,500,000
U	4,400	3,700	450
V	72,000	130,000	20,000
W	1,500	1,900	1,600
Zn	52,000	120,000	30,000

Source: Based on data compiled by Bowen (1979).

INTERACTIONS WITH METALS IN AN INSOLUBLE STATE

For microorganisms to interact with heavy metals, it is not always necessary that the metals be dissolved. Some microorganisms appear to be able to interact with metals in insoluble form, be that in the metallic state or as amorphous or crystalline oxides or salts. An example of microbial interaction with a metal in the metallic state is biocorrosion of iron or steel. According to classical corrosion theory (Iverson, 1972, 1974; Von Wohlzogen Kühr and van der Vlugt, 1934), H_2 is generated at cathodic regions at a corrosion site due to the reaction

$$2H_2O + 2e = H_2 + 2OH^-$$ (1)

Fe^{2+} is generated at the anodic region of the corrosion site due to the reaction

$$Fe^0 = Fe^{2+} + 2e$$ (2)

H_2 tends to be adsorbed by the iron or steel at the cathodic region and stop further reaction by polarization. H_2-consuming microbes such as the anaerobic sulfate-reducing bacteria may depolarize the cathodic regions by removing the adsorbed H_2 for sulfate reduction and thereby cause further Fe^{2+} to be released in the anodic regions of the corrosion site.

$$4H_2 + SO_4^{2-} + 2H^+ = H_2S + 4H_2O$$ (3)

Summers, 1978; Wong *et al.*, 1975). The marine cyanobacterium *Synechococcus* sp. has been shown to produce metallothionein when induced by Cd^{2+} and Zn^{2+} but not Cu^{2+}. This is a protein that can bind Cd^{2+}, Zn^{2+}, and Cu^{2+} and thereby render these metals less toxic to the organism (Olafsen *et al.*, 1980). Metallothioneins have a low molecular weight and a very high cysteine content, which accounts for their metal-binding capacity. They are more commonly formed by eukaryotes than prokaryotes.

Microorganisms may also develop tolerance to heavy metals by genetic modification. This modification may involve a mutation on the chromosome, or in prokaryotes it may involve the acquisition of a resistance plasmid. The latter is a more likely mechanism of heavy metal tolerance acquisition in prokaryotes than chromosome mutation. Thus, in *Pseudomonas aeruginosa*, tolerance to As(V) Cd, Hg, and U(VI) is plasmid mediated. Genetically modified resistance may manifest itself in decreased permeability, in structural modification of the reactive site so that it will not react with the metal, or in acquisition of a detoxifying enzyme system. The permeability change may involve a modification or elimination of a carrier protein in the cell envelope which in the unmutated cell is essential for uptake of the metal as an ion (Pan-Hou *et al.*, 1981; Silver *et al.*, 1981; Summers, 1978). The capacity to form mercuric ion reductases as a result of acquiring a particular resistance plasmid has been noted with *Pseudomonas* spp., *E. coli*, and *Staphlyococcus aureus* (Summers, 1978).

Some microorganisms, such as the acidophilic *T. ferrooxidans*, exhibit unusually high tolerance of metals such as Co^{2+}, Cu^{2+}, Ni^{2+}, Zn^{2+} and others (Tuovinen *et al.*, 1971) compared to bacteria that grow at more neutral pH. The resistance to heavy metals of the acidophilic organisms may in part be due to their acidophily and the consequently more electropositive character of their cell envelope as opposed to that of neutrophilic bacteria.

PASSIVE INTERACTION WITH METALS

Microorganisms may react passively with heavy metals by binding them in their cell envelope. Beveridge and Murray (1976, 1980) showed that *Bacillus subtilis* bound measurable amounts of Mn, Fe(III), Ni, Cu, and Au(III) to the peptidoglycan in its cell wall, while Beveridge *et al.* (1982) showed that *B. licheniformis* bound these metals mainly to the teichoic and teichuronic acid moieties of its cell envelope. In *B. subtilis* the amount of metals per milligram (dry weight) of native cell wall ranged from 0.107 μmol for Ni to 3.581 μmol for Fe(III), while in *B. licheniformis* the amounts ranged from 0.031 μmol for Au(III) to 0.760 μmol for Fe(III). The weight of a bacterial cell may be assumed to be in the order of 10^{-12} g, and the cell wall of a gram-positive cell can be assumed to amount to 20 to 30 percent of the dry weight of the cell. Thus the quantities of metal bound by the cell envelopes were very small.

In a more recent study, Beveridge *et al.* (1983) showed that metal-loaded *B. subtilis* cells introduced into a synthetic sediment under laboratory conditions passively nucleated the formation of a mixed assemblage of minerals, including crystalline metal phosphates, metal sulfides, and polymeric metal-complexed residues. The results may furnish a model for similar diagenetic processes in nature.

CONCLUSION

Some of the interactions cited in this paper have been or can be commercially exploited. *T. ferrooxidans* and other acidophilic bacteria are being used or have been proposed for use in leaching Cu, Zn, Ni, and U from sulfide or uraninite ores, respectively. $CrO_4{}^{2-}$ reducing bacteria may be used to remove Cr(VI) from solution by forming Cr^{3+}, which precipitates as $Cr(OH)_3$ at neutral pH. Other applications are very likely to emerge in the future.

REFERENCES

Abeles, R. H., 1971, Studies on mechanisms of action of B_{12} coenzymes, p. 346–364, *in* Gould, R. F., (Ed.), *Advances in Chemistry Series 100:* Am. Chem. Soc., Washington, D.C.

Ali, S. H., and Stokes, J. L., 1971, Stimulation of heterotrophic and autotrophic growth of *Sphaerotilus discophorus* by manganous ion: *Antonie van Leeuwenhoek J. Microbiol. Serol.*, v. 37, p. 519–528.

Arcuri, E. H., and Ehrlich, H. L., 1979, Cytochrome involvement in Mn (II) oxidation by two marine bacteria: *Appl. Environ. Microbiol.*, v. 37, p. 916–923.

Avakyan, Z. A., and Rabotnova, I. L., 1966, Determination of copper concentration toxic to microorganisms: *Mikrobiologiya*, v. 35, p. 805–811.

Babich, H., and Stotzky, G., 1980, Environmental factors that influence the toxicity of heavy metal and gaseous pollutants to microorganisms: *CRC Crit. Rev. Microbiol.*, v. 8, p. 99–145.

——, and Stotzky, G., 1981, Manganese toxicity to fungi: influence of pH: *Bull. Environ. Contam. Toxicol.*, v. 27, p. 474–480.

Balashova, V. V., Vedenina, I. Ya., Markosyan, G. E., and Zavarzin, G. A., 1974, The autotrophic growth of *Leptospirillum ferrooxidans: Mikrobiologiya*, v. 43, p. 581–585 (Engl. transl. p. 491–494).

Baross, H. A., Lilley, M. D., and Gordon, J. L., 1982, Is the CH_4, H_2, and CO venting from submarine hydrothermal system produced by thermophilic bacteria? *Nature (Lond.)*, v. 298, p. 366–368.

Bennett, J. C., and Tributsch, H., 1978, Bacterial leaching patterns on pyrite crystal surfaces: *J. Bacteriol.*, v. 134, p. 310–317.

Beveridge, T. J., and Murray, R. G. E., 1976, Uptake and retention of metals by the cell walls of *Bacillus subtilis: J. Bacteriol.*, v. 127, p. 1502–1518.

——, and Murray, R. G. E., 1980, Sites of metal deposition in the cell wall of *Bacillus subtilis: J. Bacteriol.*, v. 141, p. 876–887.

——, Forsberg, C. W., and Doyle, R. C., 1982, Major sites of metal binding in *Bacillus licheniformis* walls: *J. Bacteriol.*, v. 150, p. 1438–1448.

——, Meloche, J. D., Fyfe, W. S., and Murray, R. G. E., 1983, Diagenesis of metals chemically complexed by bacteria: laboratory formation of metal phosphates, sulfides, and organic condensates in artificial sediments: *Appl. Environ. Microbiol.*, v. 45, p. 1094–1108.

Balkwill, D. L., Maratea, D., and Blakemore, R. P., 1980, Ultrastructure of a magneto-tactic spirillum: *J. Bacteriol.*, v. 141, p. 1399–1408.

Blakemore, R. P., 1975, Magnetotactic bacteria: *Science,* v. 190, p. 377–379.

———, 1982, Magnetotactic bacteria: *Annu, Rev. Microbiol.,* v. 36, p. 217–238.

Bopp, L. H., 1980, Chromate resistance and chromate reduction in bacteria: Ph. D. thesis, Rensselaer Polytechnic Institute, Troy, N.Y.

———, and Ehrlich, H. L., 1981, Bacterial chromate reduction: a process applicable to chromate removal from waste solutions resulting from hydrometallurgical extraction of roasted chromite ore: *Abstr. 28th Congr., Int. Union Pure Appl. Chem.,* BR 127.

Bowen, J. J. M., 1979, *Environmental Chemistry of the Elements:* Academic Press, New York.

Bragg, P. S., and Rainnie, D. J., 1974, The effect of silver ions on the respiratory chain of *Escherichia coli: Can. J. Microbiol.,* v. 20, p. 883–889.

Brierley, J. A., 1978, Thermophilic iron-oxidizing bacteria found in copper leaching dumps: *Appl. Environ. Microbiol.,* v. 36, p. 523–525.

Brierley, C. L., and Brierley, J. A., 1973, A chemoautotrophic and thermophilic microorganism isolated from an acid hot-spring: *Can. J. Microbiol.,* v. 19, p. 183–188.

Brock, T. D., Cook, S., Peterson, S., and Mosser, J. L., 1976, Biochemistry and bacteriology of ferrous iron oxidation in geothermal habitats: *Geochim. Cosmochim. Acta.,* v. 40, p. 439–500.

Bromfield, S. M., 1956, Oxidation of manganese by soil microorganisms: *Aust. J. Biol. Sci.,* v. 9, p. 238–252.

———, 1978, The oxidation of manganous ions under acid condition by an acidophilous actinomycete from acid soil: *Aust. J. Soil Res.,* v. 16, p. 91–100.

———, 1979, Manganous ion oxidation at pH values below 5.0 by cell-free substances from *Streptomyces* sp. cultures: *Soil Biol. Biochem.,* v. 11, p. 115–118.

———, and David, D. J., 1976, Sorption and oxidation of manganous ions and reduction of manganese oxide by cell suspensions of a manganese oxidizing bacterium: *Soil Biol. Biochem.,* v. 8, p 37–43.

Burke, K. A., Calder, K., and Lascelles, J., 1980, Effects of molybdenum and tungsten on induction of nitrate reductase and formate dehydrogenase in wild type and mutant *Paracoccus denitrificans: Arch. Microbiol.,* v. 126, p. 155–519.

Demerec, M., and Hanson, J., 1951, Mutagenic action of manganese chloride: *Cold Spring Harbor Symp. Quant. Biol.,* v. 16, p. 215–228.

Diekert, G., and Thauer, R. K., 1980, The effect of nickel on carbon monoxide dehydrogenase formation in *Clostridium thermoaceticum* and *Clostridium formicoaceticum: FEMS Microbiol. Lett.,* v. 7, p. 187–198.

Diekert, G. B., Graf, E. G., and Thauer, R. K., 1979, Nickel requirements for carbon monoxide dehydrogenase formation in *Clostridium pasteurianum: Arch. Microbiol.,* v. 122, p. 117–120.

———, Weber, B., and Thauer, R. K., 1980, Nickel dependence of factor F_{430} content in *Methanobacterium thermoautotrophicum: Arch. Microbiol.,* v. 127, p. 273–278.

DiSpirito, A. A., and Tuovinen, O. H., 1981, Oxygen uptake coupled with uranous sulfate oxidation by *Thiobacillus ferrooxidans* and *T. acidophilus: Geomicrobiol. J.,* v. 2, p. 275–291.

———, and Tuovinen, O. H., 1982a, Uranous ion oxidation and carbon dioxide fixation by *Thiobacillus ferrooxidans: Arch. Microbiol,,* v. 133, p 33–37.

———, and Tuovinen, O. H., 1982b, Kinetics of uranous ion and ferrous iron oxidation by *Thiobacillus ferrooxidans: Arch. Microbiol.,* v. 133, p 33–37.

Douka, C. E., 1977, Study of bacteria from manganese concretions. Precipitation of manganese by whole cells and cell-free extracts of isolated bacteria: *Soil Biol. Biochem.,* v. 9, p. 89–97.

Drake, H. L., 1981, Occurrence of nickel in carbon monoxide dehydrogenase from *Clostridium pasteurianum* and *Clostridium thermoaceticum: J. Bacteriol.,* v. 149, p. 561–566.

_____ , Hu, S. -I., and Wood, H. G., 1980, Purification of carbon monoxide dehydrogenase, a nickel enzyme from *Clostridium thermoaceticum: J. Biol. Chem.,* v. 255, p. 7174–7180.

Ehrlich, H. L., 1963, Bacteriology of manganese nodules. I. Bacterial action on manganese in nodule enrichments: *Appl. Microbiol.,* v. 11, p. 15–19.

_____ , 1966, Reactions with manganese by bacteria from ferromanganese nodules: *Dev. Ind. Microbiol.,* v. 7, p. 279–286.

_____ , 1976, Manganese as an energy source for bacteria, p. 633–644, *in* Nriagu, J. O. (Ed.), *Environmental Biogeochemistry,* Vol. 2, *Metals Transfer and Ecological Mass Balances:* Ann Arbor Science, Ann Arbor, Mich.

_____ , 1978a, Inorganic energy sources for chemolithotrophic and mixotrophic bacteria: *Geomicrobiol. J.,* v. 1, p. 65–83.

_____ , 1978b, How microbes cope with heavy metals, arsenic and antimony in their environment, p. 381–408, *in* Kushner, D. J. (Ed.), *Microbial Life in Extreme Environments:* Academic Press, New York.

_____ , 1980, Bacterial leaching of manganese ores, p. 609–614, *in* Trudinger, P. A., Walter, M. R., and Ralph, B. J. (Eds.), *Biogeochemistry of Ancient and Modern Environments:* Australian Academy of Science, Canberra/Springer-Verlag, Berlin.

_____ , 1981a, *Geomicrobiology:* Marcel Dekker, New York.

_____ , 1981b, Microbial oxidation and reduction of manganese as aids in its migration in soil, p. 209–213, *in Migrations organo-minérales dans les sols tempérés:* Colloques Internationaux du CNRS, Éditions du CNRS, Paris.

_____ , 1982a, Manganese-oxidizing bacteria from a hydrothermally active area on the Galapagos Rift: *Ecol. Bull.* (Stockholm), v. 35 (in press).

_____ , 1982b, Enhanced removal of manganese (2+) from seawater by marine sediments and clay minerals in the presence of bacteria: *Can. J. Microbiol.,* v. 28, p. 1389–1395.

_____ , 1984, Different forms of bacterial manganese oxidation, p. 47–56, *in* Strohl, W. R., and Tuovinen, O. H. (Eds.), *Microbial Chemoautotrophy:* Ohio State University Press, Colombus, Ohio.

Ferguson, J. F., and Gavis, J., 1972, Review of the arsenic cycle in natural waters: *Water Res.,* v. 6, p. 1259–1274.

Fox, S. I., 1967, Bacterial oxidation of simple copper sulfides: Ph.D. thesis, Rensselaer Polytechnic Institute, Troy, N.Y.

Ghiorse, W. C., and Ehrlich, H. L., 1974, Effects of seawater cations and temperature on MnO_2-reductase activity in a marine *Bacillus: Appl. Microbiol.,* v. 28, p. 785–792.

_____ , and Ehrlich, H. L., 1976, Electron transport components of the MnO_2 reductase system and the location of the terminal reductase in a marine *Bacillus: Appl. Environ. Microbiol.,* v. 31, p. 977–985.

Golovacheva, R. S., and Karavaiko, G. T., 1978, *Sulfobacillus,* a new genus of thermophilic sporeforming bacteria: *Mikrobiologiya,* v. 47, p. 815–822 (Engl. transl., p. 658–665).

Gordon, R. M., Martin, J. H., and Knauer, G. A., 1982, Iron in north-east Pacific waters: *Nature (Lond.),* v. 299, p. 611–612.

Hallas, L. E., Means, J. C., and Cooney, J. J., 1982, Methylation of tin by estuarine microorganisms: *Science*, v. 215, p. 1505–1507.

Hanert, H., 1982, Chemolithotrophic and autotrophic life of *Gallionella ferruginea* in relation to its taxonomy: *Roundtable on Taxonomy of Budding and/or Appendaged Bacteria, 82nd Annu. Meet. Am. Soc. Microbiol.,* unpublished report.

Imai, K., Sakaguchi, H., Sugio, T., and Tano, T., 1973, On the mechanism of chalcocite oxidation by *Thiobacillus ferrooxidans: J. Ferment. Technol.* (Japan), v. 51, p. 865–870.

Ingledew, W. J., Cox, J. C., and Halling, P. J., 1977, A proposed mechanism for energy conservation during Fe^{2+} oxidation by *Thiobacillus ferro-oxidans:* chemiosmotic coupling to net H^+ influx: *FEMS Microbiol. Lett.,* v. 2, p. 193–197.

Iverson, W. P., 1972, Biological corrosion, p. 475–513, *in* Fontana, M. G., and Staehle, W. W. (Eds.), *Aqueous Corrosion Science,* Vol. 2: Plenum Press, New York.

——, 1974, Microbial corrosion of iron, *in* Neilands, J. B. (Ed.), *Microbial Iron Metabolism, A Comprehensive Treatise:* Academic Press, New York.

——, 1975, Anaerobic corrosion: metals and microbes in two worlds: *Dev. Ind. Microbiol.,* v. 16, p. 1–10.

Jensen, S., and Jernelöv, A., 1969, Biological methylation of mercury in aquatic organisms: *Nature (Lond.),* v. 223, p. 753–754.

Jernelöv, A., 1970, Release of methylmercury from sediments with layers containing inorganic mercury at different depths: *Limnol. Oceanogr.,* v. 15, p. 958–960.

Klinkhammer, G. P., and Bender, M., 1980. The distribution of manganese in the Pacific Ocean: *Earth Planet. Sci. Lett.,* v. 46, p. 361–384.

Komura, I., Izaki, K., and Takahashi, H., 1970, Vaporization of inorganic mercury by cell-free extracts of drug resistant *Escherichia coli: Agric. Biol. Chem.,* v. 34, p. 480–482.

Landing, W. M., and Bruland, K. W., 1980, Manganese in the north Pacific: *Earth Planet. Sci. Lett.,* v. 49, p. 45–56.

Lankford, C. E., 1973, Bacterial assimilation of iron: *CRC Crit. Rev. Microbiol.,* v. 2, p. 273–331.

Lazaroff, N., Sigal, W., and Wasserman, A., 1982, Iron oxidation and precipitation of ferric hydroxysulfates by resting *Thiobacillus ferrooxidans* cells: *Appl. Environ. Microbiol.,* v. 43, p. 923–938.

Lebedeva, E. V., and Lyalikova, N. N., 1979, Reduction of chrocoite by *Pseudomonas chromatophila* sp. nov.: *Mikrobiologiya,* v. 48, p. 517–522.

Lehninger, A. L., 1975, *Biochemistry,* 2nd ed.: Worth, New York.

Lewis, A. J., and Miller, J. D. A, 1977, Stannous and cuprous ion oxidation by *Thiobacillus ferrooxidans: Can J. Microbiol.,* v. 23, p. 319–324.

Lodge, J. S., Gaines, C. G., Arcenaux, J. E. L., and Byers, B. R., 1982, Heme inhibition of ferric siderophore reductase in *Bacillus subtilis: J. Bacteriol.,* v. 152, p. 943–945.

Löfroth, G., and Ames, B. N., 1978, Mutagenicity of inorganic compounds in *Salmonella typhimurium:* arsenic, chromium, and selenium: *Mutat. Res.,* v. 53, p. 65–66.

Maciag, W. J., and Lundgren, D. G., 1964, Carbon dioxide fixation in the chemoautotroph *Ferrobacillus ferrooxidans: Biochem. Biophys. Res. Commun.,* v. 17, p. 603–607,

Mann, P. J. G., and Quastel, J. H., 1946, Manganese metabolism in soils: *Nature (Lond.),* v. 158, p. 154–156.

Markosyan, G. E., 1972, A new iron-oxidizing bacterium—*Leptospirillum ferrooxidans* nov. gen. nov. sp.: *Biol. Zh. Armenii*, v. 25, p. 26–29.

Maxwell, W. A., Metzler, R., and Spoerl, E., 1971, Uranyl nitrate inhibition of transport systems in *Saccharomyces cerevisiae: J. Bacteriol.*, v. 105, p. 1205–1206.

Miller, J. D. A., 1980, Principles of microbial corrosion: *Br. Corros. J.*, v 15, p. 92–94.

Munch, J. C., and Ottow, J. C. G., 1982, Effect of cell contact and iron III oxide form on bacterial iron reduction: *Z. Pflanzenernaehr. Bodenkd.*, v. 145, p. 66–77.

Nakahara, H., Ishikawa, T., Sarai, Y., Kondo, I., Kozokue, H., and Silver, S., 1977, Linkage of mercury, cadmium, and arsenate and drug resistance in clinical isolates of *Pseudomonas aeruginosa: Appl. Environ. Microbiol.*, v. 33, p. 975–976.

National Academy of Sciences, 1971, *Marine Chemistry, A Report of the Marine Chemistry Panel of the Committee of Oceanography:* National Academy of Sciences, Washington, D.C.

Neilands, J. B., 1973, Microbial iron transport compounds (siderochromes): p. 167–202, *in* Eichhorn, G. L. (Ed.), *Inorganic Biochemistry*, Vol. 1: Elsevier, Amsterdam.

Nielsen, A. M., and Beck, J. V., 1972, Chalcocite oxidation and coupled carbon dioxide fixation by *Thiobacillus ferrooxidans: Science*, v. 175, p. 1124–1125.

Olafsen, R. W., Loya, S., and Sim, R. G., 1980, Physiological parameters of prokaryotic metallothionein induction: *Biochem. Biophys. Res. Commun.*, v. 96, p. 1495–1503.

Olson, G. J., Iverson, W. P., and Brinkman, F. E., 1981, Volatilization of mercury by *Thiobacillus ferrooxidans: Curr. Microbiol.*, v. 5, p. 115–118.

———, Porter, F. D., Rubinstein, J., and Silver, S., 1982, Mercuric reductase enzyme from a mercury-volatilizing strain of *Thiobacillus ferrooxidans: J. Bacteriol.*, v. 151, p. 1230–1236.

Pan-Hou, H. S., Nishimoto, M., and Imura, N., 1981, Possible role of membrane proteins in mercury resistance of *Enterobacter aerogenes: Arch. Microbiol.*, v. 130, p. 93–95.

Petrilli, F. L., and DeFlora, S., 1977, Toxicity and mutagenicity of hexavalent chromium on *Salmonella typhimurium: Appl. Environ. Microbiol.*, v. 33, p. 805–809.

Ramamoorthy, S., and Kushner, D. J., 1975a, Binding of mercuric and other heavy metal ions by microbial growth media: *Microb. Ecol.*, v. 2, p. 162–176.

———, and Kushner, D. J., 1975b, Heavy metal binding components of river water: *J. Fish. Res. Board Can.*, v. 32, p. 1755–1766.

Romanenko. V. I., and Koren'kov, V. N., 1977, A pure culture of bacteria utilizing chromates and bichromates as hydrogen acceptors in growth under anaerobic conditions: *Mikrobiologiya*, v. 46, p. 414–417 (Engl. transl. p. 329–332).

Rosson, R., and Nealson, K. H., 1982, Manganese binding and oxidation by spores of a marine bacillus: *J. Bacteriol.*, v. 151, p. 1027–1037.

Sadler, W. R., and Trudinger, P. A., 1967, The inhibition of microorganisms by heavy metals: *Mineralium Deposita*, v. 2, p. 158–168.

Schönheit, P., Moll, J., and Thauer, R. K., 1979, Nickel, cobalt, and molybdenum requirement for growth of *Methanobacterium thermoautotrophicum: Arch. Microbiol.*, v. 123, p. 105–107.

Silver, S., Budd, K., Leaky, K. M., Shaw, W. V., Hammond, D., Novick, R. P., Willsky, G. R., Malamy, M. H., and Rosenberg, H., 1981, Inducible plasmid-determined resistance to arsenate, arsenite, and antimony III in *Escherichia coli* and *Staphylococcus aureus: J. Bacteriol.*, v. 146, p. 983–996.

Silverman, M. P., and Lundgren, D. S., 1959, Studies on the chemoautotrophic iron bacterium *Ferrobacillus ferrooxidans*. II. Manometric studies: *J. Bacteriol.*, v. 78, p. 326–331.

Summers, A. O., 1978, Microbial transformations of metals. *Annu. Rev. Microbiol.*, v. 32, p. 637–672.

Tabillion, R., Weber, F., and Kaltwasser, H., 1980, Nickel requirement for chemolithotrophic growth in hydrogen-oxidizing bacteria: *Arch. Microbiol.*, v. 124, p. 131–136.

Temple, K. L., and Colmer, A. R., 1951, The autotrophic oxidation of iron by a new bacterium: *Thiobacillus ferrooxidans: J. Bacteriol.*, v. 62, p. 605–611.

Tributsch, H., 1976, The oxidative disintegration of sulfide crystals by *Thiobacillus ferrooxidans: Naturwissenschaften*, v. 63, p. 88.

Trimble, R. B., 1967, MnO_2 reduction by two strains of marine ferromanganese nodule bacteria: M.S. thesis, Rensselaer Polytechnic Institute, Troy, N.Y.

——, and Ehrlich, H. L., 1968, Bacteriology of manganese nodules. III. Reduction of MnO_2 by two strains of nodule bacteria: *Appl. Microbiol.*, v. 16, p. 695–702.

——, and Ehrlich, H. L., 1970, Bacteriology of manganese nodules. IV. Induction of an MnO_2-reductase system in a marine bacillus: *Appl. Microbiol.*, v. 19, p. 969–972.

Troshanov, E. P., 1968, Iron and manganese-reducing microorganisms in ore-containing lakes of the Karelian isthmus: *Mikrobiologiya*, v. 37, p. 934–940 (English transl. p. 786–791).

——, 1969, Conditions affecting the reduction of iron and manganese by bacteria in the ore-bearing lakes of the Karelian isthmus: *Mikrobiolgiya*, v. 38, p. 634–643 (English transl. p. 528–535).

Tuovinen, O. H., and Kelly, D. P., 1974, Studies on the growth of *Thiobacillus ferrooxidans*. II. Toxicity of uranium to growing cultures and tolerance conferred by mutation, other metal cations and EDTA: *Arch. Microbiol.*, v. 95, p. 153–164.

——, Niemelá, S. I., and Gyllenberg, H. G., 1971, Tolerance of *Thiobacillus ferrooxidans* to some metals: *Antonie van Leeuwenhoek J. Microbiol. Serol.*, v. 37, p. 489–496.

Wong, P. T. S., and Chau, Y. K., and Luxon, P. L., 1975, Methylation of lead in the environment: *Nature (Lond.)*, v. 253, p. 263–264.

Von Wolzogen Kühr, C. A. H., and van der Vlugt, L. S., 1934, Graphitization of cast iron as an electro-biochemical process in anaerobic soils: *Water,* v. 18, p. 147–165.

G. Stotzky and H. Babich
Laboratory of Microbial Ecology
Department of Biology
New York University
New York, New York

15

PHYSICOCHEMICAL ENVIRONMENTAL FACTORS AFFECT THE RESPONSE OF MICROORGANISMS TO HEAVY METALS:

Implications For The Application Of Microbiology To Mineral Exploration*

*The literature search and manuscript for this chapter were completed in 1983.

ABSTRACT

The physicochemical characteristics of an environment associated with mineral deposits determine the chemical speciation forms and, hence, the bioavailability and toxicity of the metals to the colonizing and indigenous microbiota. The abiotic factors involved include pH, E_h, inorganic anions and cations, clay minerals and hydrous metal oxides, particulate and soluble organic matter, temperature, and hydrostatic pressure. These environmental factors influence the speciation and toxicity of each metal differently, and therefore, it may not be possible to predict the microbial response to a specific metal in one environment from data obtained in different environments associated with other types of mineral deposits. These factors, which vary greatly from one environment to another, not only modify the toxicity to and the uptake of heavy metals by microbes, but also the selection pressures of the metals for the enrichment of heavy metal-resistant strains. Consequently, if microbiological methods are to be applied effectively to mineral exploration, the confounding influences of the physicochemical factors of environments associated with mineral deposits must be fully appreciated, both in the collection and interpretation of data. Rigorous preliminary tests should be conducted to determine the heterogeneity of the environment being explored, and caution should be used in extrapolating any microbiological findings from one environment to another that differs significantly in its physicochemical characteristics and in the type of mineral deposit associated with it. Although the use of geomicrobiological methods is attractive because of their relatively low cost, rapidity, and facility, potentially significant mineral deposits could be missed because of differences in the physicochemical characteristics between deposits if only microbiological methods are used. Consequently, microbiological methods for detecting such deposits should be used in conjunction with geochemical and geophysical techniques whenever possible. Furthermore, the composition of the laboratory media used to isolate and enumerate metal-resistant microbes from soils, waters, sediments, and plants affects the amounts of heavy metals that are actually available to interact with the microbes in vitro and, therefore, may result in erroneous conclusions about the numbers of metal-resistant microbes that are present in natural environments.

INTRODUCTION

The toxicity of heavy metals to the biota is dependent on the physicochemical characteristics of the environment in which the metals are found (Babich and Stotzky, 1980a,b, 1983e,f). The toxicity of a metal may be reduced or eliminated by the specific abiotic properties of one environment, whereas in another environment with different physicochemical characteristics, the toxicity of an equivalent dose of the same metal may be potentiated. These concepts may also be important in environments in which mineral deposits are being sought, and if microbiological methods are to be applied effectively to mineral exploration, the confounding influence of such environmental differences must be considered.

Microorganisms, as well as plants and animals, are sensitive to heavy metals. This sensitivity has been demonstrated with a variety of physiological responses at various

G. STOTZKY, H. BABICH

levels of experimental complexity, ranging from pure culture to in situ studies. The concentrations that are toxic to the microbiota vary greatly for different metals, for different groups of microorganisms, and for the physiological processes being studied. At lethal concentrations, microbes are rapidly killed, and at sublethal concentrations, the lag phase of growth may be extended, growth rates may be reduced, growth may be completely inhibited, morphological abnormalities may occur, and sporulation and spore germination of fungi may be reduced more than mycelial extension. At some concentrations, specific biochemical activities, such as photosynthesis, nitrogen fixation, nitrification, and carbon mineralization, are affected. In natural environments, species diversity may be reduced, and dominant populations may be replaced by species that were minor components before the introduction of metals from anthropogenic sources. The accumulation of metals by the microbiota can affect the survival and population dynamics of higher organisms dependent on the microbiota for food (see Babich and Stotzky, 1983e).

The preponderance of studies on the effects of heavy metals on microbes and their activities and of the influence of physicochemical factors on these effects have been conducted in pure culture or in relatively simple model systems, usually with single metals. Although such studies are necessary to clarify both the effects and mechanisms of toxicity of metals and reflect the present state of the art of studying interactions between metals, microbes, and environmental factors, caution must be exercised in extrapolating the results of such studies to natural environments, which may contain multiple metals and wherein numerous environmental factors may interact to affect their toxicity. For example, the in vitro effect of pH on the toxicity of Cd was influenced by the composition of the laboratory media in which the microorganisms were exposed: in one medium, toxicity was increased with increases in pH, whereas in another medium, toxicity was decreased with the same increases in pH. In soils and waters, however, the toxicity of Cd was usually increased with increases in pH (see Babich and Stotzky, 1983e). Consequently, the results of studies in pure culture and in model systems must be verified in natural environments, as the important point is not how pH, or any other abiotic factor, affects the toxicity of metals to microbes in vitro but how microbes respond to the same metals and physicochemical factors in situ.

This chapter summarizes data that show that the abiotic, physicochemical factors of aquatic and terrestrial ecosystems influence the toxicity of heavy metals to microbes. These environmental factors probably also influence the bioavailability and, hence, both the uptake of metals by microbes and the selection pressures exerted by metals for the enrichment of metal-resistant strains of microbes. Even though the content of metals in a surficial environment may be elevated, perhaps reflecting an underlying mineral deposit, the microbiota may respond differently to the metals in that environment than the microbiota in different metalliferous environments. The microbiota and the metals may interact to a lesser or greater extent in some environments because of complexation of the metals with organic matter, adsorption to clay minerals and hydrous metal oxides, conversion to a nontoxic or toxic speciation form, and so on.

It must be emphasized that the interests, experience, and orientation of the authors have been primarily in the ecotoxicology of heavy metals and not in mineral exploration. Hence, this chapter has been written from the former viewpoint. However, it is hoped that the data and concepts presented will have some relevance to the application of microbiology in mineral exploration.

pH

Numerous studies have demonstrated that pH influences the toxicity of heavy metals to the microbiota, with the toxicity either increasing or decreasing as the pH is altered. However, many of the data are contradictory, and mechanisms for the effects of pH have not been clearly defined, as pH influences several aspects of the cell-heavy metal system: (1) pH affects the metabolic state of the cell, and a specific biotic response to a pH–metal interaction may simply reflect the altered physiology of the cell; (2) pH (i.e., the OH⁻ concentration) affects the chemical speciation of some divalent metals (M^{2+}):

$$M^{2+} \xrightarrow{OH^-} M(OH)^+ \xrightarrow{OH^-} M(OH)_2 \xrightarrow{OH^-} M(OH)_3^- \xrightarrow{OH^-} M(OH)_4^{2-}$$

(Hahne and Kroontje, 1973; Zirino and Yamamoto, 1972; Richter and Theis, 1980), and the different hydroxylated forms of the same metal have different toxicities; and (3) pH affects the extent of complexation of metals with organic constituents, and metals complexed with organics are, in general, less toxic than the free forms of the same metals (Babich and Stotzky, 1980b, 1983e,f).

Cd is most mobile in soils of pH 5 and lower (e.g., acid peats), whereas in soils of high pH (e.g., alkaline soils rich in $CaCO_3$), Cd is less mobile, presumably as the result of the formation of insoluble $Cd(OH)_2$ and $CdCO_3$. The different chemical speciation forms of Cd, as influenced by pH, have different toxicities. For example, increasing the pH from acidic to alkaline levels (ranging from pH 5 to 9) enhanced the toxicity of Cd to various eubacteria, actinomycetes, fungi (Babich and Stotzky, 1977a; Korkeala and Pekkanen, 1978), and the green alga, *Chlorella pyrenoidosa*, in which the enhanced toxicity was correlated with an increased uptake of Cd (Gipps and Coller, 1980). The increase in Cd toxicity as a function of increased pH appeared to be related to the formation of $CdOH^+$ (which, as a monovalent cation, presumably penetrates biological membranes more readily than Cd^{2+} and, thus, would be more toxic than the divalent cation) and/or the reduced competition between H^+ and Cd for sites on the cell surface (as the pH is increased, less H^+ is available to compete with Cd for uptake by the cell) (Babich and Stotzky, 1980b). Fungi were also more tolerant of Cd when grown in a naturally acidic (pH 5.1) than in a naturally alkaline (pH 7.8) soil (Babich and Stotzky, 1978b), and when the pH of the acidic soil was adjusted to pH 7.2, the toxicity of Cd increased (Babich and Stotzky, 1977c).

Conversely, Cd toxicity has also been shown to decrease as the pH is increased. Cd was more toxic to growth of *Chlorella pyrenoidosa* at pH 7 than at pH 8, with enhanced toxicity being correlated with the greater uptake of Cd at pH 7 than at pH 8 (Hart and Scaife, 1977). Similarly, the toxicity of Cd to growth of the cyanobacterium, *Nostoc calcicola*, decreased as the pH was increased from 6 to 9 (Singh and Pandey, 1981).

This lack of uniformity in the biotic response to Cd as affected by changes in pH may be related, in part, to the composition of the growth medium. The various organic constituents commonly incorporated into microbiological media have different affinities for Cd (e.g., the order of the binding of Cd to organic substrates followed the sequence, casamino acids > proteose peptone > tryptone ≫ yeast extract, with peptone not binding any Cd) and other heavy metals (Ramamoorthy and Kushner, 1975), and the binding of metals to organics is pH dependent (Farrah and Pickering, 1978). Furthermore, the effect of pH on the toxicity of Cd is influenced by the composition of the test medium. Species of the freshwater fungi, *Saprolegnia* and *Achyla*, were exposed to 25 ppm Cd in media containing either 2.0 percent glucose and 1.0 percent neopeptone (medium 1) or 1.0 percent glucose, 0.25 percent peptone, 0.1 percent NH_4NO_3, 0.5 percent $MgSO_4 \cdot 7H_2O$, 0.02 percent $CaCl_2 \cdot 2H_2O$, and 0.02 percent yeast extract (medium 2). Increasing the pH from 5.5 to 7.5 did not affect the toxicity of Cd to the *Saprolegnia* sp. in either medium, but from pH 7.5 to 9.5, Cd toxicity was reduced in medium 1 but was potentiated in medium 2. Similarly, there were no differences in the toxicity of Cd to the *Achyla* sp. in either medium over the pH range of 5.5 to 8.5, but from pH 8.5 to 9.5, the toxicity of Cd was decreased in medium 1 and increased in medium 2 (Babich and Stotzky, 1983e).

Although hydroxylated species of Pb also form as the pH is increased from acidic to alkaline levels, increasing the pH reduced the toxicity of Pb to the alga, *Selenastrum capricornutum* (Monahan, 1976), and to the fungi, *Aspergillus niger, Trichoderma viride* (Babich and Stotzky, 1979a), an *Achyla* sp., and a *Saprolegnia* sp. (Babich and Stotzky, 1983e). The hydroxylated species of Pb are apparently either less toxic than Pb^{2+} or the different physiological states of the cells in acidic and alkaline environments influence their sensitivity to Pb differently than to Cd.

The toxicity of Zn to fungi increased as the pH was increased from 5.5 to 7.5, but thereafter, further increases in the pH to 9.5 produced no additional potentiation of Zn toxicity (Babich and Stotzky, 1983e). Similarly, increasing the pH from 4 to 8 increased the toxicity of Zn to Zn-resistant and Zn-sensitive populations of the freshwater alga, *Hormidium rivulare* (Hargreaves and Whitton, 1977; Say and Whitton, 1977). Increasing the pH from 6.5 to 8 increased the toxicity of Zn to Zn-tolerant strains of the cyanobacterium, *Anacystis nidulans*, but decreased the toxicity of Zn to Zn-sensitive strains (Shehata and Whitton, 1982). Conversely, increasing the pH from 4 to 8 progressively decreased the toxicity of Zn to growth of *Chlorella vulgaris* (Rai *et al.*, 1981). Increasing the pH from 7.1 to 7.6 decreased the toxicity of Zn to populations of the alga, *Stigeoclonium tenue*, isolated from a stream carrying a low level (i.e., 0.012 mg/liter) of Zn but had no effect on the toxicity of Zn to similar populations isolated from a stream with a high level (i.e., 2.39 mg/liter) of Zn (Harding and Whitton, 1977).

Although increasing the pH from acidic to alkaline levels had no effect on the toxicity of Hg to an *Achyla* sp. and a *Saprolegnia* sp. (Babich and Stotzky, 1983e), the toxicity of Hg to spores and mycelium of *Fusarium lycopersici* (Horsfall, 1956) and to *Chlorella vulgaris* (Rai *et al.*, 1981) was enhanced as the medium was made more alkaline. Conversely, the toxicity of Hg to growth and photosynthesis of a species of *Ankistrodesmus* was greater at pH 5 than at pH 7 (Baker *et al.*, 1983), and that of methyl Hg^+ to growth of *Chlorella vulgaris* was reduced as the pH was increased from 4 to 8 (Rai *et al.*, 1981).

There was no consistent relation between increasing the pH from 5.5 to 8.5 and the toxicity of Mn to fungi. For example, increasing the pH increased the toxicity of Mn to *Rhizopus stolonifer* and *Trichoderma viride*, decreased that to *Scopulariopsis brevicaulis*, and had no effect on that to a *Gliocladium* sp. (Babich and Stotzky, 1981a).

The pH of the environment can also affect metal toxicity even without the accompanying formation of hydroxylated species. For example, Ni occurs as Ni^{2+} over the pH range from 5.5 to 8.5 (Richter and Theis, 1980), yet increasing the pH from 5.5 to 8.5 progressively reduced the toxicity of Ni to a variety of eubacteria, actinomycetes, yeasts, and filamentous fungi (Babich and Stotzky, 1982b, 1983c–f). Furthermore, growth rates of fungi were reduced to a greater extent by 1000 ppm Ni in a naturally acidic soil (pH 4.9) than in the same soil adjusted to pH 7.1 (Babich and Stotzky, 1982d), and survival of *Serratia marcescens* and *Nocardia corallina* in the presence of 75 ppm Ni was greater in a natural lake water of pH 6.8 than in the same water adjusted to pH 5.3 (Babich and Stotzky, 1983c,d). Conversely, the acidophilic fungus, *Penicillium nigricans*, which exhibited maximum growth at pH 2.8, was more tolerant of Ni at pH 2.6 than at pH 3.5 or 5.9 (Singh, 1977). The greater tolerance to Ni at the lower pH may have reflected the enhanced physiologic state of the organism at this unusually low pH level.

The toxicity of Cu to spores of *Fusarium lycopersici* increased as the pH was increased (Horsfall, 1956), and increasing the pH from 5 to 8 increased the toxicity of Cu to growth (Steemann Nielsen and Kamp-Nielsen, 1970) and photosynthesis (Steemann Nielsen *et al.*, 1969) of *Chlorella pyrenoidosa*. Increasing the pH from 3.5 to 4.7 increased the toxicity of Cu to the fungus, *Aureobasidium pullulans*, with increased toxicity paralleling increased uptake of Cu (Gadd and Griffiths, 1980). Acid-tolerant strains of *Scytalidium* (Starkey, 1973) and *Penicillium nigricans* (Singh, 1977) tolerated Cu better in very acidic media (i.e., pH 2.6 and 2.0, respectively) than in media at pH 4.6 and 6.8, respectively.

E_h

The E_h (oxidation-reduction, or redox, potential), which is a measure of the availability of electrons, with negative E_h values being indicative of a reducing environment and positive values of an oxidizing environment, is an important factor in determining the bioavailability of metals in the environment. Reducing conditions, such as those encountered in anaerobic ecosystems, may lead to the microbial conversion of SO_4^{2-} to S^{2-}, with the subsequent precipitation of the sulfide salts of metals (e.g., NiS, HgS, CdS). The formation of insoluble S^{2-}-containing salts greatly reduces the bioavailability of metals, thereby also reducing their potential uptake by, and toxicity to, the microbiota. For example, the toxicity of Hg to fermentation by a mixed rumen microbiota (Forsberg, 1978), of Zn to photosynthesis of *Selenastrum capricornutum* (Hendricks, 1978), and of Ni to growth of *Asteromyces cruciatus* (Babich and Stotzky, 1983d,f) was reduced or eliminated by the addition of sufficient levels of S^{2-} to precipitate the metals.

The E_h of the environment also determines the valency of some metals. For example, in the oxygenated part of the fjord of Saanich Inlet, British Columbia, Cr occurs as Cr(VI), whereas in the anoxic zone, it occurs as Cr(III) (Cranston and Murray, 1978). Differentially charged forms of the same element may exert different toxicities to the microbiota. For example, Cr(VI) was more toxic than Cr(III) to the mycelial proliferation of *Rhizopus stolonifer*, an *Oospora* sp., *Trichoderma viride*, and *Penicillium vermiculatum*; to spore germination and sporulation of *Penicillium vermiculatum* and *Aspergillus giganteus* (Babich *et al.*, 1982b); to growth and survival of *Klebsiella pneumoniae* (Baldry *et al.*, 1977); and to fermentation by a mixed rumen microbiota (Forsberg, 1978). Moreover, Cr(VI) exhibited much greater mutagenicity than did Cr(III), as determined with the *Salmonella typhimurium* reverse mutation assay (Lofroth and Ames, 1978; Tso and Fung, 1981) and the *Bacillus subtilis* rec assay (Nishioka, 1975; Nakamura *et al.*, 1978).

Inorganic Cations

The cationic composition of the environment may reduce the toxicity of heavy metals to the microbiota. This reduction in toxicity is particularly significant for those metals that retain their cationic form, as competition for sites on cell surfaces between cations normally present in an environment and the cationic speciation form of the heavy metals determines the extent of uptake of the metals by the microbiota.

Mg, in particular, influences the toxicity of many heavy metals. Increasing the level of Mg decreased the toxicity of Cu to photosynthesis (Overnell, 1976) and of Zn to growth (Braek *et al.*, 1976) of *Phaeodactylum tricornutum*; of Cd to growth of *Escherichia coli* (Abelson and Aldous, 1950) and *Aspergillus niger* (Laborey and Lavollay, 1973); of Zn to growth of *Hormidium rivulare* (Say and Whitton, 1977), *Anacystis nidulans* (Shehata and Whitton, 1982), *Klebsiella pneumoniae* (Ainsworth *et al.*, 1980), *Escherichia coli* (Abelson and Aldous, 1950), and *Neurospora crassa* (Sastry *et al.*, 1962); and of Ni to growth of *Bacillus licheniformis* (Haavik, 1976), *Bacillus megaterium, Bacillus subtilis* (Webb, 1970a,b), *Escherichia coli* (Abelson and Aldous, 1950), *Klebsiella pneumoniae* (Webb, 1970a; Ainsworth *et al.*, 1980), an unidentified gram-negative bacterium isolated from a deep-sea sediment (Yang and Ehrlich, 1976), *Torulopsis utilis* (Abelson and Aldous, 1950), and numerous filamentous fungi (Babich and Stotzky, 1981c, 1982e, 1983a,f). Furthermore, the ameliorating effects of seawater or salinity (as a solution of sea salts) on Ni toxicity to marine fungi was related to the Mg rather than to the Na or Cl ions in the marine systems (Babich and Stotzky, 1983c,d,f). Similarly, the acute toxicity of Ni to *Caulobacter maris* was greater in simulated estuarine than in seawater (Babich and Stotzky, 1983c), and the chronic toxicity of Ni to coliphage T1 was less in seawater than in simulated estuarine or lake waters (Babich *et al.*, 1983a).

The mediating influence of Mg on the toxicity of Ni was probably a result of competition between these cations for sites on the cell surface, which reduced the uptake of Ni. This competition is probably related to the similarity in the size and ionic potential (charge/ionic radius) of Mg and Ni: the nonhydrated ionic radii of Mg^{2+} and Ni^{2+} are 0.066 and 0.069 nm, respectively (Webb, 1970a), and based on single-ion hydration energies (enthalpies), the hydrated ionic radii (for which there are no direct data) of Mg^{2+} (−452.0 kcal/mol) and Ni^{2+} (−495.9 kcal/mol) are also similar (Phillips and Williams, 1965), as are their ionic potentials (3.0303 for Mg^{2+} and 2.8989 for Ni^{2+}) (Babich and Stotzky, 1983a,f).

Ca also affects heavy metal toxicity toward microbes. Increasing the level of Ca decreased the toxicity of Cd to *Aspergillus niger* (Laborey and Lavollay, 1977), of Zn to *Hormidium rivulare* (Say and Whitton, 1977) and *Anacystis nidulans* (Shehata and Whitton, 1982), of Cu to *Phaeodactylum tricornutum* (Overnell, 1976), and of Zn and Hg to *Chlorella vulgaris* (Rai *et al.*, 1981). Zn, although a heavy metal pollutant when occurring in elevated levels in the environment, is also a micronutrient and has been shown to affect the toxicity of other metals. Increasing the level of Zn reduced the toxicity of Ni to an *Achlya* sp. (Babich and Stotzky, 1982e), of Cu to *Phaeodactylum tricornutum*

(Braek *et al.*, 1976), and of Cd to freshwater species of *Chlorella* (Upitis *et al.*, 1973; Gipps and Biro, 1978), to *Aspergillus niger* (Laborey and Lavollay, 1973), and to the freshwater protozoan, *Tetrahymena pyriformis* (Dunlop and Chapman, 1981).

Inorganic Anions

The type and amount of inorganic anions in an environment influence the speciation of heavy metals and, hence, their toxicities to the microbiota. Most heavy-metal cations have a coordination number from 1 to 4 and form coordination complexes not only with OH^- (see the section on pH), but also with other inorganic ligands, such as Cl^-, the dominant inorganic anion in seawater, where it occurs at an average concentration of 20,000 ppm. The various coordination complexes have different stabilities; for example, in seawater, Cd occurs as a mixture of $CdCl^+/CdCl_2/CdCl_3^-$ and Hg occurs as a mixture of $HgCl_3^-/HgCl_4^{2-}$, whereas in freshwater systems, depending on the pH, these metals occur as Cd^{2+}, $CdOH^+$, Hg^{2+}, $HgOH^+$, $Hg(OH)_2$, and so on (Hahne and Kroontje, 1973). These different speciation forms of the same metal exert different toxicities to the microbiota. For example, species of the marine bacteria, *Aeromonas* and *Acinetobacter*, the terrestrial bacteria, *Erwinia herbicola* and *Agrobacterium tumefaciens*, and the bacteriophages, $\phi11M15$ of *Staphylococcus aureus* and P1 of *Escherichia coli*, tolerated Hg better as mixtures of $HgCl_3^-/HgCl_4^{2-}$ than as mixtures of $Hg^{2+}/HgOH^+/Hg(OH)_2$. These differences in tolerance to Hg were evident also in natural ecosystems, as the *Aeromonas* sp., *Agrobacterium tumefaciens*, and the $\phi11M15$ phage tolerated Hg better in seawater than in lake water (Babich and Stotzky, 1979b).

Salinity also reduced the toxicity of Cd to the microbiota. Increasing the salinity from 5 to 15 ⁰/₀₀ reduced the uptake by and toxicity of Cd to the estuarine alga, *Chlorella salina* (Wong *et al.*, 1979), and increasing the salinity from 13.5 to 45 ⁰/₀₀ reduced the uptake by and toxicity of Cd to an unidentified marine bacterium (Gauthier and Flatau, 1980). Although seawater is a harsh environment for many terrestrial fungi, increasing the concentration of seawater in a synthetic medium decreased the toxicity of Cd to a variety of fungi, and the ameliorating effect of seawater on the toxicity was correlated with its chlorinity component, as progressively increasing the level of Cl^- (even to concentrations that themselves were inhibitory) decreased the relative toxicity of Cd (Babich and Stotzky, 1982c). The lower toxicity of negatively charged Cd–Cl complexes than of Cd^{2+} was also noted in the better growth of a mixed microbiota from activated sludge exposed to $Cd(CN)_4^{2-}$ than to equivalent concentrations of Cd as Cd^{2+} (Cenci and Morozzi, 1977). Conversely, $ZnCl_3^-/ZnCl_4^{2-}$ mixtures were highly toxic to numerous coliphages, whereas an equivalent concentration of Zn as Zn^{2+} was only slightly toxic or nontoxic. The negatively charged Zn–Cl species may have interacted with positively charged sites on the tails of the phages that are involved in recognizing complementary negatively charged sites on the host cells (Babich and Stotzky, 1978a).

Metals whose speciation is not affected by the levels of chlorinity that occur in seawater do not exhibit differences in their toxicities as a function of increasing concentrations of Cl^-. For example, Ni occurs as Ni^{2+} in both seawater and lake water (Richter and Theis, 1980), and chlorinity, at a concentration occurring in seawater, did not affect the toxicity of Ni to the marine fungi, *Dendryphiella salina, Asteromyces cruciatus*, and *Dreschlera halodes* (Babich and Stotzky, 1983a).

Other inorganic anions may also interact with heavy metals and affect their toxicity to the microbiota. As already noted (see the section on E_h), S^{2-} forms insoluble salts with metals, thereby reducing their bioavailability. Similarly, CO_3^{2-} influences metal toxicity (see the section on water hardness); for example, CO_3^{2-} decreased the toxicity of Pb and Ni to growth of fungi, presumably as the result of the formation of insoluble $PbCO_3$ and $NiCO_3$ (Babich and Stotzky, 1979a, 1981c, 1983e,f). PO_4^{3-} also reduces the toxicity of metals, again probably by the precipitation of the metals as PO_4^{3-}-containing salts. The addition of PO_4^{3-} decreased the toxicity of Pb and Ni to fungi (Babich and Stotzky, 1979a, 1983d,f), of Pb to *Chlamydomonas reinhardtii* (Schulze and Brand, 1978), and of Zn to *Chlorella vulgaris* (Rana and Kumar, 1974; Rai *et al.*, 1981), *Hormidium rivulare* (Say and Whitton, 1977), *Plectonema boryanum* (Rana and Kumar, 1974), and *Anacystis nidulans* (Shehata and Whitton, 1982).

Water Hardness

Although there has been considerable research on the effects of water hardness on the response of the macrobiota to heavy metals, few studies have evaluated the response of the microbiota. Water hardness is caused primarily by dissolved alkaline-earth ions, which in fresh waters are mainly Ca^{2+} and Mg^{2+}. Hard waters are usually alkaline and contain substantial amounts of HCO_3^- and CO_3^{2-} rather than free CO_2, which occurs primarily in soft and/or acidified waters. The major components of water hardness (i.e., Ca^{2+}, Mg^{2+}, and CO_3^{2-}) reduced the toxicity of heavy metals to the microbiota (see the sections on inorganic cations and anions). Hardness is commonly reported as an equivalent concentration of $CaCO_3$, with waters containing from 0 to 75 mg/liter $CaCO_3$ being classified as "soft," from 75 to 150 mg/liter as being "moderately hard," from 150 to 300 mg/liter as being "hard," and greater than 300 mg/liter as being "very hard" (EPA, 1976).

The toxicities of Pb (Carter and Cameron, 1973), Cd, and Zn (Chapman and Dunlop, 1981) to *Tetrahymena pyriformis*, of Cd, Pb, and Ni to a variety of fungi (Babich and Stotzky, 1981b, 1982d, 1983c-f), and of Zn and Co to *Chlorella pyrenoidosa* (Wong, 1980) were reduced in hard, as compared to soft, water. Uptake of Cd by *Nitella flexilis* was lower in hard than in soft water (Kinkade and Erdman, 1975). However, the toxicity of Hg to *Tetrahymena pyriformis* was twice as great in hard than in soft water (Carter and Cameron, 1973), and hardness did not affect the toxicity of Mn to fungi (Babich and Stotzky, 1981b).

Survival of the yeast, *Rhodotorula rubra*, was greater after 35 days in lake water containing 10 ppm Ni and amended with 200 or 400 mg/liter $CaCO_3$ than in the same lake water with a natural background of only 34 mg/liter $CaCO_3$ (Babich and Stotzky, 1983c), and the toxicity of Ni to growth of various filamentous fungi was also reduced when the nutrient-enriched lake water was amended with 400 mg/liter $CaCO_3$. The ameliorating effect of $CaCO_3$ on the toxicity of Ni to fungi was correlated with the CO_3^{2-} rather than with the Ca^{2+} component of hardness (Babich and Stotzky, 1981c, 1983f). Although in these studies hardness was achieved with amendments of $CaCO_3$, when hardness was subsequently provided as a mixture of Ca and Mg salts (as has been done in other studies; e.g., Kinkade and Erdman, 1975; Chapman and Dunlop, 1981), Mg was also a major determinant in reducing the toxicity of Ni (Babich and Stotzky, 1983f) (see the section on inorganic cations).

PHYSICOCHEMICAL ENVIRONMENTAL FACTORS

Clay Minerals

At the pH of most natural ecosystems, clay minerals possess surfaces that are predominantly negatively charged and to which charge-compensating cations (e.g., H^+, K^+, Na^+, NH_4^+, Ca^{2+}, Mg^{2+}) are adsorbed. These cations are not permanently bound to the clays and are constantly being exchanged by other cations. The total amount of cations that can be exchanged by clays is expressed in milliequivalents (meq) per 100 g of oven-dried clay and is termed the cation exchange capacity (CEC). The CEC of some of the more common clay minerals is 3 to 5 meq/100 g for kaolinite, 5 to 30 for attapulgite (palygorskite), 10 to 30 for illite, 100 to 150 for vermiculite, and 80 to 150 for montmorillonite (Baver *et al.*, 1972).

Clay minerals in soil and aquatic ecosystems sorb heavy metals and remove them, at least temporarily, from solution, thereby reducing their bioavailability and, hence, toxicity to the microbiota. The sorption of heavy metals to clay minerals is influenced by various abiotic factors, including pH [e.g., increasing the pH from 3.5 to 6.5 increased the amount of Cu, Pb, and Cd adsorbed to illite and kaolinite but had no effect on their adsorption to montmorillonite (Farrah and Pickering, 1977)]; concentration of competing cations [e.g., increasing the concentration of Ca or Mg reduced the adsorption of Cu to kaolinite (Gupta and Harrison, 1981)]; and the concentration and type of ligands present [e.g., increasing the concentration of Cl reduced the ability of Cd to adsorb to montmorillonite, presumably as a result of the conversion of Cd^{2+} to negatively charged Cd–Cl complexes, which have a lower affinity for the net negatively charged clay particles (Garcia-Miragaya and Page, 1976; Egozy, 1980)]. Consequently, the nature of the specific environment will determine the ability of heavy metals to adsorb to clay particles.

Montmorillonite and, to a lesser extent, kaolinite protected a variety of bacteria and fungi against concentrations of Cd that were inhibitory or lethal to growth in a synthetic medium. The protective ability of the clays increased as their concentration increased, and the greater protection afforded by montmorillonite than by kaolinite was correlated with the higher CEC of montmorillonite (Babich and Stotzky, 1977b, 1978b). When an acidic soil was amended with montmorillonite, *Aspergillus niger, Pencillium brefeldianum*, and *Trichoderma viride* were protected against the toxicity of Cd. Amendments with kaolinite provided only limited protection (Babich and Stotzky, 1977c). Clays probably also protect against Cd toxicity in fresh waters, but the high levels of competing cations (e.g., Na^+, Mg^{2+}) and of Cl^- may limit a similar protective effect in marine ecosystems.

Montmorillonite, attapulgite, and kaolinite protected various fungi against inhibitory or lethal levels of Pb. The sequence of this protective ability (*i.e.*, montmorillonite > attapulgite > kaolinite) followed the order of the magnitude of and was correlated with the CEC of the clays (Babich and Stotzky, 1979a). In contrast to Cd, the level of Cl^- present in seawater does not appreciably affect the speciation of Pb, as pH exerts the dominating influence and Pb occurs primarily as $PbOH^+$ (Hahne and Kroontje, 1973). Consequently, Pb has a cationic charge both in lake and seawater, and clay minerals in these two diverse ecosystems may exert a protective effect against Pb toxicity.

Various fungi were also protected against Ni in an acidic soil amended with montmorillonite; amendments with kaolinite provided less protection (Babich and Stotzky, 1982d, 1983f). Survival of bacteria in lake water contaminated with Ni was enhanced when the water was amended with montmorillonite (Babich and Stotzky, 1983c). Similarly, incorporation of freshwater sediment, composed primarily of clays, into a synthetic medium reduced the toxicity of Hg to photosynthesis of a phytoplankton community consisting primarily of diatoms (Hongve *et al.*, 1980).

Hydrous Metal Oxides

Amorphous hydrous oxides of iron, aluminum, and manganese are also capable, but to a lesser extent than are crystalline clays, of exchanging heavy metals (Murray *et al.*, 1968; Hildebrand and Blum, 1974; Kinniburgh *et al.*, 1976; Swallow *et al.*, 1980; Bowman *et al.*, 1981) and, thereby, of decreasing their potential uptake by the microbiota. Adsorption of heavy metals to these amorphous metal oxides is also dependent on several abiotic factors, such as pH [e.g., increasing the pH from 4 to 8 increased the adsorption of Pb to hydrous ferric oxides (Gadde and Laitinen, 1974)] and ligands [e.g., citrate, acting as a chelator, reduced the adsorption of Cd to hydrous aluminum and iron oxides (Chubin and Street, 1981)].

The mediating influence of hydrous metal oxides on heavy metal toxicity to terrestrial and aquatic microbes has received little attention but should be of particular interest in environments associated with deposits of iron ores. The protection against Cu toxicity toward photosynthesis of *Chlorella pyrenoidosa* by the addition of $FeCl_3$ to an alkaline medium was attributed to the adsorption of Cu to negatively charged $Fe(OH)_3$ colloids that were generated under the alkaline conditions (Steemann Nielsen and Kamp-Nielsen, 1970; Steemann Nielsen and Wium-Andersen, 1970). Increasing the concentration of hydrous aluminum oxides from 0.25 percent to 2.0 percent (w/v) or of hydrous manganese oxides from 0.1 percent to 2.0 percent progressively reduced the toxicity of Ni to *Dendryphiella salina* (Babich and Stotzky, 1983d).

Organics

The organic matter content, both in soluble and particulate forms, of an ecosystem greatly influences the mobility, bioavailability, and toxicity of heavy metals to the microbiota. Interactions between organic matter and heavy metals are also dependent on various abiotic factors of the environment, and factors that reduce or eliminate the cationic valency of the metals reduce their ability to complex with organic matter. For example, the synthetic chelators, ethylenediaminetetraacetic acid (EDTA) and nitrilotriacetic acid (NTA), reduced the uptake of Cu, Pb, Zn, and Cd by humic acids (Riffaldi and Levi-Menzi, 1975; Beveridge and Pickering, 1980); EDTA and the natural chelators, oxalate, citrate, and cysteine, reduced the adsorption of Cu, Pb, Zn, and Cd to cellulose (Farrah and Pickering, 1978); and EDTA reduced the adsorption of Cd to river mud (Gardiner, 1974b). pH also affects interactions between heavy metals and organic matter; for example, increasing the pH increased the uptake of Cu, Pb, Zn, and Cd by humic acids (Riffaldi and Levi-Menzi, 1975; Beveridge and Pickering, 1980) and cellulose (Farrah and Pickering, 1978) and of Cd by river sediment (Reid and McDuffie, 1981).

The humic acids present in soil, freshwater and marine sediments, and in the aqueous phases of aquatic environments are capable of complexing variable amounts of different heavy metals (Rashid, 1971; Stevenson *et al.*, 1973; Mishra and Kar, 1974; Gardiner, 1974a,b; Stevenson, 1976, 1977; Wilson, 1978; Beveridge and Pickering, 1980), and such complexation reduces the toxicities of the metals to the microbiota. For example, soluble humus obtained from marsh water decreased the toxicity of Cd to *Selenastrum capricornutum* (Gjessing, 1981) and of Zn, Pb, Cu, and Hg to a freshwater phytoplankton population (Hongve *et al.*, 1980). Incorporation of particulate humic acids into a synthetic medium protected a spectrum of fungi against inhibitory or lethal levels of Pb (Babich and Stotzky, 1979a, 1980a) and Ni (Babich and Stotzky, 1982e, 1983f). The toxicity of Hg to an anaerobic bacterium (a *Bacteroides* sp.) isolated from a freshwater sediment was reduced in the presence of sediment, apparently as a result of its organic matter content, as ashed sediment (i.e., sediment without the organic fraction) provided no protection against Hg (Hamdy and Wheeler, 1978).

Cu was less toxic to growth of *Thalassiosira pseudonana* in "aged" seawater (i.e., seawater that contained natural detritus and phytoplankton communities which were allowed to decompose and release their cell contents before filtration and autoclaving) than in "fresh" seawater (i.e., seawater from which the natural phytoplankton and other particulates were removed immediately after collection) (Erickson, 1972). Organic macromolecules in lake water that were not removed by ultrafiltration reduced the toxicity of Cu to phytoplankton communities (Gachter *et al.*, 1978), and treatment of seawater with ultraviolet radiation to photooxidize the soluble organics increased the toxicity of Cu to diatoms (Fisher and Frood, 1980; Sunda and Guillard, 1976), to green algae (Sunda and Guillard, 1976), and to glucose mineralization by a heterotrophic microbiota (Gillespie and Vaccaro, 1978).

Organic exudates from aquatic microbes appear to be involved in reducing the toxicity of heavy metals in natural waters. For example, various freshwater cyanobacteria (Fogg and Westlake, 1955; McKnight and Morel, 1979; Van den Berg *et al.*, 1979), freshwater algae (McKnight and Morel, 1979; Van den Berg *et al.*, 1979; Swallow *et al.*, 1978), and marine algae (McKnight and Morel, 1979; Gnassaia-Barelli *et al.*, 1978) released organic materials that complexed with and detoxified Cu. The toxicity of 0.5 to 2 ppm Zn to the diatom, *Phaeodactylum tricornutum*, was reduced by the addition of 100 to 2000 ppb polymeric polyphenols released from the marine brown algae, *Ascophyllum nodosum* and *Fucus vesiculosus* (Ragan *et al.*, 1980). Chlorophyll reduced the toxicity of Ni and Cd to terrestrial and aquatic fungi, but it has not been determined whether this effect was related to the chelating ability of chlorophyll or to the release of Mg^{2+} from chlorophyll, which antagonizes the toxicity of Ni and Cd (Babich and Stotzky, 1982e, 1983f).

Natural (e.g., dicarboxylic and amino acids) and synthetic (e.g., EDTA, NTA) soluble organics, both types acting as chelators, reduce the toxicity of heavy metals, as the chelated forms of metals are less toxic than their free, noncomplexed forms. Chelation has been suggested to be the most important abiotic factor in the reduction of Cu toxicity in aquatic ecosystems (Hodson *et al.*, 1979). The amino acids, arginine, glutamine, aspartic, and cysteine, reduced the toxicity of Cd to *Nostoc calcicola* (Singh and Pandey, 1981); aspartic acid reduced the toxicity of Ni to bacteria and fungi (Ainsworth *et al.*, 1980; Babich and Stotzky, 1983d); histidine reduced the toxicity of Cu to *Thalassiosira pseudonana* (Davey *et al.*, 1973) and *Chaetoceros socialis* (Jackson and Morgan, 1978); and cysteine reduced the toxicity of Pb to several fungi (Babich and Stotzky, 1979a) and of

Hg to marine and terrestrial bacteria and to phage ϕ11M15 of *Staphylococcus aureus* (Babich and Stotzky, 1980b). Citrate reduced the toxicity of Ni (Ainsworth *et al.*, 1980), Cd, and Zn (Pickett and Dean, 1976) to *Klebsiella pneumoniae*, of Ni to *Nocardia rhodocrous* (Babich and Stotzky, 1983d), and of Cu to *Chlorella pyrenoidosa* (Steemann Nielsen and Kamp-Nielsen, 1970); oxydisuccinate reduced the toxicity of Zn to *Microcystis aeruginosa* (Allen *et al.*, 1980); succinate reduced the toxicity of Pb to various fungi (Babich and Stotzky, 1979a); and 2,6-pyridine dicarboxylic acid (dipicolinic acid) reduced the toxicity of Ni to a yeast and a bacterium (Babich and Stotzky, 1983d).

EDTA reduced the toxicity of Cu to freshwater (Stokes *et al.*, 1973) and marine algae (Canterford and Canterford, 1980; Davey *et al.*, 1973; Schulz-Baldes and Lewin, 1976; Anderson and Morel, 1978), to glucose mineralization by a heterotrophic marine microbiota (Gillespie and Vaccaro, 1978), to cyanobacteria (Horne and Goldman, 1974; Singh and Pandey, 1981; Fogg and Westlake, 1955), and to eubacteria (Loveless and Pointer, 1968); of Zn and Cd to cyanobacteria (Allen *et al.*, 1980), eubacteria (Pickett and Dean, 1976), and marine algae (Canterford and Canterford, 1980); of Pb to marine algae (Canterford and Canterford, 1980); of Ni to an actinomycete (Babich and Stotzky, 1983d); and reduced the mutagenicity of Cr(VI) to *Bacillus subtilis* (re-assay) (Gentile *et al.*, 1981).

The rapid death of *Escherichia coli* in filter-sterilized seawater was eliminated by the addition of cysteine, EDTA, 8-hydroxyquinoline, thioglycolic acid, or *o*-phenanthroline, presumably as the result of the chelation of heavy metals (Scarpino and Pramer, 1962; Jones, 1964; Jones and Cobet, 1975). NTA reduced the toxicity of Zn to *Microcystis aeruginosa* (Allen *et al.*, 1980), of Pb to *Chlamydomonas reinhardtii* (Schulze and Brand, 1978), of Ni to an actinomycete and several fungi (Babich and Stotzky, 1983d), of Cu to an estuarine phytoplankton community (Erickson *et al.*, 1970), and of Cu, Cd, Zn, and Pb to a freshwater phytoplankton community (Hongve *et al.*, 1980).

More complex soluble organics also bind and detoxify metals, and these organics vary in their abilities to bind metals: for Hg, the sequence of binding was casamino acids $\gg$ proteose peptone $>$ yeast extract $\gg$ tryptone $>$ peptone; for Pb, it was casamino acids $\gg$ yeast extract $>$ tryptone $>$ peptone $>$ proteose peptone; and for Cu, it was casamino acids $\gg$ yeast extract $>$ tryptone $>$ proteose peptone $>$ peptone (see the section on pH for the sequence for Cd) (Ramamoorthy and Kushner, 1975). Increasing concentrations of peptone progressively reduced the toxicity of Cd to an unidentified marine bacterium (Gauthier and Flatau, 1980); yeast extract and increasing levels of neopeptone progressively reduced the toxicity of Pb to fungi (Babich and Stotzky, 1979a); tryptone reduced the toxicity of Hg to anaerobic bacteria (Hamdy and Wheeler, 1978); peptone reduced the toxicity of Ni to a gram-negative marine bacterium isolated from a ferromanganese nodule (Yang and Ehrlich, 1976); and proteose peptone reduced the toxicity of methyl Hg^+ to a protozoan (Hartig, 1971).

The toxicity of Ni to several eubacteria, actinomycetes, and yeasts was significantly reduced or eliminated in media containing yeast extract, neopeptone, casamino acids, or tryptone, whereas peptone or proteose peptone, at the same concentrations, had little or no effect on toxicity. The composition of commercial media also affected the toxicity of Ni to microbes, with the least complex media, which contained peptone as the major organic nitrogenous substrate, being least effective in reducing the toxicity. Furthermore, the toxicity of Ni to fungi was reduced on a medium solidified with Gelrite [a hetero-polysaccharide of microbial origin which requires a small amount of cation, Mg being recommended, to gel (Kelco Co.)] rather than with Bacto-agar (Difco) (Babich and Stotzky, 1983d). The toxicity of Ni to microbial populations from estuarine sediments also was less on a nutrient medium solidified with silica gel than with gelatin, and the toxicity on medium solidified with gelatin was less than with Bacto-agar or purified agar (Hallas *et al.*, 1982).

Studies to establish the concentrations of heavy metals that are inhibitory to the growth or activity of microbes, either for the purpose of enumerating heavy metal-resistant microbes in natural habitats or for laboratory studies, must be cognizant of the influence on toxicity of the components used as substrates and for solidification of the isolation or test media (Babich and Stotzky, 1974, 1978a–c, 1980a,b, 1982a,e, 1983b,d–f). This is especially important in the potential use of microbes in mineral exploration. The ability of microbes to grow on laboratory media amended with high levels of a metal does not necessarily mean that these microbes are resistant to the metal, as the microbes may not interact with the metal because of its complexation with the components of the media. Inasmuch as many microbes in natural habitats are fastidious heterotrophs, the tendency is to use complex media containing numerous growth factors to isolate as many microbes as possible from such habitats. This results in an inherent paradox, as the more complex the medium, the more it interferes with the action of the added metal on the microbes. Conversely, the less complex the medium, the greater the effect of the added metal on the microbes, but the number of microbes isolated is greatly reduced and the probability of not detecting metal-resistant strains is increased. This may be particularly important for strains whose metal resistance is a result of plasmids, as these strains probably require more energy and precursors to synthesize the additional DNA. One possible resolution of this paradox is to isolate as many microbes as possible on highly complex media not amended with metals and then to replica plate from these isolation plates to plates containing minimal media and various concentrations of the appropriate metal (Babich and Stotzky, 1982a).

Temperature

Temperature, similar to hydrostatic pressure, is an environmental factor that apparently does not influence the chemical speciation or the availability of heavy metals to microbes but, instead, affects microbial sensitivity to the metals. The data available on metal–temperature interactions are contradictory, with microbial sensitivity either increasing or decreasing as the temperature is altered. The main difficulty in interpreting these data is that the optimal temperature for the specific physiological process studied is seldom provided. The growth rates of a *Scenedesmus* sp. and a *Chlorella* sp. decreased less in the presence of Cu when exposed at $23°C$ (i.e., the optimal temperature for growth) than at 18, 29, or $35°C$ (Klotz, 1981). The toxicity of Zn^{2+} to growth of *Aspergillus niger* was unaffected by increasing the temperature from $25°C$ to $37°C$, but a $ZnCl^+/ZnCl_2/ZnCl_3^-$ mixture was more toxic at $25°C$ than at $37°C$. The greater tolerance of the fungus at $37°C$ may have reflected its enhanced physiological state, as in the absence of Zn, mycelial growth and production of conidia were greater at 37°C than at 25°C (Babich and Stotzky, 1978a). Similarly, *Aspergillus flavus* was more resistant to Ni at 33°C than at 23°C, which was related to the enhanced metabolic state of the fungus at the higher temperature (Babich and Stotzky, 1982e).

Conversely, the toxicity of Cu to *Paramecium tetraurelia* (Szeto and Nyberg, 1979), of Hg to *Scenedesmus acutis* (Huisman *et al.*, 1980), of Zn to *Nitzschia linearis* and *Nitzschia seminulum* (Cairns *et al.*, 1978), and of Ni to a marine species of *Pseudomonas* (Babich and Stotzky, 1983c) increased as the temperature was increased, and *Euglena gracilis* was more tolerant of Cr(VI) when exposed at 20°C than at the sublethal temperature of 31.5°C (Yongue *et al.*, 1979). However, the toxicity of Zn to *Chilomonas paramecium* decreased as the temperature was increased, and there were no definitive relationships between temperature and the toxicity of Cu and Cr(VI) (Honig *et al.*, 1980). Five ppm Ni stimulated the growth of a gram-negative marine bacterium isolated from a ferromanganese nodule at 18°C, whereas at 5°C, a similar concentration of Ni was neither stimulatory nor inhibitory. In contrast, growth of a gram-negative bacterium isolated from a deep-sea sediment was reduced by 1 and 5 ppm Ni at 5°C but not at 18°C (Yang and Ehrlich, 1976). Uptake of Ni by the lichen, *Umbilicaria muhlenbergii*, increased as the temperature was increased from 12°C to 22°C (Nieboer *et al.*, 1976), whereas the uptake by *Cladina rangiferina* was independent of temperature (Burton *et al.*, 1981).

Hydrostatic Pressure

There has apparently been only one study (Arcuri and Ehrlich, 1977) of the influence of hydrostatic pressure on the toxicity of heavy metals to marine microorganisms. The toxicity of heavy metals as a function of increasing hydrostatic pressure was studied with three gram-negative deep-sea bacteria: strain BIII 39, isolated from a ferromanganese nodule and characterized as a Mn^{2+} oxidizer; strain BIII 32, also isolated from a ferromanganese nodule but characterized as a Mn^{4+} reducer; and strain BIII 88, isolated from an ocean sediment and characterized as a Mn^{4+} reducer.

Increasing the hydrostatic pressure from 1 to 272, 340, and 408 atm progressively increased the toxicity of 1 ppm Ni to cell yields of strain BIII 39. Similarly, Ni had a greater toxicity toward BIII 32 and BIII 88 at 340 than at 1 atm.

Progressively increasing the hydrostatic pressure from 1 to 408 atm did not affect the toxicity of 0.1 to 10 ppm Mn to strain BIII 39, and no difference in the toxicity of Mn was noted for strain BIII 88 when exposed at 1 or 340 atm. However, 0.1 to 10 ppm Mn stimulated the growth of strain BIII 32 at 340 atm, but not at 1 atm.

Strain BIII 39 exhibited a varied response to Cu as the hydrostatic pressure was increased: at 1 atm, 0.1 to 1 ppm Cu caused a progressively increasing reduction in cell yields; at 272 atm, the toxicity was eliminated; at 340 atm, the toxicity was again apparent; and at 408 atm, the toxicity was again eliminated. With strain BIII 88, 1, but not 0.1, ppm Cu had a slight toxic effect at 1 atm, but at 340 atm, 0.1 ppm Cu had a slight stimulatory effect and 1 ppm Cu had no effect. Increasing the hydrostatic pressure from 1 to 340 atm eliminated the toxic effect of 0.1 ppm Cu on cell yields of strain BIII 32.

The microbial response to multiple heavy metals may differ from the response to stress from individual metals, as indicated by studies that have evaluated the antagonistic, synergistic, or additive interactions between multiple metals. Antagonistic interactions refer to the protective effect of one metal on the toxicity of a second metal, which is probably a reflection of competition between the two metals for sites on the cell surface, with a reduction in the subsequent uptake and accumulation of both metals. Synergistic interactions refer to the enhanced toxicity of one metal in the presence of another metal and may reflect the increased permeability of the cytoplasmic membrane when stressed by several toxicants. Additive interactions are neither antagonistic nor synergistic, and the final toxicity is simply a sum of the individual toxicities.

A specific combination of heavy metals may produce a variety of interactions. For example, the combinations of Cu + Zn was synergistic towards growth of the marine algae, *Amphidinum carteri*, *Skeletonema costatum*, and *Thalassiosira pseudonana*, but was antagonistic toward growth of *Phaeodactylum tricornutum* (Braek *et al.*, 1976). Fe(III) antagonized the toxicity of Cd, Zn, and Cu to growth of *Klebsiella pneumoniae* but acted synergistically with Co and Ni (Chapman and Dean, 1982). A specific interaction between heavy metals is also dependent on the relative concentrations of the metals and the sequence of exposure to the metals. For example, the effects of a combination of Cd + Pb on photosynthesis of a brackish-water phytoplankton community were antagonistic when the concentration of Pb was greater than that of Cd, whereas they were synergistic when the concentration of Cd was greater than that of Pb (Pietilainen, 1975). The effect of a combination of Hg + Ni on the growth rate of the cyanobacterium, *Anabaena inequalis*, was synergistic when both Hg and Ni were added simultaneously or when Hg was added first, but it was antagonistic when Ni was added before Hg (Stratton and Corke, 1979). The apparent importance of the relative concentration of the metals and the sequence of exposure to them may explain the variety of responses reported for specific combinations of heavy metals (see Babich and Stotzky, 1983e,f).

BIOTIC FACTORS

The abiotic characteristics of the environment are not the only factors that influence the toxicity of heavy metals, as the characteristics of the organisms can also influence their sensitivity to metals. Cell size may influence microbial sensitivity; for example, the concentration of Pb that caused a 50 percent reduction in photosynthesis was 15 to 18 ppm for *Chlamydomonas reinhardtii*, *Navicula pelliculosa*, and an *Anabaena* sp. but only 5 ppm for the desmid, *Cosmarium botrytis*, which has a higher surface/volume ratio than the other algae and which may have accounted for its enhanced sensitivity to Pb (Malanchuk and Gruendling, 1973). Capsulated strains of *Klebsiella pneumoniae* were more resistant to Cu and Cd than were noncapsulated strains, presumably because the polysaccharide capsule bound the metals and prevented their intracellular accumulation (Bitton and Freihofer, 1978). Extracellular polypeptide chelators produced by *Anabaena cylindrica* (Fogg and Westlake, 1955) and *Cricosphaera elongata* (Gnassaia-Barelli *et al.*, 1978) protected the cells against the toxicity of Cu.

The nutritional status of the organism can also determine its response to stress by heavy metals. For example, Mg^{2+}-, NH_4^+-, PO_4^{3-}-, and glucose-limited cells of *Klebsiella pneumoniae* were highly sensitive to Cd, and Mg^{2+}-, K^+-, and glucose-limited cells were highly sensitive to Zn, compared to cells grown in a medium containing an excess of these nutrients (Pickett and Dean, 1976). Although Mg^{2+}-limited cells of *Klebsiella pneumoniae* were more sensitive to Ni, K^+-limited cells were more resistant to Ni than were comparable cells harvested from a nutrient-sufficient medium (Ainsworth *et al.*, 1980). The uptake of Ni was reduced in PO_4^{3-}-starved cells of *Phaeodactylum tricornutum* (Skaar *et al.*, 1974) and *Saccharomyces cerevisiae* (Fuhrmann and Rothstein, 1968).

Microbial strains resistant to elevated levels of heavy metals can be isolated from contaminated environments. For example, species of *Chlorella* and *Scenedesmus* isolated from lakes in the Sudbury smelting area in Canada that were contaminated with Ni had a greater tolerance to Ni than laboratory strains of these algae (Stokes *et al.*, 1973), and gram-negative bacteria isolated from a river polluted with Cr were resistant to elevated concentrations of Cr(VI) (Simon-Pujol *et al.*, 1979). The increased tolerance of microbes to metals probably represents, in most instances, physiologic rather than genetic adaptation. For example, estuarine algae exposed in many passages to Cu or Cd (*i.e.*, "trained") at concentrations that partially inhibited growth of the original inoculants tolerated elevated levels of Cu and Cd (Wikfors and Ukeles, 1982), and *Klebsiella pneumoniae* (Ainsworth *et al.*, 1980) and *Saccharomyces ellipsoideus* (Nakamura, 1962) could be trained to tolerate elevated levels of Ni. *Klebsiella pneumoniae* trained to tolerate elevated levels of Cd produced H_2S, which formed granules of CdS on the surface of the cells (Aiking *et al.*, 1982).

However, genetic mechanisms for resistance to heavy metals can also be involved, as some bacteria contain plasmids for resistance to various metals. For example, some strains of *Escherichia coli* carry plasmids for resistance to Hg, Ni, and Co (Smith, 1967); some *Pseudomonas aeruginosa* strains have plasmids conferring resistance to Hg, organomercurials (Clark *et al.*, 1977), and Cr(VI) (Summers and Jacoby, 1978); and some strains of *Staphylococcus aureus* carry plasmids conferring resistance to Cd, Hg, Zn, Pb, and As (Novick and Roth, 1968; see Summers and Silver, 1978).

CONCLUSION

The effects of heavy metals from anthropogenic sources on the microbiota, in both aquatic and terrestrial environments, are influenced, to a large extent, by the physicochemical characteristics of the specific environment. These abiotic environmental factors, such as pH, E_h, inorganic cations and anions, organic materials, clay minerals, and hydrous metal oxides, influence the speciation of heavy metals and, hence, their bioavailability and, ultimately, their toxicity to the microbiota. In addition, other abiotic factors, such as temperature and hydrostatic pressure, as well as interactions between multiple heavy metals, influence the response of the microbiota to stress by heavy metals. The response of the microbiota to heavy metals in areas being examined for their mineral potential may also be influenced by these abiotic factors. Consequently, the mediating effects of these abiotic factors must be considered if microbes are to be used effectively in mineral exploration. Inasmuch as some of these factors may interfere with the effects on microbes of elevated concentrations of a metal in a surficial environment, potential mineral deposits may be missed if reliance is placed solely on microbiological sensitivity or resistance to that metal. Nevertheless, microbiological methods may prove to be a valuable adjunct to multidisciplinary biogeochemical and geophysical techniques for the exploration for mineral deposits. However, the value of geomicrobiological techniques in mineral exploration will not be known until they are extensively used in different environments and, ultimately, until they are successful in aiding in the discovery of economically important mineral deposits.

Although the emphasis of this chapter has been primarily on the effects of abiotic environmental factors on the toxicity to microbes of heavy metals derived primarily from anthropogenic sources (i.e., pollution), the data presented may have relevance to the potential use of microbiology in mineral exploration. Sensitivity and resistance to metals are "opposite sides of the same coin," and both are affected similarly by the physicochemical characteristics of an environment. However, caution must be exercised in extrapolating from microbial responses to relatively recent heavy metal pollution of surficial environments to responses in environments derived from metalliferous parent materials in which metals have been an intimate component for millenia and in which the microbes have either adapted to elevated concentrations of the respective metals or were selected initially for life in such environments. Little appears to be known about these and other alternatives, and this is obviously an area that demands further research— for both academic and practical reasons.

ACKNOWLEDGMENTS

Some of the research reported herein was supported, in part, by Grant R808329 from the U.S. Environmental Protection Agency. The conclusions represent the views of the authors and not necessarily those of the agency.

REFERENCES

Abelson, P. H., and Aldous, E., 1950, Ion antagonisms in microorganisms: interference of normal magnesium metabolism by nickel, cobalt, cadmium, zinc, and manganese: *J. Bacteriol.*, v. 60, p. 401–413.

Aiking, H., Kok, K., van Heerikhuizen, H., and van't Reit, J., 1982, Adaptation to cadmium by *Klebsiella aerogenes* growing in continuous culture proceeds mainly via formation of cadmium sulfide: *Appl. Environ. Microbiol.*, v. 44, p. 938–944.

Ainsworth, M. A., Tompsett, C. P., and Dean, A. C. R., 1980, Cobalt and nickel sensitivity and tolerance in *Klebsiella pneumoniae: Microbios*, v. 27, p. 175–184.

Allen, H. E., Hall, R. H., and Brisbin, T. D., 1980, Metal speciation. Effects on aquatic toxicity: *Environ. Sci. Technol.*, v. 14, p. 441–443.

Anderson, D. M., and Morel, F. M. M., 1978, Copper sensitivity of *Gonyaulax tamarensis: Limnol. Oceanogr.*, v. 23, p. 283–295.

Arcuri, E. J., and Ehrlich, H. L., 1977, Influence of hydrostatic pressure on the effects of the heavy metal cations of manganese, copper, cobalt, and nickel on the growth of three deep-sea bacterial isolates: *Appl. Environ. Microbiol.*, v. 33, p. 282–288.

Babich, H., and Stotzky, G., 1974, Air pollution and microbial ecology: *CRC Crit. Rev. Environ. Control*, v. 4, p. 353–421.

_____ , and Stotzky, G., 1977a, Sensitivity of various bacteria, including actinomycetes, and fungi to cadmium and the influence of pH on sensitivity: *Appl. Environ. Microbiol.*, v. 33, p. 681–695.

_____ , and Stotzky, G., 1977b, Reductions in the toxicity of cadmium to microorganisms by clay minerals: *Appl. Environ. Microbiol.*, v. 33, p. 696–705.

_____ , and Stotzky, G., 1977c, Effect of cadmium on fungi and on interactions between fungi and bacteria in soil: influence of clay minerals and pH: *Appl. Environ. Microbiol.*, v. 33, p. 1059–1066.

_____ , and Stotzky, G., 1978a, Toxicity of zinc to fungi, bacteria, and coliphages: influence of chloride ions: *Appl. Environ. Microbiol.*, v. 36, p. 904–913.

_____ , and Stotzky, G., 1978b, Effects of cadmium on the biota: influence of environmental factors: *Adv. Appl. Microbiol.*, v. 23, p. 55–177.

_____ , and Stotzky, G., 1979a, Abiotic factors affecting the toxicity of lead to fungi: *Appl. Environ. Microbiol.*, v. 38, p. 506–514.

_____ , and Stotzky, G., 1979b, Differential toxicities of mercury to bacteria and bacteriophages in sea and in lake water: *Can. J. Microbiol.*, v. 25, p. 1252–1257.

_____ , and Stotzky, G., 1980a Physicochemical factors that affect the toxicity of heavy metals to microbes in aquatic habitats, p. 181–203 *in* Colwell, R. R., and Foster, J. (Eds.), *Proceedings of the ASM Conference, Aquatic Microbial Ecology*, College Park, Md, University of Maryland Sea Grant Publication.

_____ , and Stotzky, G., 1980b, Environmental factors that influence the toxicity of heavy metals and gaseous pollutants to microorganisms: *CRC Crit. Rev. Microbiol.*, v. 8, p. 99–145.

_____ , and Stotzky, G., 1981a, Manganese toxicity to fungi: influence of pH: *Bull. Environ. Contam. Toxicol.*, v. 27, p. 474–480.

_____ , and Stotzky, G., 1981b, Influence of water hardness on the toxicity of heavy metals to fungi: *Microbios Lett.*, v. 16, p. 79–84.

_____ , and Stotzky, G., 1981c, Components of water hardness which reduce the toxicity of nickel to fungi: *Microbios Lett.*, v. 18, p. 17–24.

_____ , and Stotzky, G., 1982a, Gaseous and heavy metal pollutants, p. 631–670 *in* Burns, R. G., and Slater, J. H. (Eds.), *Experimental Microbial Ecology*, Blackwell Scientific, Oxford.

_____ , and Stotzky, G., 1982b, Nickel toxicity to microbes: effect of pH and implications for acid rain: *Environ. Res.*, v. 29, p. 335–350.

_____ , and Stotzky, G., 1982c, Influence of chloride ions on the toxicity of cadmium to fungi: *Zentralbl. Bacteriol. Hyg., I. Abt. Orig., C*, v. 3, p. 421–426.

_____ , and Stotzky, G., 1982d, Toxicity of nickel to microorganisms in soil: influence of some physicochemical characteristics: *Environ. Pollut.*, v. 29A, p. 303–315.

_____ , and Stotzky, G., 1982e, Nickel toxicity to fungi: influence of environmental factors: *Ecotoxicol. Environ. Safety*, v. 6, p. 577–589.

_____ , and Stotzky, G., 1983a, Nickel toxicity to estuarine/marine fungi and its amelioration by magnesium in sea water: *Water Air Soil Pollut.*, v. 19, p. 193–202.

_____ , and Stotzky, G., 1983b, Developing standards for environmental toxicants: the need to consider abiotic environmental factors and microbe-mediated ecologic processes: *Environ. Health Perspect.*, v. 49, p. 247–260.

——, and Stotzky, G., 1983c, Temperature, pH, salinity, hardness, and particulates mediate nickel toxicity to eubacteria, an actinomycete, and yeasts in lake, simulated estuarine, and sea waters: *Aquat. Toxicol.*, v. 3, p. 195–208.

——, and Stotzky, G., 1983d, Further studies on environmental factors that modify the toxicity of nickel to microbes: *Reg. Toxicol. Pharmacol.*, v. 3, p. 82–99.

——, and Stotzky, G., 1983e, Influence of chemical speciation on the toxicity of heavy metals to the microbiota, *in* Nriagu, J. O. (Ed.), *Aquatic Toxicology*, Wiley, New York, p. 1–46.

——, and Stotzky, G., 1983f, Toxicity of nickel to microbes: environmental aspects: *Adv. Appl. Microbiol.*, v. 29, p. 1–46.

——, Davis, D. L., and Trauberman, J., 1981, Environmental quality criteria: some considerations: *Environ. Manage.*, v. 5, p. 191–205.

——, Gamba-Vitalo, C., and Stotzky, G., 1982a, Comparative toxicity of nickel to mycelial proliferation and spore formation of fungi: *Arch. Environ. Contam. Toxicol.*, v. 11, p. 465–468.

——, Schiffenbauer, M., and Stotzky, G., 1982b, Comparative toxicity of trivalent and hexavalent chromium to fungi: *Bull. Environ. Contam. Toxicol.*, v. 28, p. 452–459.

——, Schiffenbauer, M., and Stotzky, G., 1983a, Sensitivity of coliphage T1 to nickel in fresh and salt waters: *Cur. Microbiol.*, v. 8, p. 101–105.

——, Bewley, R. J. F., and Stotzky, G., 1983b, Application of the "ecological dose" concept to the impact of heavy metals on some microbe-mediated ecologic processes in soil: *Arch. Environ. Contam. Toxicol.*, v. 12, p. 421–426.

Baker, M. D., Mayfield, C. I., Inniss, W. E., and Wong, P. T. S., 1983, Toxicity of pH, heavy metals, and bisulfite to a freshwater green alga: *Chemosphere*, v. 12, p. 35–44.

Baldry, M. G. C., Hogarth, D. S., and Dean, A. C. R., 1977, Chromium and copper sensitivity and tolerance in *Klebsiella* (*Aerobacter*) *aerogenes*: *Microbios Lett.*, v. 4, p. 7–16.

Baver, L. D., Gardner, W. H., and Gardner, W. R., 1972, *Soil Physics*: Wiley, New York.

Beveridge, A., and Pickering, W. F., 1980, Influence of humate-solute interactions on aqueous heavy metal ion levels: *Water Air Soil Pollut.*, v. 14, p. 171–185.

Bewley, R. J. F., and Stotzky, G., 1983a, Simulated acid rain (H_2SO_4) and microbial activity in soil: *Soil Biol. Biochem.*, v. 15, p. 425–429.

——, and Stotzky, G., 1983b, Anionic constituents of acid rain and microbial activity in soil: *Soil Biol. Biochem.*, v. 15, p. 431–437.

——, and Stotzky, G., 1983c, Effects of cadmium and zinc on microbal activity in soil; influence of clay minerals. Part I. Metals added individually: *Sci. Total Environ.*, v. 31, p. 41–55.

——, and Stotzky, G., 1983d, Effects of cadmium and zinc on microbial activity in soil; influence of clay minerals. Part II. Metals added simultaneously: *Sci. Total Environ.*, v. 31, p. 58–69.

——, and Stotzky, G., 1983e, Effects of combinations of simulated acid rain and cadmium or zinc on microbial activity in soil: *Environ. Res.*, v. 31, p. 332–339.

——, and Stotzky, G., 1983f, Effects of cadmium and simulated acid rain on ammonification and nitrification in soil: *Arch. Environ. Contam. Toxicol.*, v. 12, p. 285–291.

Bitton, G., and Freihofer, V., 1978, Influence of extracellular polysaccharides on the toxicity of copper and cadmium towards *Klebsiella aerogenes*: *Microb. Ecol.*, v. 4, p. 119–125.

Bowman, R. S., Essington, M. E., and O'Connor, G. A., 1981, Soil sorption of nickel: influence of solution composition: *Soil Sci. Soc. Am. J.*, v. 45, p. 860–865.

Braek, G. S., Jensen, A., and Mohus, A., 1976, Heavy metal tolerance of marine phytoplankton. III. Combined effects of copper and zinc ions on cultures of four common species: *J. Exp. Biol.*, v. 25, p. 37–50.

Burton, M. A. S., LeSueur, P., and Puckett, K. J., 1981, Copper, nickel and thallium uptake by the lichen *Cladina rangiferina: Can. J. Bot.*, v. 59, p. 91–100.

Cairns, J., Jr., Buikema, A. L., Jr., Heath, A. G., and Parker, B. C., 1978, Effects of temperature on aquatic organism sensitivity to selected chemicals: *Virginia Water Resources Center Bull. 106*, Virginia Polytechnic Institute, Blacksburg, VA.

Canterford, G. S., and Canterford, D. R., 1980, Toxicity of heavy metals to the marine diatom *Ditylum brightwellii* (West) Grunow: correlation between toxicity and metal speciation: *J. Mar. Biol. Assoc. U.K.*, v. 60, p. 227–242.

Carter, J. W., and Cameron, I. L., 1973, Toxicity bioassay of heavy metals in water using *Tetrahymena pyriformis: Water Res.*, v. 7, p. 951–961.

Cenci, G., and Morozzi, G., 1977, Evaluation of the toxic effect of Cd^{2+} and $Cd(CN)_4^{2-}$ ions on the growth of mixed microbial population of activated sludges: *Sci. Total Environ.*, v. 7, p. 131–143.

Chapman, R., and Dean, A. C. R., 1982, Action of iron and iron-complexes on *Klebsiella pneumoniae (Klebsiella aerogenes): Folia Microbiol.*, v. 27, p. 295–302.

Chapman, G., and Dunlop, S., 1981, Detoxication of zinc and cadmium by the freshwater protozoan *Tetrahymena pyriformis*. I. The effect of water hardness: *Environ. Res.*, v. 26, p. 81–86.

Chubin, R. G., and Street, J. J., 1981, Adsorption of cadmium on soil constituents in the presence of complexing ligands: *J. Environ. Qual.*, v. 10, p. 225–228.

Clark, D. L., Weiss, A. A., and Silver, S., 1977, Mercury and organomercurial resistances determined by plasmids in *Pseudomonas: J. Bacteriol.*, v. 132, p. 186–196.

Cranston, R. E., and Murray, J. W., 1978, The determination of chromium species in natural waters: *Anal. Chim. Acta*, v. 99, p. 275–282.

Davey, E. W., Morgan, M. J., and Erickson, S. J., 1973, A biological measurement of the copper complexation capacity of seawater: *Limnol. Oceanogr.*, v. 18, p. 993–997.

Dunlop, S., and Chapman, G., 1981, Detoxication of zinc and cadmium by the freshwater protozoan *Tetrahymena pyriformis*. II. Growth experiments and ultrastructural studies on sequestration of heavy metals: *Environ. Res.*, v. 24, p. 264–274.

Egozy, Y., 1980, Adsorption of cadmium and cobalt on montmorillonite as a function of solution composition: *Clays Clay Min.*, v. 28, p. 311–318.

Environmental Protection Agency, 1976, *Quality Criteria for Water:* EPA, Washington, D.C.

Erickson, S. J., 1972, Toxicity of copper to *Thalassiosira pseudonana* in unenriched inshore water: *J. Phycol.*, v. 8, p. 318–323.

———, Lackie, N., and Maloney, T. E., 1970, A screening technique for estimating copper toxicity to estuarine phytoplankton: *J. Water Pollut. Control Fed.*, v. 42, p. 270–278.

Farrah, H., and Pickering, W. F., 1977, Influence of clay-solute interaction on aqueous heavy metal ion levels: *Water Air Soil Pollut.*, v. 8, p. 189–197.

———, and Pickering, W. F., 1978, The effect of pH and ligands on the sorption of heavy metal ions by cellulose: *Aust. J. Chem.*, v. 31, p. 1501–1509.

Fisher, N. S., and Frood, D., 1980, Heavy metals and marine diatoms: influence of dissolved organic compounds on toxicity and selection for metal tolerance among four species: *Mar. Biol.*, v. 59, p. 85–93.

Fogg, G. E., and Westlake, D. F., 1955, The importance of extracellular products of algae in freshwater: *Verh. Int. Verein. Limnol.*, v. 12, p. 219–232.

Forsberg, C. W., 1978, Effects of heavy metals and other trace elements on the fermentative activity of the rumen microbiota and growth of functionally important rumen bacteria: *Can. J. Microbiol.*, v. 24, p. 298–306.

Fuhrmann, G. -F., and Rothstein, A., 1968, The transport of Zn^{2+}, Co^{2+}, and Ni^{2+} into yeast cells: *Biochim. Biophys. Acta*, v. 163, p. 325–330.

Gachter, R., Davis, J. S., and Mares, A., 1978, Regulation of copper availability to phytoplankton by macromolecules in lake water: *Environ. Sci. Technol.*, v. 12, p. 1416–1421.

Gadd, G. M., and Griffiths, A. J., 1980, Influence of pH on toxicity and uptake of copper in *Aureobasidium pullulans: Trans. Br. Mycol. Soc.*, v. 75, p. 91–96.

Gadde, R. R., and Laitinen, H. A., 1974, Studies of heavy metal adsorption by hydrous iron and manganese oxides: *Anal. Chem.*, v. 46, p. 2022–2026.

Garcia-Miragaya, J., and Page, A. L., 1976, Influence of ionic strength and inorganic complex formation on the sorption of trace amounts of Cd by montmorillonite: *Soil Sci. Soc. Am. J.*, v. 40, p. 658–663.

Gardiner, J., 1974a, The chemistry of cadmium in natural water. I. A study of cadmium complex formation using the cadmium specific ion electrode: *Water Res.*, v. 8, p. 23–30.

——, 1974b, The chemistry of cadmium in natural water. II. The adsorption of cadmium on river muds and naturally occurring solids: *Water Res.*, v. 8, p. 157–164.

Gauthier, M. J., and Flatau, G. N., 1980, Étude de l'accumulation du cadmium par une bactérie marine en fonction des conditions de cultures: *Chemosphere*, v. 9, p. 713–718.

Gentile, J. M., Hyde, K., and Schubert, J., 1981, Chromium genotoxicity as influenced by complexation and rate effects: *Toxicol. Lett.*, v. 7, p. 439–448.

Gillespie, P. A., and Vaccaro, R. F., 1978, A bacterial bioassay for measuring the copper-chelation capacity of seawater: *Limnol. Oceanogr.*, v. 23, p. 543–548.

Gipps, J. F., and Biro, P., 1978, The use of *Chlorella vulgaris* in a simple demonstration of heavy metal toxicity: *J. Biol. Educ.*, v. 12, p. 207–214.

——, and Coller, B. A. M., 1980, Effect of physical and culture conditions on uptake of cadmium by *Chlorella pyrenoidosa: Aust. J. Mar. Freshwater Res.*, v. 31, p. 747–755.

Gjessing, E. T., 1981, The effect of aquatic humus on the biological availability of cadmium: *Arch. Hydrobiol.*, v. 91, p. 144–149.

Gnassaia-Barelli, W., Romeo, M., Laumond, F., and D. Pesando, D., 1978, Experimental studies on the relationship between natural copper complexes and their toxicity to phytoplankton: *Mar. Biol.*, v. 47, p. 15–19.

Gupta, G. C., and Harrison, F. L., 1981, Effect of cations on copper adsorption by kaolin: *Water Air Soil Pollut.*, v. 15, p. 323–327.

Haavik, H. I., 1976, On the role of bacitracin peptides in trace metal transport by *Bacillus licheniformis: J. Gen. Microbiol.*, v. 96, p. 393–399.

Hahne, H. C. H., and Kroontje, W., 1973, Significance of pH and chloride concentration on behavior of heavy metal pollutants: mercury (II), cadmium (II), zinc (II), and lead (II): *J. Environ. Qual.*, v. 2, p. 444–450.

Hallas, L. E., Thayer, J. S., and Cooney, J. J., 1982, Factors affecting the toxic effect of tin on estuarine microorganisms: *Appl. Environ. Microbiol.*, v. 44, p. 193–197.

Hamdy, M. K., and Wheeler, S. R., 1978, Inhibition of bacterial growth by mercury and the effects of protective agents: *Bull. Environ. Contam. Toxicol.*, v. 20, p. 378–386.

Harding, J. P. C., and Whitton, B. A., 1977, Environmental factors reducing the toxicity of zinc to *Stigeoclonium tenue: Br. Phycol. J.*, v. 12, p. 17–21.

Hargreaves, J. W., and Whitton, B. A., 1977, Effect of pH on tolerance of *Hormidium rivulare* to zinc and copper: *Oecologia*, v. 26, p. 235–245.

Hart, B. A., and Scaife, B. D., 1977, Toxicity and bioaccumulation of cadmium in *Chlorella pyrenoidosa: Environ. Res.*, v. 14, p. 401–413.

Hartig, W. J., 1971, Studies on mercury toxicity in *Tetrahymena pyriformis: J. Protozool.*, v. 18, p. 26 (Abstr.).

Hendricks, A. C., 1978, Response of *Selenastrum capricornutum* to zinc sulfides: *J. Water Pollut. Control Fed.*, v. 50, p. 163–168.

Hildebrand, E. E., and Blum, W. E., 1974, Lead fixation by iron oxides: *Naturwissenschaften*, v. 61, p. 169–170.

Hodson, P. V., Borgmann, U., and Shear, H., 1979, Toxicity of copper to aquatic biota, p. 307–372 *in* Nriagu, J. O. (Ed.), *Copper in the Environment, Part II: Health Effects:* Wiley, New York.

Hongve, D., Skogheim, O. K., Hindar, A., and Abrahamsen, A., 1980, Effects of heavy metals in combination with NTA, humic acid, and suspended sediment on natural phytoplankton photosynthesis: *Bull. Environ. Contam. Toxicol.*, v. 25, p. 594–600.

Honig, R. A., McGinniss, M. J., Buikema, A. L., and Cairns, J., Jr., 1980, Toxicity tests of aquatic pollutants using *Chilomonas paramecium* Ehrenberg (Flagellata) populations: *Bull. Environ. Contam. Toxicol.*, v. 25, p. 169–175.

Horne, A. J., and Goldman, C. R., 1974, Supression of nitrogen fixation by blue-green algae in a eutrophic lake with trace additions of copper: *Science*, v. 183, p. 409–411.

Horsfall, J. G., 1956, *Principles of Fungicidal Action:* Chronica Botanica Co., Waltham, Mass.

Huismann, J., Ten Hoopen, H. J. G., and Fuchs, A., 1980, The effect of temperature upon the toxicity of mercuric chloride in *Scenedesmus acutus: Environ. Pollut.*, v. 22A, p. 133–148.

Jackson, G. A., and Morgan, J. L., 1978, Trace metal-chelator interactions and phytoplankton growth in seawater media: theoretical analysis and comparison with reported observations: *Limnol. Oceanogr.*, v. 23, p. 268–282.

Jones, G. E., 1964, Effect of chelating agents on the growth of *Escherichia coli* in seawater: *J. Bacteriol.*, v. 87, p. 483–499.

——, and Cobet, A. B., 1975, Heavy metal ions as the principal bactericidal agent in Caribbean sea water, p. 199–208 *in* Gameson, A. L. H. (Ed.), *International Symposium on Discharge of Sewage from Sea Outfalls:* Pergamon Press, Oxford.

Kinkade, M. L., and Erdman, H. E., 1975, The influence of hardness components (Ca^{2+} and Mg^{2+}) in water on the uptake and concentration of cadmium in a simulated freshwater ecosystem: *Environ. Res.*, v. 10, p. 308–313.

Kinniburgh, D. G., Jackson, M. L., and Syers, J. K., 1976, Adsorption of alkaline earth, transition, and heavy metal cations by hydrous oxide gels of iron and aluminum: *Soil Sci. Soc. Am. J.*, v. 40, p. 796–799.

Klotz, R. L., 1981, Algal response to copper under riverine conditions: *Environ. Pollut.*, v. 24A, p. 1–19.

Korkeala, H., and Pekkanen, T. J., 1978, The effect of pH and potassium phosphate buffer on the toxicity of cadmium for bacteria: *Acta Vet. Scand.*, v. 19, p. 93–101.

Laborey, F., and Lavollay, J., 1973, Sur la nature des antagonismes responsables de l'intéraction des ions Mg^{++}, Cd^{++}, et Zn^{++} dans la croissance d'*Aspergillus niger: C. R. Acad. Sci. (Paris) D*, v. 276, p. 529–532.

——— , and Lavollay, J., 1977, Sur l'antitoxicité du calcium et du magnésium à l'égard du cadmium, dans la croissance d'*Aspergillus niger: C. R. Acad. Sci (Paris) D*, v. 284, p. 639–642.

Lofroth, G., and Ames, B. N., 1978, Mutagenicity of inorganic compounds in *Salmonella typhimurium:* arsenic, chromium, and selenium: *Mutat. Res.*, v. 53, p. 65–66 (Abstr.).

Loveless, J. E., and Pointer, H. A., 1968, The influence of metal ion concentrations and pH value on the growth of a *Nitrosomonas* strain isolated from activated sludge: *J. Gen. Microbiol.*, v. 52, p. 1–14.

Malanchuk, J. L., and Gruendling, G. K., 1973, Toxicity of lead nitrate to algae: *Water Air Soil Pollut.*, v. 2, p. 181–190.

McKnight, D. M., and Morel, F. M. M., 1979, Release of weak and strong copper-complexing agents by algae: *Limnol. Oceanogr.*, v. 21, p. 823–837.

Mishra, D., and Kar, M., 1974, Nickel in plant growth and metabolism: *Bot. Rev.*, v. 40, p. 395–452.

Monahan, T. J., 1976, Lead inhibition of chlorophycean microalgae: *J. Phycol.*, v. 12, p. 358–362.

Murray, D. J., Healy, T. W., and Fuerstenau, D. W., 1968, The adsorption of aqueous metal on colloidal hydrous manganese oxide: *Adv. Chem.*, v. 79, p. 74–90.

Nakamura, H., 1962, Adaptation of yeast to cadmium. V. Characteristics of RNA and nitrogen metabolism in the resistance: *Mem. Konan Univ., Sci. Ser.*, v. 6, p. 19–31.

Nakamuro, K., Yoshikawa, K., Sayato, Y., and Kurata, H., 1978, Comparative studies of chromosomal aberration and mutagenicity of trivalent and hexavalent chromium: *Mutat. Res.*, v. 58, p. 175–181.

Nieboer, E., Puckett, K. J., and Grace, B., 1976, The uptake of nickel by *Umbilicaria muhlenbergii:* a physicochemical process: *Can. J. Bot.*, v. 54, p. 724–733.

Nishioka, H., 1975, Mutagenic activities of metal compounds in bacteria: *Mutat. Res.*, v. 31, p. 185–189.

Novick, R. P., and Roth, C., 1968, Plasmid-linked resistance to inorganic salts in *Staphylococcus aureus: J. Bacteriol.*, v. 95, p. 1335–1340.

Overnell, J., 1976, Inhibition of marine algal photosynthesis by heavy metals: *Mar. Biol.*, v. 38, p. 335–342.

Phillips, C. S. G., and Williams, R. J. P., 1965, *Inorganic Chemistry*, Vol. I: Oxford University Press, Oxford.

Pickett, A. W., and Dean, A. C. R., 1976, Cadmium and zinc sensitivity and tolerance in *Klebsiella (Aerobacter) aerogenes: Microbios*, v. 15, p. 79–91.

Pietilainen, K., 1975, Synergistic and antagonistic effects of lead and cadmium on aquatic primary production, *Int. Conf. Heavy Met. Environ., Symp. Proc.*, Toronto, Oct. 27–31, Vol. 2. p. 861–873.

Ragan, M. A., Ragan, C. M., and Jensen, A., 1980, Natural chelators in sea water: detoxification of Zn^{2+} by brown algal polyphenols: *J. Exp. Mar. Biol. Ecol.*, v. 44, p. 261–267.

Rai, L. C., Gaur, J. P., and Kumar, H. D., 1981, Protective effects of certain environmental factors on the toxicity of zinc, mercury, and methylmercury to *Chlorella vulgaris: Environ. Res.*, v. 25, p. 250–259.

Ramamoorthy, S., and Kushner, D. J., 1975, Binding of mercuric and other heavy metal ions by microbial growth media: *Microb. Ecol.*, v. 2, p. 162–176.

Rana, B. C., and Kumar, H. D., 1974, The toxicity of zinc to *Chlorella vulgaris* and *Plectonema boryanum* and its protection by phosphate: *Phykos*, v. 13, p. 60–66.

Rashid, M. A., 1971, Role of humic acids of marine origin and their different molecular weight fractions in complexing di- and trivalent metals: *Soil Sci.*, v. 111, p. 298–306.

Reid, J. D., and McDuffie, B., 1981, Sorption of trace cadmium on clay minerals and river sediments: effects of pH and Cd(II) concentrations in a synthetic river water medium: *Water Air Soil Pollut.*, v. 15, p. 375–386.

Richter, R. O., and Theis, T. L., 1980, Nickel speciation in a soil/water system, p. 189–202 *in* Nriagu, J. O. (Ed.), *Nickel in the Environment:* Wiley, New York.

Riffaldi, R., and Levi-Minzi, R., 1975, Adsorption and desorption of Cd on humic acid fraction of soils: *Water Air Soil Pollut.*, v. 5, p. 179–184.

Sastry, K. S., Adiga, P. R., Venkatasubramanyam, V., and Sarma, P. S., 1962, Interrelationships in trace-element metabolism in metal toxicities in *Neurospora crassa: Biochem. J.*, v. 85, p. 486–491.

Say, P. J., and Whitton, B. A., 1977, Influence of zinc on lotic plants. II. Environmental effects on toxicity of zinc to *Hormidium rivulare: Freshwater Biol.*, v. 7, p. 377–384.

Scarpino, P. V., and Pramer, D., 1962, Evaluation of factors affecting the survival of *Escherichia coli* in seawater. VI. Cysteine: *Appl. Microbiol.*, v. 10, p. 436–440.

Schulz–Baldes, M., and Lewin, R. A., 1976, Lead uptake in two marine phytoplankton organisms: *Biol. Bull.*, v. 150, p. 118–127.

Schulze, H., and Brand, J. J., 1978, Lead toxicity and phosphate deficiency in *Chlamydomonas: Plant Physiol.*, v. 62, p. 727–730.

Shehata, F. H. A., and Whitton, B. A., 1982, Zinc tolerance in strains of the blue-green alga *Anacystis nidulans: Br. Phycol. J.*, v. 17, p. 5–12.

Simon-Pujol, M. D., Marques, A. M., Ribera, M., and Congregado, F., 1979, Drug resistance of chromium tolerant gram-negative bacteria isolated from a river: *Microbios Lett.*, v. 7, p. 139–144.

Singh, N., 1977, Effect of pH on the tolerance of *Penicillium nigricans* to copper and other heavy metals: *Mycologia*, v. 69, p. 750–755.

Singh, S. P., and Pandey, A. K., 1981, Cadmium toxicity in a cyanobacterium: effect of modifying factors: *Environ. Exp. Bot.*, v. 21, p. 257–265.

Skaar, H., Rystand, B., and Jensen, A., 1974, The uptake of ^{63}Ni by the diatom *Phaeodactylum tricornutum: Physiol. Plant.*, v. 32, p. 353–358.

Smith, D. H., 1967, R factors mediate resistance to mercury, nickel, and cobalt: *Science*, v. 156, p. 1114–1116.

Starkey, R. L., 1973, Effect of pH on toxicity of copper to *Scytaldium* sp., a copper-tolerant fungus, and some other fungi: *J. Gen. Microbiol.*, v. 78, p. 217–225.

Steemann Nielsen, E., and L. Kamp-Nielsen, L., 1970, Influence of deleterious concentrations of copper on the growth of *Chlorella pyrenoidosa: Physiol. Plant.*, v. 23, p. 828–840.

——— , and Wium-Andersen, S., 1970, Copper ions as poison in the sea and in freshwater: *Mar. Biol.*, v. 6, p. 93–97.

——— , Kamp-Nielsen, L., and Wium-Andersen, S., 1969, The effect of deleterious concentrations of copper on the photosynthesis of *Chlorella pyrenoidosa: Physiol. Plant.*, v. 22, p. 1121–1133.

Stevenson, F. J., 1976, Binding of metal ions by humic acids, p. 519–540 *in* Nriagu, J. O. (Ed.), *Environmental Biogeochemistry*, Vol. 2, *Metals Transfer and Ecological Mass Balances:* Ann Arbor Science, Ann Arbor, Mich.

——— , 1977, Nature of divalent transition metal complexes of humic acids as revealed by a modified potentiometric titration method: *Soil Sci.*, v. 123, p. 10–17.

_____ , Krastanov, S. A., and Ardakani, M. S., 1973, Formation constants of Cu^{2+} complexes with humic and fulvic acids: *Geoderma*, v. 9, p. 129–141.

Stokes, P. M., Hutchinson, T. C., and Krauter, K., 1973, Heavy-metal tolerance in algae isolated from contaminated lakes near Sudbury, Ontario: *Can. J. Bot.*, v. 51, p. 2155–2168.

Stotzky, G., and Babich, H., 1980, Mediation of the toxicity of pollutants to microbes by the physicochemical composition of the recipient environment, p. 352–354 *in* Schlessinger, D. (Ed.), *Microbiology—1980:* American Society for Microbiology, Washington, D.C.

Stratton, G. W., and Corke, C. T., 1979, The effect of mercuric, cadmium, and nickel ion combinations on a blue-green alga: *Chemosphere*, v. 8, p. 731–740.

Summers, A. O., and Jacoby, G. A., 1978, Plasmid-determined resistance to boron and chromium compounds in *Pseudomonas aeruginosa: Antimicrob. Agents Chemother.*, v. 13, p. 637–641.

_____ , and Silver, S., 1978, Microbial transformation of metals: *Annu. Rev. Microbiol.*, v. 32, p. 637–672.

Sunda, W., and Guillard, R. R. L., 1976, The relationship between cupric ion activity and the toxicity of copper to phytoplankton: *J. Mar. Res.*, v. 34, p. 511–529.

Swallow, K. C., Westall, J. C., McKnight, D. M., Morel, N. M. L., and Morel, F. M. M., 1978, Potentiometric determination of copper complexation by phytoplankton exudates: *Limnol. Oceanogr.*, v. 23, p. 538–542.

_____ , Hume, D. N., and Morel, F. M. M., 1980, Sorption of copper and lead by hydrous ferric oxide: *Environ. Sci. Technol.*, v. 14, p. 1326–1331.

Szeto, C., and Nyberg, D., 1979, The effect of temperature on copper tolerance of *Paramecium: Bull. Environ. Contam. Toxicol.*, v. 21, p. 131–135.

Tso, W. -W., and Fung, W. -P., 1981, Mutagenicity of metallic cations: *Toxicol. Lett.*, v. 8, p. 195–200.

Upitis, V. V., Pakalne, D. S., and Nollendorf, A. F., 1973, The dosage of trace elements in the nutrient medium as a factor in increasing the resistance of *Chlorella* to unfavorable conditions of culturing: *Microbiology*, v. 42, p. 758–762.

Van den Berg, C. M. G., Wong, P. T. S., and Chau, Y. K., 1979, Measurement of complexing materials excreted from algae and their ability to ameliorate copper toxicity: *J. Fish. Res. Board Can.*, v. 36, p. 901–905.

Webb, M., 1970a, Interrelationships between the utilization of magnesium and the uptake of other bivalent cations by bacteria: *Biochim. Biophys. Acta*, v. 222, p. 428–439.

_____ , 1970b, The mechanism of acquired resistance to Co^{2+} and Ni^{2+} in gram-positive and gram-negative bacteria: *Biochim. Biophys. Acta*, v. 222, p. 440–446.

Wikfors, G. H., and Ukeles, R., 1982, Growth and adaptation of estuarine unicellular algae in media with excess copper, cadmium, or zinc, and effects of metal-contaminated algal food on *Crassostrea virginica* larvae: *Mar. Ecol. Prog. Ser.*, v. 7, p. 191–206.

Wilson, D. E., 1978, An equilibrium model describing the influence of humic materials on the speciation of Cu^{2+}, Zn^{2+}, and Mn^{2+} in freshwaters: *Limnol. Oceanogr.*, v. 23, p. 499–507.

Wong, M. H., 1980, Toxic effects of cobalt and zinc to *Chorella pyrenoidosa* (26) in soft and hard water: *Microbios*, v. 28, p. 19–25.

Wong, K. G., Chan, K. Y., and Ng, S. L., 1979, Cadmium uptake by the unicellular green alga *Chlorella salina* Cu-1 from culture media with high salinity: *Chemosphere*, v. 8, p. 887–891.

Yang, S. H., and Ehrlich, H. L., 1976, Effect of four heavy metals (Mn, Ni, Cu, and Co) on some bacteria from the deep sea, p. 867–874 *in* Miles, J. M., and Kaplan, A. M. (Eds.), *Proceedings of the Third International Biodegradation Symposium:* Applied Science Publishers, Barking, Essex, England.

Yongue, W. H., Jr., Berrent, B. L., and Cairns, J., Jr., 1979, Survival of *Euglena gracilis* exposed to sublethal temperature and hexavalent chromium: *J. Protozool.*, v. 26, p. 122–125.

Zirino, A., and Yamamoto, S., 1972, A pH-dependent model for the chemical speciation of copper, zinc, cadmium, and lead in seawater: *Limnol. Oceanogr.*, v. 17, p. 661–671.

Anne O. Summers
Department of Microbiology
University of Georgia
Athens, Georgia

16

GENETIC MECHANISMS OF HEAVY-METAL AND ANTIBIOTIC RESISTANCES

ABSTRACT

Free-living, single-cell microorganisms can be either genetically haploid (having one copy of each gene) or diploid (having two copies of each gene). The former group, called prokaryotes, includes the bacteria and the blue-green algae. The latter group, called lower eukaryotes, includes the fungi and the true algae. Resistance to toxic metals and to antibiotics has been observed in many genera of these two classes. In some cases the resistance involves the ability to alter the chemical state of either the metal or antibiotic. Until the mid-1960's it was thought that such resistance phenotypes arose only from mutations in the major cell replicon (commonly called the chromosome). However, in the mid-1960's it was discovered in both clinical and environmental studies that the procaryotes have additional, small replicons (called plasmids) which confer resistance to many antibiotics and to toxic metals. A given plasmid can carry several different resistance loci. Many plasmids are capable of moving from cell to cell by a process called conjugation; conjugation can take place even between cells of unrelated genera allowing the resistance plasmids to become widely disseminated through the bacterial population of a given niche. In addition, the resistance loci themselves are unusual in that they can recombine almost randomly with any stretch of bacterial DNA by a process called transposition. In other words, the resistance loci are "jumping genes" which move freely from a plasmid to the cell chromosome, to bacterial viruses, or to other plasmids. In several cases, the transposition process has been shown to be stimulated by exposure to the selective agent to which the transposable locus confers resistance. Transposability further enhances the dissemination of these resistance loci.

GENETIC ELEMENTS OF BACTERIA

Bacteria are the simplest single-cell organisms. Each free-living cell contains a principal replicon or chromosome. (A replicon is any self-replicating, inheritable subcellular structure.) This principal replicon, which consists of double-stranded DNA arranged in a closed circle, encodes all the information necessary for the elementary "housekeeping" functions of growth and cell division. In addition to the principal replicon, the bacterial cell can also have one or more smaller "accessory" replicons. These smaller replicons are called *plasmids* and they encode functions which allow the bacteria to adapt to specific ecological niches. It is these accessory replicons which will concern us in this discussion (Davis *et al.*, 1980, and Fig. 16-1).

It is possible for the genes of the principal replicon to mutate so that the bacterial cell can survive harsh circumstances or metabolize unusual nutrients, but such mutations are rare events. The adaptability of the bacterial population is managed in a different way. Rather than mutating the extensive but finite complement of genes carried by the principal replicon every time a new danger or opportunity arises, the bacterial population (and it is essential here to think of the population, not of an individual cell) has available a pool of genes carried by the accessory replicons. Not every cell in a given wild population carries the same set of plasmids, although there may be extensive overlap. (Only in pure cultures studied in the laboratory does each cell in the population have the

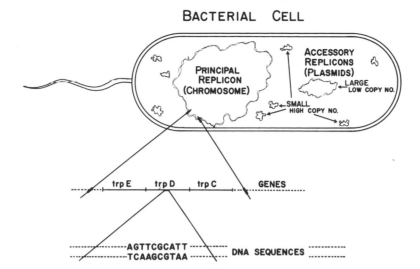

Fig. 16-1 Structures that carry genetic information in the bacterial cell. Each replicon consists of DNA. The genes depicted encode enzymes involved in tryptophan biosynthesis. The code is made up of the four nucleotide bases: adenine, cytosine, guanine, and thymine.

same genetic complement of both principal and accessory replicons.) The important thing about these plasmids is their "exchangeability." They can move very rapidly from one cell in the population to another (Falkow, 1975).

MODES OF GENE EXCHANGE IN BACTERIA

There are three different modes of gene exchange in bacteria. One mode, *conjugation*, requires cell-to-cell contact mediated by organelles specialized for the exchange of DNA. A second mode, *transformation*, involves free DNA simply being extruded from one cell into the growth medium, where it drifts until taken up by another cell. The third mode, *transduction*, is mediated by a bacterial parasite called a *bacteriophage* or *bacterial virus*. Bacteriophages are the simplest replicons, consisting only of a nucleic acid chromosome encased in a protein coat. A bacteriophage invades a bacterial cell and replicates itself 50 to 100 times before bursting out of the remnants of the destroyed cell to infect yet another hapless cell. Occasionally, via an error in its "packaging" process, a bacteriophage will enclose in its protein coat DNA from one of the host cell's resident replicons. This nonbacteriophage DNA, a kind of "sheep in wolf's clothing," can then invade another cell using the invasion mechanism built into the phage coat. However, once inside the new host, this DNA does not cause the destruction of the cell. If it is an entire replicon (e.g., a complete plasmid) which has been "transduced" this way, it will simply take up its normal, quiescent, replicative mode and be distributed to each daughter cell upon division. If a piece of DNA smaller than an entire replicon has been transduced, it must integrate into an existing replicon already in the cell by a process called *recombination* or it will not be passed on to succeeding generations of daughter cells (Davis *et al.*, 1980).

ANNE O. SUMMERS

RECOMBINATION

There are two types of recombination. The earliest described type, called *homologous recombination*, occurs only between regions of DNA which are similar over a great extent. We can tell that an exchange has taken place if the incoming DNA has a slight difference (i.e., a mutation) which would be detectable by subsequent screening of the "progeny" of the bacterial "cross" (or mating). Using a variety of techniques we can determine the location at which the recombination event took place and see that the incoming DNA physically replaces the DNA which was originally there. This process of reciprocal exchange is carried out by a group of enzymes which together are called the *rec* functions (Stahl, 1979). As an analogy, imagine that you are writing a chapter on bacterial leaching of low-grade mineral ores. You find that you would like to make a change or two in the wording of one paragraph of your text. In effect you would replace that paragraph with another which, word by word, was pretty much the same except for the small changes. The new paragraph would "fit" in the context of the chapter and would not be discontinuous with the surrounding information.

There is a second type of recombination which has been recently discovered, primarily through the study of bacterial plasmids and their genes. This type of recombination, called *transposition*, involves interactions between regions of DNA which have little or no homology. It was formerly thought that all regions of DNA were equivalent in their ability to undergo recombination (and for practical purposes this is true of homologous recombination). However, transposition involves certain unique sequences of DNA. These sequences, which are called *transposons*, occur in a variety of distinct sizes from just a few hundred to several thousand bases in length. Transposons can recombine almost randomly from one replicon to another and for this reason, they have been termed "jumping genes" (Fig. 16-2 and Kleckner, 1981).

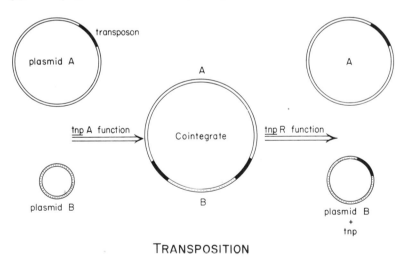

TRANSPOSITION

Fig. 16-2 Mechanism of transposition. The transposase enzyme (tnpA) effects the formation of a cointegrate structure. This cointegrate is then resolved into two separate plasmids by breakage-and-rejoining action carried out within the transposon itself. The resolvase enzyme (tnpR) effects the latter step.

To continue our analogy from above, imagine that you were proofreading your chapter on bacterial leaching and suddenly discovered in the middle of a paragraph four sentences on Early Renaissance music! This would be disturbing to the flow of information in the paragraph. Imagine your consternation when many pages later you discover in a completely different paragraph the *same* four sentences on Early Renaissance music! You might then begin to suspect that your article had suffered a couple of transposition events of information from a totally unrelated source. (Perhaps there is a problem with your word processor.) The disruption in the flow of ideas effected by these transposition events is entirely analogous to the disruption effected by transposons in genetic systems. Moreover, if you look closely at the transposition into your article, you realize that it is a brief explanation of the development of the well-tempered clavier. That is, it is not just garbled nonsense; it "does something." And this is also true of the transposons of bacteria. Most of the bacterial resistance genes which we discuss below are carried on transposons. Transposition is not limited to bacteria but, like homologous recombination, occurs in all higher genetic systems, including humans. The existence of transposition has profound implications for everything from normal development to neoplasia.

GENE EXPRESSION

Having described where genes reside and how they are exchanged, we will consider how genes are expressed. A "gene" itself does not *do* the function for which it is named; it is simply the *information* for making a protein (or enzyme) which will carry out that function. The process of converting this information (potential) into a functioning protein (actual) is called *gene expression* and it is a carefully regulated process even in a creature as simple as a bacterium.

To borrow an analogy from computer science, the information present in the linear array of the nucleotide bases adenine, guanine, cytosine, and thymine can be thought of as the "program" for the cell. Various "subroutines" of this program are called upon depending on the stage of the cell's growth, availability of certain kinds of nutrients, temperature, oxygen, and so on. Many of these "subroutines" are actually "running" at the same time, as the cell is continually metabolizing carbohydrates, synthesizing new proteins, nucleic acids, and membrane components. And although many genes are expressed simultaneously, not all genes are expressed all of the time.

A "subroutine" is "called" when an enzyme called RNA polymerase begins to copy the specific DNA sequence which encodes that gene into another polymer called *messenger RNA*. Subcellular organelles called *ribosomes* interact with the messenger RNA to "read" the message and synthesize a protein, a polymer of totally different subunits, called *amino acids*. That is, the code, which is composed of *deoxyribonucleotides*, is first "transcribed" into *ribonucleotides* and then "translated" into *amino acids*. The ribosome can be thought of as a kind of compiler (to use another computer analogy) which converts information from one form into another more usable form. The proteins, each of which has a specific activity either as an enzyme (biological catalyst) or as a structural member of the membrane or other organelle, are the *functioning* aspect of the gene. It is primarily via interactions with these proteins that toxic substances affect the cell. In other words, selective agents do not operate at the level of the gene itself but at the level of the gene's functioning "product." The exceptions to this generalization are those agents, called *mutagens*, which interact directly with the DNA (Lewin, 1974).

WAYS TO KILL BACTERIAL CELLS

There are physical, chemical, and biological ways to kill bacterial cells or to prevent their growth. Physical means include manipulation of the temperature or exposure to ionizing radiation. Chemical means include alteration of the pH of the growth medium or exposure to toxic chemical substances, including metal cations or organic solvents which disrupt the cell. Such physical and/or chemical damage to the cell is generally nonspecific. Biological agents that disrupt normal cell processes are called *antibiotics* and in contrast to the physical and chemical agents, they are very specific. Typical target processes and the antibiotics that affect them are listed in Table 16-1. The nature of the target of a given toxic agent has some bearing on the kind of resistance to that agent which it is possible for the bacterial cell to have (Davis *et al.*, 1980).

TABLE 16-1   Typical Antibiotics and Their
Target Processes in the Cell

Process	Antibiotic
DNA synthesis	Nalidixic acid
	Novobiocin
	Mitomycin C
RNA synthesis	Rifampin
Protein synthesis	Streptomycin
	Chloramphenicol
	Tetracycline
	Erythromycin
Cell envelope synthesis	Penicillin
	Bacitracin

GENETIC BASIS FOR RESISTANCE
TO HEAVY METALS AND ANTIBIOTICS

Alterations in the genetic information of a cell which confer resistance to a toxic agent can arise in two ways:

1. A mutation (a change in the DNA sequence) can take place in the genetic locus that encodes the target system. For example, the function of one of the protein components of the ribosome, which is encoded by the principal replicon, is inhibited by streptomycin. At a frequency of about one per billion cell generations, mutations arise that alter this protein, making it resistant to inhibition by streptomycin. However, the structure of each of the protein components of the ribosome has been very finely tuned by many millions of years of evolution. Any alteration in even one protein component can affect the functioning of the ribosome in its normal activity, which is the synthesis of proteins. So the cell has lost something while gaining resistance to the antibiotic (Lewin, 1974).

2. A more flexible way in which resistance can be acquired without permanently altering the function of essential enzymes in the cell is by the acquisition of a plasmid which confers resistance to that antiobiotic. As noted above, most naturally occurring bacterial populations carry more than one plasmid. Although each cell in the population may not carry a plasmid whose genes would allow it to survive, other cells in the population do (Falkow, 1975). These plasmids are continually exchanged among the members of the population by the mechanisms described above. In addition, while a given plasmid resides in a bacterial cell it may receive a transposon from another replicon within that cell. It will subsequently carry that new genetic information when it is transferred to another cell. Note that in two of the gene transfer processes, conjugation and transformation, the "donor" cell retains the "original" DNA and passes a "copy" on to another ("recipient") cell. In the case of transduction, the "donor" cell is destroyed by the bacteriophage replication, so the question of whether it retains a copy of the gene is moot. Since the replication of larger plasmids constitutes a metabolic drain on the cell carrying them, the process of gene accretion is in equilibrium with gene loss. However, when selective pressure is imposed, the accretion process dominates and more cells in the population are found to carry larger plasmids, conferring resistances to greater numbers of antibiotics and/or metals.

This process of accretion of new markers by plasmids has been demonstrated both in the bacterial normal flora of higher mammals, including humans (Richmond, 1974), and also in the bacterial flora of urban and industrial sewage (Levy et al., 1981). Procaryotes rely a great deal on plasmid-encoded genes to afford protection against deleterious agents in their environment. Essentially all strains isolated from nature which are resistant to commonly used antibiotics and toxic metal cations are resistant by virtue of plasmid-carried genes. In addition, plasmid-carried antibiotic resistance genes have all been shown to be on transposable elements (Kleckner, 1981). For the metal-resistance genes it has been shown that those for resistance to inorganic mercury compounds (Stanisich et al., 1977), organic mercury compounds (Radford et al., 1981), and tellurite (D. Taylor, personal communication, 1983, and our unpublished observations) are on transposable genetic elements.

PLASMID-DETERMINED HEAVY-METAL RESISTANCE

Resistance to toxic heavy-metal cations and to oxyanions of toxic metalloids in both bacteria and the fungi has been reported for many years. The earliest well-documented work is that of Challenger et al. (1933) on biotransformation of arsenic compounds by *Penicillium*. While plasmids have been found in the fungi (Fincham et al., 1979) they do not appear to carry resistance genes naturally. It is presently assumed that these rare occurrences of metal-resistant (and/or biotransforming) strains of yeast and molds arise from chromosomal mutations. In contrast, metal-resistant strains of bacteria have been found quite easily and in relatively large numbers in every habitat examined. In many cases it has been established that the resistance locus is plasmid-borne (Summers and Silver, 1978).

The common bacteria are divided into two large groups based on their reaction to a straining procedure devised by C. Gram. This division is corroborated by consistent differences in cell structure, biochemical pathways, DNA homology, and ability to exchange and express genetic information (Davis *et al.*, 1980). Loci conferring resistance to inorganic mercury, organomercurials, arsenate, arsenite, antimonate, tellurite, chromate, meta-borate, and silver have been found on plasmids of the gram-negative bacteria. Plasmids of the gram-positive bacteria have been found which confer resistance to inorganic mercury, organomercurials, arsenate, arsenite, antimonate, cadmium, lead, and bismuth. Extensive searches have been made among natural isolates of the gram-negative bacteria for resistances to other metals. No resistance loci other than those listed above have been clearly demonstrated by all appropriate criteria to be plasmid-determined, although variations in resistance to a number of toxic metals have been reported in many gram-negative genera in environments contaminated with such metals as cadmium (Devanas *et al.*, 1980) and uranium (DiSpirito *et al.*, 1982).

In addition to these plasmid-determined resistance mechanisms, there have been a number of reports of bacteria able to carry out transformations (solublization or volatilization) of economically and/or environmentally important metals such as gold (Korobushkina *et al.*, 1976) and lead (Wong *et al.*, 1975). Presently, little is known of the biochemistry and nothing is known of the genetics of these processes.

Only three of the plasmid-determined metal ion resistance mechanisms have been studied in any detail. These are the resistances to mercury compounds (both inorganic and organic), to cadmium, and to arsenate. The resistance mechanisms are very different from a physiological point of view. The resistance to mercury compounds involves a biotransformation of the toxic compounds to a nontoxic, volatile form. The resistances to both the cation cadmium and the oxyanion arsenate are effected by systems that eject the toxic compounds from the cell.

Resistance to Inorganic Mercury

Bacteria resistant to inorganic mercury and to a small number of organomercurials, including fluorescein mercuric acetate and merbromin, are referred to as having *narrow-spectrum resistance* (Schottel *et al.*, 1974). These strains reduce inorganic mercury to the metallic state (Hg^0), which is volatile and rapidly diffuses from the medium. The few organomercurials to which these bacteria are resistant are not detectably modified by the bacteria and presently the mechanism of this aspect of narrow-spectrum resistance is unknown.

The agent responsible for the reduction of Hg^{2+} to Hg^0 is an enzyme called *mercuric ion reductase*. Information for the synthesis of this enzyme is encoded in the plasmid-determined *mer* locus. This enzyme is a soluble, intracellular protein that uses a biological reducing agent (reduced nicotinamide adenine dinucleotide phosphate or NADPH) as a source of electrons for the reduction of inorganic mercury. The enzyme from gram-negative bacteria has been purified and characterized biochemically by several groups (Fox and Walsh, 1982; Schottel, 1978; Summers and Sugarman, 1974). It is a flavin-

There has been considerably more research on the mechanisms of plasmid-determined antibiotic resistance and there is now extensive information on all of the resistance mechanisms for the common antibiotics. Like the metal ion resistances, these mechanisms can be categorized biochemically as follows:

Resistance Involving Alteration of the Cell Membrane

Tetracycline inhibits the activity of the ribosome (i.e., it halts protein synthesis). In plasmidless wild-type bacteria, tetracycline enters the cell via an active transport system. It is rather curious that the cell should waste energy in taking up a toxic agent. However, as in the cadmium/manganese system above, many other toxic agents can "mimic" essential nutrients and thereby enter the cell. Presently, the normal function of the transport system which takes tetraycline into the cell is not known. The resistance mechanism provided by the plasmid-determined *tet* gene effects an "efflux" of tetracycline from the cell. As rapidly as the tetracycline enters the cell it is ejected back into the medium by this very specific membrane-bound molecular "sump pump." The intracellular level of tetracycline is thus maintained below the level at which it can inhibit protein synthesis. However, raising the external concentration of tetracyline above a certain level can overwhelm the capacity of the resistance mechanism which, like all of the other resistance mechanisms described here, does not confer *absolute* resistance (Chopra and Howe, 1978).

Resistance to tetracycline is inducible; the membrane protein(s) responsible for tetracycline efflux are not synthesized unless tetracycline is present. Use of this ejection mechanism for tetracycline resistance is the only such example for antibiotic resistance. It is similar in principle to the metal ion resistances for cadmium and for arsenate described above.

Resistance Involving Modification or Degradation of the Antibiotic

Chloramphenicol and the aminoglycoside antibiotics streptomycin, gentamicin, and kanamycin which affect protein synthesis, are all subject to degradation or chemical modification by plasmid-encoded enzymes. Chloramphenicol is acetylated on either of two of its hydroxyl groups by an enzyme called *chloramphenicol acetyl transferase*. The acetylated chloramphenicol derivative no longer has antibiotic activity either in vitro or in vivo (Shaw *et al.*, 1970). The aminoglycosides can be modified in a variety of ways and several different enzymes have been described which will phosphorylate, adenylylate, or acetylate the foregoing antibiotics. Although the modified forms of the antibiotics are still inhibitory to ribosomes in a cell-extract (i.e., an in vitro protein synthesizing system), it appears that they are no longer taken up by the intact cell and have therefore lost their capacity to inhibit cell processes (Falkow, 1975).

The antibiotic penicillin that affects cell wall biosynthesis has a beta-lactam ring structure. Plasmid-encoded enzymes capable of hydrolyzing this ring are called *beta-lactamases*. Once the ring is broken, the antibiotic activity of penicillin is lost (Falkow, 1975).

These plasmid-encoded enzymes, which function by degrading or modifying the antibiotic itself rather than by altering an important cell structure such as the ribosome, are synthesized continuously. The level of their synthesis can be increased, albeit gradually, by the process of gene amplification described below. It is also important to note that the benefits of these resistances, which involve large-scale alteration of the antibiotic itself (such as penicillin or chloramphenicol resistance) can be "shared" with the entire population of bacteria in a given niche. Once the antibiotic in the niche has been destroyed, both the sensitive and resistant bacteria are free to grow. This is not the case for resistance mechanisms that involve a change in the cell membrane or in the target site of the antibiotic.

Resistance Involving an Altered Target Site

Erythromycin affects protein synthesis by interacting with a specific region of one of the RNA components of the ribosome. A plasmid-encoded enzyme called a *transmethylase* adds methyl groups to two of the exposed adenylyl residues at this critical position in the RNA molecule. Erythromycin no longer binds to this altered target site. The synthesis of the transmethylase enzyme is inducible; it increases in response to the presence of erythromycin in the cell. So, although a critical component of the cell (the ribosome) is changed, the change is not permanent. New ribosomes are continually being synthesized and therefore unmodified ribosomes can slowly replace the modified ones once the erythromycin concentration has fallen below a toxic level (Lai and Weisblum, 1971).

Resistance Involving a Substitute Enzyme

Sulfadiazine and trimethoprim affect, respectively, the biosynthesis of folic acid and an enzyme that requires folic acid as a cofactor. They are not antibiotics but are synthetic chemical analogs. Plasmid-determined resistance to sulfadiazine arises from the gene encoding a sulfa-resistant dihydropteroate synthase (Wise and Abou-Donia, 1975). Resistance to trimethoprim arises from the synthesis of a trimethoprim-resistant dihydrofolate reductase (Zolg *et al.*, 1978). When the plasmid genes are present in a cell, each of these enzymes is synthesized continuously in amounts sufficient to supplant the disabled normal enzyme. The plasmid mechanism in this case provides a bypass around the blocked enzyme encoded by the principal replicon.

For a given antibiotic the mode of plasmid-determined resistance is very similar (at least at the biochemical level) in a great number of bacterial genera. For example, penicillins are inactivated by beta-lactamases in every genus so far examined. In no case is there any evidence for naturally occurring resistance to penicillins based on an ejection mechanism or on the alteration in the target enzyme. The same is essentially true for each of the other categories of antibiotic-resistance mechanisms noted above.

The concentration of a given antibiotic or other toxic agent which is required to kill bacteria can vary among bacterial genera because of intrinsic differences in their cell surfaces which affect the ability of the antibiotic to penetrate into the cell (Chakrabarty, 1976). However, even within one bacterial genus, it is possible for there to be large differences in sensitivity to a given antibiotic because of a process called *gene amplification.*

In a bacterial cell there is only one copy of the principal replicon and therefore only one copy of each gene that it carries. Many bacterial plasmids (especially the larger ones) also exist as only a single copy per cell. However, a considerable number of bacterial plasmids exist in as many as 30 to 50 copies per cell. Therefore, each gene they carry also occurs 30 to 50 times in the cell. Both types of plasmids are capable of increasing the number of copies of resistance genes in two different ways (Fig. 16-3).

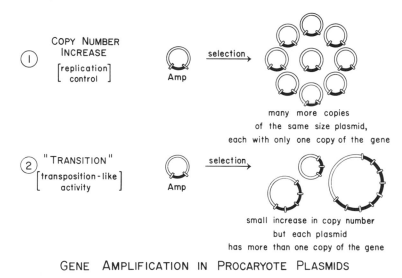

GENE AMPLIFICATION IN PROCARYOTE PLASMIDS

Fig. 16-3   Mechanisms of plasmid gene amplification in prokaryotes.

Copy Number Control

The number of copies of a given plasmid that reside in a cell is controlled by a rather complex set of interactions between several genetic loci (Helinski, 1973; Novick, 1980). In response to selective pressure, such as the administration of an antibiotic, the average number of copies of a plasmid in the cell can increase, thereby increasing the number of copies of the antibiotic resistance gene present in the cell. This mechanism of gene amplification has been found particularly in the case of resistance mechanisms, which involve inactivating enzymes such as chloramphenicol acetyl transferase and beta-lactamase. The increased number of copies of the gene in the cell can cause as much as a 10- to 25-fold increase in the concentration of the antibiotic that the strain can tolerate. Typically, upon removal of the selective agent, the copy number of the plasmid returns to its normal level. This type of gene amplification is mainly confined to the smaller plasmids, which normally tend to have a higher copy number.

The Transition

The term *transition* is used for a phenomenon in which, upon exposure to an antibiotic, tandem duplications of a set of resistance genes (now known to be on a transposable element) occur within a large, single-copy plasmid (Helinski, 1973). Although the plasmid copy number does not increase, the size of the plasmid increases as the number of tandem duplications increases. Plasmids with as many as six duplications of this "R-determinant" region have been observed. A corresponding decrease in the antibiotic susceptibility of the strain is noted. Such observations have also been made in other laboratory strains of resistant bacteria as well as in clinical isolates. Here again, upon removal of the selective agent, the plasmid population gradually reassumes its normal size and structure.

As noted, such gene amplification mechanisms appear to increase the resistance only when the antibiotic in question is one that is inactivated or degraded by the resistance mechanism. In the case of tetracycline resistance, which involves an alteration in the cell membrane, exposure to the drug does not result in *tet* locus amplification. Artificial manipulation of the *tet* locus copy number via recombinant DNA techniques has shown that strains with many copies of the locus are no more resistant than those with only one copy, and in some cases they may be slightly more sensitive (although not as sensitive as a strain with no *tet* locus at all; Coleman and Foster, 1981). The same is true for resistance to inorganic mercury compounds (our unpublished observations).

CONCLUSIONS

In this brief chapter it has only been possible to touch lightly on a very limited portion of the enormous potential of bacterial genetic systems. The flexibility provided by the system of accessory replicons and transposable genetic elements allows bacterial populations to adapt readily to toxic agents such as antibiotics and heavy-metal ions. Other genes frequently found on naturally occurring plasmids encode information for the metabolism of unusual carbon compounds such as octane, toluene, and camphor (Chakrabarty, 1976). Genes involved in pathogenesis of humans, animals, and plants (Levy *et al.*, 1981) are also carried by bacterial plasmids.

We have made use of microbes for thousands of years in such simple tasks as the making of bread, beer, and yogurt. In more recent times, it is important to note that the maintenance of the high standard of living in developed countries relies in large part on efficient sewage treatment, which is entirely based on microbial processes. Nonetheless, given the great variety of microbial abilities only recently discovered and given the power of modern genetics to understand and harness these abilities, we have only begun to exploit the potential of microbial processes.

Since this paper was presented there have been many new developments in this area. For more current reviews of these advances the reader is referred to the detailed consideration of plasmid-determined resitance factors by Tim Foster ("Plasmid-determined Resistance to Antimicrobial Drugs and Toxic Metal Ions in Bacteria," *Microbiol. Rev.* **47**:3:361-409, 1983) and the very thorough treatment of the specific area of plasmid-determined detoxification of mercury compounds (Robinson, J. H. and Tuovinen, O. H., "Mechanisms of Microbial Resistance and Detoxification of Mercury and Organomercury Compounds: Physiological, Biochemical, and Genetic Analyses." *Microbiol. Rev.* **48**:2:95-124, 1984). The most recent work on the mercury resistance systems is reviewed by Nigel Brown ("Bacterial Resistance to Mercury—Reductio ad Absurdum?", *Trends in Biochemical Sciences*, in press, 1985), and I have discussed the plasmid-determined metal resistances with a view towards their potential application in biotechnology, including mining (Summers, A. O., "Bacterial Resistance to Toxic Elements," *Trends in Biotechnology*, **3**:5:122-125, 1985).

REFERENCES

Chakrabarty, A. M., 1976, Plasmids in *Pseudomonas: Annu. Rev. Genet.*, v. 10, p. 7–30.

Challenger, F., Higginbotom, C., and Ellis, L., 1933, The formation of organo-metalloidal compounds by microorganisms. Part 1. Trimethylarsine and dimethylarsine: *J. Chem. Soc.*, 1933, p. 95–101.

Chopra, I., and Howe, T. G. B., 1978, Bacterial resistance to the tetracyclines; *Microbiol. Rev.*, v. 42, p. 707–724.

Cole, M. A., 1979, Solublization of heavy metal sulfides by heterotrophic soil bacteria: *Soil Sci.*, v. 127, p. 313–317.

Coleman, D. C., and Foster, T. J., 1981, Analysis of the reduction in expression of tetracycline resistance determined by transposon Tn10 in the multicopy state: *Mol. Gen. Genet.*, v. 182, p. 171–177.

Davis, B. D., Dulbecco, R., Eisen, H. E., and Ginsburg, H. S., 1980, *Microbiology*, 3rd ed.: Harper & Row, Hagerstown, Md.

Devanas, M. A., Litchfield, C. D., McClean, C., and Gianni, J., 1980, Coincidence of cadmium and antibiotic resistance in New York Bight apex benthic microorganisms: *Mar. Pollut. Bull.*, v. 11, p. 264–269.

DiSpirito, A. A., Silver, M., Voss, L. and Tuovinen, O. H., 1982, Flagella and pili of iron-oxidizing thiobacilli isolated from a uranium mine in northern Ontario, Canada: *Appl. Environ. Microbiol.*, v. 43, p. 1196–1200.

Falkow, S., 1975, *Infectious Multiple Drug Resistance:* Pion Press, London.

Fincham, J. R. S., Day, P. R., and Radford, A., 1979, *Fungal Genetics*, 4th ed.: University of California Press, Berkeley.

Foster, T. J., Nakahara, H., Weiss, A. A., and Silver, S., 1979, Transposon A-generated mutations in the mercuric resistance genes of plasmid R100-1: *J. Bacteriol.*, v. 140, p. 167–181.

Fox, B., and Walsh, C. T., 1982, Mercuric reductase—purification and characterization of a transposon-encoded flavoprotein containing an oxidation-reduction-active disulfide: *J. Biol. Chem.*, v. 257, p. 2498–2503.

Helinski, D. R., 1973, Plasmid determined resistance to antibiotics: molecular properties of R factors: *Annu. Rev. Microbiol.*, v. 27, p. 437–470.

Jackson, W. J. and Summers, A. O., 1982, Polypeptides encoded by the *mer* operon: *J. Bacteriol.*, v. 149, p. 479–487.

Kleckner, N., 1981, Transposable elements in procaryotes: *Annu. Rev. Genet.*, v. 15, p. 341–404.

Korobushkina, E. D., Mineev, G. G., and Praded, G. P., 1976, Mechanism of the microbiological process of dissolution of gold: *Mikrobiologiya*, v. 45, p. 535–538.

Lai, C.-J., and Weisblum, B., 1971, Altered methylation of ribosomal RNA in an erythromycin-resistant strain of *Staphylococcus aureus: Proc. Natl. Acad. Sci. (USA)*, v. 68, p. 856–860.

Legge, J. W., 1954, Bacterial oxidation of arsenite. IV. Some properties of the bacterial cytochromes: *Aust. J. Biol. Sci.*, v. 7, p. 504–514.

Levy, S. B., Clowes, R. C., and Koenig, E. L. (Eds.) 1981, *Molecular Biology, Pathogenicity, and Ecology of Bacterial Plasmids:* Plenum Press, New York.

Lewin, B., 1974, *Gene Expression*, Vol. 1: Wiley, Chichester, England.

Mobley, H. L. T., and Rosen, B. P., 1982, Energetics of plasmid-mediated arsenate resistance in *Escherichia coli: Proc. Natl. Acad. Sci. (USA)*, v. 79, p. 6119–6122.

Novick, R. P., 1980, Plasmids: *Sci. Am.*, Dec. 1980, p. 103–127.

Osborne, F. H., and Ehrlich, H. L., 1976, Oxidation of arsenite by a soil isolate of *Alcaligenes: J. Appl. Bacteriol.*, v. 41, p. 295–305.

Perry, R. D., and Silver, S., 1982, Cadmium and manganese transport in *Staphylococcus aureus* membrane vesicles: *J. Bacteriol.*, v. 150, p. 973–976.

Radford, A. J., Oliver, J., Kelly, W. J., and Reanney, D. C., 1981, Translocatable resistance to mercuric and phenylmercuric ions in soil bacteria: *J. Bacteriol.*, v. 147, p. 1110–1112.

Richmond, M. H., 1974, R factors in man and his environment, p. 27–35 in *Microbiology –1974:* American Society for Microbiology, Washington, D.C.

Schottel, J., 1978, Mercuric and organomercurial detoxifying enzymes from a plasmidbearing strain of *Escherichia coli: J. Biol. Chem.*, v. 253, p. 4341–4349.

___ , Mandal, A., Clark, D., Silver, S., and Hedges, R., 1974, Volatilisation of mercury and organomercurials determined by inducible R-factor systems in enteric bacteria: *Nature*, v. 251, p. 335–337.

Shaw, W. V., Bentley, D. W., and Sands, L., 1970, Mechanism of chloramphenicol resistance in *Staphylococcus aureus: J. Bacteriol.*, v. 104, p. 1095–1105.

___ , Budd, K., Leahy, K. M., Shaw, W. V., Hammond, D., Novick, R. P., Willsky, G. R., Malamy, M. H., and Rosenberg, H., 1981, Inducible plasmid-determined resistance to arsenate, arsenite and antimony (III) in *Escherichia coli* and *Staphylococcus aureus: J. Bacteriol.*, v. 146, p. 983–996.

Silver, S., and Keach, D., 1982, Energy-dependent arsenate efflux: the mechanism of plasmid-mediated resistance: *Proc. Natl. Acad. Sci. (USA)*, v. 79, p. 6114–6118.

Stahl, F. W., 1979, *Genetic Recombination:* W. H. Freeman, San Francisco.

Stanisich, V. A., Bennett, P. M., and Richmond, M. H., 1977, Characterization of a translocation unit encoding resistance to mercuric ions that occurs on a nonconjugative plasmid in *Pseudomonas aeruginosa: J. Bacteriol.*, v. 129, p. 1227–1233.

Summers, A. O., and Jacoby, G. A., 1977, Plasmid-determined resistance to tellurium compounds: *J. Bacteriol.*, v. 129, no. 1, p. 276–281.

___ , and Kight-Olliff, L., 1980, Tnl generated mutants in the mercuric ion reductase of the Inc P plasmid, R702: *Mol. Gen. Genet.*, v. 180, p. 91–97.

___ , and Silver, S., 1978, Microbial transformations of metals: *Annu. Rev. Microbiol.*, v. 32, p. 637–672.

———, and Sugarman, L. I., 1974, Cell-free mercury(II)-reducing activity in a plasmid-bearing strain of *Escherichia coli: J. Bacteriol.*, v. 119, p. 242–249.

Tezuka, T., and Tonomura, K., 1978, Purification and properties of a second enzyme catalyzing the splitting of carbon-mercury linkages from mercury-resistant *Pseduomonas* K-62: *J. Bacteriol.*, v. 135, 138–143.

Tynecka, Z., Gos, Z., and Zajac, J., 1981a, Reduced cadmium transport determined by a resistance plasmid in *Staphylococcus aureus: J. Bacteriol.*, v. 147, p. 305–312.

———, Gos, Z., and Zajac, J., 1981b, Energy-dependent efflux of cadmium coded by a plasmid resistance determinant in *Staphylococcus aureus: J. Bacteriol.*, v. 147, p. 313–319.

Wise, E. M., Jr., and Abou-Donia, M. M., 1975, Sulfonamide resistance mechanism in *Escherichia coli:* R plasmids can determine sulfonamide-resistant dihydropteroate snythases: *Proc. Natl. Acad. Sci. (USA)*, v. 72, p. 2621–2625.

Wong, P. T. S., Chau, Y. K., and Luxon, P. L., 1975, Methylation of lead in the environment: *Nature*, v. 253, p. 263–264.

Zolg, J. W., Hanggi, U. J. and Zachau, H. G., 1978, Isolation of a small DNA fragment carrying the gene for a dihydrofolate reductase from a trimethoprim resistance factor: *Mol. Gen. Genet.*, v. 164, p. 15–29.

Rita R. Colwell
Department of Microbiology
University of Maryland
College Park, Maryland

Dawn Allen-Austin
Department of Brewing and Biological Sciences
Heriot-Watt University
Edinburgh, Scotland

Tamar Barkay
U.S. EPA
Sabine Island
Gulf Breeze, Florida

Juan Barja
Departmento de Microbiologia
Universidad de Santiago
Santiago de Compostela
Spain

J. D. Nelson, Jr.
Olin Research Center
Cheshire, Connecticut

17

ANTIBIOTIC RESISTANT BACTERIA ASSOCIATED WITH ENVIRONMENTAL HEAVY-METAL CONCENTRATIONS

The incidence of bacterial resistance to several antibiotics commonly prescribed, including ampicillin, chloramphenicol, gentamicin, kanamycin, streptomycin, and tetracycline, and paired combinations of these antibiotics was surveyed in the upper Chesapeake Bay and in offshore waters of the east coast of the United States. Water and sediment samples collected in areas receiving domestic and industrial wastes were found to contain greater numbers of antibiotic-resistant and heavy-metal-resistant bacteria than similar samples collected from unpolluted locations. Seasonal distribution of bacterial species playing important roles in the nutrient cycles of estuarine and coastal waters had been recorded for the areas included in this study. Antibiotic-resistant bacterial populations in these same areas, however, did not demonstrate seasonal variation, although maximum numbers were frequently recovered during the spring months. The conclusion drawn was that heavy metal concentrations in the water and sediment can be directly related to metal and antibiotic resistance, more so than seasonal parameters, such as temperature and salinity.

Characterization and identification of several hundred isolates obtained to date revealed antibiotic- and heavy-metal-resistant bacteria represented a diverse range of gram-negative and gram-positive bacteria, similarly distributed among the sample sites. An exception were *Bacillus* spp., which were isolated from sediment samples, predominantly. Plasmids were demonstrated in metal- and antibiotic-resistant strains, with up to 68 to 70 percent of mercury-resistant strains and 30 to 40 percent of heavy-metal- and antibiotic-resistant strains, overall possessing one or more plasmids. The taxonomic groups demonstrated heterogeneity with respect to the number of strains carrying plasmids, as well as plasmid size. A number of different metals can contribute to resistance to one or more antibiotics, offering the possibility of a nonspecific geochemical screening tool for assaying large numbers of samples in an extensive mineral lands reconnaissance study.

INTRODUCTION

About a decade ago, the mechanisms and extent of influence of bacterial transformations of metals that take place in the estuarine environment were only beginning to be understood. Isolated reports by several investigators of bacterial synthesis and metabolism of mercury compounds, coupled with a serious question about mercury concentrations in Chesapeake Bay, stimulated our interest and resulted in the initiation of an extensive research program resulting in a better understanding of the role of bacteria in cycling of metals in Chesapeake Bay and coastal waters of the U.S. east coast.

The first major premise was that bacterial resistance to mercury (and other heavy metals) was a function of ability to detoxify mercury by metabolic transformation. A corollary was that measurement of numbers of mercury-resistant bacteria in situ should provide an index of real, or potential, microbiological mercury transformation in Chesapeake Bay, the "field laboratory" chosen for our studies.

Initial experiments were designed to measure the resistance of natural populations of aerobic, heterotrophic bacteria by their ability to form colonies on a simple solid-growth medium supplemented with selected inorganic and organic mercury compounds. The concentration of resistant bacteria in each sample was expressed as a percent of the "total viable, aerobic, heterotrophic bacterial count" (TVC) obtained by counting colonies appearing after incubation for 7 days at 25°C. The comparative effects of selected concentrations of mercury and concentrations approximating environmental levels were employed. Concentrations as low as 1.2 ppb of phenylmercuric acetate (PMA), or 1.2 ppb of HgCl$_2$, showed measurable effects on the bacterial populations in water and sediment samples. When the effect of incubation temperature on numbers of mercury-resistant bacteria was examined, the maximum TVC was obtained at those temperatures most closely approximating in situ temperatures.

Since an estuary is subjected to extensive changes in salinity, among other parameters, the effect of salt content in the growth medium used for routine isolation of mercury-resistant bacteria was evaluated for samples collected from sites encompassing a wide range of salinities. Media of three salinities (2.66, 11.38, and 26.59°/oo), including a routine isolation medium, were compared. Media of the highest (26.5°/oo) and the lowest (2.66°/oo) salinities consistently yielded higher proportions of mercury-resistant bacteria than an 11.38°/oo medium. It was concluded that the effect of salt concentration is related to selective effects on the bacterial population in the samples tested (Colwell and Nelson, 1975).

The natural habitat of mercury-resistant bacteria had not, at the time our studies began in 1971, been defined. Estuarine samples were fractionated into filtered water, surface sediment, interstitial water, and plankton. The results showed that a nonuniform distribution of both mercury-resistant bacteria and total mercury concentration existed among water, plankton, and sediment samples. The relative enrichment (i.e., greater concentrations) of mercury in sediments and living organisms, in comparison to water, was consistent with data published at that time (Klein and Goldberg, 1970; Williams and Weiss, 1973) and since—agreeing with the hypothesis that planktonic forms of life are influential in the transport of mercury, as well as the introduction of mercury, into food chains. A definite trend observed in the data was that mercury-resistant bacterial populations found associated with plankton were relatively larger than those in water or sediment. Observations of water and sediments indicated that in most cases, mercury-resistant bacteria were also distributed nonuniformly between different particle size fractions, with bacterial populations of interstitial water containing a larger fraction of mercury-resistant forms than filtered water and somewhat larger fractions than in sediment (see Table 17-1). Thus the evidence suggested that examination of mercury concentrations and bacterial populations of individual microenvironments would be the most logical approach to the problems of defining the relationship of mercury (and other heavy-metal)-resistant bacterial population size to environmental metal concentrations.

MULTI-RESISTANT BACTERIA IN THE ENVIRONMENT

When metal-resistant bacteria are examined for a given metal resistance, cross-resistance to other metals is the norm. Furthermore, mercury-resistant bacterial populations of Chesapeake Bay samples have been shown to degrade oil (Walker and Colwell, 1974). The concentration of mercury in water and sediment and in the oil extracted from water and

TABLE 17-1 Distribution of Bacterial HgCl$_2$ Resistance[a] and Total Mercury Concentrations among Water and Sediment Fractions

Source

Sample	Surface Water % Resistant HgCl$_2$	Surface Water % Resistant PMA	Surface Water [Hg] (ppb)	Filtered Water[b] (% Resistant HgCl$_2$)	Surface Sediment % Resistant HgCl$_2$	Surface Sediment % Resistant PMA	Surface Sediment [Hg] (ppm)	Interstitial Water[c] (% Resistant HgCl$_2$)	Plankton[d] % Resistant HgCl$_2$	Plankton[d] % Resistant PMA	Plankton[d] [Hg] (ppm)
A2, 3/29/72	6.5	–	<0.2 ± 0.02	3.1	7.0	–	0.27 ± 0.03	33.6	11.4	–	0.77 ± 0.04
B2, 4/03/72	–	–	–	7.5	29.8	–	0.82 ± 0.02	22.5	31.0	–	0.68 ± 0.03
Rhode River, 4/13/72	4.8	–	<0.2 ± 0.02	7.3	23.8	–	0.10 ± 0.01	5.9	–	–	–
B1, 5/15/72	4.2	–	<0.2 ± 0.02	–	8.2	–	0.17 ± 0.01	–	10.9	–	–
A2, 6/01/72	0.3	–	–	<0.4	1.5	–	0.37 ± 0.02	1.7	2.9	–	3.0 ± 0.90
B2, 10/05/72	13.2	–	0.09 ± 0.01	35.0	6.3	–	0.97 ± 0.04	35.7	–	–	–
EB1, 10/06/72	–	–	–	–	4.8	–	0.04 ± 0.00	31.0	–	–	–
A2, 12/05/72	6.0	–	0.04 ± 0.01	8.1	1.5	–	0.20 ± 0.08	1.8	–	–	–
B2, 1/04/73	8.9	22.4	–	–	3.3	5.5	0.67 ± 0.01	–	18.8	–	0.09 ± 0.01
B2, 5/24/73	8.3	19.6	–	–	6.0	8.2	–	–	6.1	31.8	0.05 ± 0.00
EB1, 5/25/73	0.08	0.14	–	–	0.31	–	–	–	0.23	53.5	0.06 ± 0.05

[a]Dilutions of material were spread on basal medium with and without 6 ppm of HgCl$_2$ or 0.3 ppm of PMA added. The percent of the total, viable, aerobic, heterotrophic bacterial population resistant to mercury was calculated.
[b]Water was filtered through sterile 8-μm-pore-size Millipore filters.
[c]Supernatant solution resulting from centrifugation of sediment at $1610 \times g$ for 20 min.
[d]Collected with No. 20 nylon mesh plankton net.

sediment was determined, with the oil found to contain 4000 times higher a concentration than water samples (Walker and Colwell, 1974).

Thus, in heavy-metal-enriched environments, mercury and other heavy metals soluble in oil may concentrate in the oil, and the metal-resistant microorganisms present in the metal-rich, oil-laden samples will be capable of utilizing the oil and perhaps playing very significant roles in the biological transfer of both metals and oil in the food chain. Principal among the metals greatly concentrated in oils are chromium, copper, nickel, lead, and zinc (Brinckman and Iverson, 1975).

METAL-RESISTANT BACTERIA AS AN INDEX OF POTENTIAL METAL MOBILIZATION

A variety of bacterial cultures were isolated and characterized. Figure 17-1 shows the average comparative population distribution of $HgCl_2$-resistant and total populations of bacteria (Nelson et al., 1972). A greater diversity of genera was noted for the total culturable population. A relative enrichment for *Pseudomonas* spp. was found among the resistant population.

Subsequently, a total of 230 strains of metal-resistant bacteria from water and sediment samples collected in Chesapeake Bay were isolated on a medium containing cobalt, lead, mercury, or molybdenum. As a control, a set of 71 cultures were isolated simultaneously on the medium without metals (Austin et al., 1977). Including 12 reference strains, the total number of strains analyzed by numerical taxonomy was 324 with 112 biochemical, cultural, morphological, and physiological characters tested. Interestingly, *Pseudomonas* spp. were found to be the predominant organisms among the metal-tolerant organisms. There was no demonstration of specificity of taxa to water or sediment or to

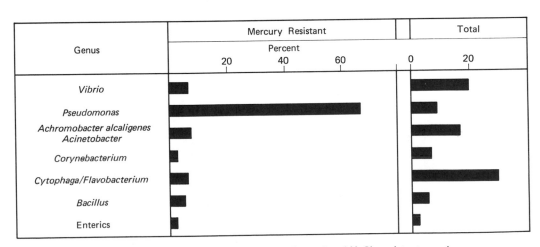

Fig. 17-1   Average distribution of genera in total and $HgCl_2$-resistant populations.

areas of the Chesapeake Bay, such as Colgate Creek or Chesapeake Beach, two of the stations sampled (i.e., a lack of geographical specificity). The distribution of metal-tolerant bacterial taxa was observed to be similar in the polluted and unpolluted areas of Chesapeake Bay. However, *Bacillus* and coryneforms resistant to heavy metals were recovered predominantly from sediment. *Erwinia carotovora*, a representative of the Enterobacteriaceae, was identified among the metal-tolerant strains. Thus the metal-tolerant bacteria found in Chesapeake Bay water and sediment comprise only a few taxa (Austin *et al.*, 1977).

The heavy-metal-resistant bacteria were subsequently also found to be antibiotic resistant; ampicillin and chloramphenicol were the antibiotics to which resistance was most common (Allen *et al.*, 1977). Studies of clinical isolates of *Staphylococcus aureus* had previously demonstrated a correlation between resistance to penicillin, erythromycin, and tetracycline and tolerance to mercury, lead, cadmium, and zinc (Hall, 1970; Novick and Roth, 1968). In the studies of Chesapeake Bay metal-resistant bacteria, resistance to antibiotics, including ampicillin, chloramphenicol, gentamicin, kanamycin, streptomycin and tetracycline, was demonstrated (Table 17-2). Multiple-antibiotic resistance [i.e., resistance to two or more antibiotics (usually ampicillin and chloramphenicol)] was found to occur in 16 percent of the strains tested. Resistance to ampicillin (24 percent) and to chloramphenicol (32 percent) were relatively common, whereas few strains were resistant to gentamicin (1 percent), kanamycin (2 percent), or tetracycline (3 percent).

Strains isolated from media containing cobalt, lead, mercury, and molybdenum demonstrated resistance to ampicillin and chloramphenicol, whereas only a few of the strains isolated from media containing lead or molybdenum were resistant to gentamicin (Table 17-2). Most of the strains in taxa comprising the metal-resistant strains were resistant to one or more antibiotics (Table 17-3). Bacteria resistant to both antibiotics and heavy metals were present in greater abundance in polluted sites, notably Colgate Creek in Baltimore Harbor of Chesapeake Bay.

Taking a different approach, antibiotic-resistant bacteria present in the surface water and sediment samples of Chesapeake Bay were enumerated and identified. The media employed included ampicillin, chloramphenicol, kanamycin, tetracycline, gentamicin, and streptomycin (Allen and Colwell, unpublished data) and the number of antibiotic resistant, colony-forming units (Cfu) were calculated from plate counts. The percentage of the total viable aerobic heterotrophic bacterial population that was antibiotic resistant was determined for each station and sample. Statistical analyses of the data were performed by computer using statistical routines of SAS-76 and BMDP (Dixon, 1975). The antibiotic-resistant strains were also tested for resistance to heavy metals, including cadmium, chromium, cobalt, lead, mercury, and tin.

As in the case of metal-resistant isolates, a polluted site in Chesapeake Bay, Baltimore Harbor, yielded the consistently greater percentage of antibiotic-resistant bacteria, compared to Chesapeake Beach and Eastern Bay. The smaller populations of antibiotic-resistant bacteria were similar at Chesapeake Beach and Eastern Bay (Tables 17-4, 17-5, and 17-6).

Interestingly, the proportion of the bacterial populations resistant to kanamycin was much higher than that resistant to other antibiotics. In Baltimore Harbor, kanamycin-

TABLE 17-2 Antibiotic Resistance of Metal-Tolerant Bacteria

Isolation Medium	Number of Strains Examined[a]	Percent Resistant to:					
		Ampicillin	Chloramphenicol	Gentamicin	Kanamycin	Streptomycin	Tetracycline
Cobalt	52 (21)	33	67	0	5	14	5
Lead	82 (81)	10	21	1	0	4	3
Mercury	60 (57)	56	25	0	4	7	5
Molybdenum	36 (33)	12	36	3	3	15	0
Control	71 (56)	18	41	0	0	25	2
Total number of strains	301 (248)	24	32	1	2	10	3

[a]Number in parentheses refers to number of strains growing on Mueller–Hinton medium and, thereby, tested for antibiotic resistance.

TABLE 17-3 Metal Resistance of Genera Identified during the Study

Phenon[a]	Presumptive Identification	Number of Strains[b]	Percent Resistant to:[c]					
			Cd^{2+}	Cr^{3+}	Co^{2+}	Pb^{2+}	Hg^{2+}	MoO_4^{2-}
1a	Pseudomonas fluorescens	8 (8)	100	88	87	87	75	100
1b	P. alcaligenes	51 (50)	16	68	39	100	82	100
1c	Pseudomonas spp.	3 (3)	100	100	100	100	33	100
1d	Pseudomonas spp.	2 (2)	100	100	50	100	100	100
2	P. maltophilia	43 (39)	2	32	9	88	88	98
3	Erwinia carotovora	2 (2)	50	100	0	100	50	100
4a	Unidentified yellow chromogens (gram-negative rods)	52 (15)	48	60	73	56	60	76
4b	Unidentified yellow chromogens (gram-negative rods)	30 (21)	20	60	0	30	3	50
5	Unidentified gram-negative rods	2 (2)	50	0	100	100	50	100
6	Coryneforms	19 (18)	10	42	37	100	100	100
7	Coryneforms	7 (7)	0	57	14	100	100	100
8	Mycobacterium spp.	12 (12)	0	75	17	100	33	83
9	Bacillus megaterium	53 (53)	38	91	58	100	91	100
10	Bacillus spp.	2 (2)	50	50	100	100	100	100
11	Bacillus spp.	6 (6)	17	33	17	100	83	100
12	Bacillus spp.	2 (2)	50	50	50	100	100	100

[a]Identification of the metal-tolerant bacteria was achieved using numerical taxonomy, the results of which have been published separately (Austin et al., 1977).
[b]Numbers in parentheses refer to the number of strains capable of growth on Mueller–Hinton medium and included in the study.
[c]Concentration of heavy metals used in the study were: 100 ppm (Cd^{2+}, Cr^{3+}, Co^{2+}, Pb^{2+}, and MoO_4^{2-}) and 10 ppm (Hg^{2+}).

resistant bacteria comprised more than 25 percent and 50 percent of the total bacterial population in sediment and surface water, respectively. However, in contrast to single antibiotic-resistant, multiple-antibiotic resistance did not show significant specificity to sampling site.

The majority of the antibiotic-resistant bacteria were identified as *Pseudomonas* spp. Other groups included *Aeromonas, Vibrio, Alcaligenes, Acinetobacter, Bacillus, Flavobacterium, Micrococcus, Staphylococcus,* coryneforms, and Enterobacteriaceae (Fig. 17-1).

The gram-positive, antibiotic-resistant strains were predominantly from sediment, with the exception of *Micrococcus*, which, with *Flavobacterium*, was isolated exclusively from surface water samples. *Acinetobacter* isolates were recovered entirely from Jones Falls in Baltimore Harbor, as were the majority of the Enterobacteriaceae isolates.

Multiple resistance (i.e., resistance to two or more antibiotics, up to 11) occurred in the majority of isolates (76 percent). Of these 18 percent were resistant to five antibiotics, and 15 percent to six.

When tested for heavy-metal resistance, the antibiotic-resistant isolates were also resistant to heavy metals (Table 17-7). Resistance to chromium, cobalt, lead and tin was most prevalent. Among the ampicillin-containing combinations, resistance to chromium was demonstrated by 96 to 100 percent, cobalt 94 and 96 percent, lead 98 and 100 percent, and tin 98 and 96 percent, of isolates resistant to ampicillin–kanamycin and

TABLE 17-4 Enumeration of Antibiotic-Resistant Bacteria in Surface Water Collected at Jones Falls

						Percent of TVC							
Antibiotic	Apr. 1976	May 1976	June 1976	July 1976	Sept. 1976	Oct. 1976	Nov. 1976	Dec. 1976	Jan. 1977	Feb. 1977	Mar. 1977	Apr. 1977	May 1977
Ampicillin	84	18	8	5	4	2	0.9	40	1	11	2	3	6
Chloramphenicol	54	31	0.9	3	4	3	4	15	1	11	3	13	4
Gentamicin	54	8	41	11	8	6	4	11	6	41	6	23	23
Kanamycin	100	37	53	18	21	29	3	21	1	50	80	92	100
Streptomycin	—	21	26	12	9	24	4	9	5	1	29	31	23
Tetracycline	78	9	2	4	2	1	9	10	0.2	1	1	2	9
Ampicillin–chloramphenicol	38	5	0.2	0.5	0.4	0.9	0.2	5	0.2	2	0.5	<0.1	0.9
Ampicillin–gentamycin	8	1	0.3	2	<0.1	0.1	0.1	1	<0.1	0.8	0.5	0.2	0.4
Ampicillin–kanamycin	35	7	3	2	0.2	0.3	0.1	1	<0.1	1	0.5	0.2	0.4
Ampicillin–streptomycin	26	3	1	2	0.4	0.7	0.3	2	<0.1	3	2	0.2	0.7
Ampicillin–tetracycline	18	2	0.5	0.7	0.8	0.5	0.7	4	0.1	4	2	<0.1	2
Chloramphenicol–gentamycin	100	6	0.7	0.8	5	<0.1	3	2	<0.1	1	0.2	0.1	0.6
Chloramphenical–kanamycin	17	3	3	1	2	0.9	0.2	11	<0.1	1	0.2	0.4	0.2
Chloramphenicol–streptomycin	24	4	0.2	0.8	1	0.4	0.2	3	0.2	3	1	2	1
Chloramphenicol–tetracycline	11	1	0.5	0.1	1	0.1	0.6	13	0.3	1	0.6	0.3	1
Gentamicin–streptomycin	33	5	38	14	9	6	1	5	2	6	11	4	5
Gentamicin–tetracycline	5	0.2	0.1	0.1	0.4	0.1	0.3	0.1	0.1	1	0.1	<0.1	0.6
Kanamycin–streptomycin	34	5	13	0.1	7	7	1	13	9	8	18	4	17
Kanamycin–tetracycline	13	0.9	0.9	12	1	0.5	9	3	<0.1	2	<0.1	<0.1	3
Streptomycin–tetracycline	8	0.3	0.1	0.3	0.3	0.5	0.5	20	<0.1	0.1	0.1	0.2	1

TABLE 17-5 Enumeration of Antibiotic Resistant Bacteria in Surface Water Collected at Chesapeake Beach

Antibiotic	Percent of TVC										
	Apr. 1976	May 1976	June 1976	Aug. 1976	Oct. 1976	Nov. 1976	Dec. 1976	Feb. 1977	Mar. 1977	Apr. 1977	May 1977
Ampicillin	1	2	–	2	0.1	1	0.2	<0.1	0.1	<0.1	0.3
Chloramphenicol	2	2	–	<0.1	<0.1	0.4	0.2	<0.1	0.1	<0.1	<0.1
Gentamicin	0.5	2	–	0.8	<0.1	0.2	0.2	<0.1	<0.1	0.2	0.5
Kanamycin	1	13	–	4	0.1	0.4	2	<0.1	0.1	0.2	8
Streptomycin	0.4	13	–	1	0.1	0.1	9	<0.1	0.2	1	3
Tetracycline	2	0.5	–	3	<0.1	<0.1	0.2	<0.1	0.2	0.2	<0.1
Ampicillin–chloramphenicol	1	–	<0.1	1	<0.1	<0.1	0.7	<0.1	<0.1	<0.1	<0.1
Ampicillin–gentamicin	<0.1	<0.1	<0.1	0.9	<0.1	<0.1	8	<0.1	0.1	<0.1	<0.1
Ampicillin–kanamycin	<0.1	5	<0.1	1	<0.1	<0.1	1	<0.1	0.1	<0.1	<0.1
Ampicillin–streptomycin	<0.1	<0.1	<0.1	0.8	<0.1	<0.1	7	<0.1	0.1	<0.1	<0.1
Ampicillin–tetracycline	1	<0.1	<0.1	0.6	<0.1	0.4	0.2	<0.1	0.1	<0.1	<0.1
Chloramphenicol–gentamicin	<0.1	<0.1	<0.1	4	<0.1	<0.1	0.3	<0.1	0.1	<0.1	<0.1
Chloramphenicol–kanamycin	0.2	3	<0.1	<0.1	<0.1	<0.1	0.2	<0.1	0.1	<0.1	<0.1
Chloramphenicol–streptomycin	<0.1	<0.1	<0.1	<0.1	<0.1	<0.1	0.4	<0.1	<0.1	<0.1	<0.1
Chloramphenicol tetracycline	2	<0.1	<0.1	<0.1	0.1	0.2	0.1	<0.1	0.1	<0.1	<0.1
Gentamicin–streptomycin	2	<0.1	0.1	0.2	0.1	3	2	<0.1	0.2	0.2	<0.1
Gentamicin–tetracycline	<0.1	<0.1	0.1	0.3	<0.1	0.1	0.2	<0.1	<0.1	<0.1	<0.1
Kanamycin–streptomycin	0.6	4	2	12	<0.1	0.2	8	<0.1	<0.1	0.1	0.5
Kanamycin–tetracycline	<0.1	0.1	<0.1	1	0.4	0.2	0.4	<0.1	<0.1	<0.1	<0.1
Streptomycin tetracycline	<0.1	<0.1	<0.1	0.3	<0.1	<0.1	0.4	<0.1	<0.1	<0.1	<0.1

TABLE 17-6 Enumeration of Antibiotic-Resistant Bacteria in Surface Water Collected at Eastern Bay

	Percent of TVC											
Antibiotic	Apr. 1976	May 1976	June 1976	July 1976	Sept. 1976	Oct. 1976	Nov. 1976	Dec. 1976	Feb. 1977	Mar. 1977	Apr. 1977	May 1977
Ampicillin	4	2	—	6	0.1	<0.1	<0.1	0.1	<0.1	<0.1	0.1	0.3
Chloramphenicol	5	0.3	—	1	0.1	<0.1	<0.1	0.1	<0.1	<0.1	<0.1	<0.1
Gentamicin	0.3	0.6	—	4	0.2	<0.1	<0.1	0.1	<0.1	0.2	0.2	0.7
Kanamycin	1	8	—	22	0.1	0.3	0.1	0.1	0.2	0.3	3	15
Streptomycin	0.3	34	—	8	4	0.2	<0.1	0.3	0.1	0.4	4	16
Tetracycline	6	0.2	—	2	0.6	<0.1	0.2	0.1	<0.1	<0.1	<0.1	<0.1
Ampicillin–chloramphenicol	6	0.3	0.3	1	1	<0.1	<0.1	0.1	<0.1	0.1	<0.1	<0.1
Ampicillin–gentamycin	<0.1	<0.1	2	1	0.2	<0.1	<0.1	0.1	<0.1	<0.1	<0.1	<0.1
Ampicillin–kanamycin	0.6	0.4	0.5	0.4	2	<0.1	<0.1	0.1	<0.1	<0.1	<0.1	<0.1
Ampicillin–streptomycin	—	0.1	0.3	0.4	0.3	<0.1	<0.1	0.1	<0.1	<0.1	<0.1	<0.1
Ampicillin–tetracycline	6	2	<0.1	2	0.3	<0.1	0.2	0.1	<0.1	<0.1	<0.1	<0.1
Chloramphenicol–gentamycin	<0.1	0.2	0.1	0.2	0.2	<0.1	<0.1	0.1	<0.1	<0.1	<0.1	<0.1
Chloramphenicol–kanamycin	<0.1	0.4	<0.1	0.4	2	<0.1	<0.1	0.1	<0.1	<0.1	<0.1	<0.1
Chloramphenicol–streptomycin	<0.1	0.3	<0.1	0.8	4	<0.1	<0.1	0.1	<0.1	0.1	<0.1	<0.1
Chloramphenicol–tetracycline	5	<0.1	<0.1	0.4	2	<0.1	<0.1	0.4	<0.1	<0.1	<0.1	<0.1
Gentamicin–streptomycin	0.3	0.1	0.3	10	2	<0.1	<0.1	0.1	<0.1	<0.1	0.1	<0.1
Gentamicin–tetracycline	<0.1	<0.1	0.3	0.6	2	<0.1	<0.1	0.3	<0.1	<0.1	<0.1	<0.1
Kanamicin–streptomycin	0.1	2	1	1	2	<0.1	0.2	0.1	<0.1	0.2	0.2	<0.1
Kanamycin–tetracycline	<0.1	<0.1	0.7	5	2	<0.1	0.1	0.1	<0.1	<0.1	<0.1	<0.1
Streptomycin–tetracycline	<0.1	<0.1	0.2	0.6	2	<0.1	<0.1	0.3	<0.1	<0.1	<0.1	<0.1

TABLE 17-7 Resistance to Heavy Metals Demonstrated by Antibiotic-Resistant
Bacteria Isolated from Chesapeake Bay

Antibiotic Combination in Isolation Medium	Number of Strains Selected for Testing	Heavy Metal[a]					
		Cd^{2+}	Cr^{3+}	Co^{2+}	Pb^{2+}	Hg^{2+}	Sn^{2+}
Ampicillin–kanamycin	123	76	96	94	98	20	98
Ampicillin–streptomycin	136	80	100	96	100	32	96
Chloramphenicol–gentamicin	42	52	79	71	78	40	78
Chloramphenicol–kanamycin	87	41	80	73	77	9	79
Gentamicin–tetracycline	85	55	81	69	85	14	85
Streptomycin–tetracycline	158	45	75	70	96	20	94
Total number of strains	631	60	86	81	92	22	91

[a]Concentration of heavy metals used were: 10 ppm (Cd^{2+}, Cr^{3+}, Co^{2+}, Pb^{2+}, Hg^{2+}, and Sn^{2+}). Numbers given are percent of strains tested and found to be resistant to the heavy metal indicated.

ampicillin–streptomycin, respectively. Strains resistant to streptomycin–tetracycline combinations showed high resistance to lead (96 percent) and tin (94 percent), with lower percentages to chromium (75 percent) and cobalt (70 percent).

Interestingly, only 9 and 14 percent of isolated strains resistant to chloramphenicol–kanamycin and gentamicin–tetracycline, respectively, were resistant to mercury. The penicillin–streptomycin and chloramphenicol–gentamicin isolates were markedly resistant to mercury (32 and 42 percent).

A large proportion of antibiotic-resistant bacteria (94 percent) were multiply resistant to heavy metals (i.e., to two and up to six heavy metals). Whereas no consistent pattern of multiple resistance to antibiotics was observed, a pattern for metal resistance was noted, the most common of which was resistance to cadmium–chromium–cobalt–lead–tin (39 percent). A number of isolates (18 percent) showed resistance to all six of the metals tested.

A corollary of the working hypothesis related to heavy-metal indicator bacteria was that the presence of bacteria in the autochthonous microbial population which demonstrated resistance to mercury could be used as an index of potential microbial activity in the generation of Hg^0 from inorganic or organic sources of mercury (Colwell and Nelson, 1975). The hypothesized relationship between mercury resistance and metabolism was suggested early on by experiments in which Hg^0 was generated by a mixture of bacterial colonies growing on an $HgCl_2$-containing nutrient agar. Fogging of areas on an x-ray film exposed to the inoculated mercury-containing agar plates resulted from reduction of the Ag^+ emulsion by Hg^0 (Nelson et al., 1973). Reduction (detoxification) of $HgCl_2$ by individual mercury-resistant colonies, randomly selected from platings of sediment and water, was assayed by measuring the disappearance of ^{203}Hg from aerated, $HgCl_2$-containing, buffered salts suspensions of bacteria (Nelson et al., 1973). A large proportion of the isolates (77 percent) evolved ^{203}Hg in amounts that exceeded those from killed and uninoculated controls. Thus examination of isolates corroborated the relationship between resistance and metabolism. Subsquent work on the morphology, including ultrastructure, of mercury-resistant bacteria revealed a variety of adaptive mechanisms evolved by these bacteria, culminating in resistance to and/or metabolism of heavy metals.

SEASONAL VARIATION IN POPULATIONS
OF METAL AND ANTIBIOTIC RESISTANT BACTERIA

Studies of the seasonal incidence of mercury-resistant bacteria were undertaken in 1972–1973. A distinct periodicity was evident for both water and sediment (Figs. 17-2 and 17-3). A major peak was noted in mercury-resistant bacteria in the spring months of both 1972 and 1973 and, possibly, a secondary peak in the fall of 1972. Several physical parameters were found to demonstrate a seasonal periodicity (Fig. 17-4) characteristic of the Chesapeake Bay estuary (Schubel, 1972).

The data were subjected to multiple stepwise regression analysis to detect possible relationships between the selected physical parameters and number of resistant organisms and to establish a basis for comparison of the three sites. Resistance in sediments was found to be correlated positively with water transparency (Secchi disk), dissolved oxygen, and total mercury concentration in the sediment. Resistance was not observed to be related to total viable counts.

Interestingly, seasonal fluctuations in the population of metal-resistant bacteria were also observed in subsequent studies done in 1975. The antibiotic-resistant populations did not demonstrate a marked seasonality, but it should be noted that the winter of 1976–1977, when the antibiotic-resistant populations of bacteria were being analyzed, was one of the most severe on record, with the entire Chesapeake Bay under ice cover. Also, the antibiotic-resistant bacteria included a larger number of genera, including more gram-positive genera, and hence was less species specific.

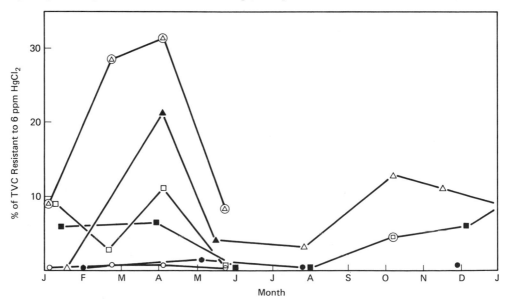

Fig. 17-2   Seasonal variation in percent of HgCl$_2$-resistant bacteria in water. The percent of total viable, aerobic, heterotrophic bacterial count (TVC) capable of growth on a solid medium containing 6 ppm HgCl$_2$ was determined. Water samples from Station B-1, 1972 (▲); Station B-2, 1972 (△); Station A-1, 1972 (◉); Station A-2, 1972 and 1973 (■, □); and Station EB-1, 1972 and 1973 (●, ○) were plated and incubated at 25°C for 1 week.

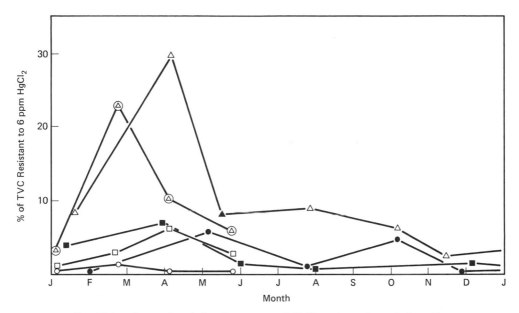

Fig. 17-3   Seasonal variation in percent of HgCl₂-resistant bacteria in sediment. The percent of total, viable, aerobic heterotrophic bacterial count (TVC) capable of growth on a solid medium containing 6 ppm HgCl₂ was determined. Sediment samples from Station B-1, 1972 (▲); Station B-2, 1972 and 1973 (△, ⊕); Station A-2, 1972 and 1973 (■, □); and Station EB-1, 1972 and 1973 (●, ○) were plated and incubated at 25°C for 1 week.

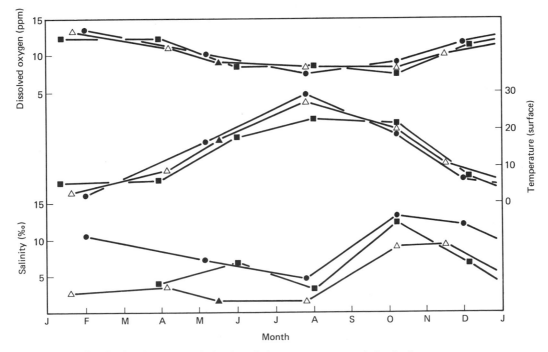

Fig. 17-4   Seasonal variation in salinity, temperature, and dissolved oxygen concentration in surface waters from three locations in Chesapeake Bay. (▲), Station B-1; (△), Station B-2; (■) Station A-2; (●), Station EB-1.

RELATIONSHIP OF TOTAL METAL CONCENTRATIONS AND PERCENTAGE OF THE TOTAL VIABLE BACTERIAL COUNT RESISTANT TO METALS

A working hypothesis underlying the microbial heavy-metal-resistance studies was that the percentage of bacteria in the autochthonous microbial population which demonstrated resistance to metals or antibiotics could be used as an index of the concentration of heavy metals in the environment, or, conversely, that metals in the environment will have an observable effect on the bacterial community, in terms of predisposition to metal resistance and metabolism.

A plot of $HgCl_2$ resistance versus total mercury in sediments, shown in Figure 17-5, suggested a relationship between resistance and in situ mercury concentration. Data obtained for six stations sampled routinely, as well as stations sampled on an irregular basis are shown. A positive correlation coefficient of 0.58 was calculated, when average percent resistance and mercury concentration were compared, indicating that mercury resistance is related, at least in part, to the mercury concentration in the environment. The plankton data, although limited, indicate a similar relationship. A stricter relationship may be blurred by background fluctuation, due to temporal and positional variation in sample populations, as well as to the probable variety of chemical forms of mercury present in the given environment. Chi-square analysis of resistant populations from six stations in three areas showed significant differences at $\alpha = 0.1$. The observed variation in mercury resistance from site to site thus may have been the result of population differences, of mercury-mediated selection, or a combination of both. Clearly, sites encompassing a complete range of mercury concentrations must be surveyed to separate out these effects.

The distribution of total mercury concentrations (Fig. 17-6) in surface sediments

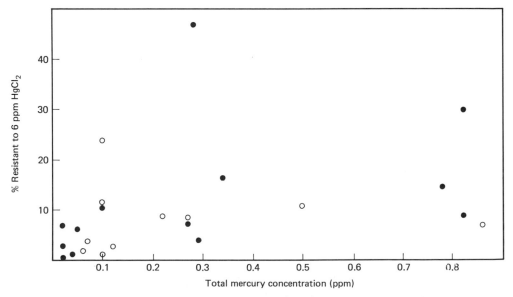

Fig. 17-5   Percent of total viable count (TVC) resistant to 6 ppm $HgCl_2$ and total concentration of mercury in sediment. (●), regularly sampled; (○) spot checked.

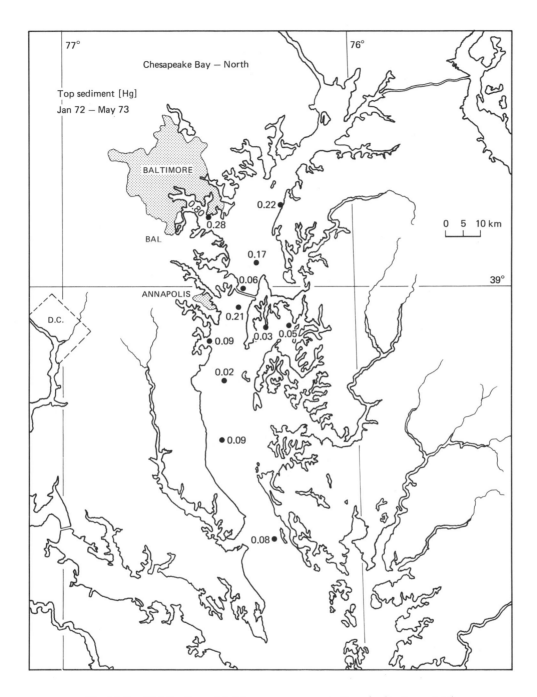

Fig. 17-6   Distribution of total mercury concentrations (μg/g, dry weight) in surface (0 to 5 cm) sediments of the northern reach of the Chesapeake Bay collected over an annual period. In general, those locales subject to heavy commercial traffic and industry show a 10- to 50-fold increase in mercury loading as compared with "background" levels (0.01 to 0.02 ppm) observed at pristine sites.

collected over an annual period (Brinckman and Iverson, 1975), when compared with the distribution and incidence of mercury-resistant bacteria, permit the conclusion that the proportion of mercury- and other heavy metal-resistant bacteria in the total viable, aerobic, heterotrophic bacterial population serves as a valid index of potential mercury mobilization in Chesapeake Bay water and sediment.

PLASMID-MEDIATED HEAVY-METAL AND ANTIBIOTIC RESISTANCE IN AQUATIC BACTERIA

The location of heavy-metal and antibiotic resistance genes on plasmids (R factors) was documented in the literature at the time when our studies were carried out (Novick, 1969). It was, therefore, interesting to determine the extent of this phenomenon in natural bacterial populations. Plasmids were subsequently found in mercury-resistant strains isolated from the Chesapeake Bay and deep ocean (Cayman and Puerto Rico Trenches). Further, mercury volatilization was demonstrated to be plasmid mediated, after appropriate curing experiments were conducted (Olson *et al.*, 1978, 1979).

More recent data were obtained using antibiotic and heavy-metal-resistant strains described in the studies cited above (Allen and Colwell, *manuscript in preparation*; Barja and Colwell, *manuscript in preparation*). Methods specifically designed to detect large plasmids (Portnoy *et al.*, 1981; Kado and Liu, 1981) were employed in the isolation and characterization of plasmid DNA. A total of 154 metal- and antibiotic-resistant strains isolated from water and sediment samples, which represented three levels of pollution in the Chesapeake Bay, were examined. Forty-eight strains (31 percent) harbored at least one plasmid and 27 were multi-plasmid-bearing strains. Plasmid carriage was higher among strains isolated from the polluted location, Jones Falls (59 percent), compared to Chesapeake Beach isolates (17 percent) and Eastern Bay isolates (14 percent). In addition, 21 of 33 Jones Falls strains harbored more than a single plasmid and all strains with more than two plasmids were from this site. The difference in plasmid carriage between water (27 percent) and sediment (32 percent) strains was not significant.

Of the 48 strains that harbored plasmids, 23 had plasmids with a mass of > 20 megadaltons (Mdal), 12 strains contained plasmids with a mass < 20 Mdal, and 13 strains had both plasmid types. Nearly all plasmids of > 50 Mdal mass were detected in strains isolated at Jones Falls. Plasmids larger than 20 Mdal are commonly those which can mediate their own conjugal transfer (Novick, 1969). This process may explain the increased occurrence of resistant strains in polluted locations of Chesapeake Bay (see above).

When strains selected on mercury-emended media were tested, 26 of 38 (68 percent) harbored plasmids (Barkay, Barja, and Colwell, *manuscript in preparation*). Plasmids were detected in 80 percent of the gram-negative bacteria and in 59 percent of the gram-positive bacteria isolated from cadmium-laden sediment using a cadmium-enriched medium (Litchfield *et al.*, 1982).

Thus, bacterial strains carrying heavy-metal-resistance plasmids, or strains able to gain these plasmids via genetic transfer, thrive in heavy-metal-enriched habitats. These plasmids may play a role in the observed multiple-antibiotic and metal-resistance patterns among the Chesapeake Bay bacterial populations. Furthermore, a possible relationship between antibiotic resistance and heavy-metal resistance among bacteria in the natural environment is indicated.

Heavy-metal- and antibiotic-resistant bacteria were found to be associated with metal-laden sediments and waters. It is possible that indicator bacteria useful in exploration for mineral resources can be identified. However, more promising are biotechnological approaches, such as gene cloning and related genetic engineering methods providing "gene probes" for metal-deposit exploration and for recovery of metals from wastes, mining tailings, and low concentration sources of elements in critically short supply.

REFERENCES

Allen D. A., Austin, B., and Colwell, R. R., 1977, Antibiotic resistance patterns of metal-tolerant bacteria isolated from an estuary: *Antimicrob. Agents Chemother.*, v. 12, p. 545–547.

Austin, B., Allen, D. A., Mills, A. L., and Colwell, R. R., 1977, Numerical taxonomy of heavy metal-tolerant bacteria isolated from an estuary: *Can. J. Microbiol.*, v. 23, p. 1433–1447.

Brinckman, F. E., and Iverson, W. P., 1975, Chemical and bacterial cycling of heavy metals in the estuarine system: *American Chemical Society Symposium on "Marine Chemistry in the Coastal Environment," 169th ACS Natl. Meet.*, Philadelphia, Apr. 8-10, 1975.

Colwell, R. R., and Nelson, J. D., Jr., 1975, Metabolism of mercury compounds in micro-organisms: *U.S. EPA Ecol. Res. Ser. Rep. EPA-600/3-75-007*, National Technical Information Service, Springfield, Va., 84 p.

Dixon, W. J., 1975, *BMDP Biomedical Computer Programs:* University of California Press, Berkeley.

Hall, B. M., 1970, Distribution of mercury resistance among *Staphylococcus aureus* isolated from a hospital community: *J. Hyg., Comb.*, v. 68, p. 111–119.

Kado, C. I., and Liu, S., 1981, Rapid procedure for detection and isolation of large and small plasmids: *J. Bacteriol.*, v. 145, p. 1365–1373.

Klein, D., and Goldberg, E., 1970, Mercury in the marine environment: *Environ. Sci. Technol.*, v. 4, p. 765–768.

Litchfield, C. D., Devanas, M. U., Zindulis, J., Meskill, M., Freedman, J., and McLean, C., 1982, Influence of cadmium on the microbial populations and processes in sediments from the New York Bight apex, p. 587–604 in Mayer, G. F. (Ed.), *Ecological Stress in the New York Bight: Science and Management:* Estuarine Res. Fed., Columbia, S.C.

Nelson, J. D., Jr., McClam, H. L., and Colwell, R. R., 1972, The ecology of mercury-resistant bacteria in Chesapeake Bay: *Proc. 8th Annu. Conf. Mar. Technol. Soc., Washington, D.C., Mar. Technol. Soc. Publ.*, p. 303–312.

—— , Blair, W., Brinckman, F. E., Colwell, R. R., and Iverson, W. P., 1973, Biodegradation of phenylmercuric acetate by mercury-resistant bacteria: *App. Microbiol.*, v. 26, p. 321–326.

Novick, R. P., 1969, Extrachromosomal inheritance in bacteria: *Bacteriol. Rev.*, v. 33, p. 210–235.

—— , and Roth, C., 1968, Plasmid-linked resistance to inorganic salts in *Staphylococcus aureus: J. Bacteriol.*, v. 95, p. 1335–1342.

Olson, B. H., Barkay, T., Nies, D., Bellama, J. M., and Colwell, R. R., 1978, Plasmid mediation of mercury volatilization and methylation by estuarine bacteria: *Dev. Ind. Microbiol.*, v. 20, p. 275–284.

_____ , Barkay, T., and Colwell, R. R., 1979, The role of plasmids in mercury transformation by bacteria isolated from the aquatic environment: *Appl. Environ. Microb.*, v. 38, p. 478–485.

Portnoy, D. A., Moseley, S. L., and Falkow, S., 1981, Characterization of plasmids and plasmid-associated determinants of *Yersinia enterocolitica* pathogenesis: *Infect. Immun.*, v. 31, p. 775–782.

Schubel, J. R., 1972, The physical and chemical conditions of the Chesapeake Bay: *J. Wash. Acad. Sci.*, v. 62, p. 56–87.

Walker, J. D., and Colwell, R. R., 1974, Mercury-resistant bacteria and petroleum degradation: *Appl. Microbiol.*, v. 27, p. 285–287.

Williams, P. M., and Weiss, H. V., 1973, Mercury in the marine environment: concentration in sea water and in a pelagic food chain: *J. Fish. Res. Board Can.*, v. 30, p. 293–295.

G. B. Michaels
School of Natural and Social Sciences
Western State College
Gunnison, Colorado
and
Gainesville College
Gainesville, Georgia

J. J. Hill and D. J. Schneck
Department of Biology
Western State College
Gunnison, Colorado

18

INCIDENCE OF ANTIBIOTIC AND HEAVY METAL-RESISTANT BACTERIA IN A HEAVILY MINERALIZED HIGH-ALTITUDE WATERSHED: A CASE HISTORY

ABSTRACT

Antibiotic resistance has usually been associated with antibiotic use, either for medical purposes or in animal feeds. Recently, however, there has been an increasing awareness of antibiotic resistance apparently associated with other factors. In a diversity study of stream populations, we found 17 of 256 isolates exhibited multiple antibiotic resistance; 15 of these were also resistant to at least one heavy metal. During 1981, nine stream sites were surveyed for antibiotic- and heavy metal-resistant bacteria. The highest incidence of chloramphenicol and kanamycin resistance was found at control sites remote from human activity while streptomycin and tetracycline resistances were highest in populated areas. Few heavy-metal-resistant strains were found in the initial screen, although 50 percent of randomly selected multiply antibiotic-resistant isolates were resistant to cadmium, lead, or mercury. Similar data were obtained from the same sites in 1982. When randomly selected resistant isolates were screened for multiple resistance by a disk method, 95 percent were resistant to five or more antibiotics. More than 50 percent of these were resistant to arsenate, cadmium, mercury or lead, with multiple resistance more common than single.

A preliminary survey of Tomichi Creek in the Tomichi mining district indicates that there may be a correlation between the incidence of antibiotic resistance in the stream bacteria and minerals in the area.

INTRODUCTION

With increasing use of antibiotics for treatment of disease and in feeds and drinking water for improvement of animal production, we have become increasingly aware of the incidence of antibiotic-resistant bacteria. Many reports in the literature document an increase in the incidence of antibiotic-resistant bacteria in environments where antibiotics are present (Huber et al., 1971; Feary et al., 1972; Richmond, 1972; Fontaine and Hoadley, 1976; Levy et al., 1976; Austen and Trust, 1981). Antibiotic-resistant bacteria are also found, however, in environments where antibiotics are not used (Maré, 1968; Huber et al., 1971; Cooke, 1976; Kelch and Lee, 1978). In recent years, increasing attention has been directed toward the relationship between antibiotic resistance and resistance to a variety of heavy metals. Environments containing or contaminated with heavy metals often have a high incidence of antibiotic-resistant bacteria (Timoney et al., 1978; Lighthart, 1979; Devanas et al., 1980; Sjogren and Port, 1981). Use of silver compounds for treatment of burns has also led to an increase of heavy-metal- and antibiotic-resistant bacteria in the hospital environment (McHugh et al., 1975). This correlation is not surprising since genes conferring resistance to antibiotics and heavy metals are often found on the same plasmid (Novick and Roth, 1968; Summers and Silver, 1972; Varma et al., 1976; Nakahara et al., 1977; Smith et al., 1978; Summers et al., 1978). Two other factors that may increase the incidence of antibiotic resistance are the presence of antibiotic-producing organisms in the environment (Soulides, 1965) and ultraviolet irradiation, since genes for ultraviolet resistance may be found on the same plasmids with antibiotic-resistance genes (Howarth, 1965; Drabble and Stocker, 1968; Marsh and Smith, 1969; Siccardi, 1969; Frazier and Zimmerman, 1980).

In this paper we report an unusually high incidence of antibiotic-resistant bacteria in both the streams and the native animal populations of a high-altitude watershed with low population density and limited use of antibiotics.

The sampling sites used in this study are in the watersheds of the Gunnison River and Tomichi Creek in central Colorado, an area bounded by 38°17'N–38°57'N latitude and 106°26'W–107°05'W longitude. The stream sites were selected to represent areas minimally affected by human or domestic animal activity, areas affected by domestic animals (primarily cattle) but not human activity, and areas affected by both human and animal activity. These sites are described in Table 18-1. A preliminary survey was also carried out along Tomichi Creek from its headwaters near Tomichi Pass to the regular sampling

TABLE 18-1 Description of Stream Sampling Sites

Site Designation	Description
TC-1	Tomichi Creek above old town of White Pine, approximately 7 miles from headwaters. Control site (minimal human/animal activity upstream); site lies between two highly mineralized areas which have been extensively mined for silver, gold, lead, and zinc
TC-3	Tomichi Creek 3 miles downstream from its confluence with Marshall Creek near the community of Sargents (approximately 13 miles from TC-1)
MC-1	Marshall Creek above the Homestake Pitch Project (uranium mine); grazing allotments (cattle) but minimal human activity; no dwellings
ER	East River, approximately 15 miles below town of Crested Butte; the area above Crested Butte is highly mineralized and has been extensively mined in the past
CmC	Cement Creek, tributary of East River, at the Cement Creek campground (U.S. Forest Service); grazing areas and dwellings upstream
WFC	Waterfall Creek, tributary of Cement Creek. Control site (no grazing allotments or dwellings, very little human activity: no known mines or evidence of mining.
TR	Taylor River, 2 miles north of community of Almont; dwellings and grazing areas upstream; areas around the headwaters of the Taylor and tributary streams have been mined in the past; no current mining activity
GR-A	Gunnison River at Almont campground (U.S. Forest Service), 1 mile downstream of the confluence of the East and Taylor Rivers at Almont, which forms the Gunnison River
GR-C	Gunnison River downstream of the city of Gunnison and below the sewage outfall

site below the community of Sargents (site TC-3) to determine possible effects of known metallogenic zones on the incidence of antibiotic resistance in the stream flora.

Fecal materials were collected from native animal species throughout the watershed area, with particular attention to locations potentially affecting stream sites.

METHODS

Water samples were collected in sterile Whirl-Pak bags and transported to the laboratory. The samples were diluted with sterile distilled water if necessary and plated by the spread plate method. Casein–peptone–starch (CPS) agar (Collins and Willoughby, 1962) as recommended by Staples and Fry (1973) for unpolluted waters was used for all water samples. For detecting antibiotic resistance, chloramphenicol, kanamycin, streptomycin, and tetracycline (Sigma Chemical Company) were incorporated into the CPS agar at a concentration of 25 μg/ml. The plates were incubated at 20°C for 72 hours. Antibiotic resistance was confirmed by reinoculation into the same medium.

Randomly selected antibiotic resistant isolates from the stream samples were screened for multiple resistance using antibiotic sensitivity disks (Difco Laboratories) as follows: ampicillin, 10 μg; chloromycetin (chloramphenicol), 30 μg; erythromycin, 15 μg; kanamycin, 30 μg; neomycin, 30 μg; streptomycin, 10 μg; tetracycline, 30 μg. These isolates were also tested for heavy-metal resistance by a stab method (A. Summers, University of Georgia, personal communication) using CPS agar containing 0.1 mM HgCl$_2$, 5 mM AgNO$_3$, 1 mM CdCl$_2$, 5 mM Pb (C$_2$H$_3$O$_2$)$_2$, 80 mM NaHAsO$_4$ · 7H$_2$O. A group of these multiply-resistant isolates were stored in CPS agar at 4°C and retested at 6 months for loss of resistance.

Animal fecal samples were collected using sterile forceps and placed in Whirl-Pak bags for transport. Sterile distilled water was added to each sample bag and the sample was thoroughly agitated. The samples were diluted, plated on MacConkey's agar (Difco Laboratories), and incubated at 35°C for 24 to 48 hours. Those plates with viable flora were then replica plated onto MacConkey's agar containing 25 μg/ml chloramphenicol, kanamycin, streptomycin, and tetracycline. Randomly selected fecal isolates were tested for heavy-metal resistance using MacConkey's agar containing metals as described for CPS agar above. Resistant isolates were confirmed by reinoculation into the same antibiotic- or heavy-metal-containing medium.

RESULTS

The incidence of antibiotic resistance among the aquatic bacterial populations at the stream sites during the summers of 1981 and 1982 is shown in Table 18-2. No tetracycline resistance was observed at any stream site in 1982. Table 18-3 gives the incidence of antibiotic resistance found in a preliminary survey along Tomichi Creek from its headwaters through a highly mineralized and previously mined area (designated as Tomichi town site) to two sites downstream.

When 94 resistant stream isolates were screened by the disk method for multiple antibiotic resistance using ampicillin, chloramphenicol, erythromycin, kanamycin, neomycin, streptomycin, and tetracycline, 95 percent were resistant to five or more anti-

TABLE 18-2 Average Incidence of Antibiotic-Resistant Bacteria
among Stream Populations in 1981 and 1982

Site	*Percent of Cultivable Heterotrophic Population Resistant to:*			
	Chloramphenicol	*Kanamycin*	*Streptomycin*	*Tetracycline*
TC-1	13.37	21.43	1.05	<0.01
TC-3	9.63	11.27	3.42	0.200
MC-1	11.93	10.35	0.72	<0.01
GR-A	8.56	15.17	0.93	<0.01
GR-C	6.88	9.59	1.25	<0.01
WFC	17.66	28.30	0.55	0
CmC	8.07	15.21	0.83	<0.01
ER	11.69	17.65	0.84	0.012
TR	9.05	16.80	0.55	0.013
$\overline{X}$, all sites	10.76	16.20	1.13	0.027
$\overline{X}$, control sites (TC-1, WFC)	15.52	24.86	0.80	<0.01
$\overline{X}$, human/animal-influenced sites	9.40	13.72	1.22	0.034

TABLE 18-3 Survey of Antibiotic Resistance along Tomichi Creek

Site	*Percent of Cultivable Heterotrophs Resistant to:*			
	Chloramphenicol	*Kanamycin*	*Streptomycin*	*Tetracycline*
Headwaters ponds	21.0	27.0	3.20	0
Tomichi town site	37.0	37.0	3.20	0
TC-1	18.0	16.0	1.0	0
TC-3	13.0	11.0	2.0	0

biotics. At least 50 percent of the multiply-resistant strains were resistant to one or more heavy-metal ions (arsenate, cadmium, lead, mercury, silver). Of 60 resistant strains stored at 4°C for 6 months, 40 percent had lost multiple-antibiotic resistance.

One hundred twenty-two samples of animal feces were obtained which contained viable flora. These samples were distributed among six animal groups as shown in Table 18-4. The percentage of samples with antibiotic resistance is shown in Table 18-5. Among the animal samples the highest incidence of resistance was to chloramphenicol (51 per-

TABLE 18-4 Animal Feces Analyzed for
Antibiotic-Resistant Bacteria

Animal	*Number of Specimens*
Elk (*Cervus canadensis*)	65
Deer (*Odocoileus hemionus*)	27
Black bear (*Ursus americanus*)	13
Marmot (*Marmota flaviventris*)	10
Rodent, species unknown	4
Coyote (*Canis latrans*)	3
	122

TABLE 18-5 Percent of Fecal Isolate Samples Showing Resistance
to Antibiotics Tested

	Number of Samples	Percent
No resistance	41	34
Resistance to one	18	15
Chloramphenicol 13		
Kanamycin 3		
Streptomycin 2		
Tetracycline 0		
Resistance to two (any combination)	17	14
Resistance to three (any combination)	25	21
Resistance to four (any combination)	21	17
Resistance to two or more	63	52

cent) followed by streptomycin (48 percent), kanamycin (39 percent), and tetracycline (34 percent). When 145 isolates were selected at random from the resistant strains in the fecal flora and screened against the other antibiotics used in the initial screen, 57 percent were resistant to two or more. Of these multiply resistant strains, 6 percent were resistant to mercury. No assays were made for resistance to other heavy metals.

DISCUSSION

Little information is available in the literature on the incidence of antibiotic resistance in the overall heterotrophic bacterial populations of freshwater systems, most studies having focused on the coliform bacteria which are used as indicators of water quality. Because the majority of native aquatic bacteria are gram negative, resistance to a number of antibiotics, such as the penicillins, would be anticipated, but resistance to broad-spectrum antibiotics might or might not be present. Antibiotic resistance for some aquatic species has been determined by Kelch and Lee (1978), but they did not determine the overall incidence of resistance in the aquatic population. Limited data on some aquatic species (using pure cultures) were published by van Dijck and van de Voorde (1976). It has also been shown that the incidence of plasmids in aquatic populations is widespread (Burton et al., 1982). All of these data suggest that the native aquatic species are generally more resistant than coliforms if there has been no previous exposure to antibiotics.

If antibiotic use (i.e., the presence of antibiotics) is the major selection pressure leading to increased antibiotic resistance it would be anticipated that bacterial populations at locations remote from human dwellings and domestic animals would show low levels of antibiotic resistance. Bacterial populations in populated areas would be expected to show higher levels of resistance. This has been demonstrated in a number of environments (Huber et al., 1971; Feary et al., 1972; Richmond, 1972; Fontaine and Hoadley, 1976; Levy et al., 1976; Austen and Trust, 1981). Our data on streptomycin and tetracycline resistance in the stream populations (Table 18-2) tend to support this hypothesis.

The data on chloramphenicol and kanamycin resistance, however, are entirely different. For these two antibiotics, the incidence of resistance was highest at the control sites and lower in the populated areas. This may have been due in part to the types of organisms present. *Pseudomonas* spp., commonly found in aquatic populations, have a high level of resistance to chloramphenicol; while the *Flavobacterium–Cytophaga* group is highly resistant to kanamycin (Kelch and Lee, 1978). Although we have not done positive identifications, *Flavobacterium*-type organisms are quite common at the control sites, especially Waterfall Creek. We feel, however, on the basis of present data that this cannot entirely explain the levels of resistance observed.

The data indicating a relatively high incidence of antibiotic resistance in the fecal flora of the native animal populations were also in marked contrast to the literature (Maré, 1968; Huber *et al.*, 1971). This antibiotic resistance was not due to high levels of antibiotic use in domestic animals with which the native animal populations might then associate. The cattle industry in this area involves strictly rangeland grazing; there are no feed lots, and the ranchers do not use feed supplements that contain antibiotics. Antibiotic use in domestic animals is strictly for treatment of disease and therefore quite limited.

One hypothesis that could explain the levels of antibiotic resistance observed in the Gunnison watershed is the concomitant occurrence of genes for antibiotic and heavy-metal resistance on the same plasmids (Novick and Roth, 1968; Varma *et al.*, 1976; Smith *et al.*, 1978; Summers *et al.*, 1978; Tomoney *et al.*, 1978). This is a highly mineralized area, parts of which were extensively mined in the late nineteenth and early twentieth centuries, and the water has a high mineral content. Silver, lead, zinc, cadmium, and molybdenum are known to be present. We have not yet confirmed that the antibiotic resistance we observed is plasmid-borne, but the presence of multiple-antibiotic and heavy-metal resistances and the loss of these resistances upon storage of the cultures would suggest that this may be the case.

When resistant isolates were screened for multiple antibiotic and metal resistance, 95 percent of the isolates were multiply antibiotic resistant and at least 50 percent were resistant to one or more heavy metals. The incidence of heavy-metal resistance should probably be higher, since we used only four metals in the screen. Also, we have recently recognized that the concentrations of heavy-metal ions generally used in screening media that contain more complex organic compounds are too high for use in CPS agar with its very low organic content. Even strains known to be resistant often fail to grow. Hence the 50 percent incidence of heavy-metal resistance represents a minimum value; the actual incidence may well be much higher.

The preliminary data from the Gunnison–Tomichi watersheds suggest that there may be a definite correlation between environmental factors other than antibiotic use and the observed incidence of antibiotic resistance in these bacterial populations. The presence of heavy metals in an environment represents a significant selection pressure for heavy-metal resistance and increases the number of resistant organisms present. Due to the presence of antibiotic resistance and metal resistance genes on the same plasmids, the incidence of antibiotic resistant bacteria would also increase. We cannot, however, correlate the presence of resistance to a specific antibiotic or antibiotics with specific metals. Plasmids are unstable and genes are frequently exchanged among them (Koch, 1981; Chau *et al.*, 1982). Thus we cannot say that chloramphenicol resistance, for example, correlates with the presence of any specific metal. Chloramphenicol resistance does appear to be one of those most frequently associated with heavy-metal resistance (Nakahara *et al.*, 1977; Smith *et al.*, 1978). This is consistent with our data, which indicated high levels of chloramphenicol resistance in mineralized areas.

The data presented here suggest a positive correlation between antibiotic resistance, especially multiple antibiotic resistance, and the presence of heavy metals in the environment. If this correlation is confirmed, it would suggest that the presence of multiply antibiotic resistant and/or heavy metal-resistant bacteria could be used as an exploration tool.

CONCLUSION

The incidence of antibiotic resistance among both aquatic bacterial populations and the fecal flora of native animals in the Gunnison watershed area is generally higher than that which has been reported in the literature. Random sampling of resistant isolates indicates a high incidence of multiple antibiotic resistance, and heavy-metal resistance is common among these multiply-resistant isolates. Data from the upper Tomichi Creek drainage, a highly mineralized area, would suggest a relationship between resistance to chloramphenicol and kanamycin and the presence of heavy metals. However, a high incidence of chloramphenicol and kanamycin resistance was also found in Waterfall Creek, an area with no known heavy-metal deposits. Both of these areas are minimally affected by human and/or domestic animal activity. This would suggest that perhaps more than one environmental factor is contributing to the high levels of resistance observed. Two such possible factors are ultraviolet radiation and the presence of antibiotic-producing organisms in the soil.

The high incidence of antibiotic resistance in the fecal flora of the native animal populations indicates that care must be taken in interpreting results involving antibiotic-resistant organisms in the soil (and also in the streams). The resistant forms might be indigenous or might have been transported into the area by animals.

If bacteriological systems for mineral exploration are to be developed, use of heavy-metal resistance rather than antibiotic resistance may be the better choice. This would reduce error due to antibiotic production in the soil or to antibiotic-resistant organisms deposited in the feces of animals. However, since the genes for antibiotic and metal resistance are often linked, not all such errors can be eliminated. For field use, heavy-metal resistance studies would also be advantageous because of the greater stability of the media. Metal-containing agars can be stored for some time, whereas antibiotics readily break down even at refrigeration temperatures and antibiotic-containing media should be prepared no more than 24 hours before use.

Our data do not indicate a single antibiotic which is correlated well enough with the presence of heavy metals to serve as a pathfinder. High incidence of multiple antibiotic resistance may be relevant, however, especially in combination with metal resistance.

ACKNOWLEDGMENTS

The contributions of Susan Engel, who did the original diversity study, and Timothy Flynn, who assisted with the field work, are gratefully acknowledged. We would also like to express our appreciation to John Castle for allowing us to establish sampling stations on his ranch. This work was supported in part by a grant from the Western State College Foundation.

REFERENCES

Austen, R. A., and Trust, T. J., 1981, Plasmid specification of resistance to antibacterial compounds in environmental *Citrobacter freundii: Can. J. Microbiol.*, v. 27, p. 343–349.

Burton, N. F., Day, M. J., and Bull, A. T., 1982, Distribution of bacterial plasmids in clean and polluted sites in a South Wales River: *Appl. Environ. Microbiol.*, v. 44, p. 1026–1029.

Chau, P. Y., Ling, J. Threlfall, E. J., and Im, S. W. K., 1982, Genetic instability of R plasmids in relation to the shift of drug resistance patterns in *Salmonella johannesburg: J. Gen. Microbiol.*, v. 128, p. 239–245.

Collins, V. G., and Willoughby, L. G., 1962, The distribution of bacterial and fungal spores in Blelham Tarn with particular reference to an experimental overturn: *Arch. Mikrobiol.*, v. 43, p. 294–307.

Cooke, M. D., 1976, Antibiotic resistance among coliform and fecal coliform bacteria isolated from the freshwater mussel *Hyridella menziesii: Antimicrob. Agents Chemother.*, v. 9, p. 885–888.

Devanas, M. A., Litchfield, C. D., McClean, C., and Gianni, J., 1980, Coincidence of cadmium and antibiotic resistance in New York Bight apex benthic microorganisms: *Mar. Pollut. Bull.*, v. 11, p. 264–269.

Drabble, W. T., and Stocker, B. A. D., 1968, R (Transmissible drug-resistance) factors in *Salmonella typhimurium:* pattern of transduction by phage P22 and ultraviolet-protection effect: *J. Gen. Microbiol.*, v. 53, p. 109–123.

Feary, T. W., Sturtevant, A. B., and Lankford, J., 1972, Antibiotic-resistant coliforms in fresh and salt water: *Arch. Environ. Health*, v. 25, p. 215–220.

Fontaine, T. D., III, and Hoadley, A. W., 1976, Transferable drug resistance associated with coliforms isolated from hospital and domestic sewage: *Health Lab. Sci.*, v. 13, p. 238–245.

Frazier, M. L., and Zimmerman, L. N., 1980, Plasmid-borne resistance to ultraviolet light and phage in *Streptococcus faecalis* ssp. *zymogenes: Can. J., Microbiol.*, v. 26, p. 1253–1255.

Howarth, S., 1965, Resistance to the bactericidal effect of ultraviolet radiation conferred on enterobacteria by the colicine factor col I: *J. Gen Microbiol.*, v. 40, p. 43–55.

Huber, W. G., Korica, D., Neal, T. P., Schnurrenberger, P. R., and Martin, R. J., 1971, Antibiotic sensitivity patterns and R factors in domestic and wild animals: *Arch. Environ. Health*, v. 22, p. 561–567.

Kelch, W. J., and Lee, J. S., 1978, Antibiotic resistance patterns of gram-negative bacteria isolated from environmental sources: *Appl. Environ. Microbiol.*, v. 36, p. 450–456.

Koch, A. L., 1981, Evolution of antibiotic resistance gene function: *Microbiol. Rev.*, v. 45, p. 355–378.

Levy, S. B., Fitzgerald, G. B., and Macone, A. C., 1976, Changes in intestinal flora of farm personnel after introduction of a tetracycline-supplemented feed on a farm: *N. Engl. J. Med.*, v. 295, p. 583–588.

Lighthart, B., 1979, Enrichment of cadmium-mediated antibiotic-resistant bacteria in a Douglas fir (*Pseudotsuga menziesii*) litter microcosm: *Appl. Environ. Microbiol.*, v. 37, p. 859–861.

Maré, I. J., 1968, Incidence of R factors among gram negative bacteria in drug-free human and animal communities: *Nature,* v. 200, p. 1046–1047.

Marsh, E. B., and Smith, D. H., 1969, R factors improving survival of *Escherichia coli* K-12 after ultraviolet irradiation: *J. Bacteriol.*, v. 100, p. 128–139.

McHugh, G. L., Moellering, R. C., Hopkins, C. C., and Swartz, M. N., 1975, *Salmonella typhimurium* resistant to silver nitrate, chloramphenicol, and ampicillin: *Lancet*, v. i, p. 235–240.

Nakahara, H., Ishikawa, T., Sarai, Y., Kondo, I., Kozukue, H., and Silver, S., 1977, Linkage of mercury, cadmium, and arsenate and drug resistance in clinical isolates of *Pseudomonas aeruginosa: Appl. Environ. Microbiol.*, v. 33, p. 975–976

Novick, R. P., and Roth, C., 1968, Plasmid-linked resistance to inorganic salts in *Staphylococcus aureus: J. Bacteriol.*, v. 95, p. 1335–1342.

Richmond, M. H., 1972, Some environmental consequences of the use of antibiotics: or "What goes up must come down": *J. Appl. Bacteriol.*, v. 35, p. 155–176.

Siccardi, A. G., 1969, Effect of R factors and other plasmids on ultraviolet susceptibility and host cell reactivation property of *Escherichia coli: J. Bacteriol.*, v. 100, p. 337–346.

Sjogren, R. E., and Port, J., 1981, Heavy metal-antibiotic resistant bacteria in a lake recreational area: *Water Air Soil Pollut.*, v. 15, p. 29–44.

Smith, H. W., Parsell, Z., and Green, P., 1978, Thermo-sensitive antibiotic resistance plasmids in enterobacteria: *J. Gen. Microbiol.*, v. 109, p. 37–47.

Soulides, D. A., 1965, Antibiotics in soils: VII. Production of streptomycin and tetracyclines in soil: *Soil Sci.*, v. 100, p. 200–206.

Staples, D. G., and Fry, J. C., 1973, A medium for counting aquatic heterotrophic bacteria in polluted and unpolluted waters: *J. Appl. Bacteriol.*, v. 36, p. 179–181.

Summers, A. O., and Silver, S., 1972, Mercury resistance in a plasmid-bearing strain of *Escherichia coli: J. Bacteriol.*, v. 112, p. 1228–1236.

——, Jacoby, G. A., Swartz, M. N., and McHugh, G., and Sutton, L., 1978, Metal cation and oxyanion resistances in plasmids of gram-negative bacteria: *Microbiology 1978*, p. 128–131.

Timoney, J. F., Port, J., Giles, J., and Spanier, J., 1978, Heavy-metal and antibiotic resistance in the bacterial flora of sediments of New York Bight: *Appl. Environ. Microbiol.*, v. 36, p. 465–472.

Van Dijck, P., and van de Voorde, H., 1976, Sensitivity of environmental microorganisms to antimicrobial agents: *Appl. Environ. Microbiol.*, v. 31, p. 332–336.

Varma, M. M., Thomas, W. A., and Prasad, C., 1976, Resistance to inorganic salts and antibiotics among sewage-borne Enterobacteriaceae and Achromobacteriaceae: *J. Appl. Bacteriol.*, v. 41, p. 347–349.

B. H. Olson and T. Barkay
Program in Social Ecology
University of California
Irvine, California

19

FEASIBILITY OF USING BACTERIAL RESISTANCE TO METALS IN MINERAL EXPLORATION

ABSTRACT

Metal-resistance patterns of bacterial populations in polluted soil environments closely paralleled the metal content of the associated soil for cadmium, copper, and zinc. Lead produced a clear but less closely correlated association. Although higher levels of resistance were noted in areas where the source metal was currently being introduced (Avonmouth, smelting), weathering of mineralogical material produced the most stable and well-delineated bacterial resistance patterns. On a population basis, patterns of bacterial resistance to metals for the sites investigated significantly correlated only with metals that were elevated in the associated soils. The findings suggest that heavy-metal resistance in naturally occurring bacteria may indicate the presence of heavy metals in the soil and could provide an excellent tool for the detection of pathfinder elements employed in mineral exploration. The applicability of this technique to mineral exploration is discussed.

INTRODUCTION

Biogeochemistry and hence biogeoprospecting has its origin in the beginning of the twentieth century. The field was introduced by Vernadskii, who was concerned with the study of living organisms from a geochemical point of view (Kovalevskii, 1979). This early work and much subsequent research have focused primarily on the use of plants to indicate the presence of mineral deposits. A number of orebodies have been discovered with the aid of this method, including deposits rich in such elements as Mo, Cu, U, Pb, and Zn. The depth of these orebodies has varied from surface deposits to those with 35 m of overburden (Cannon, 1960, 1964; Talipov, 1966; Tkalich, 1970; Viotkevich *et al.*, 1970; Warren and Delavault, 1970; Warren *et al.*, 1953). The work to be presented here is based on the concept of using bacteria to indicate the presence of an orebody. As yet, this technique is only beginning to be applied to mineral prospecting. The data presented in this article show the promise of its application to mineral exploration but draw from environmental pollution data.

Bacteria respond rapidly to the presence of metals in the environment. This response is based on their close interaction and in certain instances the mediation of metal chemistry in soils, and stream and lake sediments. This dynamic association between the microbial population and the soil/sediment milieu results in adaptation within bacterial populations and selection for those organisms best able to survive in a given environment. Many metals, especially those that are present in low concentrations in the earth's crust, have the ability to rapidly inhibit bacterial metabolism or growth. Thus such elements as Hg and Ag have been used as bactericidal agents for centuries (Summers and Silver, 1978). The ability of bacteria to adapt through mutation and selection to the presence of toxic substances in the environment can be used as a tool to indicate whether certain metals are present or have been present in an environment. To present why biogeochemical mineral exploration using bacterial populations is feasible, it will be useful to discuss in a limited manner some of the factors that govern metal toxicity to bacteria.

In any environment, toxicity of a metal to bacterial populations is dependent on both biotic and abiotic factors (Sterritt and Lester, 1980; Babich and Stotzky, 1977). These factors probably act by altering the availability of the metal to the numerous microorganisms in soil environments. Bacterial populations demonstrate a different toxicity response to a variety of metals and also react distinctly to the chemical species of a metal (Summers and Silver, 1978). Few studies, however, have delineated metal species in the soil environment. Mattigod and Sposito (1977), have described a number of physical and chemical factors that affect metal speciation in the soil environment, while Sposito and Bingham (1981) have examined differential metal uptake into plants depending on the metal complex present. The importance of metal speciation in the bioavailability of metals is currently gaining acceptance and its relationship to metal toxicity in microbial populations is beginning to be more fully understood. A broader knowledge of factors governing bioavailability of metals in soil environments would aid in understanding gas geochemistry in temperate climates as well as in the application of bacterial metal resistance to mineral exploration.

Intracellular factors that mediate bacterial response to metals are also of the utmost importance in understanding how endogenous bacterial populations may be used in mineral exploration. One of the major characteristics governing bacterial response to metals in an environment is the organism's genetic makeup. In bacteria the genes coding for metal resistance have been shown to be carried by extrachromosomal elements (plasmids). Some plasmids which are readily transferred among different species and even genera of bacteria may constitute an important means in the adaptation of natural bacterial populations to heavy-metal stress (Maynell, 1972). In addition, plasmid heavy-metal resistance genes may be duplicated to yield tandem repeats, which result in the amplification of the genes products, thus enhancing bacterial adaptability to metal stress (Summers, Chapter 16, this volume). Therefore, certain segments of bacterial populations will demonstrate the ability to tolerate high concentrations of a given metal based on genetic adaptation, the stimulus for that adaptation being triggered by the environment (Summers and Silver, 1978). The result of this genetic adaptation is that bacterial metal resistance at the population level is a mirror of metals that are or were present in the associated environment.

The present investigation was undertaken to examine bacterial resistance patterns in a number of polluted soil environments in order to determine if the source of the metal pollutant or the time the pollutant had resided in the soil environment affected the metal tolerance patterns of bacterial populations. This chapter discusses these findings in relation to their possible application to mineral exploration.

MATERIALS AND METHODS

A total of 47 soil samples (0 to 10 cm depth) were collected during January and February 1981, and January 1982, from seven different metalliferous locations in the United Kingdom (Tables 19-1 and 19-2). Ten soils were collected in 1981 and 1982 near a current zinc/lead smelter located in Avonmouth. An additional 10 soils were obtained, five in 1981 and five in 1982 in Swansea near a zinc/lead smelter that had discontinued operations during the early 1970s. In 1981, 10 soil samples were collected from four sites that had been amended with sewage sludge. A further 10 soils were sampled, six at historic arsenic and/or copper mining sites and four from adjacent control areas in Cornwall. A total of seven soil samples were obtained from two locations overlying marine black shales in central England. In 1979, 10 soil samples were collected at historic Zn/Pb mining and smelting areas and three adjacent control areas in Somerset, and 11 soil samples were taken from Pb mining and smelting areas as well as control areas in Derbyshire.

The "total" metal content of the soils was determined after digestion with HNO_3 using a Perkin-Elmer 403 flame atomic absorption spectrophotometer (AAS) measuring for arsenic, cadmium, copper, lead, and zinc, as described by Troyer *et al.* (1980). Acidity and organic content of soil samples were measured in terms of pH and loss on ignition (LOI), respectively.

From each site, 10 g of soil were placed into 90 ml of buffer (Troyer *et al.*, 1980), and using standardized microbiological techniques, appropriate dilutions of each sample were spread plated in triplicate onto plate count agar (Difco) amended with various concentrations of the four metals (Fig. 19-1). Sterile metal solutions were prepared from a 1 percent solution of each metal in deionized water filtered through a 0.45-μm cellulose-acetate membrane filter (Gelman GN-6). The final pH of each medium was adjusted to approximately 6.8 with the addition of NaOH or HCl.

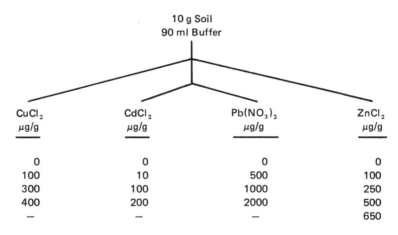

Fig. 19-1 Diagram of the experimental procedure utilized in the study indicating the concentration of the various metals in the media.

Bacterial colonies were enumerated on all plates after 48 to 96 hours of incubation at room temperature. Percent bacterial resistance was calculated by comparing mean bacterial numbers on each of the metal-amended media with the appropriate unamended control medium.

Values for "total" metal and percent bacterial resistance were transformed into their log forms. Subsequent statistical analyses were performed on log-transformed data, utilizing the BMDP-6D scattergrams and correlations program at the University of California, Irvine, computing facility (Dixon and Brown, 1977).

TABLE 19-1 Descriptive Statistics for Metals, pH, and Loss on Ignition (LOI) for Soils Downwind of Current and Recent Smelting Operations

	Cd	Cu	Pb	Zn	pH	LOI (%)
Avonmouth total metal (μg/g)(10)[a]						
Mean	30.5	77.2	662.6	1792.8	6.0	16.3
Range	5–50	28–259	106–1622	380–3540	4.4–7.1	6.3–24.4
Swansea total metal (μg/g)(10)[a]						
Mean	24.1	210	797	944	4.8	13.6
Range	3–81	84–452	120–1552	540–2300	3.9–5.9	6.8–22.5

[a]( ), sample number

TABLE 19-2 Historical Mining, Sewage-Amended Soils, and Geologically Enriched Areas[a]

	Cd	Cu	Pb	Zn	pH	LOI
Historical mining, Derbyshire total metal (μg/g)						
Mean	14	30	5441	683	6.0	13.4
Range	2–102	17–55	120–29,544	75–1259	5.0–6.5	5.0–25.5
n	11	11	11	11	10	10
Historical mining, Somerset Total metal (μg/g)						
Mean	197	44	5192	27,217	6.5	7.9
Range	2–468	22–84	96–13,200	220–66,400	6.1–7.2	5.3–10.2
n	14	14	14	14	3	3
Historical mining, Cornwall Total metal (μg/g)						
Mean	3	140	136	191	6.2	12.8
Range	0–12	37–432	76–240	58–500	5.5–7.6	7.1–19.4
n	10	10	10	10	10	10
Marine black shale, Derbyshire Total metal (μg/g)						
Mean	4	96	243	423	4.9	24.8
Range	2–9	25–212	180–320	116–644	3.9–5.7	17.3–37.1
n	6	6	6	6	7	7
Sewage sludge Total metal (μg/g)						
Mean	30	215	226	648	6.3	13.2
Range	1–157	17–806	84–480	60–2918	5.3–7.2	3.5–43.4
n	9	9	9	9	10	10

[a]n, number of samples.

B. H. OLSON AND T. BARKAY

RESULTS

Soil Characteristics

Descriptive statistics for the levels of the four metals, pH and loss on ignition (LOI) values for current and recent smelting areas, Avonmouth and Swansea, are presented in Table 19-1, while the same data are presented for historical mining areas, sewage-sludge-amended soils and soils derived from marine black shales in Table 19-2. The total metal concentrations shown in Tables 19-1 and 19-2 indicate that metal concentrations in the soil vary dramatically over the seven locations and within each location. Each site was selected to include for a given metal or series of metals concentrations ranging from background to anomalous. Data for the metals of interest in British soil are described below. The copper content of normal British soils varies from 5 to 50 μg/g with a mean concentration of 20 μg/g (Swain and Mitchell, 1960). Background cadmium levels in British soils range from 1 to 5 μg/g, while lead values are normally less than 150 μg/g in most of the United Kingdom except for the Derbyshire soils, which may contain up to 350 μg/g lead (Archer, 1977; Colburn and Thornton, 1978). In England and Wales, Zn concentrations range from 10 to 150 μg/g. Utilizing these data, it can be seen that Cd is enriched in current and recent smelting sites, the historical mining location in Somerset, and to some small degree, in Derbyshire. Copper is significantly elevated in Cornwall, recent smelting sites, and in sewage-sludge-amended areas. Only in Cornwall were Pb levels within the normal limits. In both the sewage-sludge-amended areas and in soils derived from marine black shales, Pb levels were only slightly elevated above those reported as normal values and would not be considered anomalous. Even though Zn has an elevated mean value for the Derbyshire area, only four of the 11 samples taken were highly enriched in Zn, whereas Zn was elevated in all Somerset soil samples.

Soil pH varied from acidic to neutral for all the areas examined. The highest mean soil pH value was 6.5, while all others ranged between 4.8 and 6.3. Soils near the recent smelting site and the soils derived from marine shales were the most acidic soils, with mean pH values of 4.8 and 4.9, respectively. Mean values for loss on ignition, an indication of the organic content in the soil samples, were similar at five of the seven locations, with Somerset containing considerably less organic material and marine shale soils considerably higher than average levels.

Bioassay System

The bioassay system (Troyer *et al.*, 1980) employed in this study is capable of distinguishing metal-tolerant or metal-resistant bacteria in metal-polluted soils, as exemplified by the cadmium data shown in Figure 19-2. The assay utilizes three metal test concentrations to distinguish metal-tolerant from metal-sensitive bacteria. The Cd example shown in Figure 19-2 depicts a strong layering pattern among the Cd concentrations used when log percent bacterial resistance on each test medium concentration (designated as A, B, or C in the figure) was plotted against log "total" soil cadmium concentration. These data indicated that the correct Cd concentrations were employed to show resistance (low or no tolerance 10 μg/g, moderate tolerance 100 μg/g, and high tolerance 200 μg/g). Thus,

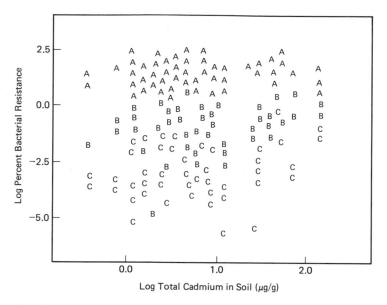

Fig. 19-2   Relationship between log total cadmium and log percent bacterial resistance on media amended with different levels of cadmium for the seven study locations. A, 10 μg Cd/g; B, 100 μg Cd/g; C, 200 μg Cd/g; $n = 255$.

as would be expected, increasing the metal content of the medium lowered the numbers of bacteria capable of growth (percent resistance) at a particular concentration of metal. This layering pattern was demonstrated for all metals studied. The concentrations of copper (50, 100, 200 μg/g) previously used in the assay described by Troyer *et al.*, (1980) were altered to the concentrations shown in Figure 19-1 to better delineate the copper-resistant population.

Sampling regimes are extremely important in determining the validity of the data. The seven locations (each encompassing four to seven sites) discussed in this chapter represent large tracts of land containing from background to anomalous metal concentrations for a given element or set of elements over a defined area.

Therefore, a few samples can be collected over a transect that represents a gradient from high-to-low metal levels in the soil. A typical sampling regime used in this investigation is shown for the current smelting location in Figure 19-3. The samples were collected along the path of aerial fallout, which was determined in an earlier study (Thornton, personal communication 1980), and up to a distance of 8 km from the plant. As shown in Figure 19-3, two samples were collected approximately one year apart at each site within the location. The data indicated that the relative relationships between bacterial metal resistance patterns and soil concentrations remained very stable (Table 19-3). Thus, as shown in Table 19-3, relatively few soil samples are needed to establish a relationship between soil metal concentration and bacterial resistance. However, as with any sampling technique, the more samples that are collected, the higher the assurance in the results.

To establish the bioassay method, it was necessary to compare the soil metal concentrations with the corresponding bacterial metal resistances. Soils often have a hetero-

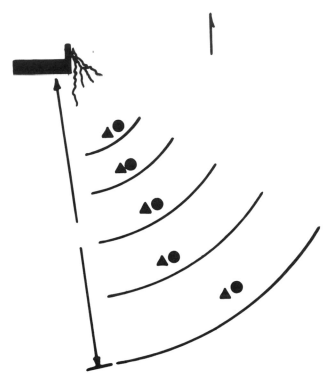

▲-- Location of sample collected, 1981
●-- Location of sample collected, 1982

Fig. 19-3   Diagram of sampling regime at the current smelting location, Avonmouth. ▲, location of sample collected,
1981; ●, location of sample collected, 1982.

geneous distribution of metals in the surface layers, as was shown in this investigation. Therefore, although previously reported measurements are useful in determining the concentrations of metals that should be used in the assay, one set of values may not be useful for all soils collected even within a local area. Optimum test concentrations can, however, usually be applied to an entire study area once they have been established. However, with each new element to be tested in an area, concentrations must first be established with preliminary trials.

To establish the specificity of the bacterial indicator system for the presence of metals, it was necessary to select sites that contained metals of interest over a significant range. Each site was selected for the metal(s) of interest, such as Somerset for Cd and Zn, Derbyshire for Pb, Cornwall for Cu, and so on. Therefore, sites at each location contain levels varying from background to highly anomalous concentrations for certain metals and within normal limits for others. The data in Tables 19-3, 19-4 and 19-5 present clear evidence that bacterial resistance to specific metals is demonstrated at these locations and reflects the metal content in the soil.

The ability of the bioassay to produce an index of the metals that are present in the soil in elevated concentrations was assessed using Pearson r product-moment correlations

TABLE 19-3 Correlational Relationships between Log Percent Bacterial Resistance on Amended Media and Log Total Metal Soil Levels for Current and Recent Smelting Locations in the United Kingdom

Location	Metal Tested	Metal Elevated in Soil[a]	r value, Media Concentration of Metal[b,c,d] (µg/g)			
			A	B	C	D
			Smelting, current			
Avonmouth	Cd	+ to ++	0.34 (55)	0.70 (47)	0.85 (39)	NA
	Cu	0 to ++	−0.47 (52)	0.07 (41)	0.53 (28)	NA
	Pb	0 to ++	0.09 (58)	0.55 (52)	0.50 (41)	NA
	Zn	++ to +++	ND	0.75 (55)	0.68 (60)	0.57 (33)
			Smelting, recent			
Swansea	Cd	+ to ++	0.28 (36)	0.28 (36)	0.18 (30)	NA
	Cu	+ to +++	0.10 (57)	−0.39 (35)	0.08 (30)	NA
	Pb	0 to ++	0.08 (58)	0.21 (38)	−0.03 (22)	NA
	Zn	++ to +++	ND	0.24 (44)	0.14 (39)	0.63 (26)

[a]0, background level; +, slightly elevated over background levels; ++, moderately elevated over background levels; +++, highly elevated over background levels.
[b]$r \geq 0.5$; $p \leq 0.05$; r = correlation between log percent bacterial resistance and log total soil metal concentration.
[c]A: Cd = 10 µg/g; Cu = 100 µg/g; Pb = 500 µg/g; Zn = 100 µg/g.
B: Cd = 100 µg/g; Cu = 300 µg/g; Pb = 1000 µg/g; Zn = 250 µg/g.
C: Cd = 200 µg/g; Cu = 400 µg/g; Pb = 2000 µg/g; Zn = 500 µg/g.
D: Cd = NA; Cu = NA; Pb = NA; Zn = 650 µg/g.
[d]( ), number of samples; NA, not applicable; NC, not correlated; ND, not determined.

TABLE 19-4 Correlational Relationships between Log Percent Resistant Bacteria and Log Total Soil Metal Levels for Historical Mining Locations in the United Kingdom

Location	Metal Tested	Metal Elevated in Soil[a]	r value Media Concentration of Metal[b,c,d] (µg/g)		
			A	B	C
Derbyshire	Cd	0 to +	0.19 (49)	0.77 (28)	NC
	Pb	0 to +++	0.30 (50)	0.47 (29)	0.17 (13)
	Zn	0 to ++	ND	0.57 (43)	0.83 (36)
Somerset	Cd	0 to +++	0.11 (67)	0.52 (45)	0.94 (22)
	Zn	+ to +++	0.53 (32)	0.77 (32)	ND
Cornwall	Cd	0 to +	0.38 (28)	−0.28 (18)	−0.13 (17)
	Cu	0 to +++	0.15 (35)	0.64 (28)	0.38 (14)
	Pb	0 to +	0.01 (35)	−0.28 (35)	−0.55 (32)
	Zn	0 to ++	−0.02 (28)	0.01 (27)	−0.15 (29)

[a]0, background level; +, slightly elevated over background levels; ++, moderately elevated over background levels; +++, highly elevated over background levels.
[b]$r \geq 0.5$; $p \leq 0.05$; r, correlation between log percent bacterial resistance and log total soil metal concentration.
[c]A: Cd = 10 µg/g; Cu = 100 µg/g; Pb = 500 µg/g; Zn = 100 µg/g.
B: Cd = 100 µg/g; Cu = 300 µg/g; Pb = 1000 µg/g; Zn = 250 µg/g.
C: Cd = 200 µg/g; Cu = 400 µg/g; Pb = 2000 µg/g; Zn = 500 µg/g.
[d]( ), number of samples; ND, not determined; NC, not correlated.

TABLE 19-5 Correlation Relationships between Log Percent Resistant Bacteria and Log Total Soil Metal Levels for Marine Black Shales and Sewage Sludge Locations in the United Kingdom

Location	Metals Tested	Metals Evaluated in Soil[a]	r_2 Value, Media Concentration of Metal[b,c,d] $(\mu g/g)$		
			A	B	C
		Marine black shales			
Derbyshire	Cd	0 to +	0.16 (25)	0.55 (14)	−0.32 (8)
	Cu	0 to +	0.68 (25)	0.29 (17)	−0.41 (16)
	Pb	0 to +++	0.17 (27)	0.35 (13)	−0.22 (8)
	Zn	0 to ++	0.59 (23)	0.45 (26)	0.66 (21)
		Sewage sludge			
	Cd	0 to +++	0.16 (32)	0.68 (20)	0.56 (18)
	Cu	0 to +++	0.51 (30)	0.39 (25)	0.32 (20)
	Pb	0 to +	0.46 (32)	0.38 (18)	0.37 (13)
	Zn	0 to ++	0.28 (33)	0.31 (23)	0.45 (25)

[a]0, background level; +, slightly elevated over background levels; ++, moderately elevated over background levels; +++, highly elevated over background levels.
[b]r J 0.5; p F 0.05; r, correlation between log percent bacterial resistance and log total soil metal concentration.
[c]A: Cd = 10 $\mu g/g$; Cu = 100 $\mu g/g$; Pb = 500 $\mu g/g$; Zn = 100 $\mu g/g$.
  B: Cd = 100 $\mu g/g$; Cu = 300 $\mu g/g$; Pb = 1000 $\mu g/g$; Zn = 250 $\mu g/g$.
  C: Cd = 200 $\mu g/g$; Cu = 400 $\mu g/g$; Pb = 2000 $\mu g/g$; Zn = 500 $\mu g/g$.
[d]( ), number of samples; ND, not determined; NC, not correlated.

between log percent bacterial resistance and log total metal concentrations. The results of these analyses, shown in Tables 19-3, 19-4, and 19-5, demonstrate that bacterial metal resistance can be used to index the relative content of a metal at the seven locations. How the bio-assay technique is applied to develop an index of anomalous metals is discussed by location and metal source.

At the current smelting sites statistically significant positive correlations between soil metal and percent metal-resistant bacteria were observed at the highest concentrations for each amended medium. Bacteria from Avonmouth are highly resistant to the four metals examined, and the metal resistance patterns exhibited by these bacteria are greater than for all other locations, although the soil metal concentrations here are less than those in some other test areas. An explanation of this observation is that the availability of the given metal is higher here. Metals recently having entered the environment may be more available, and thus in this instance cause a greater bacterial response, in spite of lower total concentrations of the metal. Metal-uptake studies with plants using both metal-amended soils and metal-enriched sludge soils indicated that the highest levels of translocation into plants occur when the metal source had most recently entered the soil system (Sposito and Bingham, 1981), and therefore support the present findings. We interpret these data to indicate that in the case of artificially and sludge-contaminated soils, the time a metal has resided in an environment will influence its availability to microorganisms, as well as to other life forms; this idea is also supported by data reported

gold deposits. Nichols (presented at Applied Geochemistry in the 1980s, Imperial College, London, April 1983) has stated that Au particles are often distributed in a random fashion in soil samples, that is, its presence may not be detected by using ordinary sample sizes (Harris, 1981). This is a recognized difficulty in the exploration for most placer and certain Carlin-type deposits. Because the bioassay is less affected by dilution, which traditional digestion and atomic absorption analysis face, it could be of greater value. For example, if a sample were to have only 10 resistant bacteria in relation to a total number of 10^6 bacteria per gram, resistance corresponding to the presence of unevenly distributed gold could perhaps be detected in a 10-g sample even if all the Au were located in only one fraction of that sample. Further, advantages may exist in the determination of elements such as As, Te, and Se which are difficult to analyze. All of the above-mentioned metals should be readily detectable by bacterial bioassay systems (Olson, unpublished data; Watterson, personal communication, 1982).

Geochemical data on volatile elements have proven to be variable and unreliable in wet climates. Vapors of Hg or CO_2 can be produced by microbial activity, in addition to abiotic processes, and these microbial processes are well documented in the literature (Olson *et al.*, 1979; Spangler *et al.*, 1973; Vonk and Sijpesteijn, 1973). Many naturally occurring bacteria are able to volatilize mercury, and thus, depending on the temperature, soil moisture content, and season, bacteria will play a dramatic role in the conversion of Hg^{2+} to volatile Hg vapor and short-chain alkyl mercurials. Therefore, an understanding of seasonal microbial activity and of the numbers of organisms resistant to Hg would help provide information on the role of bacteria in the overall production of volatile Hg compounds. The same considerations are applicable to As (Summers and Silver, 1978).

Further, bacteria have the ability to degrade a number of organic compounds that are common in the soil environment with the end products of CO_2 and water. Thus if carbon dioxide is being measured, bacterial activity in the soil could play a significant role and should be taken into consideration in interpreting results. Microbial data may result in better interpretation of geochemical reconnaissance data on volatile elements in temperate climates.

Certain current experimental exploration techniques are based on the presence of volatile elements over different deeply concealed ore deposits. One of the difficulties associated with these methods is the uncertain spatial relationship of volatile elements to the hidden ore. If these volatile elements have followed a tortuous path from a deposit through a significant depth of overburden to the surface, then volatile element data alone may not be precise enough to justify expensive drilling, especially in situations where volatile elements may travel through different local pathways at different times. Bacterial data may be more precise and presumably could be used as a tracer (provided that the volatile elements of interest had toxic effects on bacteria). Some of these gases and volatile elements are CO_2, COS, H_2S, H_2, Hg^0, and CS_2: they all would modify the bacterial population but not necessarily through toxic effects. Those bacteria that are exposed to the main pathway of the gas would adapt, while those that are not in contact as often would not. Thus the assay could presumably be used to pinpoint the average pathway of the gas as it traversed the soil column to the surface. This information could be used to better define the area in which to carry out core drilling.

Perhaps the hardest step in the introduction of a new technique is to obtain opportunities to test its applicability to that field. For the bioassay technique described here to have any future, it must be applied to a number of known mineral deposits and be shown to verify existing finds. A still better test is for it to be used in conjunction with current methods in mineral exploration so that its merits can be evaluated in a "real-world" situation. Currently, both phases of the above-mentioned testing are either in the planning stages or already under way. It is hoped that within relatively few years this technique, if proved successful, will be available.

ACKNOWLEDGMENTS

We wish to acknowledge the Applied Geochemistry Research Group, Imperial College, for analytical analysis of the soil samples.

REFERENCES

Archer, F. C., 1977, Trace elements in soils in England and Wales: *Proc. ADAS Conf. Inorg. Pollut. Agric.,* London (in press).

Babich, H., and Stotzky, G., 1977, Sensitivity of various bacteria, including actinomycetes. and fungi to cadmium and the influence of pH on sensitivity: *Appl. Environ. Microbiol.,* v. 33, p. 681–695.

Barkay, T., and Olson, B. H. 1986, Phentotypic and Genetypic Adaptation of Aerobic Heterotrophic Sediment Bacterial Communities to Mercury Stress: *Appl. Environ. Microbiol.* (in press).

Brownlow, A. H., 1977, *Geochemistry:* Prentice-Hall, Englewood Cliffs, N.H., 498 p.

Cannon, H. L., 1960, Botanical prospecting for ore deposits: *Science,* v. 132, no. 3427, p. 591–598.

___ , 1964, Geochemistry of rocks and related soils and vegetation in the Yellow Cat area, Grand Country, Utah: *U.S. Geol. Surv. Bull. 1176*, p. 1–127.

Colburn, P., and Thornton, I., 1978, Lead pollution in agricultural soils: *J. Soil Sci.,* v. 29, p. 513–526.

Dixon, W. J., and Brown, M. B., 1977, *Biomedical Computer Programs P-Series:* University of California Press, Berkeley.

Duxbury, T., 1981, Toxicity of heavy metals to soil bacteria: *FEMS Microbiol. Lett.,* v. 11, p. 217–220.

Harris, J. F., 1982, Sampling and analytical requirements for effective use of geochemistry in exploration for gold, p. 53–67 in Levinson, A. A. (Ed.), *Precious Metals in the Northern Cordillera:* Assoc. Explor. Geochem., Chicago.

Kovalevskii, A. L., 1979, *Biogeochemical Exploration of Mineral Deposits:* Nedra, Moscow, 1974; transl. into English by V. S. Kothekar (Ed.), Oxonian Press, New Delhi, 1979, 136 p.

Mattigod, S. V., and Sposito, G., 1977, Chemical modelling in aqueous systems—speciation, sorption, solubility and kinetics, p. 837–856 in Jenne, E. (Ed.), *ACS Symposium Series No. 93*, American Chemical Society, Washington, D.C.

Maynell, G. G., 1972, *Bacterial Plasmids—Conjugation, Colicinogeny and Transmissable Drug-Resistance:* Macmillan Press, London.

Olson, B. H., Barkay, T., and Colwell, R. R., 1979, Role of plasmids in mercury transformation by bacteria isolated from the aquatic environment: *Appl. Environ. Microbiol.,* v. 38, p. 478–485.

Shearer, D., and Olson, B. H., 1983, The role of chemical speciation on bacterial resistance to cadmium: *Abstr. Annu. Meet. Am. Soc. Microbiol.,* p. 260.

Spangler, W. J., Spigarelli, J. L., Rose, J. M., Flippin, R. S., and Miller, H. M., 1973, Degradation of methylmercury by bacteria isolated from environmental sample: *Appl. Microbiol.,* v. 25, p. 448–493.

Sposito, G., and Bingham, F., 1981, Computer modeling of trace metal speciation in soil solution: correlation with trace metal uptake by higher plants: *J. Plant Nutr.,* v. 3, p. 35–49.

Sterritt, R. M., and Lester, J. N., 1980, Interactions of heavy metals with bacteria: *Sci. Total Environ.,* v. 14, p. 5–17.

Summers, A. O., and Silver, S., 1978, Microbial transformations of metals: *Annu. Rev. Microbiol.,* v. 32, p. 637–672.

Swain, D. J., and Mitchell, R. L., 1960, Trace element distribution in soil profiles: *J. Soil. Sci.,* v. 11, p. 347–368.

Talipov, R. M., 1966, *Biogeokhimicheskie poiski polimetallicheskikh i mednykh mestorozhdenii v usloviyakh Uzbekistana (Biochemical prospecting of polymetallic and copper deposits in the conditions of Uzbekistan):* FAN, Tashkent, 105 p. with illustrations.

Tkalich, S. M., 1970, *Fitogeokhimicheskii metod poiskov mestorozhdenii poleznyky iskopaemykh (Phytogeochemical prospecting of ore deposits):* Nedra, Leningrad, 175 p. with illustrations.

Troyer, L., Olson, B. H., Hill, D. C., Thornton, I., and Matthews, H., 1980, Assessment of metal availability in soil through the evaluation of bacterial metal resistance: p. 129–141 *in* Hemphill, D. D. (Ed.), *Trace Substances in the Environment,* Vol. 14: University of Missouri, Columbia, Mo.

Viotkevich, G. V., Alekseenko, V. A., and Aleksenko, V. A., 1970, Opyt primeneniya biogeokhimicheskikh metodov pri poiskakh nekotorykh rudnykh elementov v Dzhungarskom Alatau (Experimental application of biogeochemical techniques in prospecting of some ore elements in the Zhungar Alatau): *Izv. Vuzov Geol. Razved.,* no. 2, p. 64–69.

Vonk, S. W., and Sijpesteijn, A. K., 1973, Studies on the methylation of mercuric chloride by pure cultures of bacteria and fungi: *Antonie van Leeuwenhoek, J. Microbiol. Serol.,* v. 39, p. 505–513.

Warren, H. V., and Delavault, R. E., 1970, Biogeochemistry in Canada, p. 1–14 *in Background Papers on the Earth Sciences in Canada,* Montreal.

—— , Delavalut, R. E., and Routley, D. G., 1953, Preliminary studies of the biogeochemistry of molybdenum: *Trans. R. Soc. Can.,* v. 47, p. 71–75.

John R. Watterson
U.S. Geological Survey
Denver Federal Center,
Denver, Colorado

Laslo A. Nagy
Aquatech
Page, New South Wales, Australia

David M. Updegraff
Department of Chemistry and Geochemistry
Colorado School of Mines
Golden, Colorado

20

PENICILLIN RESISTANCE IN SOIL BACTERIA IS AN INDEX OF SOIL METAL CONTENT NEAR A PORPHYRY COPPER DEPOSIT AND NEAR A CONCEALED MASSIVE SULFIDE DEPOSIT

Geochemical and microbiological tests have been conducted on soils collected above the Poorman Creek porphyry copper deposit, Lewis and Clark County, Montana, and above the Keystone massive sulfide deposit, Shasta County, California, to determine the possible influence of natural metal enrichments on the penicillin resistance of a normally penicillin-sensitive group of soil bacteria.

Over both deposits, penicillin resistance in *Bacillus* spp. correlates significantly with soil metal content. At the Poorman Creek deposit, the metalliferous soils contain increased numbers of one naturally penicillin-resistant organism, *Bacillus cereus*. Statistical analysis reveals a close correlation between soil metal content, percent *B. cereus*, and percent penicillin-resistant *Bacillus* spp. These findings, supported by recent information on the metal-binding capacity of penicillamine, a penicillin hydrolysate, suggest that the production of penicillin and hence penicillin resistance may confer an ecological advantage on the microflora of metalliferous soils.

Either penicillin resistance in *Bacillus* spp. or increased numbers of *B. cereus* in soils or stream sediments may characterize similar ore deposits and be an aid in prospecting.

INTRODUCTION

Mineral deposits studied by the U.S. Geological Survey over the past 30 years have been found to display surface indications of their presence, commonly the dispersion in water, vegetation, or surface materials of ore-associated "indicator elements." One of the tasks of the U.S. Geological Survey has long been to try new methods of detecting known, but concealed, mineral deposits. The reason for this effort is that economic mineral deposits that remain to be discovered are deeply concealed and thus increasingly difficult to locate. The intimate association of soil bacteria with their geochemical environment and their ability to adapt rapidly to one or more of the toxic ore indicator elements have led us to begin investigations of microbial heavy metal and antibiotic resistance as it may relate to concealed mineral deposits and to ore-associated geochemical dispersion patterns in soil.

Evidence has accumulated that the genetic resistance features of soil microorganisms faithfully reflect the chemical constituents, including the concentrations of heavy metals, in their soil environment. The monograph of Letunova and Koval'sky (1978), summarizing their research since 1954, presented a great deal of evidence that microbial adaptation to natural heavy metal accumulations is a predictable phenomenon. The work of other researchers (Marques *et al.*, 1979; Timoney *et al.*, 1978; Austin *et al.*, 1977; Nelson and Colwell, 1975; Kendrick, 1962; Olson and Thornton, 1981; Troyer *et al.*, 1980, 1981; Ehrlich, Chapter 14, this volume; Colwell *et al.*, Chapter 17, this volume; Michaels *et al.*, Chapter 18, this volume; and Olson and Barkay, Chapter 19, this volume) supports the conclusion that microorganisms reproducing in a metal-enriched environment can be expected to carry genetic traits or consist of distinct populations suited to that environment. Our investigations of microorganisms associated with naturally metalliferous soils in the vicinity of metal deposits are based on the apparent trustworthiness of this phenomenon, which we have come to refer to as the predictability of adaptation.

An obvious approach in investigating the possible utility of microorganisms in geochemical studies is to determine the response of soil or stream sediment microflora to growth media containing different test concentrations of the various metals of interest. This approach, which is being taken by Olson and her group at the University of California at Irvine, has much to recommend it, especially if a rapid, multielement metal-resistance test perhaps similar to that of Bauer *et al.* (1966) can be exploited. The potential advantage of such a test is that microbial resistance to any toxic metal or other agent can, in theory, be tested with equal facility. In some cases this bioassay approach could serve as a screening tool and circumvent analytical difficulties involved in direct chemical determination of such elements as tellurium and mercury.

Another approach to using microorganisms as metal sensors, which we believe may have equal merit in mineral exploration, is that of using a single test to indicate the presence of any combination of several toxic metals. Evidence that such a test is feasible is presented here.

Although numerous medical and environmental studies have noted the coincidence of various antibiotic and heavy-metal resistances in bacteria, there has apparently been little interest in discovering whether ecological reasons exist for the coevolution of these resistances. Timoney *et al.* (1978), in a pioneering study, were the first to conclude forcefully that heavy metals can result in a selection pressure for antibiotic resistance in bacteria in that system. Unfortunately for our purposes, the conclusion of Timoney *et al.* (1978) was compromised by the unknown origin and composition of the New York Bight dumping ground sediments tested in their study. It could be argued that ampicillin or other semisynthetic penicillins in hospital sewage may have provided a direct selective pressure for the ampicillin resistance observed in *Bacillus* spp. examined. However, because of their conclusion we became interested in learning if antibiotic resistance in soil bacteria could be found to increase with natural increases in soil metal content. We therefore began to look for variations in antibiotic resistance in soil bacteria in pristine environments affected only by natural variations in soil geochemistry. Preliminary disk assay tests of *Bacillus* spp. in heat-treated samples of soil collected over and adjacent to a vein deposit in Montana indeed indicated that large variations in ampicillin resistance in soil bacteria coincided with and were apparently conditioned by the metal content of the soil. Bacteria in soils with high metal content were far more ampicillin resistant than those in control soils.

The genus *Bacillus* was of particular interest because their durable endospores are present in most soils (Holding *et al.*, 1965; Mishustin and Mirsoeva, 1968; Gray and Williams, 1971) and stream sediments (Watterson, unpublished data) and they are ideally suited, because of their longevity, for use in potential geochemical assays. Because penicillin (a natural product of many common soil molds) is twice as active against *Bacillus* spp. as ampicillin and the resistance mechanism is the same (Davis *et al.*, 1980), penicillin was chosen for this study.

MATERIALS AND METHODS

Ore Deposits

Poorman Creek deposit The Poorman Creek deposit is marked by an oval treeless zone about 80 m at its maximum (north-south) diameter. The deposit is located on a 20 to 30° south-facing slope just north of Poorman Creek about 11 miles southeast of

Lincoln, Montana, and can be seen from the road. It lies at the common boundary of the Swede Gulch and Nevada Mountain 7.5-minute quadrangles (USGS topographic series, 1968). The soils in the treeless zone contain substantially higher values of copper, lead, silver, and other metals than do soils in the surrounding forest. The predominant vegetation consists of sparse, presumably metal-tolerant grasses and a luxuriant field of the copper-indicator plant, *Eriogonum ovalifolium* (Cannon, 1960). The deposit is of the porphyry copper type, in chloritized quartz monzonite. This and neighboring deposits are related to the Tertiary Silver Bell stock which intrudes Precambrian igneous and metamorphic rocks, and these deposits occur at the southwest edge of the Big Belt Mountain thrust fault zone (Kleinkopf and Mudge, 1972). Components of Tertiary gravels have been interpreted (R. G. Schmidt, oral communication, 1983) as indicating that the Silver Bell stock was exposed as far back as the Eocene. Accordingly, similar deposits in the Poorman Creek area may have been weathering since Eocene time (50 m.y.). Individual deposits in this area may have been continuously exposed since the end of Pliocene volcanism (2 m.y.). The Poorman Creek deposit may or may not have been exposed prior to valley glaciation approximately 10,000 years ago. The microflora colonizing the Poorman Creek deposit, however, have had at least 2,000,000 years to adapt to soils derived from other similar deposits nearby.

Keystone deposit The Keystone mine is in Sec. 14, T. 33 N., R. 6 W., Shasta County, California. The Keystone deposit is one of nine massive pyrite sulfide deposits in the West Shasta copper–zinc district that produced approximately half of California's copper through 1946. Kinkel *et al.* (1956) have interpreted the base metal mineralization in this district to be Late Jurassic or Early Cretaceous (150 m.y.). Later investigations have suggested that this is a typical massive sulfide deposit formed as a submarine exhalative deposit and of the same age as the surrounding Devonian rocks (360 m.y.). All of the massive pyrite deposits in the district occur in the Balaklala Rhyolite, of Middle Devonian age, and contain chalcopyrite, sphalerite, and minor quantities of gold and silver minerals. The most striking features of the ore are its uniform, dense, pyritic character, and its sharp boundary with barren or weakly pyritized wall rocks. Most of the ore in the mine stopes occurs as horizontal lenses about 15 ft in thickness with a maximum thickness of 50 ft. The lenticular ore bodies are within the middle unit of the Balaklala Rhyolite. A total of 200 ft of flat-lying middle and upper units of the Balaklala Rhyolite presently overlies the ore bodies. Most of the ore bodies dip 5 to $10°$ more steeply than the contact between the middle and upper units of the Balaklala Rhyolite. The ore bodies are intersected in various places by vertical faults.

The Balaklala Rhyolite capping this ore deposit, judging by exposures in other places, is broken and faulted and therefore probably not impervious to solutions or gases from the weathering sulfide minerals. The thickness of the rhyolite, however, despite the presence of faults, apparently serves as a barrier to the surface expression of the target minerals. If, as has been suggested, this is a typical massive sulfide deposit, this deposit and the Poorman Creek copper porphyry deposit have a fundamental difference in origin (and exposure) that could explain differences in the soil expression of the two deposits. The difference in time of formation may also have a strong influence. Indeed, it can be plausibly argued that the minor variations in soil metal content over the Keystone deposit are not directly attributable to subjacent ore.

Sampling Procedures

Sampling and soil-handling procedures at the Poorman Creek and Keystone deposits differed. Undifferentiated samples were collected at the Keystone deposit, refrigerated, and the −1-mm fraction milled to about −80 mesh prior to chemical and microbiological study. The more satisfactory procedure of sieving to −30 mesh at the sampling site was adopted for the Poorman Creek study. The latter procedure has the advantage that no further preparation is required prior to microbiological and chemical analysis. In addition, the −30-mesh samples require little or no centrifuging after the shaking step in the microbiological procedure and thus are easier to work with than milled samples.

Poorman Creek deposit Sampling was carried out under dry conditions, August 21, 1982. Twenty-six samples were taken along a north-south line at 10-ft intervals. The first 11 sample sites were upslope of the deposit, in the forest, which consisted of both deciduous and coniferous trees. Samples 12 through 26 were in the barren *Eriogonum* zone. A control sample was taken several hundred yards west of the deposit. At each sample site any loose forest litter was removed, and soil from an area of about 100 cm^2 to a depth of about 10 cm was dug up with a stainless steel garden trowel, sieved through a 30-mesh Tyler equivalent stainless steel sieve and transferred to four factory-closed plastic bags. A total of about 1.5 kg of sieved soil was taken from each site. No sterile precautions were taken during sampling, but visible soil was brushed from the equipment between sites. The sieved material was a light grayish tan in color and did not change appreciably in color or texture between the wooded and the barren zone. Samples were stored in the dark at ambient temperatures for about 10 days prior to division into two sample sets using a cross-channel sample splitter. One-half of each sample was stored at −15°C. Bulk samples for chemical and microbiological testing were kept at room temperature in the dark; tests were conducted on 150-g subsamples kept at room temperature in sterile, wide-mouth glass jars.

Keystone deposit Samples were collected under dry conditions on September 21, 1981, at 50-ft intervals in two parallel traverses that intersected mapped underground workings of the Keystone mine in mixed coniferous and deciduous forest. After loose forest litter was removed, organic-rich soil was collected within 5 to 10 cm of the surface. Unsieved samples were taken. Soils for microbiological study were kept as cool as possible for a week until they could be refrigerated at 6°C.

As needed for chemical and microbiological study, several grams of sample were sieved through a 1-mm sieve and ground to approximately −80 mesh in a porcelain mortar. The 80-mesh sample was stored in sterile sputum jars at 6°C between operations.

Analysis of soil All chemical analyses on Poorman Creek soils were carried out at least three times; all chemical analyses on Keystone soils were carried out at least five times. Cd, As, and Ag were determined by a modification of the method of Viets (1978). Al, Ca, Fe, Mg, Ba, Be, Ce, Co, Cu, La, Mn, Mo, Nb, Ni, P, Pb, Sr, Ti, V, Y, and Zn were determined by inductively coupled argon plasma-atomic emission spectrometry (ICP) (Motooka and Sutley, 1982) following overnight digestion in concentrated nitric acid, evaporation to dryness, and dissolution in 20 percent HCl. Hg was determined according to an unpublished in-house procedure as follows: 0.1-g samples were placed in fritted Pyrex tubes connected to a Jerome[1] mercury vapor collection unit and heated to red heat with a propane torch. Collected mercury was then revolatilized and passed through a glass chamber in the light path of a Perkin-Elmer 303 atomic absorption instrument with a chart recorder. Rare earth elements in the Keystone soils were determined by the procedure of Crock and Lichte (1982); pH values were obtained by mixing 2.5 g of either −1-mm (Keystone) or −30-mesh (Poorman) soils with 5 ml of deionized water and testing with an Orion model 201 digital pH meter after about 10 minutes of equilibration. Percent moisture was calculated from the weight loss occurring in a 3.00-g sample after heating at 90°C overnight and temperature equilibration in a desiccator containing $CaCl_2$. Percent organic matter was calculated from the weight loss occurring in desiccated, approximately 3-g samples after heating at 525°C overnight and temperature equilibration in a 90°C drying oven and drying over $CaCl_2$.

Microbiological Tests

Test medium A minimal medium similar to one suggested by Stanier *et al.* (1976) was used alone or amended with one-tenth its volume of appropriate concentrations of filter-sterilized potassium penicillin G (United States Biochemical). The minimal medium consisted of 1 g of K_2HPO_4, 0.2 g of $MgSO_4 \cdot 7H_2O$, 0.01 g of $FeSO_4 \cdot 7H_2O$, 0.01 g of $CaCl_2$, 1 g of glucose, 1 g of NH_4Cl, 0.1 g of yeast extract, 0.07 g of cycloheximide, and 15 g of agar per liter of deionized water. In the Poorman Creek study, appropriate volumes of fresh penicillin solution were filter sterilized and added to sterilized medium at 47°C to make media containing 0, 0.05, 0.1, 0.2, 0.5, 1.0, 2.0, 5.0, 10., 20., 50., 100, 200, and 500 μg/ml penicillin immediately prior to use. Final test concentrations of penicillin in the Keystone study were 0, 0.001, 0.002, 0.005, 0.01, 0.02, 0.05, 0.1, 0.2, 0.5, 1.0, 2.0, 5.0, 10., 20., 50., and 100 μg/ml.

Procedure One gram of soil was added to 9 ml of sterile 33 mM K_2HPO_4 in 16 × 150 mm screw-cap culture tubes, shaken 10 minutes in a mechanical shaker, diluted to three successive 1/10 dilutions in 33 mM K_2HPO_4, submerged in circulating 68°C water for 12 minutes, and cooled in cold tap water. Prior to the dilution step, Keystone samples were centrifuged twice for 1 minute at 1200 rpm in a benchtop centrifuge (56 × g) and the supernatant cloudy suspension immediately transferred by Pasteur pipette to sterile 16 × 150 mm tubes; this served to separate most of the spores from soil debris that interfered with colony counting. This step was unnecessary with the −30-mesh Poorman Creek samples, as sediment from those samples settled adequately without centrifuging. Within a few hours of heat treatment, 0.5-ml portions of the four dilutions of each sample were used to make standard 10-ml pour plates with unamended or penicillin-amended medium in the concentrations noted above. Plates were incubated inverted at 30°C for 72 hours and counted at 9x with a stereo binocular microscope. Only typical bacterial colonies were counted. Randomly selected colonies from the Keystone soils were transferred by loop from the surface of spread plates made with unamended medium and isolated with at least three successive streakings. Approximately 400 Keystone isolates were tested for gram reaction and slides prepared for phase-contrast microscopic examination; approximately 80 percent were clearly gram positive. The presence of rod-

---

[1]Use of commercial names in this report does not constitute product endorsement by the U.S. Geological Survey.

shaped vegetative cells, spores, and sporangia (typical of virtually all isolates) was considered adequate to classify the bacteria as *Bacillus* spp. All sterile operations were carried out in a laminar flow hood.

B. cereus enumeration procedure The *B. cereus* enumeration procedure was based on the development of a circular, white, opaque zone around *B. cereus* colonies in overnight pour plates made by amending our medium with 2.5 percent (v/v) egg yolk as a 1:1 dilution in deionized water and 0.5% (w/v) trisodium citrate as recommended by Donovan (1958). According to McGaughey and Chu (1948), only two *Bacillus* species (according to the then current classification) give positive results with this test, *B. cereus* and *B. anthracis*. The anthrax bacillus, apart from being rare enough to be dismissed a priori at most sites, gives a much weaker reaction than *B. cereus* with egg-yolk agar (the opaque zone not extending beyond the colony in the case of *B. anthracis*). McGaughey and Chu (1948) tested 80 different strains of *B. cereus* and obtained positive results with all 80 strains. The insect pathogen, *B. thuringiensis*, otherwise identical to *B. cereus* (Kreig, 1981) which has subsequently been given species status, also gives a positive egg-yolk reaction and is penicillin resistant (Norris *et al.*, 1981). The insect-pathogenic habit of all but one strain (DeBarjac, 1981), however, probably limits the distribution of *B. thuringiensis.*

Statistical analysis The percent of *Bacillus* spp. resistant to each test concentration of penicillin was calculated according to the formula

$$\text{percent of organisms resistant to penicillin} = \frac{(\text{no. colony-forming units in medium amended with penicillin}) \times 100}{\text{no. colony-forming units in control plate (medium not amended with penicillin)}}$$

Data were standardized by the following procedure. All percent values were plotted on the y (four-cycle) axis against test concentrations of penicillin on the x (six-cycle) axis of 4×6 cycle log paper for each soil site. Points for all sites were then grouped into either two or three sets forming approximately straight lines. Linear regression lines were then fitted to each set of points after converting coordinates to their logarithms. The antilogs of values read off the regression lines were used for statistical analysis. The U.S. Geological Survey STATPAC program library (VanTrump and Miesch, 1977) was used to generate additional statistics.

RESULTS

Poorman Creek deposit Figure 20-1 shows the relationship between the percent *Bacillus* spp. resistant to the seven lowest test concentrations of penicillin and the location of the soil sample sites with respect to the Poorman Creek deposit. Sample sites 0 through 11 are in the forested zone, and sites 12 through 26 are in the *Eriogonum* or "barren" zone directly over the deposit. A general increase of penicillin resistance in *Bacillus* spp. at all test concentrations of penicillin was observed in the soils over the barren zone. No significant correlations were found between the bioparameters tested and percent moisture content or pH (both deposits).

Statistical analyses of the Poorman Creek soil data are summarized in Tables 20-1 and 20-2. Scatter diagrams were printed and inspected for all significant correlations shown

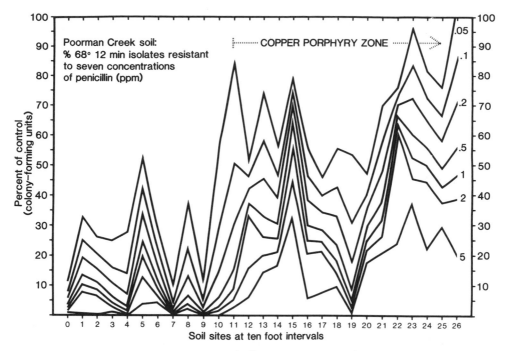

Fig. 20-1   Percent *Bacillus* spp. (68°C 12-minute isolates) resistant to the seven lowest test concentrations of penicillin. A general increase in percent penicillin resistant *Bacillus* spp. coincides with the treeless zone over the porphyry copper deposit.

TABLE 20-1   Correlation Coefficients, Poorman Creek Soils[a]

	Cu	Percent of Bacillus spp. Resistant to 2 ppm Penicillin	Percent of Bacillus spp. Resistant to 5 ppm Penicillin	Total B. cereus	Percent B. cereus
Total *B. cereus*	0.77	0.80	0.78	1.00	0.86
Percent *B. cereus*	0.74	0.91	0.89	0.86	1.00
Percent organic matter	0.68	0.55	0.62	0.61	0.60
Al	0.75	0.41	0.43	0.63	0.54
Fe	0.64	0.25	0.26	0.50	0.44
As	0.81	0.58	0.50	0.66	0.65
Ba	0.64	0.38	0.42	0.50	0.47
Be	0.94	0.57	0.60	0.74	0.73
Ce	0.99	0.67	0.68	0.78	0.76
Co	0.69	0.56	0.60	0.67	0.69
Cu	1.00	0.66	0.67	0.77	0.74
La	0.98	0.61	0.62	0.75	0.70
Mo	0.63	0.37	0.39	0.41	0.51
Nb	0.74	0.46	0.49	0.61	0.60
Ni	0.87	0.49	0.50	0.65	0.64
P	0.92	0.57	0.50	0.60	0.56
Sr	0.87	0.49	0.55	0.71	0.64
V	0.84	0.56	0.56	0.72	0.71
Y	0.97	0.69	0.69	0.77	0.76

[a]$N = 27$, $P \leqslant 0.05$ for $\pm r \geqslant 0.37$, $P \leqslant 0.01$ for $\pm r \geqslant 0.47$.

TABLE 20-2 Correlations between Numbers of *Bacillus* spp. Resistant to Various Test Concentrations of Penicillin and Total Numbers of *B. cereus* per Gram of Soil or Percent of Total *Bacillus* spp. Consisting of *B. cereus*[a]

Test Concentrations Of Penicillin	Correlation Coefficients ($n = 27$) between Percent Penicillin-Resistant Bacteria and:	
	Total *B. cereus*	Percent *B. cereus*
0.05	0.69	0.72
0.1	0.77	0.82
0.2	0.79	0.86
0.5	0.79	0.88
1.0	0.80	0.89
2.0	0.80	0.91
5.0	0.78	0.89
10.0	0.64	0.76
20.0	0.58	0.70

[a]$N = 27$, $P \leqslant 0.05$ for $\pm\, r \geqslant 0.37$, $P \leqslant 0.01$ for $\pm\, r \geqslant 0.47$. The correlations indicate that the majority of penicillin-resistant soil *Bacillus* spp. at these sites are *Bacillus cereus*. The correlations indicate that *B. cereus* may be selected in these soils by their penicillin-resistance trait.

to ensure that the assumptions underlying correlation analysis (using Pearson's r) were not violated. Although many significant correlations between various metals and percent penicillin-resistant bacteria occurred at other test concentrations of penicillin in the Poorman Creek data set, the correlations became maximum at either 2 or 5 μg/ml penicillin, with most occurring at 5 μg/ml penicillin. Only soil samples 7 and 9 contained no spores able to form colonies in 5 μg/ml penicillin. We conclude from this that the maximum penicillin test concentrations that permit the growth of some *Bacillus* spp. from all, or nearly all, sample sites would be the most diagnostic of soil metal content.

Keystone deposit Figure 20-2 shows the relationship between *Bacillus* spp. at the Keystone sample sites resistant to various test concentrations of penicillin in an early trial. The extreme penicillin resistance recorded at sites 14 and 15 was due to one organism (not *B. cereus*), which was an unidentified, small, highly motile, aerobic, spore-forming rod that formed tiny, amber-colored, spindle-shaped subsurface colonies. In two separate tries we were not able to repeat the data at sites 14 and 15 six months after the first resistance trials, although some tiny amber colonies were still present. The correlations

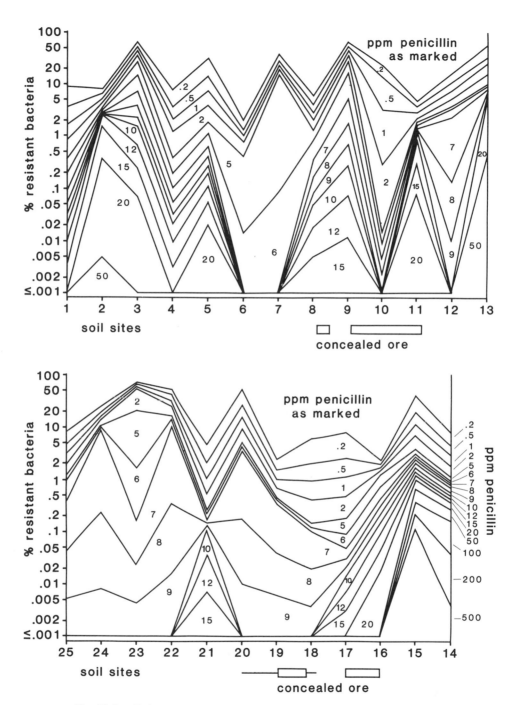

Fig. 20-2   Early Keystone data composites showing the regular collapse of penicillin resistance in *Bacillus* spp. with increased penicillin through both traverses. Each line traces the percent of *Bacillus* organisms (from site to site) able to form colonies in minimal agar amended with the indicated test concentrations of penicillin. Note difference in scale between Figures 20-1 and 20-2.

JOHN R. WATTERSON, LASLO A. NAGY, DAVID M. UPDEGRAFF

recorded in Table 20-3 are based on a more exhaustive range of penicillin test concentrations in a repeat experiment conducted about 6 months after the first one.

Correlation and stepforward regression analysis of the Keystone chemical and penicillin resistance data, based on 24 samples, are summarized in Table 20-3. Site number six, which was suspected of having been disturbed, was eliminated from the data set. Scatter diagrams for all single correlations shown in Table 20-3 were printed and inspected to ensure that assumptions underlying regression and correlation analyses were not violated. Superscripts in two cases indicate that the correlation is questionable because of being based on too few points.

TABLE 20-3 Percent Penicillin Resistant *Bacillis* spp. at Various Test Concentrations of Penicillin versus Element Content of Soil at the Keystone Deposit[a]

Dependent Variable (ppm penicillin)	Correlation Analysis		Stepforward Regression Analysis	
	Indep. Var. (Soil Conc.)	Cor. Coef. (Range)	Indep. Var.	Mult. Coef.
0.001	Al Fe	0.63 0.50	Al Be	0.75 (57%)
	Ca	−0.44		
0.002	Bact.	−0.49	−	
0.005	Be Ni Bact.	−0.56 −0.52	Be	−0.56 (31%)
	V Ba Nb Ti	−0.51 −0.46		
	Al Mn	−0.42 −0.39		
0.01	Be Nb Al	−0.67 −0.59	Be	−0.67 (45%)
	Ni Ba Bact.	−0.58 −0.49		
	La V[b]	−0.48 −0.46		
	Ca/La	0.49		
0.02	Be Al La Ni	−0.65 −0.52	Be	−0.65 (42%)
	Nb Ba Bact.	−0.51 −0.47		
	Ce Tb Co	−0.45 −0.42		
	Ca/La	0.50		
0.05	Be La Ce Ba	−0.64 −0.52	Be Fe	0.74 (54%)
	Al Co Tb	−0.50 −0.49		
	Ni Pr Nb	−0.47 −0.41		
	Nd Bact.	−0.40		
	Ca/La	0.47		
0.1	Be La Ce Ba	−0.60 −0.50	Be Tb	0.70 (49%)
	Tb Co Al Ni	−0.48 −0.44		
	Pr Nd Y	−0.43 −0.40		
0.2	Be La Ce Ba Tb	−0.54 −0.44	Be	−0.54 (29%)
	Al Pr Co Ni[b]	−0.40 −0.39		
0.5	Be La Ba Tb	−0.49 −0.41	Be	−0.49 (24%)
1	Be La Ba	−0.46 −0.40	Be	−0.46 (21%)
2	Be La Ba	−0.44 −0.39	Be	−0.44 (20%)
5	Be	−0.39	−	

[a]$N = 24$; $P \leqslant 0.05$ for $\pm r \geqslant 0.39$, $P \leqslant 0.01$ for $\pm r \geqslant 0.50$.
[b]Poor scattergram.

PENICILLIN RESISTANCE IN SOIL BACTERIA

Poorman Creek Deposit Interelement Correlations

Of the elements listed in Table 20-1, copper correlates most strongly with the other elements for which we have analyses. Of the transition metals, copper has the highest correlation with penicillin resistance and *B. cereus* counts, with a maximum $r = 0.77$ for total *B. cereus*. The high correlations between copper and a number of other elements caused concern about the possibility of a spectral interference by Cu in the ICP procedure. Repeat ICP runs spiked with similar concentrations of copper, however, revealed no spectral interferences with any of the elements correlating with copper in Table 20-1. A somewhat bimodal distribution of copper and other metal values in the soil traverse, in conjunction with the substantial spread of copper values (141 to 2253 ppm) appears to adequately explain the rather high correlations between copper and other elements. The high positive correlations between Ce and Y and the bioparameters may be ascribed to mutual correlations with copper rather than to any biological effects caused by Ce and Y.

Metal–Bioparameter Correlations

An apparent trend exists between the magnitude and sign of the copper–metal correlation coefficients and that of bioparameter–metal correlation coefficients. That is, the decrease in correlation between the bioparameters and the elements can be seen to parallel the decrease in correlation between copper and those elements. This is consistent with our hypothesis that more than one metal is involved in determining penicillin resistance in these soils. However, with the possible exception of lead, the correlation of the various toxic metals other than copper with penicillin resistance or *B. cereus* counts could thus be satisfactorily explained by their correlation with copper. We note that, in contrast to the Keystone deposit, the soil content of the rare earth elements (REE) correlates positively with percent penicillin resistance (*pen^r*) (and *B. cereus* counts) at this deposit.

B. Cereus versus Metals and Penicillin Resistance versus Metals

B. cereus counts and percent *B. cereus* (i.e., *B. cereus* × 100/total *Bacillus* spp.) both correlate substantially better with all the metals than do the percentages of penicillin-resistant *Bacillus* spp. (Table 20-1). This, in concert with the strong agreement between *B. cereus* counts and penicillin resistance at all test concentrations (Table 20-2), leads us to believe that *B. cereus* is an intervening variable by which the penicillin-resistance data can be explained. That is, the percentage of *B. cereus* in the *Bacillus* population accounts for the percent penicillin-resistant *Bacillus* spp. This idea is supported by the work of Curran and Evans (1946) (summarized in Table 20-4), which shows that *B. cereus* is the least sensitive of the *Bacillus* species to penicillin. The fact that *B. cereus* counts correlate better with the metals than do the percentages of penicillin-resistant *Bacillus* spp. may

TABLE 20-4 Relative Sensitivity of *Bacillus* spp. to Penicillin

		Number of Strains
Most sensitive	*B. alvei*	4
	B. brevis	3
	B. circulans	1
	B. firmus	1
	B. laterosporus	2
	B. mascerans	1
	B. megaterium	5
	B. polymyxa	1
	B. pumilus	1
	B. sphaericus	2
	B. subtilis	5
Intermediately sensitive	*B. alvei*	2
	B. subtilis	4
	B. subtilis (anaerobic)	1
	B. subtilis var. *aterrimus*	1
	B. subtilis-niger	1
Least sensitive	*B. albolactis*	1
	B. cereus	7
	B. metiens[a]	1
	B. mycoides[a]	2

[a]Now classified as *B. cereus*.
Source: Data from Curran and Evans (1946, p. 89).

indicate that the metals are the primary selective determinants and that penicillin is a secondary or intermediate selective determinant.

That metals, and not penicillin, could be *directly* selecting for *B. cereus* via its penicillin-resistance trait, however, appears unlikely. The only means by which metals are known to exert a direct genetic selective pressure for increased antibiotic resistance is through selection for plasmids containing both antibiotic- and metal-resistance genes (Summers, Chapter 16, this volume). Nearly all specific metal resistances studied so far appear to be carried on plasmid DNA. Because *B. cereus* penicillinase genes are exclusively chromosomal (Pollock, 1967), there appears to be little possibility that metals could *directly* (genetically) select for penicillin resistance in this species.

Percent Organic Matter versus Copper

It is interesting and perhaps paradoxical in view of the apparent toxicity of the soils for trees at the Poorman Creek deposit that copper should correlate significantly with percent organic matter in the soils. The primary explanation for this may be a type of "reverse-rain forest" effect. Copper is a well-known inhibitor of fungi—the principal agents that decompose soil organic matter. It may thus be that copper and other available metals in the more metalliferous soils act as natural preservatives of soil organic matter. A minor, contributing explanation may lie in a well-known experimental phenomenon. Powell (1950), Levinson and Sevag (1953), and Krishna Murty and Halvorson (1957) have noted the ability of the cations of mercury, copper, chromium, and iron to inhibit completely the germination of *Bacillus* spores. The experiments of Krishna Murty and Halvorson (1957) indicate that this inhibition occurs through the binding of metal cations to the bacterial spore coat, similar to the effect observed with fungal spores. It therefore seems probable that in soils with substantial concentrations of these cations, a large percentage of *Bacillus* spores and fungal spores would be inhibited, i.e., unable to germinate. Since these soils apparently have active spore-forming populations more or less continuously forming spores, it appears possible that organic matter in the form of metal-inhibited spores may accumulate in proportion to the presence of the inhibitory cations and contribute to the organic matter present. The correlation (Table 20-1) between percent organic matter and the bioparameters may be adequately explained by their common correlation with soil copper.

Keystone Deposit

The correlations shown in Table 20-3 are more complex. An explanation for the significant negative correlations between penicillin resistance and the rare-earth elements may lie in the quite different effects rare-earth elements are known to have on the calcium-transport systems of *Bacillus* spp. and higher organisms such as fungi that contain mitochondria. Lanthanum and all other rare-earth cations are potent inhibitors of respiration-dependent calcium transport in mitochondria, extrapolating to complete inhibition at less than 0.1 ηmol of lanthanide per milligram of protein (Mela, 1969; Vainio *et al.*, 1970). The experiments of Eisenstadt and Silver (1972) with sporulating cells of *B. subtilis*, on the other hand, show that lanthanum does not inhibit calcium accumulation in these prokaryotes (which lack mitochondria), but, in fact, stimulates it. This markedly different effect may explain our negative correlations. Where the rare-earth elements increase in the traverse, the penicillin production of the molds would suffer; as a consequence, there would be fewer penicillin-resistant *Bacillus* spp. Conversely, penicillin production and percent penicillin-resistant *Bacillus* spp. would increase where the rare-earth elements decrease. That this effect occurs despite the ordinary content of rare-earth elements in Keystone soils (La varies between 4 and 22 ppm, Y between 17 and 120 ppm) suggests to us that the ecology of molds and *Bacillus* spp. may be quite sensitive to low-level variations in the rare-earth elements. In a geologic setting similar to that of the Keystone deposit (the Kuroko massive sulfide deposit in Japan), the rare-earth elements, which may normally be tied up in biologically unavailable rutile, are thought to be present as phosphate inclusions in readily weathering ilmenite (S. E. Church, oral communication, 1983). At the Keystone deposit, the correlations suggest that the rare earths must be present in a biologically available form. The considerable differences in solubility among the major rare-earth-containing minerals suggests that the observed effect must be heavily dependent on the soil mineralogy of these elements.

The consistently positive correlations between the Ca/La ratio and the percent penicillin-resistant *Bacillus* spp. (Table 20-3) is consistent with the interpretation above that the lanthanide elements are interfering with calcium transport, the idea being that high Ca/La ratios are favorable to penicillin production.

The correlations with beryllium are puzzling. The beryllium content of these soils is low and the variation minor (0.25 to 0.49 ppm). On this basis alone, one might suppose that these correlations are fortuitous. But even lacking an organic rationale, it is most improbable that beryllium should correlate so well with penicillin resistance throughout the range of test concentrations. We tentatively conclude, therefore, that beryllium in some way frustrates penicillin production even more effectively than the rare-earth elements, although we have no idea by what means this might occur.

Ecology of Penicillin Resistance

Selman Waksman (1947, 1951) was of the opinion that antibiotic antagonism and hence antibiotic production, despite its evident usefulness in nature, is not the decisive factor in determining the numerical hierarchy of microorganisms in most settings. Otherwise, he reasoned, the antibiotic-producing organisms would now dominate the microbial world, and clearly they do not. Brian (1957), although he stated that "the capacity to produce antibiotics is a character conducing to fitness," nevertheless concluded in his review of experiments attempting to measure the production of antibiotics in soils that most antibiotics are produced in extremely minor amounts under natural conditions and even then, in such restricted microenvironments as to be virtually without effect in "the bulk soil as a whole." This conclusion appears to be the consensus of most of the investigators who have undertaken the difficult task of demonstrating the presence of antibiotics in soil. But U.S. patent laws may also be of significance to the existence of this opinion because they discourage the patenting of natural substances. It may thus be partly a consequence of these laws that the view that antibiotics, by and large, are produced only with the aid of clever biochemists has its adherents and has so often been put forward.

In a review of the literature on antibiotic production and antibiotic resistance in soils, we have found only a single study (Hill, 1972) on both antibiotic production and antibiotic resistance in soil and none on either antibiotic production or antibiotic resistance in naturally metalliferous soils. The lack of such studies is interesting in view of the well-known coincidence of metal- and antibiotic-resistance genes on bacterial plasmids. It seems rather elementary, in view of this genetic coincidence, to ask whether bacteria encounter unusual concentrations of metals and antibiotics in the same environmental settings. The genetic knowledge, however, came somewhat after the excitement about antibiotics had peaked.

A basic assumption of our investigation was that the documented coincidence of metal- and penicillin-resistance genes on bacterial plasmids may be evidence that these metals and penicillin are commonly encountered by bacteria in the same natural settings and perhaps at a similar frequency. Other persuasive, strictly genetic reasons for the common occurrence of antibiotic and heavy-metal-resistance genes on plasmids can be forwarded, but we do not think they invalidate our assumption.

We began this study partly on the premise that plasmid-borne penicillin resistance might serve as a general index of plasmid-borne metal resistance on the basis of published observations that metal- and antibiotic-resistances are commonly linked (Novick and Roth, 1968) and can be co-selected by antibiotics (Summers, Chapter 16, this volume) or metals alone (Timoney et al., 1978). As applied to natural geologic settings and geologic

time, however, such a premise is apparently false. According to modern gene theory (Koch, 1981), any gene whose product is not required in a given environment, will inevitably be lost because of competition for the energy necessary to maintain DNA. Even though penicillin resistance in certain strains of bacteria can be temporarily increased by exposure to metals alone (assuming a common plasmid), if there were no selective pressure by penicillin itself, the penicillinase gene would be inevitably lost over a long period of time. Long periods of time for bacteria are virtually instantaneous geologically. The premise that metals alone might continuously select for penicillin resistance in the absence of penicillin in plasmid-bearing strains, therefore was not logical. However, the looked-for phenomenon, penicillin resistance related to natural metal occurrences, was observed. Evidently, the penicillin resistance observed had been selected for by penicillin produced naturally in the soil.

Our conclusions about the ecology of penicillin resistance and soil metal are summarized in the flow diagram in Figure 20-3. The diagram is based on a conjectural interpretation of our data. What follows is evidence from the literature for this interpretation.

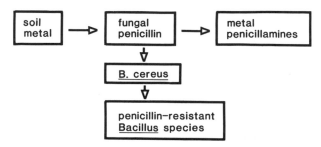

Fig. 20-3   Flow diagram of ecological causality according to our interpretation. Our data furnish evidence for the soil metal–*B. cereus*–penicillin resistance relationship. The soil metal–fungal penicillin–penicillamines relationship is conjectural.

Penicillin production in nature is not limited to the organisms used for its commercial production, but is an ability common to dozens of different species of molds belonging to at least nine genera (Pollock, 1967). Penicillin-producing molds of the two genera that account for most of the organisms that had been identified by the end of World War II are listed in Table 20-5. In trying to gain insights into the possible coincidence of penicillin production and metal resistance in these molds, we looked at the documented metal resistances of the known penicillin-producing organisms. The result of cross-referencing penicillin production to metal resistance in a cursory study of the subject is shown in Table 20-6. A systematic literature search would undoubtedly result in a larger list. The impression gained from this exercise is that the penicillia and aspergilli most noted for their resistance to several toxic metals share in common the ability to produce penicillin. We note that copper resistance is something of a common denominator in this group. Why might penicillin-production confer resistance to heavy metals, particularly copper? We will now attempt to answer this question.

TABLE 20-5 Fungi That Produce Penicillin[a]

Penicillia
P. *turbatum* Westling
P. *cinerascens* Biourge
P. *rubens* Biourge
P. *chloro-leucon* Biourge
P. *steckii* Zaleski
P. *chrysogenum* Thom
P. *chlorophaeum* Biourge
P. *griseo-roseum* Dierckx
P. *notatum* Westling
P. *citreo-roseum* Dierckx
P. *cyaneo-fulvum* Biourge
P. *brunneo-rubrum* Dierckx
P. *baculatum* Westling
P. *fluorescens* Laxa
P. *asperulum* Bainier
P. *meleagrinum* Biourge
P. *raciborskii* Zaleski
P. *roseo-citreum* Biourge
P. *griseo-fulvum* Dierckx
P. *avellaneum* Thom et Turesson
P. *crateriforme* Gilman and Abbott
P. *baarnense* von Beyma
P. *euglaucum* von Beyma
    Aspergilli
A. *giganteus* Wehmer
A. *nidulans* Eidam
A. *flavipes* Thom and Church
A. *niger* van Tieghem
A. *oryzae* Cohn
A. *parasiticus* Speare
A. *flavus* Link

[a]Original references in Florey *et al.*, (1949, Vol. 1, p. 647, ff.).

Penicillin–Metal Interactions

While trying to prepare salts of penicillin, early investigators noticed that the substance was inactivated by a large number of metal cations (Abraham and Chain, 1942). The metals with the most powerful inactivating effect on penicillin were copper, lead, zinc, and cadmium; but other heavy metals, such as nickel, mercury, and uranium, also exerted the inactivating effect. They did not understand the mechanism of this inactivation. Abraham and Chain (1942) noted that no antibiotic activity could be recovered by decomposing the inactivated penicillin with acid and extracting with ether. Although they did not deduce from this that the (still unresolved) molecule was broken, they guessed that complex formation was taking place between the metals and the penicillin. Within the year Abraham *et al.* (1943) had purified the primary penicillin hydrolysate and named it penicillamine. They did not, apparently, associate this compound with their previous observations of the chemical reactivity of penicillin toward metals.

TABLE 20-6   Metal Resistance of the Penicillin-Producing Fungi

Aspergillus nidulans (Eidam)	Mn^{2+}, Zn^{2+} Co^{2+}, Ni^{2+} Fe^{3+}, Ba^{2+} Sn^{2+}	Elorza, V., 1969, *Microbiologia Espanola*, v. 22, p. 131–138.
A. niger (Thom)	Cu^{2+}	Starkey, R. L., 1973, *J. gen. microbiol.*, v. 78, p. 217–225.
	Hg^{2+}	Ashworth, L. J., Jr., and Admin, J. V., 1964, *Phytopathology*, v. 54, p. 1459–1463.
	Pb^{2+}	Zlochevskaya, I. V., 1968, *Microbiology*, v. 37, p. 709–714.
	$AsO_3{}^{3-}$	Terui, G., and others, 1960, *Tech. Rept.* *Osaka Univ.* v. 10, p. 279–290.
	Au^{3+}	Mineyev, G. G., and others, 1975, cited in: Mineyev, G. G., 1976, *Geochem. Inter-* *national* v. 13, no. 2, p. 164–168.
Penicillium chrysogenum (Thom)	Cu^{2+}	Jarvis, F. G.., and Johnson, M. J., 1950, *J. Bact.*, v. 59, p. 51–60.
	$AsO_3{}^{3-}$	Challenger, F., 1944, *Chem. Rev.*, v. 36, p. 315–361.
	$SeO_3{}^{2-}$ $TeO_3{}^{2-}$	
P. notatum (Westling)	Cu^{2+} Hg^{2+} $AsO_3{}^{3-}$ $SeO_3{}^{2-}$ $TeO_3{}^{2-}$	Partridge, A. D., and Rich, A. E., 1962, *Phytopathology*, v. 52, p. 1000–1004. Challenger, op cit.
A. oryzae (Ahlburg) Cohn	Cu^{2+} $AsO_3{}^{3-}$ Au^{3+}	Starkey, op. Terui, op cit. Mineyev, op cit.
A. flavus (Link)	Cu^{2+}	Starkey, op cit.
P. thomii (Maire)	Cu^{2+}	Kendrick, W. B., 1962, *Can. J. Microbiol.*, v. 8, p. 639–647.

The presently understood route of penicillin hydrolysis, omitting an unstable inter-mediate in the conversion of penicilloic acid to penicillamine, penamaldate, is shown in Figure 20-4 (Davis *et al.*, 1980).

In 1955, it occurred to an English physician, J. M. Walshe (Walshe, 1981), that the penicillamine molecule might have a suitable structure to act as a copper-binding agent

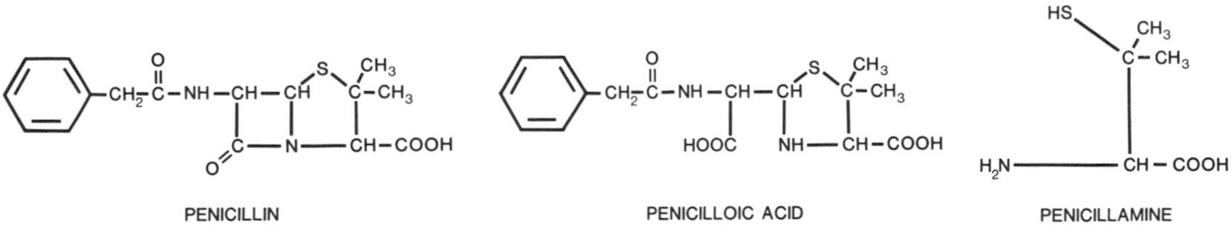

PENICILLIN   PENICILLOIC ACID   PENICILLAMINE

Fig. 20-4   Steps in the hydrolysis of penicillin. A more elaborate account is given in Davis *et al.* (1980).

for use in the treatment of Wilson's disease, a toxic buildup of copper that causes destructive lesions in the brain and liver. He obtained a few grams of the scarce compound from Merck and administered 0.5 g to himself. The next morning he gave a similar dose to a terminal patient suffering from Wilson's disease. These early experiments showed that administration of 1 g of penicillamine by mouth led to a 10- to 20-fold increase in the urinary excretion rate of copper. By the late 1950s, the use of penicillamine had revolutionized the prognosis of this hitherto invariably fatal disease. Since then, penicillamine, which also forms strong complexes in vitro with a wide range of heavy metals, has been used experimentally in treating poisoning due to lead, mercury, cadmium, and gold (Gergely and Sovago, 1979; Lyle, 1981). The various metal complexes that form with penicillamine, as listed by these authors, are shown in Figure 20-5. It now appears to us that the formation of these strong penicillamine chelates adequately explains the well-documented destruction of the penicillin molecule by these metals. We thus believe that penicillin may have at least two roles in nature, that of an antibiotic and that of a heavy-metal detoxifying agent. In the highly metalliferous soils of the Poorman Creek deposit, the heavy-metal detoxifying role must predominate.

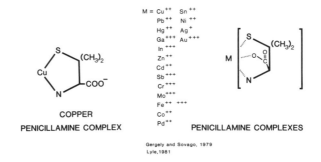

Fig. 20-5 Copper penicillamine complex and a general form of other known heavy-metal penicillamine complexes.

CONCLUSION

In response to a conclusion by Timoney *et al.*, (1978) that metals can create a selective pressure for antibiotic resistance, we tested the penicillin resistance of a normally penicillin-sensitive group of bacteria (*Bacillus* spp.) in the naturally metalliferous soils associated with two ore deposits. In both cases we found significant correlations between soil metal content and penicillin resistance. Further investigation of the Poorman Creek soils revealed that the numbers of one organism, *B. cereus*, the most naturally penicillin resistant of the group, explains the correlations between soil metal content and penicillin resistance. We conclude from this and from other circumstantial evidence having to do with the metal resistance of the penicillin-producing molds, the chemistry of the penicillin molecule, and the extraordinary metal-binding properties of its most characteristic hydrolysate, penicillamine, that penicillin production confers a selective advantage on penicillin-producing molds in metalliferous soils, particularly those containing copper. Although this mechanism for increased penicillin resistance in bacteria is fundamentally

different from the direct genetic selection invoked by Timoney *et al.* (1978), we think that these findings may bear on the evolution of the common plasmid loci of metal- and antibiotic-resistance genes in other bacteria in which plasmid-borne resistances appear to be more common. We conclude that *B. cereus* may be an authentic heavy-metal indicator microbe whose presence in soil, water, and stream-sediment samples should be investigated further for its potential as an indicator of mineralization. The ease with which *B. cereus* can be enumerated in egg-yolk agar commends it for investigation as a rapid geochemical assay.

Note added during text editing Between submission of this manuscript in 1983 and the present (late 1985), additional facts bearing on the above interpretation have come to light:

(1) During the last two and half years we have studied the populations of microfungi inhabiting barren soils over the Cotter Basin Prospect, Lewis and Clark County, Montana (a deposit similar to the Poorman Creek deposit with respect to metal, penicillin-resistance, and *B. cereus* anomalies). These fungal populations consist almost exclusively of penicillin-producing *Penicillium* molds, as determined by microscopic examination and *beta*-lactamase-induction bioassays using an inducible strain of *B. cereus* (modified after Hill, 1972). Fungal populations adjacent to the deposit, in contrast, were taxonomically heterogenous and contained *Penicillium* spp. less adept at penicillin production.

(2) Purified *B. cereus* strains isolated from metalliferous soils over the Cotter deposit and identified to species level have been found to exhibit dramatically enhanced penicillin (and other antibiotic) resistances relative to strains isolated from adjacent, unmineralized soils as well as to American Type Culture Collection strains 13061 and 14579. In addition, large percentages (up to perhaps 40 percent) of the *B. cereus* populations inhabiting the Cotter deposit soils are constitutive (deregulated) for *beta*-lactamase production, as shown by penicillin-hydrolysis assay (modified from Hill, 1972); whereas this mutation occurs in only about 3×10^{-8} wild-type cells (Sneath, 1955). Thus, penicillin resistance occurs as a premium trait at both the genus and species levels over the deposit. It would appear that penicillin produced by the *Penicillium* sp. dominating this deposit can only have brought about these effects.

(3) A paper by Mitchell and Alexander (1963) came to hand which documented the discovery of a lytic strain of *B. cereus* which could digest members of 11 out of 12 genera of fungi tested—including a *Penicillium* sp. Tests in our laboratory then indicated that strains of *B. cereus* isolated from the Cotter soils share this lytic trait: they can grow in ultrapure water with no nutrient other than intact mycelium tufts of the dominant *Penicillium* sp. Elevated penicillin resistance in the *B. cereus* populations over the Cotter deposit may thus have resulted from successful inter- and intra-species competition for one of its major nutrient sources—the penicillin-producing molds.

ACKNOWLEDGMENTS

G. H. Allcott, P. K. Theobald, Jr., J. H. McCarthy, Jr., A. P. Pierce, W. L. Campbell, D. J. Grimes, J. M. Motooka, J. G. Viets, J. D. Sharkey, R. F. Sanzolone, D. R. Rice, J. G. Crock, J. M. Hoffman, R. M. O'Leary, S. S. Elliott, S. Leatham, N. L. Parduhn, and

R. E. Tucker, all USGS colleagues, contributed generously to this study in various capacities. H. W. Lakin, who opened the door for this research, cannot be adequately acknowledged. Part of this work was done while J. R. Watterson was a guest at the laboratory of B. H. Olson at the University of California at Irvine.

J. R. Watterson gratefully acknowledges the kind interest of three of his fellow convenors, Don Carlisle, Ian Kaplan, and Wade Berry (who helped collect the Poorman Creek samples). These three, along with Paul Coleman, made possible the early presentation of this data under the best possible circumstances.

REFERENCES

Abraham, E. P., and Chain, E., 1942, Purification and some physical and chemical properties of penicillin: *Bri. J. Exp. Pathol.*, v. 23, no. 3, p. 8–115.

___ , Chain, E., Baker, W., and Robinson, R., 1943, Penicillamine, a characteristic degradation product of penicillin: *Nature*, v. 151, p. 107.

Austin, B., Allen, D. A., Mills, A. L., and Colwell, R. R., 1977, Numerical taxonomy of heavy metal-tolerant bacteria isolated from an estuary: *Can. J. Microbiol.*, v. 23, p. 1433–1477.

Bauer, A. W., Kirby, W. M., Sherris, J. C., and Turck, M., 1966, Antibiotic susceptibility testing by a standardized single disc method: *Am. J. Clin. Pathol.*, v. 45, p. 493–496.

Brian, P. W., 1957, The ecological significance of antibiotic production, p. 168–188 *in* Williams, R. E. O., and Spicer, C. C. (Eds.), *Microbial Ecology:* Cambridge University Press, Cambridge.

Cannon, H. L., 1960, Botanical prospecting for ore deposits: *Science*, v. 132, no. 3427, p. 591–598.

Crock, J. G., and Lichte, F. E., 1982, Determination of rare earth elements in geological materials by inductively coupled argon plasma/atomic emission spectroscopy: *Anal. Chem.*, v. 54, p. 1329–1332.

Curran, H. R., and Evans, F. R., 1946, The activity of penicillin in relation to bacterial spores and the preservation of milk: *J. Bacteriol.*, v. 52, p. 89–98.

Davis, B. D., Dulbecco, R., Eisen, H. N., and Ginsberg, H. S., 1980, *Microbiology:* Harper & Row, New York, 1355 p.

DeBarjac, H., 1981, Insect pathogens in the genus *Bacillus*, p. 241–250 *in* Berkeley, R. C. W., and Goodfellow, M. (Eds.), *The Aerobic, Endospore-Forming Bacteria: Classification and identification:* Academic Press, New York, 373 p.

Donovan, K. O., 1958, A selective medium for *Bacillus cereus* in milk: *J. Bacteriol.*, v. 21, no. 1, p. 100–103.

Eisenstadt, E., and Silver, S., 1972, Calcium transport during sporulation in *Bacillus subtilis*, p. 180–186 *in* Halvorson, H. O., Hensen, R., and Campbell, L. L. (Eds.), *Spores V:* American Society for Microbiology, Washington, D.C.

Florey, H. W., and six coauthors, 1949, *Antibiotics*, 2 Vols.: Oxford University Press, Oxford.

Gergely, A., and Sovago, I., 1979, The coordination chemistry of L-cysteine and D-penicillamine, Chap. III *in* Sigel, H. (Ed.), *Metal Ions in Biological Systems*, Vol. 9., *Amino Acids and Derivatives as Ambivalent Ligands:* Marcel Dekker, New York, 277 p.

Gray, T. R. G., and Williams, S. T., 1971, Microbial productivity in soils, p. 255–286 *in*

Hughs, D. E., and Rose, A. H. (Eds.), *Microbes and Biological Productivity:* Cambridge University Press, Cambridge.

Hill, P., 1972, The production of penicillins in soils and seeds by *Penicillium chrysogenum* and the role of penicillin beta-lactamase in the ecology of soil *Bacillus: J. Gen. Microbiol.*, v. 70 no. 20, p. 243–252.

Holding, A. J., Franklin, D. A., and Watling, R., 1965, The microflora of peat–podzol transitions: *J. Soil Sci.*, v. 16, p. 44–59.

Kendrick, W. B., 1962, Soil fungi of a copper swamp: *Can. J. Microbiol.*, v. 8, p. 639–647.

Kinkel, A. R., Hall, W. E., and Albers, J. P., 1956, Geology and base-metal deposits of West Shasta copper–zinc district, Shasta County, California: *U.S. Geol. Surv. Prof. Pap. 285.*

Kleinkopf, M. D., and Mudge, M. R., 1972, Aeromagnetic, bouger gravity, and generalized geologic studies of the Great Falls-Mission Range area, northwestern Montana: *U.S. Geol. Surv. Prof. Pap. 726-A.*

Koch, A. L., 1981, Evolution of antibiotic resistance gene function: *Microbiol. Rev.*, v. 45, no. 2, p. 355–378.

Krieg, A., 1981, The genus *Bacillus:* insect pathogens, p. 1743–1755 *in* Starr, M. P., Stolt, H., Truper, H. G., Balows, A., and Schlegel, H. G. (Eds.), *The Prokaryotes*, Vol. II: Springer-Verlag, New York, 2440 p.

Krishna Murty, G. G., and Halvorson, H. O. 1957, Effect of enzyme inhibitors on the germination of and growth from *Bacillus cereus* var. *terminalis* spores: *J. Bacteriol.*, v. 73, p. 230–234.

Letunova, S. V., and Koval'sky, V. V., 1978, *Geochemical Ecology of Microorganisms* (in Russian): Publishing House 'Science', Moscow, 148 p.

Levinson, H. S., and Sevag, M. G., 1953, Stimulation of germination and respiration of the spores of *Bacillus megatherium* by manganese and monovalent cations: *J. Gen. Physiol.*, v. 36, p. 617–629.

Lyle, W. H., 1981, Penicillamine in metal poisoning: *J. Rheumatol.* (Suppl. 7), no. 8, p. 96–99.

Marques, A. M., Congregado, F., and Simon-Pujol, D. M., 1979, Antibiotic and heavy-metal resistance of *Pseudomonas aeruginosa* isolated from soils: *J. Appl. Bacteriol.*, v. 47, p. 347–350.

McGaughey, C. A., and Chu, H. P., 1948, The egg-yolk reaction of aerobic sporing bacilli: *J. Gen. Microbiol.*, v. 2, p. 334–340.

Mela, L., 1969, Inhibition and activation of calcium transport in mitochondria; effect of lanthanides and local anesthetic drugs: *Biochemistry*, v. 8, p. 2481–2486.

Mishustin, E. N., and Mirsoeva, V. A., 1968, Spore-forming bacteria in the soils of the USSR, p. 458–473 *in* Gray, T. R. G., and Parkinson, D. (Eds.), *The Ecology of Soil Bacteria:* Liverpool University Press, Liverpool, England.

Mitchell, R., and Alexander, M., 1963, Lysis of soil fungi by bacteria, *Can. J. Microbiol.*, v. 9, p. 169–177.

Motooka, J. M., and Sutley, S. J., 1982, Analysis of oxalic acid leachates of geologic materials by inductively coupled plasma-atomic emission spectroscopy: *Appl. Spectrosc.*, v, 36, no. 5, p. 524–533.

Nelson, J. D., Jr., and Colwell, R. R., 1975, The ecology of mercury-resistant bacteria in Chesapeake Bay: *Microb. Ecol.*, v. 1, p. 191–218.

Norris, J. R., Berkeley, R. C. W., Logan, N. A., and O'Donnell, A. G., 1981, The genera *Bacillus* and *Sporolactobacillus*, p. 1711–1742 *in* Starr, M. P., Stolt, H., Truper, H. G., Balows, A., and Schlegel, H. G. (Eds.), *The Prokaryotes*, Vol. II: Springer-Verlag, New York, 2440 p.

Novick, R. P., and Roth, C., 1968, Plasmid-linked resistance to inorganic salts in *Staphylococcus aureus, J. Bacteriol.*, . 95, no. 4, p. 1335–1342.

Olson, B. H., and Thornton, I., 1981, The development of a bacterial indicator system to assess bioavailability of metals in contaminated land, p. 254–258 *in Management and Control of Heavy Metals in the Environment:* CEP Consultants Ltd., Edinburg.

Pollock, M. R., 1967, Origin and function of penicillinase: a problem in biochemical evolution: *Brit. Med. J.*, v. 4, p. 71–77.

Powell, J. F., 1950, Factors affecting the germination of thick suspensions of *Bacillus subtilis* spores in L-alanine solution: *J. Gen. Microbiol.*, v. 4, p. 330–338.

Sneath, P. H. A., 1955, Proof of the spontaneity of a mutation to penicillinase production in *Bacillus cereus, J. Gen. Microbiol.*, v. 13, p. 561–568.

Stanier, R. Y., Adelberg, E. A., and Ingraham, J. L., 1976, *The Microbial World*, 4th ed.: Prentice-Hall, Englewood Cliffs, N.J., 871 p.

Timoney, J. F., Port, J., Giles, J., and Spanier, J., 1978, Heavy metal and antibiotic resistance in the bacterial flora of sediments of New York Bight: *Appl. Environ. Microbiol.*, v. 36, p. 465–472.

Troyer, L. S., Olson, B. H., Hill, D. C., Thornton, I., and Matthews, H., 1980, Assessment of metal availability in soil through the evaluation of bacterial metal resistance, p. 129–141, *in* Hemphill, D. D. (Ed.), *Trace Substances in Environmental Health*, Vol. 14: University of Missouri, Columbia, Mo.

_____ , Thornton, I., and Olson, B. H., 1981, Evaluation of cadmium pollution in British soils by natural bacterial populations: *Dev. Ind. Microb.*, v. 22, p. 537–542.

Vainio, H., Mela, L., and Chance, B., 1970, *Eur. J. Biochem.*, v. 12, p. 387–391.

VanTrump, G., Jr., and Miesch, A. T., 1977, The U.S. Geological Survey RASS-STATPAC system for management and statistical reduction of geochemical data: *Comput. Geosci.*, v. 3, no. 3, p. 475–488.

Viets, J. G., 1978, Determination of silver, bismuth, cadmium, copper, lead, and zinc in geologic materials by atomic absorption spectrometry with tricaprylylmethylammonium chloride: *Anal. Chem.*, v. 50, no. 8, p. 1097–1101.

Waksman, S. A., 1947, *Microbial Antagonisms and Antibiotic Substances*, 2nd ed.: Commonwealth Fund, New York, 415 p.

_____ , 1951, Biological aspects of antibiotics, *in*: Anon., *Frontiers in Medicine: the March of Medicine 1950:* Columbia University Press, New York.

Walshe, J. M., 1981, Discovery of therapeutic use of D-penicillamine, Int. Symp. Penicillamine, Miami, Fla., May 8–9, 1980: *J. Rheumatol.* (Suppl. 7), no. 8, p. 3–8.

Sue Tripp, Tamar Barkay, and Betty H. Olson
Program in Social Ecology
University of California
Irvine, California

21

RELATIONSHIP BETWEEN HEAVY-METAL CONTAMINATION AND SOIL BACTERIA

The relationship between heavy-metal contamination and the community structure of soil bacteria was examined. Six sewage-sludge-amended sites, which varied according to their contamination by Cu, Cd, Pb, and Zn, the duration of the sludge amendment, and certain physicochemical parameters, were selected in England. The sites were sampled twice, in February 1981 and August 1982. Bacteria were isolated on media supplemented with the following: 100, 300, and 400 ppm Cu; 10, 100, and 200 ppm Cd; 500, 1000, and 2000 ppm Pb; 250, 500, and 650 ppm Zn; and control plates of 0 ppm metal. Percent bacterial resistance was defined as: 100 X (no. colonies on metal-supplemented plate/no. colonies on control plate), and was determined for each metal concentration. A comparison of the resistance patterns determined for each site at the two sampling times indicated changes in resistance. In certain instances this was in response to the recent addition of sludge to the site. A more detailed study of the August 1982 samples was performed using the isolates from the 0, 10, 100 ppm Cd-supplemented plates. Fifty strains from each of the three concentrations at the six sites were purified and identified to the genus level. Results indicated that increased Cd supplementation of the media (1) increased the proportion of gram-negative to gram-positive bacteria isolated, and (2) decreased the number of different genera isolated. Both of these results were most pronounced for the sites with the lowest levels of Cd contamination. Our results indicate that for the sludge-amended sites studied, bacterial resistance may be affected by seasonality, that a background percentage of resistant bacteria are present in the soil regardless of the season or degree of heavy-metal contamination, and that this percentage does increase as the available metal in the soil increases.

Isaac R. Kaplan
Department of Earth and Space Sciences
University of California — Los Angeles
Los Angeles, California

III

AN INTRODUCTION TO ORGANIC MATTER AND GEOCHEMICAL EXPLORATION

The relationship between naturally occurring organic matter and metals is at best poorly understood. In part this is due to poor documentation and in part to the impossibility, up to the present time, to solve the various controls. From the time that organic carbon analysis of soils and sediments began to be routine, some 60 years ago, the association of certain authigenic minerals (e.g., pyrite) with organic-rich sediments was recognized. Such associations had long ago been attributed by Winogradsky to microbiological metabolism within the sediment. Thus organic matter was thought to support metabolism and not to influence metal binding directly. Perhaps the first legitimate example of an interaction between a well-recognized organic moiety and metals came from the classic work of A. Treibs in the early 1930s in which he isolated vanadium and nickel petroporphyrins. These were interpreted as arising from the reaction of diagenetically altered chlorophyll or haem pigments (which contain magnesium or iron, respectively) with nickel or vanadium present within sedimentary minerals. We still do not understand the mechanism by which some transition metals chelate with tetrapyrrole ligands.

Perhaps the clearest examples of the role of humus in the concentration of metals came from the intensive exploration for uranium during the 1950s. It became recognized that roll-front urananite deposits are closely associated with transitional oxidizing-reducing environments in sandstone aquifers and also that uranium-enriched minerals are concentrated in humus within deposition zones. A more recent example was the realization about 20 years ago that gold may also be associated with organic-rich rocks. Here the association is more poorly understood in that the organic matter is finely dispersed and is carbon rich, with no recognizable morphologic structures. Deposits such as those at Carlin, Nevada, or Olympic Dam, South Australia, also contain a small amount of extractable bitumen. However, this bitumen is highly altered and cannot easily be ascribed to any single source or origin. The source of the gold is probably hydrothermal, but the role of the organic matter in concentrating the gold (as well as uranium and copper at Olympic Dam) is not understood.

Although metals are notably enriched in the organic facies of soils and sediments, they are also enriched in the clay components. As the organic component most frequently correlates with the clay content of soils and sediment, gross total analyses of metals and organic carbon do not allow us to determine in which component of the sediment the metal has been concentrated.

The four chapters in this section each deal with some aspect of metal–organic association. The first, by Curtin and King, considers several examples for the concentration of a variety of metals by soil mull. The three areas studied were Empire District, Colorado; Chief Mountain, Colorado; and Stibnite, Idaho. The studies show that copper, cobalt, chromium, magnesium, barium, tin, strontium, and yttrium are highly mobile in soils and that the mull horizon is the most favorable for the accumulation of metals.

Baker, who has performed most of his research in Tasmania, Australia, combines laboratory experiments using humic acids and metals with observations in nature. He concludes that the primary nature of the humic matter in soils and sediments is not colloidal but is a partially soluble material whose ligands can form metal complexes. He suggests that the kinetics of humic material reacting with metals follows a first-order reaction. Dry metal humates did not readily release metals when washed with distilled water, but metals were readily solubilized in the presence of a humic acid solution. In another experiment he showed that gold salts are complexed to humic acid and that they are not converted to colloidal gold as some authors have stated. In conclusion, Baker discusses the transport of metals by humic-rich aqueous solutions and attempts to relate the presence or absence of gossans to the nature of the soil, vegetation, and climatic conditions in any area as an aid in exploration.

The third chapter in this section is written by M. Schnitzer, an international authority on the chemistry of humic acids. He defines the three organic fractions in soils as humic acids (HA), fulvic acids (FA), and humins. The latter are not soluble in acid or base. HA contain more C, H, N, and S than FA, but contain less O. FA are richer in phenolic OH, alkoholic OH, ketonic $C=O$, and carboxylic COOH groups than HA, but the latter contain more quinonoid $C=O$. All the COOH groups in HA are associated with aromatic compounds. Nonaromatic compounds such as proteins, polysaccharides, alkanes, or fatty acids are thought to be loosely associated to the humic acids by H bonds. The author discusses metal- and mineral-humic interactions and concludes that six types of association can be recognized, ranging from formation of water-soluble metal complexes to adsorption on external mineral surfaces or in clay interlayers. Schnitzer discusses the efficiencies of HA and FA reactions with different metals and minerals at different acidities (pH values). This brief summary chapter clearly suggests the importance of humic substances in the weathering of minerals and solubilization of metals by organic complexation.

The last chapter in this section, by Perdue and Lytle, is a physical chemical exposition of the thermodynamics of interaction of humic material with metals. The authors discuss the pitfalls in using simple equilibria relationships to explain relationships between minerals and humic material. In particular, they stress that certain equations, such as the Scatchard equation, used widely in biochemical research when dealing with polymers, is inapplicable when explaining humic substance reactions. One major reason is that there are multiple sites of complexation, and these may vary among humic acids derived from different soil extracts. They conclude with some advice to explorationists, stating that the largest amounts of metals available to plants will be in soils poor in clays and humus (mull).

Isaac R. Kaplan

G. C. Curtin
U.S. Geological Survey
National Center
Reston, Virginia

H. D. King
U.S. Geological Survey
Denver Federal Center
Denver, Colorado

22

UTILITY OF MULL IN GEOCHEMICAL EXPLORATION

ABSTRACT

Mull (forest humus) has been used successfully as a sample medium in geochemical exploration in various climatologic and physiographic terrains since the 1960s. Our investigations in the Rocky Mountains, in study areas in Colorado and in Idaho, show that the ash of mull contains anomalous amounts of many metals. These include gold, silver, copper, lead, zinc, molybdenum, and bismuth in the Colorado area and gold, antimony, arsenic, silver, zinc, and molybdenum in the Idaho area. Both areas are near sites of previous mining activity; the anomalous metal values are presumed to outline mineralized bedrock.

Leaching studies of reconstructed soil profiles collected in these areas show that copper, cobalt, chromium, magnesium, barium, tin, strontium, and yttrium are highly mobile in the soils; zirconium, lanthanum, nickel, manganese, arsenic, and scandium are moderately mobile; and titanium is rather immobile. Furthermore, the results indicate that in these soils, the mull horizon is a more favorable zone for the accumulation of metals and other elements than are the mineral (A, B, C) soil horizons.

INTRODUCTION

The role of humus as a concentrating medium was shown first by Goldschmidt (1937, p. 670-671) when he found that the ash of humus from an old forest of beech and oak in central Germany contained much more of some elements, including gold, than did the underlying mineral soil. He suggested that these elements enter the plant through the roots in soil solution and become enriched in the leaves, which are eventually deposited on the ground; as the leaves decay, the elements are immobilized and concentrated in the humus layer.

In the mid-1960s, we were testing various sample media both for their value in mineral exploration and for ease of collection and analysis. In part because of Goldschmidt's (1937) findings, we experimented with mull (forest humus layer) and found that high metal values in the ash of mull outlined ore zones in bedrock much better than did high metal values in the underlying soil. In general, the metal anomalies in mull ash were continuous and correlated well with concealed ore deposits, whereas anomalies in the underlying mineral soil horizons were scattered and discontinuous and did not correlate well with the deposits. Similar results using mull have been described by Hoffman and Booker (Chapter 9, this volume) for the Abitibi and Timmins–Porcupine greenstone belts of northern Quebec and Ontario; by Dunn (Chapter 7, this volume) for north-central Canada; by Boyle and Dass (1967) for Cobalt, Ontario; by Banister (1970) for Stibnite, Idaho; by Rice (1970) for the Orogrande district, Idaho; by Toverud (1979) for Sweden; and Kokkola (1977) and Nuutilainen and Peuraniemi (1977) for the Soviet Union and Finland.

This paper summarizes the results of our studies (1) comparing the value of mull in mineral exploration to that of the underlying soil in areas where the soil is derived mainly from transported materials such as glacial debris or colluvium, and (2) comparing the concentration and mobility of various elements in reconstructed soil profiles from two areas in the Rocky Mountains.

G. C. CURTIN, H. D. KING

357

GEOCHEMICAL STUDIES

We sampled and analyzed mull, the underlying soil, and other media in a number of areas, mostly in the Rocky Mountains of the western United States, and the results of our studies in three areas are given here. Two of the areas are in the Colorado Front Range (Fig. 22-1), and one is in central Idaho (Fig. 22-2). In these areas, we compared the abundance and distribution of elements in mull ash with that in the mineral soil.

This comparison of element concentrations in media of such varied composition as vegetation, soil rich in organic matter, and mineral soil is not completely satisfactory because these media cannot be reduced to a similar compositional state for analysis and still accurately represent their role in the geochemical cycle. Ashing of the samples rich in organic matter (vegetation and organic-matter-rich soil), however, produces material that is roughly similar in composition to the mineral soil horizons in that all samples are largely inorganic and essentially dry. Thus a comparison of the relative amounts of elements in the inorganic phases of the different media gives an estimate of where certain elements are being concentrated as they are cycled through the mineral soil, the vegetation, and the mull.

Sample Preparation and Analysis

Soil samples were sieved, and the fraction consisting of particles smaller than 2 mm was ground and pulverized for analysis. Mull samples were sieved to remove rock fragments and large pieces of litter, and the fraction consisting of particles smaller than 2 mm was ashed. A small split was taken from the ashed sample and sieved through a 0.25-mm sieve to remove sand-sized grains, and the fraction consisting of particles smaller than 0.25 mm was analyzed. A 10-g split from the sieved and ashed mull sample was analyzed for gold. The metal content of the mull samples is reported on an ashed-weight basis.

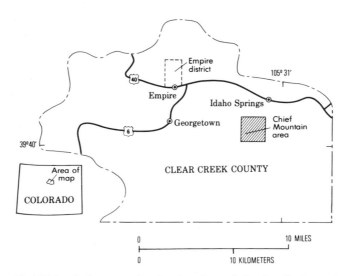

Fig. 22-1   Index map showing locations of Empire district and Chief Mountain area, Colorado.

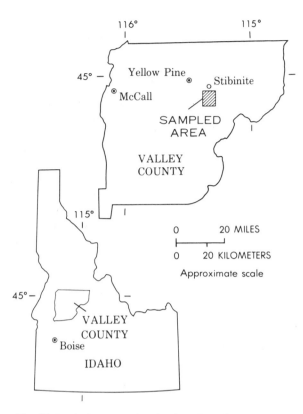

Fig. 22-2   Index map showing location of study area near Stibnite, Idaho.

Several analytical procedures were used. Gold was determined by an atomic-absorption method (Thompson *et al.*, 1968). The other elements were determined in mull ash by a semiquantitative spectrographic method for organic materials (Mosier, 1972), and in soils and rocks by another semiquantitative spectrographic method (Grimes and Marranzino, 1968).

Empire District, Colorado

The first example of our studies, that made in the Empire district (Curtin *et al.*, 1968, 1971), an old gold district in the Colorado Front Range about 48 km west of Denver (Fig. 22-1), demonstrated the usefulness of mull as a sample medium in areas of gold mineralization.

The area is mountainous and forested with lodgepole pine (*Pinus contorta latifolia* Engelm.), Douglas fir [*Pseudotsuga menziesii* (mirb.) Franco], Engelmann spruce [*Picea engelmannii* (Parry) Engelm.], and aspen (*Populus tremuloides* Michx.). The bedrock, primarily Precambrian granodiorite and biotite gneiss, is mostly concealed by an extensive cover of colluvium and glacial debris. The principal lodes in the district are a series of gold-bearing, pyritic quartz veins that surround a small Tertiary granodiorite stock on three sides and possibly on the fourth beneath a cover of glacial debris (Fig. 22-2). The mull, consisting of mixed organic and mineral soil material, usually is present as well-formed pads 2.5 to 10 cm thick under individual conifer trees and forms a rather even blanket in groves of aspen.

Analysis of the samples showed that high gold values in the ashed mull (Fig. 22-3) defined the distribution of gold occurrences in bedrock, known from underground mining operations, beneath the extensive cover of colluvium and glacial drift. In general, these high gold values in mull ash are continuous and correlate well with concealed gold deposits. In contrast, relatively high gold values in the underlying soil display patterns that, in part, outline the gold deposits but that also outline areas of barren bedrock (Fig. 22-4).

The soil is derived from colluvium or glacial debris, and in this setting, gold-bearing-vein material is dispersed and diluted by the admixture of barren soil. The process would result in the pattern that is apparent in Figure 22-4. The mull, however, is a relatively young material that is concentrating elements that apparently have been accumulated from the bedrock by the parent trees. Tree roots were observed in bedrock at several localities and were observed at least 15 m below the surface in one of the mine workings. This ability of trees to concentrate elements obtained from bedrock is likely a major reason for the difference in the geochemical anomalies between the mull ash and the soil.

Chief Mountain Area, Colorado

The Chief Mountain area in Colorado (Fig. 22-1) is about 6 km east of the Empire district and about 50 km west of Denver. The area ranges in elevation from about 2600 to 3500 m and is forested with species of pine, spruce, and fir.

Our attention was drawn to this area by a strong negative gravity anomaly similar to gravity lows that are associated with mineralized Tertiary hypabyssal intrusive stocks and plugs in the Front Range Mineral Belt. The anomaly we investigated could also reflect a concealed mineralized intrusive body. The area is underlain by schist and gneiss that have been intruded mainly by granodiorite and pegmatite, all of Precambrian age. These rocks, in turn, have been intruded by a few small Tertiary bostonite dikes.

Bedrock surface exposures are generally unaltered and appear to be barren. The only exceptions are a few scattered prospect pits in small shear zones and two small mine workings in the northeast part of the area. For our studies, we collected samples of the A and B horizons of the soil in addition to the mull, which was collected mainly beneath conifer trees. The soil is derived from an extensive cover of colluvium.

The following examples from this area demonstrate the enrichment of elements in mull ash over that in the soil in a setting where the mineralized rock may be concealed by many tens to hundreds of meters of barren rock. The geochemical anomalies in the mull ash are thought to outline the primary halo over a mineralized zone. The distributions of gold and silver in the ash of mull (Fig. 22-5) outline an anomalous area slightly offset from the center of the gravity low. In contrast, the few gold and silver values for the mineral soil that are within the limits of detection and that we consider to be reproducible (Fig. 22-6) are for scattered samples and do not show any definite trends. Nor do they correspond to the anomalous areas outlined by the high gold and silver contents in mull ash. Similarly, high zinc and cadmium contents in mull ash (Curtin and King, 1974a,

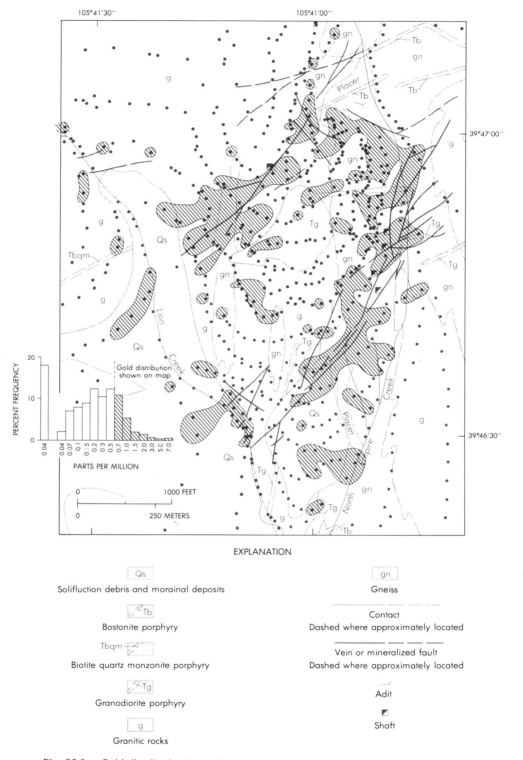

Fig. 22-3　Gold distribution in mull ash, Empire district. Dots show sample
localities. (Base modified from Braddock, 1969.)　　　　　　　361

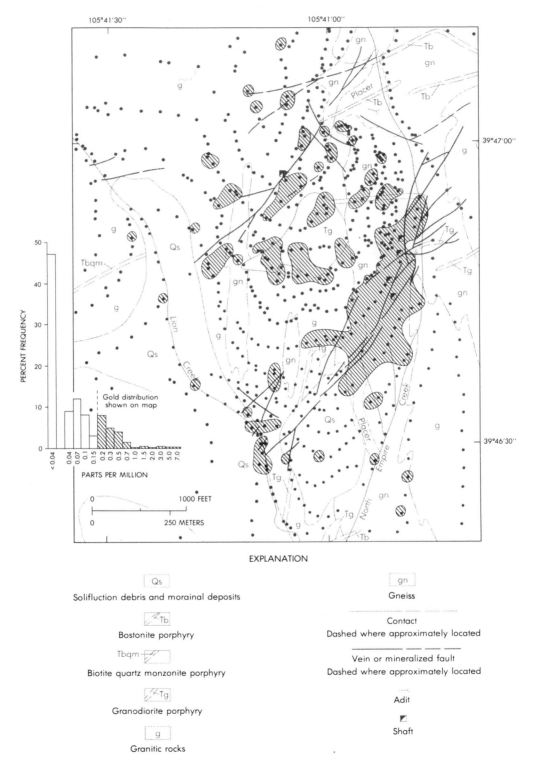

Fig. 22-4 Gold distribution in mineral soil, Empire district. Dots show sample localities. (Base modified from Braddock, 1969.)

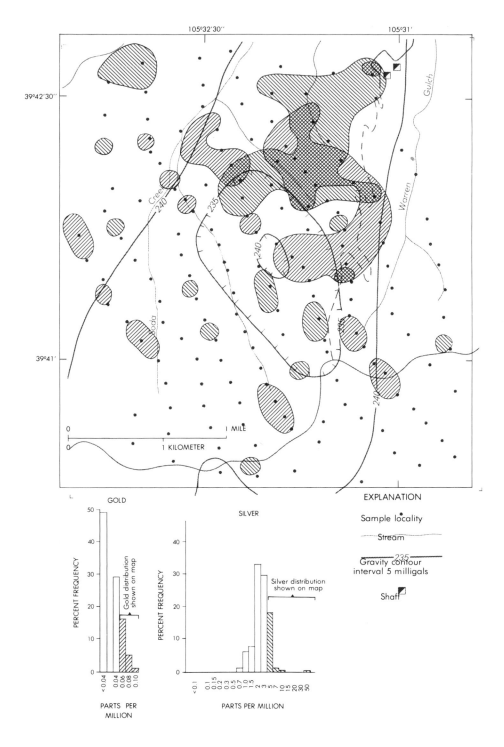

Fig. 22-5   Gold and silver distribution in mull ash, Chief Mountain area.

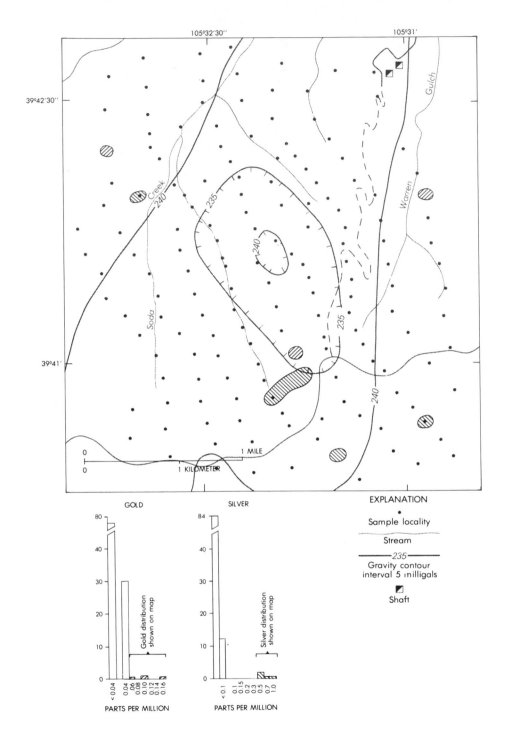

Fig. 22-6   Gold and silver distribution in mineral soil, Chief Mountain area.

UTILITY OF MULL IN GEOCHEMICAL EXPLORATION

p. 585–586) are from samples slightly offset from those having high gold and silver contents but outline the same general area. On the other hand, the anomalous zinc values in A horizon soil are scattered and do not form definitive patterns, and cadmium was not detected in any of the soil samples.

High lead values and moderately high bismuth values in mull ash (Fig. 22-7) provide a pattern different from that of gold and silver in mull ash in that it outlines an anomalous zone in the south and west parts of the area. In addition, the correlation of the moderately high bismuth values and lead values of 500 parts per million (dashed line in Fig. 22-7) forms a rather extensive pattern that partly encircles the center of the gravity low. The high lead values in the underlying soil (Fig. 22-8) also correlate, in part, with the bismuth values in the mull ash, but the high bismuth values in the underlying soil show the scatter typical of the soil anomalies.

One interesting note is that the distributions of the moderately high contents of both lead and bismuth are continuous across changes in species of vegetation. In the southern part of the area, the mull samples were derived from a subalpine fir (*Abies lasiocarpa*) and Englemann spruce forest, whereas in the northern part of the area, at a lower elevation, mull samples were collected from beneath lodgepole pine, Douglas fir, and ponderosa pine (*Pinus ponderosa*). Other metals, such as gold and silver, also concentrated equally in the mull under different species. This continuity suggests that the distribution and abundance of these metals in mull ash are not noticeably affected by changes in coniferous species.

Copper is the only metal of interest that was not more abundant in mull ash than in soil. The high copper values in mull ash (Curtin and King, 1974a, p. 589) formed only scattered anomalies. The range of copper values is also very narrow. High copper values in the B horizon of the soil, however, outlined two areas that correlate with the central part of the gravity low.

In the bedrock samples from the area, with few exceptions, the elements gold, silver, zinc, cadmium, lead, bismuth, copper, tin, and molybdenum were present in very small amounts or were not detected. Conversely, all these elements were detected in the mull ash. The exceptions were samples of bedrock from three prospect pits in small shear zones and one sample from an outcrop of the vein of one of the small mines. These samples contained anomalous amounts of gold, silver, zinc, lead, bismuth, copper, and mercury. The results of the Chief Mountain studies show that mull ash is a very useful anomaly-enhancement tool where one is trying to detect subtle metal halos over concealed mineralized zones, and where the soil is derived from transported colluvial material.

Stibnite, Idaho

Another area we investigated is just south of Stibnite, Idaho, an old antimony, gold, and tungsten mining area in the central part of the state (Fig. 22-2). In this area, we also collected samples of the A and B horizons of the soil in addition to the mull. The mull samples were derived from coniferous trees at all but about 5 percent of the sites. We selected this area to test the relative merits of the sample media because of the possibility that the Meadow Creek fault zone (Figs. 22-9 and 22-10), which bisects this area and is mineralized to the north at Stibnite, may also be mineralized here. The country rock is

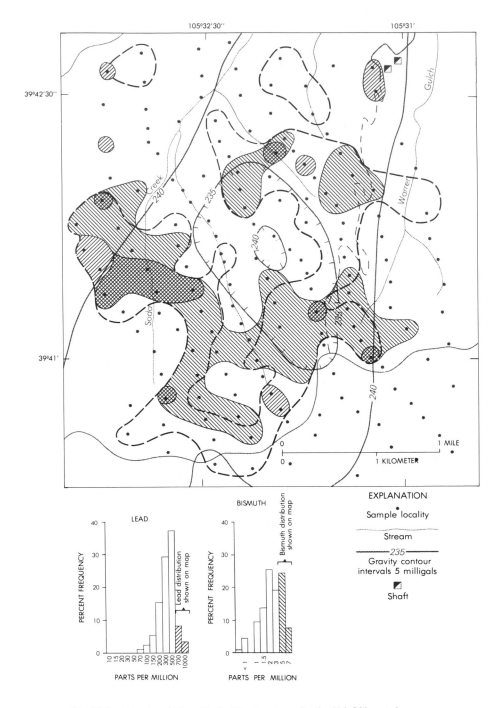

Fig. 22-7 Lead and bismuth distribution in mull ash, Chief Mountain area.

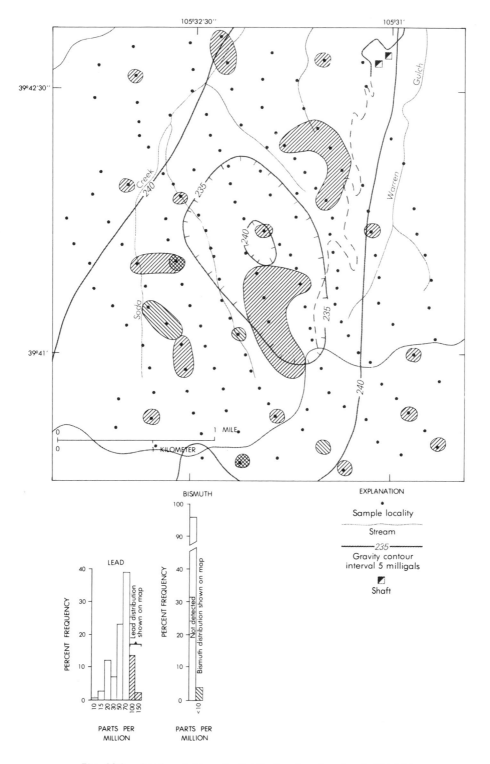

Fig. 22-8 Lead and bismuth distribution in mineral soil, Chief Mountain area.

367

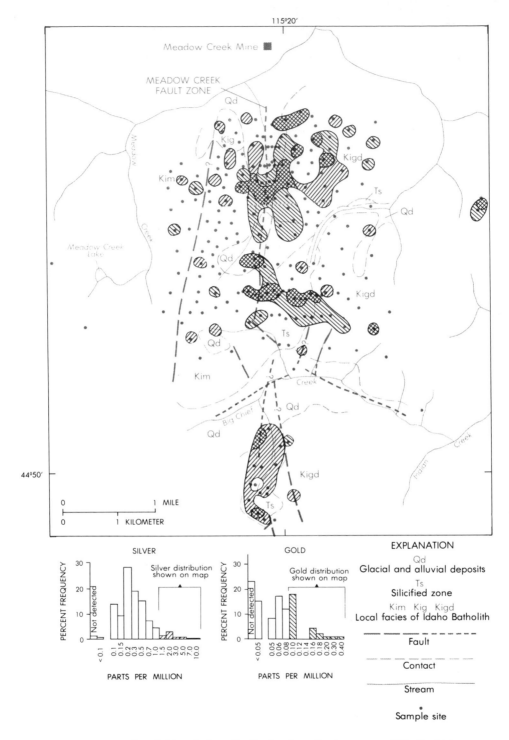

Fig. 22-9　Gold and silver distribution in mull ash, area near Stibnite, Idaho. (Modified from Curtin and King, 1974b.)

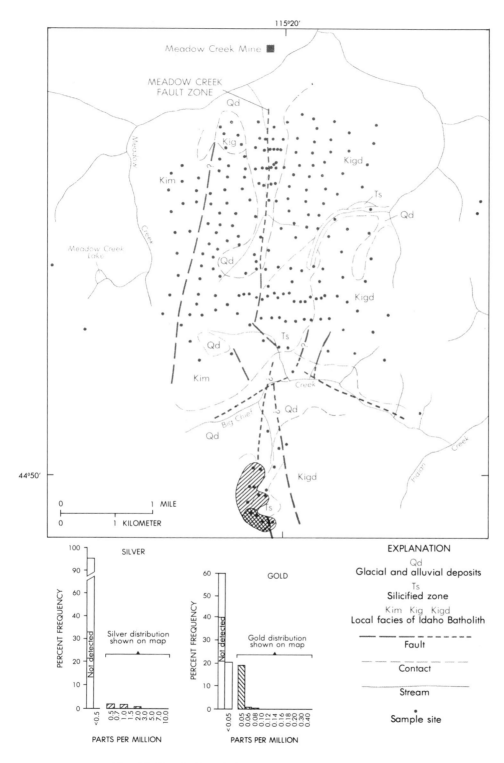

Fig. 22-10   Gold and silver distribution in mineral soil, area near Stibnite, Idaho. (Modified from Curtin and King, 1974b.)   369

mainly the granitic units of the Idaho batholith. These rocks are altered and sheared along a wide zone on either side of the Meadow Creek fault. The silicified parts of the altered zone are shown by the symbol "Ts" in Figures 22-9 and 22-10.

Two examples of the results (Curtin and King, 1974b)—the gold–silver pair in mull ash and A horizon soil and the zinc–molybdenum pair in both media—are discussed here. The distributions of gold and silver in the ash of the mull (Fig. 22-9) outline three distinct anomalous areas along the Meadow Creek fault zone. The northernmost of the three areas is partly in glacial cover, suggesting that gold and silver have migrated upward from mineralized rock beneath the glacial drift and have been concentrated in the ash of mull. The central and southern zones roughly correlate with the silicified zones (Ts, Fig. 22-9) and probably outline mineralized parts of these zones. Conversely, gold values that were reproducible by the analytical method used and measurable silver values in the A horizon soil (Fig. 22-10) outline only the anomalous area in the southern part of the Meadow Creek fault zone. Here the bedrock is relatively well exposed and the soil cover has not been displaced far from its bedrock source. Similarly, the distributions of anomalous amounts of zinc and molybdenum in mull ash (Curtin and King, 1974b, p. 43, 51–53) show similar patterns to those of gold and silver. By contrast, no zinc was detected in the soil by the spectrographic method used (Grimes and Marranzino, 1968) (lower limit of determination, 200 ppm). Molybdenum was detected in soil samples from only two places and therefore could not be used as a good guide to mineralized rock.

The distribution and abundance of other metallic elements, including antimony, reflected the patterns of gold and silver just described (Curtin and King, 1947b). The pattern for antimony may have been modified to some unknown extent by contamination from smelting operations at the Meadow Creek mine during the years 1949–1952, although the prevailing wind direction and other factors, such as the large amount of antimony (60 metric tons) estimated in the mull and soil (Curtin and King, 1974b), argue against smelting making a significant contribution of antimony or other metals to the mull.

STUDIES OF SOIL PROFILES

We also studied seven typical subalpine soil profiles in order to get a better understanding of the distribution, abundance, and mobility of elements in the mull and mineral-soil horizons. The studies and their results are described in detail in Curtin *et al.*, (1979).

The studies of these profiles outline the general distribution and mobility of 25 to 29 elements within the soils and the associated vegetation. The studies presumably approximate conditions within vegetation, mull, and the mineral soil in the spring and early summer when the soils are saturated, or at least are very wet, because of snowmelt and frequent rainstorms. Also, the rate of uptake by vegetation of soil water, and presumably of elements in solution, is highest at this time (Swanson, 1967). We found that many elements can be mobilized in all horizons of subalpine forest soils, but that the mull horizon is better suited to the fixation and concentration of elements in the biogeochemical cycle than are the mineral horizons.

UTILITY OF MULL IN GEOCHEMICAL EXPLORATION

Fieldwork

We sampled three profiles in the Chief Mountain area of Colorado (Fig. 22-1). In Idaho, we sampled four soil profiles; three were south of Stibnite, where we made our original survey, and the fourth was about 5 km northeast of Stibnite. Both areas are forested primarily with lodgepole pine, limber pine (*Pinus flexilus* James), subalpine fir, and Douglas fir. The ground cover consists of grasses and myrtle blueberry (*Vaccinium myrtillus* L.).

The soil profiles were collected in areas where anomalous amounts of many metals were present in samples of the mineral soil, ash of mull, and needles and twigs of the coniferous vegetation. We sampled the soil profiles by digging pits 0.6 to 1 m deep and separating the soil horizons at the sample sites. Bedrock was reached at only one of the sites in Colorado. The other profiles are in areas where the bedrock is covered by about 1 m to several tens of meters of colluvial and/or glacial material.

The soils are classified as inceptisols (USDA, 1960, p. 136) because they are poorly developed and lack a well-defined illuvial, clay-rich horizon. They are well drained, with the exception of two of the Idaho soils. Three of the Idaho soils are developing from glacial debris or glacial debris and alluvium; the rest are developing from colluvium.

Sample Preparation

The soils were prepared by sieving each horizon, including the mull, through a screen having 2-mm openings and discarding the fraction that did not pass the screen. The fraction that passed the screen was split several times to provide material for analysis and for reconstruction of the soil profiles. The profiles were reconstructed in polyethylene columns by using varying amounts of the fraction of the soil horizons that passed the screen. The soil material was packed into the columns, giving it a consistency similar to that of the original soil, and was supported by nylon bolting cloth, which was attached to the bottoms of the columns. Next, each reconstructed profile was saturated with demineralized water, and two additional charges (1200 ml) were added; they percolated through the profiles and collected in glass containers. Each percolation took several hours.

The resulting suspensions/solutions were filtered through 0.2-μm membrane filters to remove the suspended particles larger than 0.2 μm. The filtrates were evaporated to dryness, and the filtrate residues were ashed in a muffle furnace at 500°C. The residues contained 20 to 25 percent ash, which probably indicates the presence of colloidal or clay-size material, although the residues were clear and light brown in color.

Analytical Procedures

All element contents in the ash of vegetation and mull were determined by the semiquantitative spectrographic method of Mosier (1972). Soil contents of elements except antimony, cadmium, molybdenum, and zinc were determined by the semiquantitative spectrographic method of Grimes and Marranzino (1968). Antimony and molybdenum contents in soil were determined by colorimetric methods (Ward *et al.*, 1963). Cadmium and zinc contents in soil were determined by atomic absorption (Ward *et al.*, 1969).

Element Concentration

We compared element concentrations among the mineral soil horizons and the ash of the vegetation, mull, and leachate residue. Figure 22-11 shows the location of the highest element concentrations in the ash of vegetation and mull and in the mineral soil prior to leaching with demineralized water. The results show that most elements are concentrated in the ash of the mull and vegetation as opposed to the mineral-soil horizons; the results also indicate the enrichment of many of the elements of interest in mineral exploration—silver, copper, lead, zinc, cadmium, molybdenum—in the mull ash, which to a great extent reflects the enrichment of these elements in the ash of the vegetation.

The element abundances of a "typical" soil profile—that is, the abundances most common to these profiles (Fig. 22-12)—further demonstrate the enrichment of elements in mull relative to the soil. The locations of the highest concentration of 25 elements are shown in Figure 22-12. The element abundances in this "typical" profile illustrate that one can expect to find relatively high concentrations of silver, cadmium, cobalt, copper, molybdenum, nickel, lead, antimony, tin, and zinc in the ash of mull, whereas they may not be detected in the soil. Also, the geochemical maps (Figs. 22-3 through 22-10) demonstrated that one could also expect high concentrations of additional elements such as gold and bismuth in the mull ash. The soil horizons, on the other hand, apparently concentrate elements such as iron, beryllium, cobalt, chromium, lanthanum, and vanadium and may be better sample media than the mull for elements such as beryllium, cobalt, or chromium.

Other elements that show a tendency to be enriched in the mull are magnesium, titanium, manganese, boron, barium, strontium, and zirconium. Other elements, although not present in mull in the highest concentrations, are present in noticeable amounts.

The relationship of element concentration in the mull to that in the soil can probably best be explained by considering the nature of these soils. Because the soil profiles below the mull horizon are poorly developed and lack a well-defined illuvial, clay-rich B horizon, they lack the capacity to fix elements that are in the soil solutions. The mull horizon has the capacity to fix elements because it contains insoluble organic compounds—principally humic acid components such as porphyrins (Goodman and Cheshire, 1976) that are insoluble under natural conditions but that form complexes with various elements. Insoluble humic acid components are able to fix iron, manganese, zinc, molybdenum (Salazy, 1969, p. 34), and presumably, other elements.

Element Mobility

Elements are mobilized as well as fixed in the mull and mineral-soil horizons. To estimate the mobility of elements in subalpine forest soils, we assumed that element values in the ash of the leachate residue represented elements in the soil that had gone into solution or colloidal suspension in the demineralized water as it percolated through the soil profile. The concentrations of certain elements in the ash of the leachate residue were compared to those of the same elements in the mineral-soil horizons and ash of the mull horizons. This comparison allowed us to determine the amount of an element leached from the mineral-soil horizons and mull horizon relative to the amount of that element potentially available in these horizons. Thus we were able to estimate the relative mobilities of the

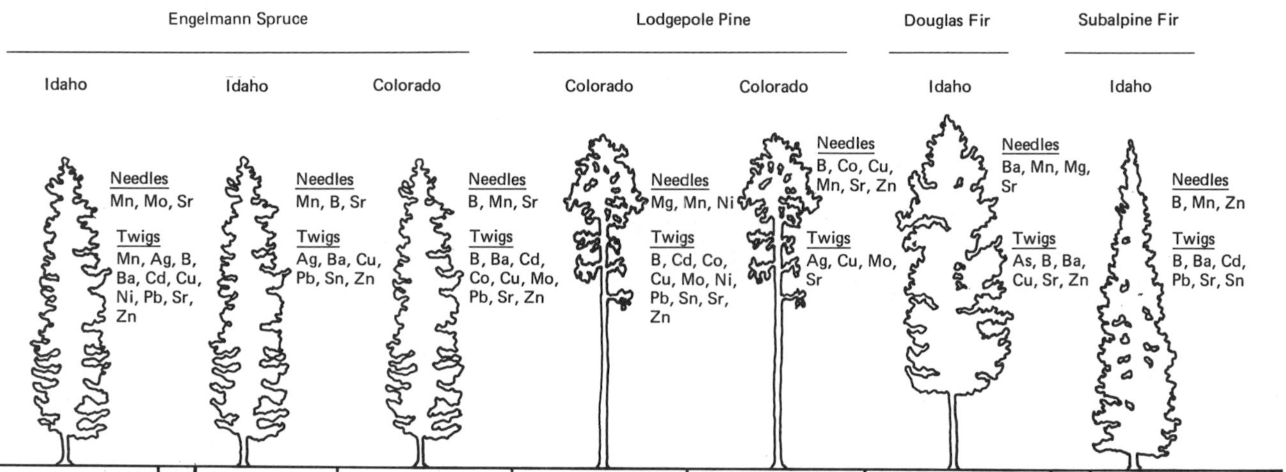

Fig. 22-11 Location of highest element concentration in the soil profiles and ash of vegetation.

	Engelmann Spruce			Lodgepole Pine		Douglas Fir	Subalpine Fir
	Idaho	Idaho	Colorado	Colorado	Colorado	Idaho	Idaho
Needles	Mn, Mo, Sr	Mn, B, Sr	B, Mn, Sr	Mg, Mn, Ni	B, Co, Cu, Mn, Sr, Zn	Ba, Mn, Mg, Sr	B, Mn, Zn
Twigs	Mn, Ag, B, Ba, Cd, Cu, Ni, Pb, Sr, Zn	Ag, Ba, Cu, Pb, Sn, Zn	B, Ba, Cd, Co, Cu, Mo, Pb, Sr, Zn	B, Cd, Co, Cu, Mo, Ni, Pb, Sn, Sr, Zn	Ag, Cu, Mo, Sr	As, Ba, Cu, Sr, Zn	B, Ba, Cd, Pb, Sr, Sn
O2 (Mull Ash)	Co, Mn, Sb, Sn, Ti, Zr	021 Mn, Ag, Cd, Sb / 022 Mn, Cd, Ni / 023 Cd, Co, Mn, Mo, Sc, Ti, Zr	Ag, Cd, Co, Mg, Ni, Sn, Ti, Zr	Ag, Ba, Be, Co, Ni, Sn, Ti, Zr	Ba, Be, Cd, Co, Mo, Ni, Pb, Sc, Sn, Ti, V, Y, Zr	Ag, Ba, Cd, Co, Fe, Mn, Mo, Ni, Pb, Sb, Sc, Sn, Ti, Zr	021 Cu, Sb / 022 Be, Cu, Fe, La, Mo, Ni, Sb, Sc, V, Zr
A	Be, Cr, Y	Be, Cr, Mo	Be, Co, Fe, La, Sc, V, Y	Be, Ni	No high values	Be	A horizon not recognized
B	Be, Cr, Fe, La, Sc, V, Zr	Be, Cr, Fe, La, V, Y	Be, Co, Cr, La, V	Be, Ni, Sc, Y	No high values	Cr, Fe, La, V, Y	Cr
C	C horizon not sampled	C horizon not sampled	C horizon not sampled	Be, Co, Cr, Fe, La, Ni, Sc, V	Be, Co, Cr, Fe, La, Sc	C horizon not sampled	Ag, As, Be, Co, Y

Profile

Soil

various elements in the soil horizons. Elements that were more highly concentrated in the ash of the leachate residue than they were in the soil horizons we judged to be highly mobile, whereas elements that showed a relatively low concentration in the residue ash were judged to have a low mobility.

In most or all of the soil profiles, the elements copper, cobalt, chromium, magnesium, barium, tin, strontium, and yttrium were more concentrated in the ash of the leachate residue than they were in the mineral-soil horizons or in the ash of the mull (Fig. 22-12), indicating that they are mobile in these soils. The elements manganese, zirconium, lanthanum, nickel, scandium, and arsenic (where they were detected) were somewhat more concentrated in the ash of the leachate residue than they were in the soil, indicating that they are at least moderately mobile in these soils. The indicated mobility of zirconium, generally considered a relatively immobile element, is supported by data from previous studies (Curtin et al., 1974, p. 254–259). Titanium is the only element that appeared to be relatively immobile in these soils.

Individual soil horizons were also leached (Curtin et al., 1979), and the results indicated that elements are mobile in all horizons, including the mull. The percentage of elements leached from the mull relative to the total element content was noticeably less than that in the mineral-soil horizons, reflecting the capacity of the mull to fix elements.

The mobilization of elements in these soils is most likely accomplished primarily by the complexing action of (1) the fulvic acid component of soil organic matter (Schnitzer, 1969, p. 80), which constitutes as much as 87 percent of the water-soluble organic matter in some soils (Schnitzer and Desjardins, 1969, p. 154); and (2) oxalic acid, which is an exudate of forest tree roots (Smith, 1976, p. 327) and is apparently produced by fungi in forest soils (Graustein and Cromack, 1977, p. 1253). Fulvic acid presumably can be generated in the humus (Davies, 1971, p. 82), can form water-soluble chelates with elements in the humus and in the A horizon and can transport them downward in the profile. Similarly, oxalic acid can mobilize elements within the soil. Polyphenols, organic acids, and other organic compounds from canopy drip also can mobilize iron, aluminum, and presumably other elements in the soil (Malcolm and McCracken, 1968, p. 838). In these studies, the elements presumably were mobilized primarily as complexes with fulvic acid, oxalic acid, and other water-soluble organic compounds.

Horizon	Highest element concentration
02 (mull ash)	Ag, B, Ba, Cd, Co, Cu, Mg, Mn, Mo, Ni, Pb, Sb, Sn, Sr, Ti, Zn, Zr
A	Be
B	Be, Cr, La, V
C (three profiles)	Be, Co, Cr, Fe, La, V, Y
Leachage residue (ash)	Ba, Co, Cr, Cu, Mg, Sn, Sr, Y (As, La, Mn, Ni, Sc, Zr)

Fig. 22-12   Location of highest element concentration in a "typical" soil profile and ash of leachate residue.

The results indicate a process in which relatively small amounts of many elements are leached from the mull and migrate downward in the soil. These elements are primarily recycled back into the vegetation and, subsequently, the mull, thus leaving the mineral-soil horizons depleted in elements relative to the mull horizon. Under these conditions, the mull would act primarily as a trap for elements accumulated by the trees from the bedrock and soil, whereas the mineral soil would act mainly as a medium through which elements travel in the biogeochemical cycle. Furthermore, this process would explain why the mineral horizons of nonresidual subalpine forest soils yield only weak and erratic anomalous patterns. The process would also explain why mull is a medium for element accumulation and is possibly the most reliable indicator of concealed mineral deposits in mountainous areas where the soil geochemical anomalies are dispersed and diluted by barren material.

REFERENCES

Banister, D. P., 1970, Geochemical investigations for gold, antimony, and silver at Stibnite, Idaho: *U.S. Bur. Mines Rep. Invest. 7417*, 7 p.

Boyle, R. W., and Dass, A. S., 1967, Geochemical prospecting—use of the A horizon in soil surveys: *Econ. Geol.*, v. 62, p. 274–276.

Braddock, W. A., 1969, Geology of the Empire quadrangle, Grand, Gilpin, and Clear Creek Counties, Colorado: *U.S. Geol. Surv. Prof. Pap. 616*, 56 p.

Curtin, G. C., and King, H. D., 1974a, The association of geochemical anomalies with a negative gravity anomaly in the Chief Mountain–Soda Creek area, Clear Creek County, Colorado: *U.S. Geol. Surv. J. Res.*, v. 2, no. 5, p. 581–592.

——, and King, H. D., 1974b, Antimony and other metal anomalies south of Stibnite, Valley County, Idaho—*with a section on* Geology, by B. F. Leonard: *U.S. Geol. Surv. Open-File Rep. 74-111*, 54 p.

——, Lakin, H. W., Hubert, A. E., Mosier, E. L., and Watts, K. C., 1971, Utilization of mull (forest humus layer) in geochemical exploration in the Empire district, Clear Creek County, Colorado: *U.S. Geol. Surv. Bull. 1278-B*, 39 p.

——, Lakin, H. W., Neuerburg, G. J., and Hubert, A. E., 1968, Utilization of humus-rich forest soil (mull) in geochemical exploration for gold: *U.S. Geol. Surv. Circ. 562*, 11 p.

——, King, H. D., and Mosier, E. L., 1974, Movement of elements into the atmosphere from coniferous trees in subalpine forests of Colorado and Idaho: *J. Geochem. Explor.*, v. 3, p. 245–263.

——, King, H. D., and Nishi, J. M., 1979, The concentration and mobility of elements in subalpine forest soils in Colorado and Idaho: *U.S. Geol. Surv. Open-File Rep. 79-710*, 40 p.

Davies, R. I., 1971, Relation of polyphenols to decomposition of organic matter and to pedogenetic processes: *Soil Sci.*, v. 111, p. 80–85.

Goldschmidt, V. M., 1937, The principles of distribution of chemical elements in minerals and rocks: *J. Chem. Soc. (London)*, p. 655–673.

Goodman, B. A., and Cheshire, M. V., 1976, The occurrence of copper–porphyrin complexes in soil humic acids: *J. Soil Sci.*, v. 27, p. 337–347.

Graustein, W. C., and Cromack, K., Jr., 1977, Calcium oxalate: occurrence in soils and effect on nutrient and geochemical cycles: *Science*, v. 198, p. 1252–1254.

Grimes, D. J., and Marranzino, A. P., 1968, Direct-current arc and alternating-current spark emission spectrographic field methods for the semiquantitative analysis of geologic materials: *U.S. Geol. Surv. Circ. 591*, 6 p.

Kokkola, M., 1977, Application of humus to exploration, *in Prospecting in Areas of Glaciated Terrain 1977:* Institution of Mining and Metallurgy, London, 140 p.

Malcolm, R. L., and McCracken, R. J., 1968, Canopy drip: a source of mobile soil organic matter for mobilization of iron and aluminum: *Soil Sci. Soc. Am. Proc.*, v. 32, no. 6, 834–838.

Mosier, E. L., 1972, A method for semiquantitative spectrographic analysis of plant ash for use in biogeochemical and environmental studies: *Appl. Spectrosc.*, v. 26, p. 636–641.

Nuutilainen, J., and Peuraniemi, V., 1977, Application of humus analysis to geochemical prospecting: some case histories, *in Prospecting in Areas of Glaciated Terrain 1977:* The Institution of Mining and Metallurgy, London, 140 p.

Rice, W. L., 1970, Investigation of a low-grade gold deposit in Orogrande district, Idaho: *U.S. Bur. Mines Rep. Invest. 7425*, 14 p.

Salazy, A., 1969, Accumulation of uranium and other micrometals in coal and organic shales and the role of humic acids in these geochemical enrichments: *Ark. Mineral. Geol.*, v. 5, no. 3, p. 23–36.

Schnitzer, M., 1969, Reactions between fulvic acid, a soil humic compound and inorganic soil constituents: *Soil Sci. Soc. Am. Proc.*, v. 33, no. 1, p. 75–81.

____ , and Desjardins, J. G., 1969, Chemical characteristics of natural soil leachate from a humic podzol: *Can. J. Soil Sci.*, v. 49, p. 151–158.

Smith, W. H., 1976, Character and significance of forest tree root exudates: *Ecology*, v. 57, p. 324–331.

Swanson, R. H., 1967, Seasonal course of transportation of Lodgepole pine and Engelmann spruce, *in* Sopper, W. E. (Ed.), *Forest Hydrology:* Pergamon Press, Elmsford, N.Y., 519 p.

Thompson, C. E., Nakagawa, H. M., and VanSickle, G. H., 1968, Rapid analysis for gold in geologic materials, in Geological Survey research 1968: *U.S. Geol. Surv. Prof. Pap. 600-B*, p. B130–B132.

Toverud, O., 1979, Humus: a new sampling medium in geochemical prospecting for tungsten in Sweden, *in Prospecting in Areas of Glacial Terrain 1979:* The Institution of Mining and Metallurgy, London, 109 p.

U.S. Department of Agriculture, 1960, *Soil Classification—A Comprehensive System, 7th Approximation:* U.S. Department of Agriculture Soil Conservation Service, 265 p.

Ward, F. N., Lakin, H. W., Canney, F. C., and others, 1963, Analytical methods used in geochemical exploration by the U.S. Geological Survey: *U.S. Geol. Surv. Bull. 1152*, 100 p.

____ , F. N., Nakagawa, H. M., Harms, T. F., and VanSickle, G. H., 1969, Atomic-absorption methods of analysis useful in geochemical exploration: *U.S. Geol. Surv. Bull. 1289*, 45 p.

W. E. Baker
Department of Mines
Rosny Park
Tasmania, Australia

23

HUMIC SUBSTANCES
AND THEIR ROLE
IN THE SOLUBILIZATION
AND TRANSPORT OF METALS

ABSTRACT

In wet temperate climates, humic substances influence the solubilization of metals at the soil-rock interface and play a continuing role in the migration of metals in the hydrosphere.

Literature and field data indicate that for experimental investigations of the role of humic substances in the solubilization of metals, a reasonable upper concentration limit is 500 μg/ml. Minerals of ore deposits, particularly those of the oxidized zone, are easily dissolved by solutions of these substances. Experiments on the leaching of minerals by solutions of humic substances demonstrated solubilization of metals which ranged from the release of 6.9 mg of lead from anglesite in 3 hours to the release of 6.6 μg of gold from the native metal in 50 days. Where the minerals are well crystallized, the rate at which metals are solubilized from sulfides by humic substances generally follows their relative stability. Once formed, the metal-humic substances association may migrate directly into the drainage system, or climatic factors may temporarily impede this progress.

In the study of the nature of the metal-humic substances association, potentiometric titration, electrophoresis, polarography, and x-ray diffraction are useful approaches. Studies utilizing these techniques indicate that the association is one of complex formation.

Due to the activity of humic substances under wet temperate climatic conditions, metals become highly mobile and relative mobilities predicted on the basis of solubilities of inorganic species are no longer relevant. Transport of gold as a complex with humic substances appears to be the most satisfactory means of explaining the migration of this noble metal. The activity of the humic substances may be a factor in the lack of development of significant gossan over some sulfides in wet temperate zones.

The considerable quantities of metals leached from the earth's surface over wide areas and generally deposited in localized near-shore marine environments could be one of the many factors involved in the ultimate production of an orebody. Nature has provided an all-pervading, effective leaching agent and it should be possible for exploration geochemists to make use of this at both regional and local scales in the search for ore deposits.

INTRODUCTION

The possibility that humic substances might be agents of solubilization of silicate minerals was commented on early in the history of soil science by Sprengel (1826), Berzelius (1839), Thenard (1870), and Senft (1871). In an extensive review of the geological activity of humic substances, Julien (1879) suggested that they were involved in a wide range of processes ranging from the solubilization of silica to the formation of diamond. Clarke (1911) rejected the early claims of activity of humic substances, as he considered them not supported by experimental evidence. Guillin (1928) found that feldspar had a higher solubility in solutions of humic substances than in water in equilibrium with atmospheric carbon dioxide. Blanck (1933) maintained that this observed solubilization was due to the high concentration of carbon dioxide produced by the decay of the humic substances. Krauskopf (1967) and Loughnan (1969) also questioned the activity of humic substances in weathering. Kononova *et al.* (1964) and Ponomareva and Ragim-Zade (1969) demonstrated that silicates were solubilized by solutions of humic substances and these results have been confirmed in other studies (e.g., Schnitzer, Chapter 24, this volume).

The literature concerning interaction between humic substances and minerals which are commonly the concern of exploration geologists is not extensive. Gruner (1922) found iron to be leached from magnetite, siderite, and pyrite by water drawn through a bed of peat. Freise (1931) found that humic substances solubilized gold, whereas Fetzer (1934, 1946) concluded that these substances were ineffective as solvents of minerals. Shcherbina (1956) and Steelink (1963) suggested that gold solubilization by humic substances should be possible, whereas Ong and Bisque (1968) found no evidence for the process. The contradictory views expressed in the literature led the writer (Baker, 1973, 1978) to commence experimental studies of the role of humic substances in the solubilization of minerals of ore deposits. These continuing studies and their possible relevance to mineral exploration are discussed in following sections.

EXPERIMENTAL STUDIES OF THE SOLUBILIZATION OF MINERALS

In the design of experimental procedures that attempt compatibility with the natural environment, the type of weathering conditions must be taken into account. In a wet, cool temperate climate, weathering of minerals takes place at an irregular interface between rock and soil, where the constituents of soil water react with solid mineral surfaces. The experimental arrangement should thus attempt to simulate this situation.

Experimental Constraints

The early studies, such as those of Freise (1931) and Fetzer (1934, 1946), made use of concentrations of humic substances that ranged from 1 percent to 5 percent (w/v) and it is not likely that such levels would occur in natural systems. The range found in several creeks in western Tasmania was 5 to 40 mg/liter (spectrophotometric determination by the method of Welte, 1956). Since the creeks carry a large volume of direct runoff, it would be reasonable to expect the concentration of humic substances in soil water to considerably exceed that in the creeks. Support for this view was found in a tunnel seepage sample in western Tasmania which contained 430 mg/liter of humic substances. Confirmation that like concentrations occur elsewhere is given by Fotiyev (1971), who has recorded values of up to 300 mg/liter for marsh waters of the USSR. Mineral preparations used in early studies produced materials which, like the humic substances concentrations, deviated widely from those occurring in nature. For example, Fetzer (1946) made use of the etching of polished mineral surfaces and dissolution of finely pulverized or precipitated samples to measure the effectiveness of humic substances as solvents.

The use of recently developed analytical techniques makes it possible to conduct experiments under conditions that approximate those existing in nature. Five samples of humic substances were extracted from podzolic soils, carrying differing vegetation, by a method based on the work of Kononova and Bel'chikova (1960) and Hori and Okuda (1961). For these soils (Table 23-1), humic acids contribute about 85 to 98 percent of

TABLE 23-1 Location Data, Soil Type, Vegetation Cover, Humic Acid Content, and Composition for Some Tasmanian Soils

Sample	Location	Soil Type	Predominant Vegetation Cover	Humic Acid Content (% of total soluble organic fraction)	Humic Acid Composition[a] (%)			
					C	H	N	O
TS-1	Oliver Hill, 41°31'N 146°09'E	Podzol	White topped stringy-bark (*Eucalyptus delegatensis*) Bracken fern (*Pteridium aquilinum*)	97.7	59.9	2.9	1.9	34.1
TS-2	Machinery Creek, 41°30'N 146°10'E	Swamp soil	Moss (*Sphagnum* spp.)	92.1	56.8	3.6	2.4	37.0
TS-3	Mt. Roland 41°26'N 146°17'E	Podzol	Tea tree (*Melaleuca ericifolia*)	97.5	54.0	4.0	3.3	38.4
TS-4	Savage River 41°30'N 145°14'E	Groundwater podzol	Button grass (*Gymnoschoenus sphaerocephalus*)	96.0	55.5	4.2	3.6	36.6
TS-5	Paradise 41°26'N 146°20'E	Gray-brown podzol	Pine (*Pinus radiata*)	83.3	52.8	4.7	3.2	38.9

[a]Composition given on an ash-free basis from analyses by CSIRO Microanalytical Service, Melbourne, Australia.

the soluble humic substances present and display a mobility approaching that of the fulvic acids of more mature soil types. A concentration of these substances of 500 mg/liter was used in most experimental studies described in this chapter. Given the wide variability in mineral cleavage and fracture, it is not possible to produce equivalent surfaces of different minerals for solubilization studies. As a compromise, coarsely crystalline mineral samples were crushed and sieved to a grainsize range of 300 to 600 μm. This procedure avoids the problems associated with the high activity of extremely large surface areas and the mechanics of dealing with fine powders.

Experimental Arrangement for Solubilization by Humic Substances

The mineral samples were exposed to the action of the humic substances in the device shown in Figure 23-1. An inner glass tube covered at one end with nylon sieve cloth supports a bed of mineral grains within an outer tube containing 50 ml of 500-mg/liter solutions of humic substances. Continuous circulation of the solution was achieved by use of a peristaltic pump. A group of sulfide, oxide, secondary, and carbonate minerals found associated with ore deposits were subjected to the action of the solutions of humic substances for a period of 24 hours. Equal bed volumes were approximated by taking a weight in grams equal to the density of the various minerals. The averaged results of duplicate runs are given in Table 23-2. Variation between duplicates ranged from about 10 to 40 percent and this is considered reasonable given the dynamic nature of the ex-

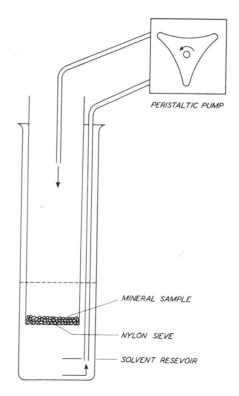

PERISTALTIC PUMP

MINERAL SAMPLE

NYLON SIEVE

SOLVENT RESEVOIR

Fig. 23-1 Mineral solubilization equipment.

TABLE 23-2 Solubilization of Minerals by H_2O/atmospheric CO_2, and 500-mg/liter Solutions of Humic Substances

Mineral	Metal Determined	H_2O/atmos. CO_2	Metal Solubilized (mg) after 24 Hours Extraction by:				
			THA-1[a]	THA-2	THA-3	THA-4	THA-5
Galena	Pb	0.06[b]	2.62 (13)	1.51 (7)	0.39 (2)	1.28 (6)	0.96 (5)
Bornite	Cu	0.10	1.21 (19)	0.98 (15)	1.07 (17)	1.27 (20)	1.20 (19)
Sphalerite	Zn	0.08	0.13 (2)	0.11 (2)	0.12 (2)	0.10 (2)	0.10 (2)
Pyrite	Fe	0.04	0.24 (4)	0.59 (11)	0.12 (2)	0.10 (2)	0.36 (6)
Loellingite	Fe	0.06	5.07 (91)	3.03 (54)	1.28 (23)	1.20 (21)	2.10 (38)
Bismuthinite	Bi	0.04	2.06	2.53	1.13	2.03	1.95
Stibnite	Sb	0.03	0.14	0.48	0.07	0.20	0.04
Hematite	Fe	0.06	0.58	0.19	0.08	0.09	0.34
Magnetite	Fe	0.10	0.54	1.05	0.14	0.34	0.37
Pyrolusite	Mn	0.17	1.47	0.78	0.57	1.25	0.58
Azurite	Cu[c]	0.21	22.05 (347)	12.10 (190)	10.25 (161)	14.60 (230)	8.10 (127)
Anglesite	Pb[c]	0.15	55.60 (268)	25.30 (122)	55.60 (268)	29.50 (142)	16.65 (80)
Calcite	Ca[c]	0.65	17.40	11.40	10.75	12.50	10.20
Dolomite	Ca	0.25	7.50	5.00	4.00	5.60	3.40
	Mg	0.14	5.65	3.75	2.46	4.95	3.00
Magnesite	Mg	0.22	1.07	0.93	0.43	0.89	0.56

[a]Humic substances extracted from Tasmanian podzolic soils (described in Table 23-1).
[b]Values are averages of duplicate runs. Bracketed figures are metal solubilized in micromoles.
[c]Metal solubilized calculated from 3-hour run.

periments, the uncertainty of surface areas, and difficulty experienced in completely re-covering reaction products from some samples prior to analysis by flame atomic absorption.

To study the possible effect of humic substances on gold, 2-g quantities of 70- to 150-μm grain-size commercial gold powder were placed in capped tubes with 20 ml of distilled water and 20 ml of 500-mg/liter solutions of THA-1 and THA-4. These tubes were turned end for end for a period of 50 days by a slow-revolution (100 rph) motor after which their contents were filtered through a 0.4-μm membrane and the filtrates analyzed for gold by carbon furnace atomic absorption spectrophotometry. The results of triplicate runs are given in Table 23-3.

TABLE 23-3 Solubilization of Gold (ng/ml) by H_2O/atmospheric CO_2 and 500-mg/liter Solutions of Humic Substances

Run No.	H_2O/atmos. CO_2	THA-1	THA-4
1	8	200	300
2	11	170	360
3	10	210	320
Average	10	190	330

Discussion of Results of Solubilization by Humic Substances

The data in Table 23-2 leave no doubt that ore-related minerals are solubilized much more effectively by solutions of humic substances than by water in equilibrium with atmospheric carbon dioxide. This is in marked contrast to the view expressed by Fetzer (1946) that these substances are ineffectual solvents of minerals. For the sulfides it can be seen that galena, bornite, and bismuthinite are extensively solubilized, whereas sphalerite, pyrite, and stibnite are less so. Some correlation between relative bond strength and de-gree of solubilization is apparent. According to Bachinski (1969), the order of bond strengths is pyrite and sphalerite > galena > copper sulfides and this is generally reflected in the solubility data. Nockolds (1966) showed the bond strength in bismuthinite and stibnite to be similar and hence the widely differing solubilization data must be due to some other factor. This may be the ease of oxidation of antimony to the pentavalent state, allowing formation of an oxyanion for which the humic substances have little affinity. Nickel (1968) has shown that the arsenides of iron have lower stability than the sulfides, and this is reflected in the high solubilization of loellingite relative to pyrite. Secondary minerals and carbonates with calcium are extensively solubilized while magnesite and oxides are affected to a lesser extent.

The solubilization data indicate that THA-1 is the most effective extract and this is compatible with the view expressed by Birrell *et al.* (1971) that humic substances devel-oped under *Pteridium aquilinum* accelerate soil profile development. Detailed comparative analysis of the effectiveness of the several extracts is not possible and, beyond the struc-tural controls already noted, no systematic variation in the reaction with the mineral species studied is apparent. This is not surprising since the humic substances are virtually a sink of functional groups in structurally indeterminate arrays. The overall stability of the metal–humic substances association is a composite of the stability of the numerous contributing metal complexes, some of which may well be of selective nature.

The data given in Table 23-3 indicate that considerable solubilization of gold has been achieved with humic substances solution strengths of 500 mg/liter. These results are in disagreement with those of Fetzer (1934, 1946) and Ong and Swanson (1969), who maintained that humic substances were not solvents of gold. Since the particle size of humic substances reaches colloidal dimensions, it is not possible to readily assess the degree of physical and chemical interaction occurring in solubilizaton experiments. Studies described later in this chapter are concerned with the nature of the association.

Solubilization in Presence of Excess Carbon Dioxide

It has been claimed by Blanck (1933) that the activity of humic substances in weathering is due to carbon dioxide generated by their decay. This postulate was examined by subjecting several minerals to the action of water under a carbon dioxide-enriched atmosphere. On the basis that humic substances are approximately 60 percent carbon, it follows that 50 ml of a 500-mg/liter solution could theoretically generate 28 ml of carbon dioxide. The experimental arrangement is shown in Figure 23-2. With this device, galena, pyrolu-

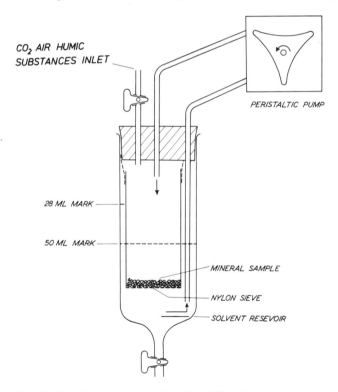

Fig. 23-2   Equipment for mineral solubilization under carbon dioxide–enriched atmosphere.

site, calcite, dolomite, and magnesite were exposed to the action of water in equilibrium with atmospheric carbon dioxide, water under a carbon dioxide–enriched atmosphere, and to 50 ml of a 500-mg/liter solution of THA-1. For the carbon dioxide–enriched studies, the device was completely filled with distilled water with the mineral sample in place. Carbon dioxide was introduced to the 28-ml mark, then air to the 50-ml mark. The THA-1 was introduced as 0.5 ml of concentrate to yield the 500 mg/liter required.

The averaged results of duplicate runs are given in Table 23-4. It can be seen that the THA-1 solution is far more active in solubilizing minerals than is the carbon dioxide that would result from its decay. This is contrary to the views of Blanck (1933) and it would appear that the effectiveness of humic substances as agents of weathering is not due to their being a source of carbon dioxide enrichment.

TABLE 23-4   Solubilization of Minerals by H_2O/atmospheric CO_2, H_2O/CO_2-Enriched Air, and a 500-mg/liter THA-1 Solution

Mineral	Metal Determined	Metal Solubilized (mg) after 24 Hours Extraction by:		
		H_2O/atmos. CO_2	H_2O/CO_2 Enriched Air	THA-1
Galena	Pb	0.06[a]	0.24	2.62
Pyrolusite	Mn	0.17	0.21	1.47
Calcite	Ca	0.65	2.75	17.40
Dolomite	Ca	0.25	0.45	7.50
	Mg	0.14	0.33	5.67
Magnesite	Mg	0.22	0.28	1.07

[a] Average of duplicate runs.

Solubilization by Simple Organic Substances

In a number of reports (Murray and Love, 1929; Fetzer, 1946; Krauskopf, 1967) that question the role of humic substances in weathering, the view has been advanced that it is the simple organic compounds in soils that are likely to be the active substances. To investigate the validity of this view, the group of minerals subjected to the action of humic substances were exposed to the action of 500-mg/liter solutions of salicylic acid, citric acid, glycine, analine, catechin, and fructose. The experiments were carried out in duplicate using the arrangement shown in Figure 23-1 and the averaged results are given in Table 23-5. The values indicate that with the exception of the attack on copper minerals by citric acid and glycine, the simple organic compounds are marginally less effective than humic substances. In view of the fact that these compounds seldom provide more than 5 percent of the total organic reserves of soil, their contribution to weathering must be considered limited. Furthermore, Barber (1968) has drawn attention to the fact that the simple organic substances do not long survive in the soil since they are rapidly converted to carbon dioxide by the microorganisms present.

TABLE 23-5 Solubilization of Minerals by H_2O/atmospheric CO_2 and 500-mg/liter Solutions of Simple Organic Compounds

Mineral	Metal Determined	Metal Solubilized (mg) after 24 Hours Extraction by:						
		H_2O/atmos. CO_2	Salicylic Acid	Citric Acid	Glycine	Alanine	Catechin	Fructose
Galena	Pb	0.06[a]	0.73	0.92	0.09	0.11	0.26	0.08
Bornite	Cu	0.10	1.48	3.02	3.90	0.36	0.49	0.12
Sphalerite	Zn	0.08	0.08	0.07	0.07	0.07	0.08	0.08
Pyrite	Fe	0.04	0.06	0.08	0.05	0.05	0.05	0.04
Loellingite	Fe	0.06	3.45	3.14	0.07	0.08	0.45	0.09
Bismuthinite	Bi	0.04	0.13	1.30	1.10	0.69	1.81	0.42
Stibnite	Sb	0.03	0.01	0.02	0.04	0.02	0.05	0.02
Hematite	Fe	0.06	0.08	0.08	0.06	0.08	0.07	0.06
Magnetite	Fe	0.10	0.46	1.81	0.10	0.11	0.18	0.12
Pyrolusite	Mn	0.17	3.02	2.47	0.38	0.18	0.41	0.27
Azurite	Cu	0.21	28.80	75.90	39.75	6.75	9.50	7.15
Anglesite	Pb	0.15	13.60	23.85	9.00	9.50	11.55	2.75
Calcite	Ca	0.65	24.25	48.90	4.15	2.80	5.45	0.90
Dolomite	Ca	0.25	10.35	20.80	0.95	0.90	2.35	0.90
	Mg	0.14	9.62	17.00	0.80	0.70	2.10	0.55
Magnesite	Mg	0.22	1.12	2.30	0.35	0.28	0.40	0.22

[a] Average of duplicate runs.

386

The uncertain character of humic substances has resulted in a number of interpretations being placed on their association with metals. Originally, Sprengel (1826) regarded the association as simple salt formation, whereas Tarkhov (1881) believed it to be the result of colloidal adsorption phenomena. The adsorption of metals by humic substances of peat deposits provides a mechanism for the enrichment of trace elements in coal. In one of the classical studies of geochemistry by Goldschmidt (1930), this mechanism was proposed to explain the presence of up to 11,000 μg/g germanium in the ash of coals. Szalay (1964) reported that a number of polyvalent cations, including UO_2^{2+}, were enriched in peat by factors up to 10,000 times the content in associated waters. Halbach et al. (1980) have made a detailed study of the uptake of uranium by a peat bog overlying granitic bedrock. They found that the action of humic substances reduced the uranium content from 14 μg/g in the fresh rock to 2 to 3 μg/g in the weathered surface. Szalay and Szilagyi (1967) found enrichment of vanadium in peat of the order of 50,000 times the content in associated waters. Radtke and Scheiner (1970) proposed that the concentration of gold in the Carlin, Nevada, deposits was in part due to adsorption of gold from hydrothermal solutions by organic matter similar to humic substances. In an examination of the sorption of a variety of metals on humic substances, Kerndorff and Schnitzer (1980) developed empirical equations to describe the sorption but were unable to find any correlation with such properties as atomic weights, valency, or ionic radii.

The colloidal activity of humic substances in metal transport has been discussed by Ong and Bisque (1968) and Ong et al. (1970). These studies suggested wide application of a colloidal model of transport. The colloidal properties displayed by humic substances in rivers and estuaries have been studied by Sholkovitz and Copland (1981), Tipping and Cooke (1982) and Tipping and Heaton (1983). These studies indicate that the presence of humic substances modifies the surfaces of particulate absorbents and leads to variable precipitation behavior not predictable on the basis of normal chemical models.

Bremner et al. (1946) suggested that the metal–humic substances association involved metal complexation. From this date, the literature of soil science and more recently geochemistry contains numerous studies of complexation by humic substances (e.g., Broadbent and Ott, 1957; Lewis and Broadbent, 1961; Schnitzer and Skinner, 1963, 1967; Saar and Weber, 1980; Sohn and Hughes, 1981). The possible reducing properties of humic substances have been invoked in their reaction with metals by Garrels and Pommer (1959), Szalay and Szilagyi (1967), Ong and Swanson (1969), Ong et al. (1970), Wilson and Weber (1979), and Templeton and Chasteen (1980).

The metal–humic substances association may be investigated by a number of procedures, and several of these that yield information relevant to mineral exploration are described below.

Potentiometric Titration

When titrations of solutions of metal ions with a base are monitored by pH measurements, the formation of metal hydroxides is indicated by inflections in the titration curves. In the presence of an excess of a complexing agent no hydroxides are formed and, as a result, there are no inflections.

This technique was used to examine the reaction of the humic substances extracts

THA-1 and THA-4 with copper, nickel, lead, and zinc ions. A study was made of the titration of 500 μ-equivalents of these ions in 50 ml of 0.1 M potassium chloride by a solution of 0.1 meq/ml sodium hydroxide. The titrations were repeated in the presence of 0.2 g THA-1 and THA-4 to examine the effect of humic substances. Adsorption effects were monitored by repeating the titrations in the presence of finely precipitated manganese dioxide. The titration curves are shown in Figure 23-3.

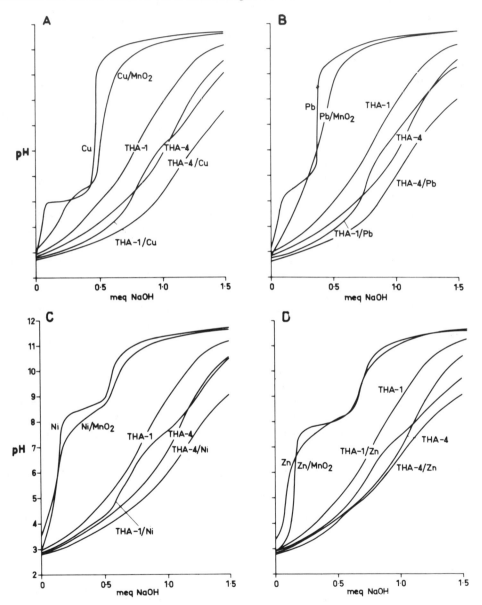

Fig. 23-3   Potientiometric titration curves for 500μ-equivalents of (A) copper, (B) lead, (C) nickel, and (D) zinc, alone and in the presence of 0.2 g of THA-1, THA-4, and MnO$_2$.

In the absence of additives, the metal solutions exhibit inflections for copper, nickel, lead, and zinc at pH values of 5.0, 5.8, 8.5, and 7.8 corresponding to the respective hydroxide precipitation. No inflections are apparent in the presence of THA-1 or THA-4, and the titration curves are displaced to the right of those for the humic substances extracts alone. This, as discussed by Martell and Calvin (1952), is as would be expected in the event of complex formation.

In comparison with the effects of THA-1 and THA-4, the effect of manganese dioxide is weak and only slightly modifies the titration paths. It would appear that adsorption processes may be of lesser consequence than complexation where humic substances are present in environments that are likely to produce hydroxides. The evidence suggests, in fact, that the formation of hydroxides is unlikely in the presence of humic substances and does not support transport mechanisms involving colloidal associations of hydroxides and these substances, as suggested by Ong et al. (1970).

Polarographic Studies

Lingane (1941) demonstrated that polarographic analysis could be applied to the study of metal complexation. Changes in the state of a metal from ionic to complexed, or from one complex to another, are generally evident in the characteristics of the relevant polarograms. Such changes are accompanied by variation in the position of the half-wave potential ($E_{1/2}$) for a particular reduction and in the value of the diffusion current (I_d) associated with this event. Since reduction processes are involved, polarographic studies may also be used to examine the reducing characteristics of humic substances.

Polarograms were recorded for Pb(II), Cu(II), Au(III), Fe(III), and U(VI) by means of direct-current pulsed polarography. The high electrolyte concentrations necessary to carry the current in polarographic analysis caused problems in maintaining solubility of the humic substances. This problem can be avoided by use of electrolytes consisting of alkalis and alkali metal salts of organic acids, although comparisons are then between various complexes rather than between complexes and ions.

The polarograms for 1-μmol/ml concentrations of Pb(II) and Cu(II) in a 1 M alkaline citrate electrolyte are shown in Figure 23-4. For Pb(II), the $E_{1/2}$ values in the absence of humic substances is −0.66 V, whereas in the presence of 50 mg of both THA-1 and THA-4 it is −0.65 V. The associated I_d values are 3.7 and 3.0 μA, respectively. These minor variations in the characteristics of the polarographic waves for Pb(II) suggest that the stabilities of the citrate and humic substances complexes with the metal are of the same order of magnitude. For Cu(II) the $E_{1/2}$ and I_d values in the absence of humic substances are −0.35 V and 2.7 μA, respectively. In the presence of THA-1 and THA-4 no polarographic wave appears and this indicates that the complexes formed with the humic substances are capable of preventing the polarographic reduction of Cu(II).

The polarographic reduction of Au(III) was carried out in a supporting electrolyte of 1 M sodium hydroxide. The aurihydroxide complex present under these conditions was stable for several hours and polarograms were recorded within a half hour of preparation of the solutions and again after a lapse of 24 hours. In Figure 23-5A the polarographic reduction of Au(III) is seen to occur with an $E_{1/2}$ value of −0.53 V and I_d value of 8.2 μA. The residual current is near zero. In the presence of THA-1 (Fig. 23-5B) the $E_{1/2}$ and I_d values are −0.75 V and 3.6 μA, respectively, which is clear evidence of strong complexa-

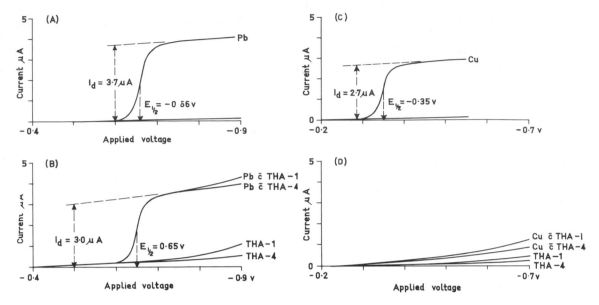

Fig. 23-4   Polarograms of 1μ mol/ml copper and lead in absence and presence of 50-mg humic substances: (A) lead; (B) lead with THA-1 and THA-4; (C) copper; and (D) copper with THA-1 and THA-4. Supporting electrolyte 1*M* sodium citrate/0.1*M* sodium hydroxide.

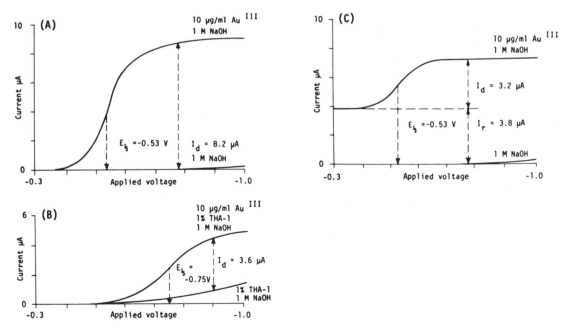

Fig. 23-5   Polarograms of 1μ mol/ml Au(III) in absence and presence of 50-mg humic substances: (A) Au(III): (B) Au(III) with THA-1; (C) Au(III) 24 hours after preparation. Supporting electrolyte 1*M* sodium hydroxide.

tion of Au(III). The results in the presence of THA-4 were almost identical, with $E_{1/2}$ and I_d values of −0.71 V and 3.6 μA.

The polarograms for Fe(III) in 0.5 M ammonium tartrate adjusted to pH 7 are shown in Figure 23-6. In the absence of humic substances (Fig. 23-6A) the reduction of Fe(III) and Fe(II) occur at $E_{1/2}$ values of −0.30 and −0.88 V, respectively, with associated I_d values of 2.8 and 1.1 μA. In the presence of THA-1 (Fig. 23-6B) there is no sign of the wave for the reduction of Fe(III) to Fe(II), while the reduction of the latter to Fe(0) occurs at an $E_{1/2}$ value of −0.94 V with an I_d value of 1.8 μA. With THA-4 (Fig. 23-6C) the reduction of Fe(II) to Fe(0) occurs at an $E_{1/2}$ value of −10.5 V with an I_d value o. 1.9 μA. The fact that the humic substances are capable of reducing Fe(III) indicates that their standard reduction potential is less than that for the reaction Fe(III) + e = Fe(II), +0.77 V. The strong shifts in position of the $E_{1/2}$ values indicate that iron complexes with humic substances are more stable than with tartrate.

For the study of U(VI) the polarograms were recorded in a supporting electrolyte of 1 M sodium citrate/0.1 M sodium hydroxide and are shown in Figure 23-7. In the absence of humic substances (Fig. 23-7A) the $E_{1/2}$ value occurs at −0.93 V with an I_d value of 2.0 μA. In the presence of THA-1 and THA-4 the respective $E_{1/2}$ values are −1.11 and −1.25 V with I_d values of 0.7 and 0.5 μA. As with iron, the large shift in $E_{1/2}$ positions is indicative of the formation of strong complexes. Since U(VI) persists in the presence of humic substances, their standard reduction potential is greater than for the reaction $UO_2^{2+} + 4H^+ = U^{4+} + 2H_2O$, +0.62 V. Thus on the basis of the polarographic studies the standard reduction potential for THA-1 and THA-4 appears to be between +0.62 and +0.77 V, which is in agreement with a value of +0.7 V calculated by Szilagyi (1971).

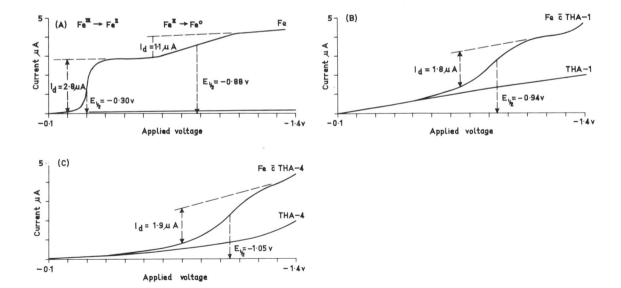

Fig. 23-6   Polarograms of 1μ mol/ml Fe(III) in absence and presence of 50-mg humic substances: (A) Fe(III); (B) Fe(III) with THA-1; (C) Fe(III) with THA-4. Supporting electrolyte 0.5M ammonium tartrate (pH 7.0).

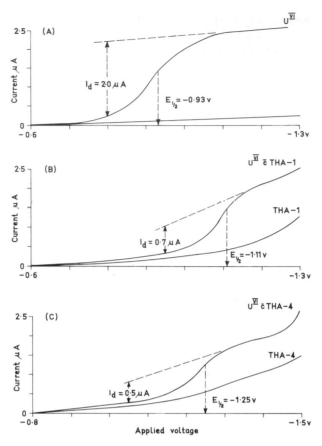

Fig. 23-7   Polarograms of 1μ mol/ml U(VI) in absence and presence of 50-mg humic substances: (A) U(VI); (B) U(VI) with THA-1; (C) U(VI) with THA-4. Supporting electrolyte $0.1M$ sodium citrate/$0.1M$ sodium hydroxide.

Electrophoresis Studies

Electrophoresis is an extremely useful technique for the examination of metal–humic substances associations. Drozdova and Emel'yanova (1960), for example, used this technique to study the interaction of copper with humic acids. In the investigations reported in this section counter-current electrophoresis was developed as a semiquantitative technique to study the interaction between humic substances, metal ions, sulfides, and gold.

The electrophoresis was carried out on 100 × 20 mm cellulose acetate membranes in a 0.05 M buffer of ammonium acetate (pH 6.0). Preliminary studies indicated that under a potential of 200 V the humic substances extract THA-1 migrated about 40 mm from the point of application toward the cathode in 15 minutes. In the same period cations moved about 30 mm toward the anode. The reactants under study were placed on the membrane by means of a slotted applicator, which was filled from a 5-μl micropipette. By this procedure, 250 μg of various metals were applied to the membranes. The general experimental arrangement of the membranes and range of substances applied is shown in Figure 23-8. The first series of runs was made with the membranes impregnated with ionic copper, cobalt, lead, nickel, and silver. A second series of runs was carried out to examine the degree of interaction of humic substances with insoluble compounds in the short time span of these types of experiments. A wide range of metals was applied to the acetate strips and immobilized as sulfides by exposure to hydrogen sulfide. Metallic gold was applied by first placing stannous chloride on the strip and then adding 500 μg of gold as the aurichloride complex. Excess reagents were washed from the strips before electrophoresis.

After electrophoresis had been continued for 15 minutes, the humic substances fronts were cut from the acetate strips and the metal leached with 0.5 ml of ammoniacal ethylenediaminetetraacetic acid. The metal content of the leachate was determined by carbon furnace atomic absorption. The results, which are given in Table 23-6, indicate that in general substantial quantities of metals have migrated toward the cathode and could only achieve this as complexes of the humic substances. Under the conditions of electrophoresis copper appears to be the most reactive and strongly bound metal, which is in agreement with its behavior during polarography. The sulfides of nickel and cobalt show anomalies relative to their ionic behavior and the sulfides of bismuth and antimony both have low solubilization. The reasons for these results are not clear and more detailed electrophoresis studies of humic substances–metal interactions appear warranted.

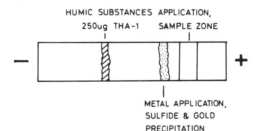

HUMIC SUBSTANCES APPLICATION, 250ug THA-1   SAMPLE ZONE

METAL APPLICATION, SULFIDE & GOLD PRECIPITATION

Fig. 23-8   Application and sampling points for electrophoresis experiments.

TABLE 23-6   Metals Mobilized during Electrophoresis

Metal	State	Solubilization [μg Metal/g THA-1 (ppm)]
Antimony	Sb_2S_3	60
Bismuth	Bi_2S_3	80
Cobalt	Co^{2+}	3,200
	CoS	40
Copper	Cu^{2+}	16,000
	CuS	2,000
Gold	Au	220
Lead	Pb^{2+}	3,200
	PbS	2,400
Nickel	Ni^{2+}	5,600
	NiS	3,600
Silver	Ag^+	2,560
	AgS	640

X-Ray Diffraction

X-ray diffraction procedures may be used to examine the products of reaction between humic substances, minerals, and compounds with crystalline structure. The fate of metallic lead, cerussite, anglesite, and galena was studied by comparing the x-ray diffractograms of 5 mg of the minerals in physical mixture with 50 mg of THA-1 and the reaction products from the mixing of the same quantities under aqueous conditions for a period of 5 days. The diffractograms are shown in Figure 23-9. For lead (Fig. 23-9A) the diffraction peak from the (111) planes at a 2θ value of 31.3° is strong in the physical mixture but absent from the reaction products. A similar result is to be seen for the (111) and (021) planes of cerussite at 2θ values of 24.8 and 25.5°, respectively (Fig. 23-9B). It is thus apparent that the result of reaction of lead and cerussite with humic substances is the formation of an amorphous lead humate. For anglesite (Fig. 23-9C), the comparative diffraction from the (210) and (102) planes at 2θ values of 26.7 and 27.7°, respectively, indicate that this mineral has been substantially attacked during the 5-day exposure to the solution of THA-1. The diffractograms for galena (Fig. 23-9D) show, from the diffraction peak for the (111) plane, that this mineral has been the least affected, although a considerable proportion has been converted to the humate. These results are in keeping with the relative values for lead solubilized by 500-mg/liter solutions of THA-1 in 24 hours, of 173.2, 127.3, 55.6, and 2.6 mg from lead, cerussite, anglesite, and galena.

A study similar to that with the lead minerals was carried out for a number of metal hydroxides. For $Cu(OH)_2$ (Fig. 23-10A) the (020) diffraction peak at a 2θ value of 16.7° is apparent in the physical mixture but is far weaker in the aqueous reaction product. A similar situation exists for $Pb(OH)_2$ (Fig. 23-10B), $Ni(OH)_2$, (Fig. 23-10C), and $Zn(OH)_2$ (Fig. 23-10D). Thus the ultimate result of interaction between humic sub-

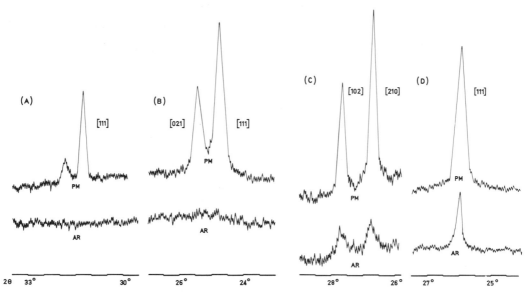

Fig. 23-9 X-ray diffraction of lead minerals as physical mixture (PM) and after aqueous reaction (AR) with jumic substances. (A) lead, (B) cerrusite, (C) anglesite and (D) galena.

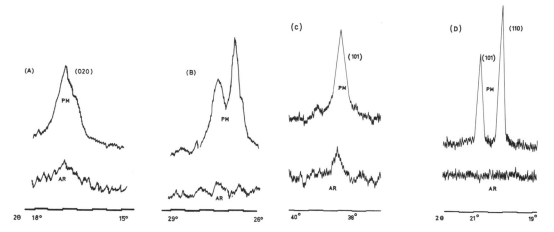

Fig. 23-10　X-ray diffraction of hydroxides as physical mixture (PM) and after aqueous reaction (AR) with humic substances. (A) copper hydroxide, (B) lead hydroxide, (C) nickel hydroxide, and (d) zinc hydroxide.

stances and metal hydroxides appears likely to be the formation of amorphous humates. The x-ray diffraction results are contrary to those that should prevail if transport models involving colloidal mixing of metal hydroxides and humic substances are correct.

The x-ray diffraction technique may also be used to examine the association of gold with humic substances. The diffractogram resulting from the product of drying 1 mg of colloidally dispersed gold in the presence of 100 mg of humic substances (THA-1) is shown in Figure 23-11A. The (111) planes produce a sharp diffraction peak at a 2θ value of 38.3°. An association with the same relative quantities of gold and humic substances was produced by adjusting a solution containing 2 mg of gold as the chloroauric acid to pH 5 and mixing this with 20 ml of a one percent THA-1 solution. After shaking for 50 hours the humic substances were coagulated with aluminum sulfate and removed by centrifuging. The residual solution contained less than 0.1 $\mu g/ml$ gold, whereas the dried humic substances carried 0.93 percent of the metal. The x-ray diffractogram of the dried product is shown in Figure 23-11B, in which it is apparent, from the absence of the (111) peak, that the association is, as with other minerals, an amorphous humate.

Reaction Kinetics

In view of the rapid reaction of many minerals with humic substances, it would be of interest to know if these reactions are of first order. Throughout the progress of a first-order reaction, the rate of change in concentration at any given time is directly proportional to the concentration at that time. If a is the initial concentration, $a - x$ the concentration after a lapse of time t, and k is the velocity constant, a first-order reaction obeys the equation

$$k = \frac{1}{t}\ln\frac{a}{a-x}$$

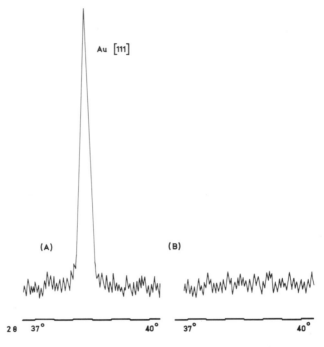

Fig. 23-11   X-ray diffraction of gold with humic substances. (A) physical mixture of 1 mg colloedal gold and 100 mg THA-1. (B) 2 mg gold as chloroauric acid reacted 50 hours with 20 ml of 1% (w/v) aqueous solution of THA-1.

If the time $t_{0.5}$ is taken as the time required for the initial concentration of a reactant to be reduced by half, in which case $x = a/2$, substitution in the equation above results in

$$t_{0.5} = \frac{\ln 2}{k}$$

Hence the important feature of a first-order reaction is that the time required to complete a definite proportion is independent of the initial concentration.

Thus a satisfactory test for a first-order reaction would be to monitor its progress after starting with two differing initial reactant concentrations. This was applied to humic substances (as 500-mg/liter and 1-g/liter solutions of THA-1) and their reaction with lead, cerussite, and anglesite. The depletion of the active sites in the humic substances was measured by determining the amount of lead entering solution as a result of reaction. The experimental arrangement was similar to that in Figure 23-1, with the outer vessel increased in capacity to allow use of 100-ml volumes of humic substances. This was placed over a magnetic stirrer to enable continuous agitation of the solvent phase. Monitoring of the reactions was carried out by flame atomic absorption analysis of 1-ml aliquots over a period of 12 hours. The results of reaction of lead, cerussite, and anglesite with THA-1

HUMIC SUBSTANCES IN THE SOLUBILIZATION AND TRANSPORT OF METALS are given in Figure 23-12. It is evident that in all cases there is rapid depletion of the active sites of the humic substances with a resultant decrease in the rate of reaction. For all practical purposes the reactions have almost ceased after 12 hours. The final lead concentrations and $t_{0.5}$ values obtained from half the final values are shown on the diagrams. From the close agreement in relevant pairs of values it is apparent that the reaction of humic substances with minerals is of first order. This is an important result, in that it validates the extrapolation of laboratory experiments, which are often of necessity car-

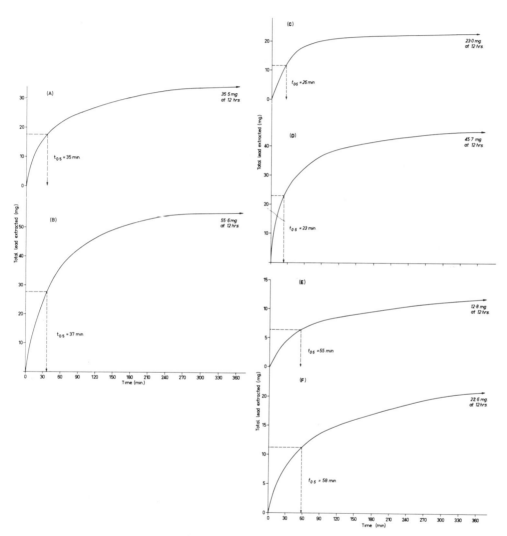

Fig. 23-12   Reaction rate curves for the dissolution of lead minerals by humic substances. Lead with THA-1. (A) 500 mg/l and (B) 1 g/l. Cerrusite with THA-1. (C) 500 mg/l and (D) 1 g/l. Anglesite with THA-1. (E) 500 mg/l and (F) 1 g/l.

ried out at high concentrations of humic substances, to the natural environment, where these materials are of lower concentration.

The experimental studies described all point to the considerable importance of complex formation in the reaction between minerals and humic substances. These reactions occur early in the erosional cycle, with the humic substances in soil water forming complexes at the soil–rock interface. The complexes formed are likely to have sufficient stability to survive the long interplay of physical and chemical factors operating between the soil–rock interface and the sea.

HUMIC SUBSTANCES–METAL ASSOCIATION AS A MOBILE PHASE IN THE WEATHERING CYCLE

The concept that metal humates, other than those formed with the alkali metals, are of low solubility originated in early studies, such as those of Odén (1919), which remain the basis of classification of these substances. This low solubility is with respect to distilled water and the actual values found are very much dependent on the means of preparation of the humates and time of aging of the products (Kononova, 1966).

In the natural environment of humid climates, rainfall supplies water to a soil layer in which bacterial action is continually producing humic substances. These substances, together with numerous other biochemicals, are potential solvents at the soil–rock interface and throughout the soil profile. The mobility of the metal humates is largely dependent on their history subsequent to formation. If long, dry periods occur, the metal humates together with metals adsorbed on clay minerals and iron oxides may be slow to migrate. Where rainfall is more abundant, however, the soil profile is frequently subjected to the passage of soil waters rich in solvents generated in the organic layers.

Experiments were carried out to examine the behavior of dry metal humates toward fresh solutions of humic substances. An excess of aluminum, calcium, cobalt, copper, ferric iron, magnesium, manganese, nickel, lead, and zinc were mixed with 25 mg of aqueous humic substances (5 ml of 50-mg/liter THA-1) and the metal humates were separated by centrifuging. The coagulated humates were mixed with macerated filter paper and allowed to dry for 3 days at 30°C. The dried pads of humate were confined between 2-cm-deep pads of macerated filter paper in glass columns sealed at the base with sintered glass. Excess inorganic salts were washed from the columns with up to 100 ml of distilled water followed by ten 10-ml aliquots of distilled water and ten 10-ml aliquots of 500-mg/liter THA-1. The leachate aliquots were analyzed by use of flame atomic absorption. The procedure was repeated for a second preparation of humic substances, which made use of THA-4. The results are shown in Figure 23-13. From these it can be seen that metals as dry humates display low mobility in pure water, whereas they become highly mobile in the presence of water containing dissolved humic substances. The humate of aluminum, which remains immobile throughout the leaching, is the one exception.

The humates of calcium, cobalt, magnesium, manganese, and nickel display some mobility in distilled water. The metal in solution shows correlation with the amount of

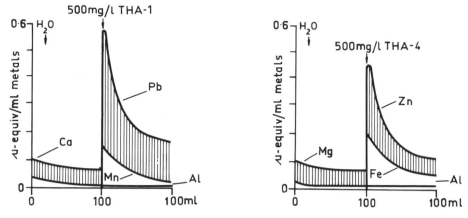

Fig. 23-13   Leaching of metals precipitated as humates (with 25 mg of THA-1 and THA-4) by 100 ml of distilled water followed by 500 mg/ml THA-1 and THA-4. Leaching data for all metals falls within the shaded fields. Maximum and minimum mobility observed for the metal humates in water and solutions of humic substances indicated as —Ca etc.

humic substances in the leachate. This is shown in Figure 23-14, where the data for THA-4 are seen to have a steeper gradient than those for THA-1. This is indicative of a higher mobility for the humates formed from **THA-4**.

Since solutions of humic substances have a considerable effect on the mobilization of metals from dry humates, a study was made of metals adsorbed on pulverized hematite and montmorillonite. Cobalt, copper, nickel, lead, and zinc were adsorbed on 25 mg of the former minerals and leaching carried out as for the metal humates. The results,

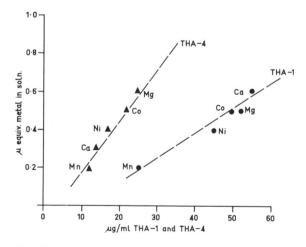

Fig. 23-14   Correlation of metal humates mobilized in distilled water with amount of humic substances present.

which are shown in Figure 23-15, exhibit the same pattern as seen with the metal humates. The metals are relatively immobile in the presence of pure water but migrate rapidly when humic substances are in solution.

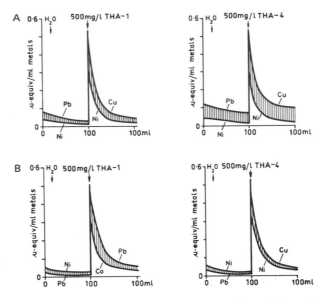

Fig. 23-15   Leaching of metals adsorbed on 25 mg of hematite (A) and montmorillonite (B). Same details as for Figure 23-13.

SOME GEOCHEMICAL ASPECTS OF HUMIC SUBSTANCES–METAL ASSOCIATIONS

There is a growing awareness that the humic substances are involved in various processes occurring at the surface of the earth. An appreciation of the significance of these substances in waters draining vegetated areas may be gained by considering the situation on the western coast of the small island of Tasmania. The average annual total discharge from the major rivers of western Tasmania is about 3×10^{10} m^3 from a catchment of approximately 2×10^4 km^2. Even with an estimate for the average content of humic substances as low as 5 mg/liter, the resulting quantity delivered to the sea annually is of the order of 1.5×10^5 metric tons. This figure does not take into account the additional contribution from the numerous minor rivers. Draeger and Lauer (1967) have estimated that almost 70 percent of the earth's land surface is covered by forest or grassland. This area, of approximately 3.6×10^8 km^2, has a potential to produce humic substances, and if the drainage conditions average to the situation observed in Tasmania, then the amount of these substances moving to the sea annually would be of the order of 2.7×10^9 metric tons. For the exploration geologist, the movement of this considerable quantity of humic substances is of importance because of its role in the weathering of ore mineral occurrences, its influence on metal transport, and its effect on geochemical exploration methods.

Orebody Weathering

The data presented on mineral solubilization show conclusively that humic substances can play an important role in the weathering cycle. Most sulfide, oxide, and secondary minerals are rapidly degraded, and this may influence the surface expression of an orebody. This could be particularly significant in gossan development because of the vulnerability of iron and manganese oxides to the action of humic substances. Some Tasmanian examples may illustrate this point.

At Oliver Hill in northwestern Tasmania (146°09′E, 41°31′S) an extensive gossan carrying up to 7 percent lead can be traced for about 1 km from its point of emergence from overlying Tertiary basalt. Analyses of gossan–soil pairs for this body indicate a rapid depletion of manganese, with iron/manganese ratios increasing from gossan to soil by factors of up to 40. This is compatible with the higher solubilization of pyrolusite compared with hematite as found in the experimental studies. As the gossan is located beneath Tertiary basalt, it is a remnant of an early Tertiary or Cretaceous land surface which is now reappearing with progress of the current erosion cycle. It is likely that the earlier climatic conditions were more arid, with consequential retardation of extensive vegetation growth and the associated activity of humic substances.

By contrast with this situation, the site of a recently developed base metal mine (Que River 145°42′E, 41°36′S), when cleared of vegetation, was found to have massive sulfides within 0.5 m of the surface. The capping was strongly leached and there was only minor development of iron and manganese oxides. It is possible that the abundance of humic substances in the current cycle of erosion, by providing a rapid rate of solution of oxidizing minerals, has been an important factor in preventing the formation of gossan at Que River. Many of the mineral discoveries in Tasmania, and no doubt elsewhere, were the result of the sighting of gossans during gold prospecting activities. The Que River deposits remained concealed for 50 years in an area of intensive prospecting and the question remains: How many others are defying our current geochemical exploration methods?

Metal Transport

The experimental studies have a substantial bearing on the mobilities of metals in wet temperate climates. Suggested relative mobilities in many reference works only take into account the solubilities of inorganic species, which results in metals such as lead, for example, being classed as immobile. Under the influence of humic substances, lead is highly mobile, as are most other metals. The leaching experiments indicate that even where metals are immobilized by the drying out of humates or by adsorption on clays or oxides the incursion of further humic substances in the groundwater will cause remobilization and continued transport. The one exception to this process appears to be aluminum, and this may be of significance in the formation of bauxite in temperate climatic regions.

The mechanism of transport by humic substances has been a contested issue. Schatz *et al.* (1954) suggested that complexation was an active weathering process and that possibly humic substances were complexing agents. De Long and Schnitzer (1955) suggested that the solution chemistry of soil organic substances could be explained in part in terms of colloidal theory. These early papers were the beginning of an extensive literature, mainly in soil science, on metal transport which largely supported the complexation interpretation of the activity of humic substances. More recently, the colloidal theory has reappeared and has been applied to the transport of gold and of metal hydroxides (Ong and Swanson 1969; Ong *et al.*, 1970). These papers have refueled the fires of controversy.

The possible means of gold transport in the weathering cycle include ionic, colloidal, and organic complexation, Krauskopf (1967) has shown that transport is possible as the aurichloride ion, which can be produced during the weathering of a gold-bearing orebody. The conditions necessary for this to occur are that both hydrogen and chloride ions are available, with one in high concentration and in addition an oxidizing agent, such as manganese dioxide, must be present. As Krauskopf (1967) points out, such conditions occur rarely and result only locally in appreciable transport of gold. Lakin *et al.* (1974) have indicated that the uptake of gold by plants could occur as the aurocyanide complex resulting from the activity of cyanoglycosides. They comment, however, that this complex would have limited mobility due to its rapid hydrolysis and enzymic destruction. Thus the mobility of gold in the weathering cycle does not appear to be adequately explained in terms of inorganic complexes of gold.

As proposed by Ong and Swanson (1969), the colloidal theory of gold transport requires the presence in solution of a gold ion species that is reduced to colloidal gold by humic substances. The mobility of this colloidal gold is then maintained by the humic substances acting as protective colloids. One problem with the colloidal theory is the provision of a gold ion species at a concentration that will enable the attainment of observed soluble gold values. For river waters Fischer (1966) gives a value of 0.002 ng/ml (ppb) and Gosling *et al.* (1971) a value of 0.010 ng/ml. If western Tasmania is a typical example of a wet temperate climate, the river and marsh waters generally range in acidity from pH 6 to 4. According to Ong and Swanson (1969), the dominant ionic gold species would be the aurochloride complex, with a solubility of 0.006 ng/ml, which agrees well with the required concentration. Baes and Mesmer (1976), however, state that the auro complexes are unstable and disproportionate as follows:

$$3AuX_2^- = AuX_4^- + 2Au^0 + 2X^-$$

This being the case, the only soluble gold species that is stable is the aurichloride complex, and this has a solubility of 0.001 ng/liter, which is far too low to support the colloidal model.

A further problem involved in the colloidal theory of gold transport is the reduction of ionic gold to colloidal by the humic substances. This mechanism is based on the premise that humic substances reduce U(VI) to U(IV) (Garrels and Pommer, 1959) and hence the reduction potential of these substances would be less than +0.447 V. If this were the case, the reduction $AuCl_4^- + 3e^- = Au^0 + 4Cl^-$, with a potential of +1.00 V, could be produced by humic substances. There are two factors which cast doubt upon this postulate. As Szilagyi (1971) has shown, and the polarographic studies support, the reduction potential of humic substances is on the order of +0.7 V. Furthermore, the reduction potentials of complexes decrease with their increasing stability. For gold, Shcherbina (1956) noted that for the reaction $AuX_4^- + 3e^- = Au^0 + 3X^-$, where X is chloride, bromide, and thiocynate, the respective reduction potentials are +1.00, +0.86, and +0.66 V. If, as suggested by Shcherbina (1956), the stability of the gold–humic substances complex is of the same order of magnitude as the thiocyanate complex, the reduction of gold by these substances is unlikely.

The studies of gold solubility, polarography, electrophoresis, and x-ray diffraction support the view that complexation of gold by humic substances is possible. These substances are well suited for a role in the transport of gold (and other metals) since they are being continuously produced by microbial action in the soil and are present at the soil–rock interface where weathering is initiated. A gold–humic substances complex formed at this location would be free to migrate readily in the hydrosphere because it is not subject to the solubility and stability problems associated with transport as ionic and colloidal gold.

During migration through the hydrosphere, some metal–humic substances associations may encounter conditions that result in destruction of the complexes. Such events may explain the occurrence of native copper in bogs (e.g., Lovering, 1927). They could also be responsible for the formation and regeneration of some placer gold deposits (Friese, 1931; Steelink, 1963). As already indicated, the quantity of humic substances migrating on the earth's surface is very large and the solubilization studies show that it is capable of complexing considerable amounts of metals. The ultimate destination of a large proportion of these substances is the sea. With their associated metals they are likely to accumulate in estuarine, lagoonal, and near-shore marine environments. Such accumulations localize metals once widely dispersed and these concentrations may be an early stage in the multicycle process which produces that rare occurrence—an orebody.

Humic Substances and Exploration Geochemistry

At the present stage of our expertise, humic substances are not very good news for the exploration geochemist. As far as an orebody is concerned these substances are likely to ensure that visible attributes, such as gossans, will have been removed. Where the chemical activity exceeds the physical removal of weathered products, the soil profile may be strongly leached and secondary dispersion patterns will be weakened or entirely obliterated by removal of trace metals as mobile humates. This is the case in western Tasmania, where companies now have to resort to C-horizon or bedrock sampling in difficult terrain and meet the escalating costs associated with these procedures.

There is some hope that the activity of humic substances might be turned to the advantage of the exploration geochemist. If the movement of metal humates on the earth's surface can be subjected to simple and rapid sampling, the potential exists for a very powerful means of assessing metal migration paths at both a regional and local level. Stream sediment sampling includes a proportion of the migrating humates due to their tendency to adsorption on clay mineral and iron oxide surfaces. The possible use of organic matter in stream sediments has been studied by Tessier *et al.* (1982), who found use of this fraction could enhance contrast. In many cases, however, particularly in areas of high runoff as in Tasmania, the fine-grained sediment of use in exploration is either nonexistent or difficult to sample. Concentration of metal humates from water is an attractive proposition, if currently not feasible. Lee and Jonassen (1983) investigated the impact of dissolved organic matter on hydrogeochemical surveys and noted that it was of particular importance where nickel, copper, and cobalt were being studied.

For geochemical exploration, the problem is largely that the laboratory research methods are far too time consuming to be applied economically. Very elegant procedures have been developed for the recovery of organics from natural waters by means of cross-linked polystyrene beads (e.g., Aiken *et al.*, 1979). Unfortunately, these techniques require operation at low pH, with consequent destruction of metal–humic substances bonding, or make use of high ionic strengths, which introduces contamination problems. There is no requirement in exploration geochemistry to reclaim humic substances from an adsorbent. All that is necessary is that the metals associated with these substances be easily eluted. Investigations are in progress in Tasmania to assess the potential of some form of activated carbon for concentration of metals associated with humic substances from natural waters. Research in this area is highly desirable and the development of successful methods that make use of the metal–humic substances association should prove very useful in the increasingly more difficult task of locating ore occurrences.

REFERENCES

Aiken, G. R., Thurman, E. M., and Malcolm, R. L., 1979, Comparison of XAD macroporous resins for the concentration of fulvic acid from aqueous solution: *Anal. Chem.*, v. 51, p. 1799–1803.

Bachinski, D. J., 1969, Bond strength and sulfur isotopic fractionation in coexisting sulfides: *Econ Geol.*, v. 64, p. 56–65.

Baes, C. F., and Mesmer, R. E., 1976, *The Hydrolysis of Cations:* Wiley, New York.

Baker, W. E., 1973, The role of humic acids from Tasmanian podzolic soils in mineral degradation and metal mobilization: *Geochim. Cosmochim. Acta*, v. 37, p. 269–281.

——, 1978, The role of humic acid in the transport of gold: *Geochim. Cosmochim. Acta*, v. 42, p. 645–649.

Barber, D. A., 1968, Microorganisms and the inorganic nutrition of higher plants: *Annu. Rev. Plant Physiol.*, v. 19, p. 71–78.

Berzelius, J. J., 1839, Untersuchung des Wassers der Porla-Quelle: *Annu. Phys. Chem.*, v. 105, p. 1–37.

Birrell, K. S., Pullar, W. A., and Heine, J. C., 1971, Pedological, chemical and physical properties of organic horizons of palaeosols underlying the Tarawera Formation: *N.Z. J. Sci.*, v. 14, p. 187–218.

Blanck, E., 1933, Die sogenannte "Humussäureverwitterung" im Lichte neuester Bodenforschung: *Ernaehr. Pfl.*, v. 29, p. 41–43.

Bremner, J. M., Mann, P. J. G., Heintze, S. G., and Lees, H., 1946, Metallo-organic complexes in soil: *Nature*, v. 158, p. 790–791.

Broadbent, F. E., and Ott, J. B., 1957, Soil–organic matter–metal complexes: 1. Factors affecting retention of various cations: *Soil Sci.*, v. 83, p. 419–427.

Clarke, F. W., 1911, The data of geochemistry, 2nd ed.: *U.S. Geol. Surv. Bull. 491.*

De Long, W. A., and Schnitzer, M., 1955, Investigations on the mobilization and transport of iron in forested soils: I. The capacities of leaf extracts and leachates to react with iron: *Soil Sci. Soc. Am. Proc.*, v. 19, p. 360–363.

Draeger, W. C., and Lauer, D. T., 1967, Present and future forestry applications of remote sensing from space: *4th Meet. Am. Inst. Aeronaut. Astronaut.*, Anaheim, p. 67–765.

Drozdova, V. T., and Emel'yanova, M. P., 1960, Copper complexes with humic acids: *Dokl. Akad. Nauk SSSR.*, v. 131, p. 3–7.

Fetzer, W. G., 1934, Transportation of gold by organic solutions: *Econ. Geol.*, v. 29, p. 599–604.

———, 1946, Humic acids and true organic acids as solvents of minerals: *Econ. Geol.*, v. 41, p. 47–56.

Fischer, K. W., 1966, Edelmetalle in der Saale und in ihrem Einzugsgebiet: *Geologie*, v. 15, p. 550–561.

Fotiyev, A. V., 1971, The nature of aqueous humus: *Dokl. Earth Sci. Sect.* (AGI Trans. 1972), v. 199, p. 193–195.

Freise, F. W., 1931, The transportation of gold by organic underground solutions: *Econ. Geol.*, v. 26, p. 421–431.

Garrels, R. M., and Pommer, A. M., 1959, Geochemistry and mineralogy of the Colorado Plateau uranium ores. Part 14. Some quantitative aspects of the oxidation and reduction of the ores: *U.S. Geol. Surv. Prof. Pap. 320*, p. 157–164.

Goldschmidt, V. M., 1930, Über das Vorkommen des Germaniums in Steinkohlen and Steinkohlen-produkten: *Nachr. Ges. Wiss. Goettingen, Math.-Phys.*, v. 1, p. 398–401.

Gosling, A. W., Jenne, E. A., and Chao, T. T., 1971, Gold content of natural waters in Colorado: *Econ. Geol.*, v. 66, p. 309–313.

Gruner, J. W., 1922, The origin of sedimentary iron formations: the Biwabik Formation of the Mesabi Range: *Econ. Geol.*, v. 17, p. 407–460.

Guillin, R., 1928, Dissociation intégrale des silicates par l'acide carbonique et les acides humiques et réactions annexes: *Comp. Rend.*, v. 187, p. 673–675.

Halbach, P., von Borstel, D., and Gundermann, K. D., 1980, The uptake of uranium by organic substances in a peat bog environment on a granite bedrock: *Chem. Geol.*, v. 29, p. 117–138.

Hori, S., and Okuda, A., 1961, Purification of humic acid by the use of exchange resin: *Soil. Sci. Plant. Nutr. Tokyo*, v. 7, p. 4.

Julien, A. A., 1879, On the geological action of the humus acids: *Proc. Am. Assoc. Adv. Sci.*, v. 28, p. 311–410.

Kerndorff, H. and Schnitzer, M., 1980, Sorption of metals on humic acid: *Geochim. Cosmochim. Acta*, v. 44, p. 1701–1708.

Kononova, M. M., 1966, *Soil Organic Matter*, 2nd English ed., Pergamon Press, Oxford.

———, and Bel'chikova, N. P., 1960, Investigations of the nature of soil humus substances by fractionation: *Sov. Soil Sci. 1960*, p. 1149–1155.

———, Alexandrova, I. V., and Titova, N. A., 1964, Decomposition of silicates by soil organic matter: *Sov. Soil Sci. 1964*, p. 1005–1014.

Krauskopf, K. B., 1967, *Introduction to Geochemistry:* McGraw-Hill, New York.

Lakin, H. W., Curtin, G. C., and Hubert, A. E., 1974, Geochemistry of gold in the weathering cycle: *U.S. Geol. Surv. Bull. 1330.*

Lee, J., and Jonasson, I. R., 1983, Contribution of organic complexation to Ni, Co and Cu speciation in surface waters: implications for hydrogeochemical surveys: *J. Geochem. Explor.*, v. 18, p. 25–48.

Lewis, T. E., and Broadbent, F. E., 1961, Soil–organic matter–metal complexes. 4. Nature and properties of exchange sites: *Soil Sci.*, v. 91, p. 393–399.

Lingane, J. J., 1941, Interpretation of the polarographic waves of complex metal ions: *Chem. Rev.* (ACS), v. 29, p. 1–35.

Loughnan, F. C., 1969, *Chemical Weathering of Silicate Minerals:* Elsevier, New York.

Lovering, T. S., 1927, Organic precipitation of metallic copper: *U.S. Geol. Surv. Bull. 795-C.*

Martell, A. E., and Calvin, M., 1952, *Chemistry of Metal Chelate Compounds:* Prentice-Hall, Englewood Cliffs, N.J.

Murray, A. N., and Love, W. W., 1929, Action of organic acids on limestone: *Bull. Am. Assoc. Petrol. Geol.*, v. 13, p. 1467–1475.

Nickel, E. H., 1968, Structural stability of minerals with the pyrite, marcasite, arsenopyrite and löllingite structures: *Can. Mineral.*, v. 9, p. 311–321.

Nockolds, S. R., 1966, The behavior of some elements during fractional crystallization of magma: *Geochim. Cosmochim. Acta*, v. 30, p. 267–278.

Odén, S., 1919, Die Huminsäuren: *Kolloidchem. Beih.*, v. 11, p. 75–82.

Ong, H. L., and Bisque, R. E., 1968, Coagulation of humic colloids by metal ions: *Soil Sci.*, v. 106, p. 220–224.

____ , and Swanson, V. E., 1969, Natural organic acids in the transportation, deposition and concentration of gold: *Q. Colo. Sch. Mines*, v. 64, p. 395–425.

____ , Swanson, V. E., and Bisque, R. E., 1970, Natural organic acids as agents of chemical weathering: *U.S. Geol. Surv. Prof. Pap. 700-C*, p. 130–137.

Ponomareva, V. V., and Ragim-Zade, A. I., 1969, Comparative study of fulvic and humic acids as the agents of silicate mineral decomposition: *Sov. Soil Sci.*, 1969, p. 157–166.

Radtke, A. S., and Scheiner, B. J., 1970, Studies of hydrothermal gold deposition (I). Carlin gold deposit, Nevada: the role of carbonaceous materials in gold deposition: *Econ. Geol.*, v. 65, p. 87–102.

Saar, R. A., and Weber, J. H., 1980, Lead(II) complexation by fulvic acid: how it differs from fulvic acid complexation of copper(II) and cadmium(II): *Geochim. Cosmochim. Acta*, v. 44, p. 1381–1384.

Schatz, A., Cheronis, N. D., Schatz, V., and Trcwlawney, G. S., 1954, Chelation (sequestration) as a biological weathering factor in pedogenesis: *Penn. Acad. Sci.*, v. 28, p. 44–51.

Schnitzer, M., and Skinner, S. I. M., 1963, Organo–metallic interactions in soils. 1. Reactions between a number of metal ions and the organic matter of a podzol B_h horizon: *Soil Sci.*, v. 96, p. 86–93.

____ , and Skinner, S. I. M., 1967, Organo–metallic interactions in soils. 7. Stability constants of Pb^{2+}, Ni^{2+}, Mn^{2+}, Co^{2+}, Ca^{2+} and Mg^{2+}–fulvic acid complexes: *Soil Sci.*, v. 103, p. 247–252.

Senft, H., 1871, Vorläufige Mittheilungen über die Humussubstanzen und ihr Verhalten zu den Mineralien: *Z. D. Geol. Ges.*, v. 23, p. 665–669.

Shcherbina, V. V., 1956, Complex ions and the transfer of elements in the supergene zone: *Geochemistry* (1956), p. 486–493.

Sholkovitz, E. R., and Copland, D., 1981, The coagulation, solubility and adsorption properties of Fe, Mn, Cu, Ni, Cd, Co and humic acids in a river water: *Geochim. Cosmochim. Acta*, v. 45, p. 181–189.

Sohn, M. L., and Hughes, M. C., 1981, Metal ion complex formation constants of sedimentary humic acids with Zn(II), Cu(II) and Cd(II): *Geochim. Cosmochim. Acta*, v. 45, p. 2393–2396.

Sprengel, C., 1826, Ueber Pflanzenhumus, Humussäure and humussäure Salze: *Kastners Arch. Ges. Naturlehre*, v. 8, p. 145–220.

Steelink, C., 1963, What is humic acid? *J. Chem. Educ.*, v. 40, p. 379–384.

Szalay, A., 1964, Cation exchange properties of humic acids and their importance in the geochemical enrichment of UO_2^{2+} and other cations: *Geochim. Cosmochim. Acta*, v. 28, p. 1605–1614.

Szalay, A., and Szilagyi, M., 1967, The association of vanadium with humic acids: *Geochim. Cosmochim. Acta*, v. 31, p. 1–6.

Szilagyi, M., 1971, Peat–water heterogeneous system and the determination of the standard redox potential: *Mag. Kem. Foly.*, v. 77, p. 172–175.

Tarkohv, K., 1881, The effect of humus on mineral salts: *Izv. Petrov. s-kh. Akad.*, v. 4, no. 1.

Templeton, G. D., III, and Chasten, D. N., 1980, Vanadium–fulvic acid chemistry: conformational and binding studies by electron spin probe techniques: *Geochim. Cosmochim. Acta*, v. 44, p. 741–752.

Tessier, A., Campbell, P. G. C., and Bisson, M., 1982, Particulate trace metal speciation in stream sediments and relationships with grain size: implications for geochemical exploration: *J. Geochem. Explor.*, v. 16, p. 77–104.

Thenard, P., 1870, Observations sur le mémoire de M. Friedel: *Comp. Rend.*, v. 70, p. 1412–1414.

Tipping, E. and Cooke, D., 1982, The effects of adsorbed humic substances on the surface charge of goethite (α-FeOOH) in freshwaters: *Geochim. Cosmochim. Acta*, v. 46, p. 75–80.

——, and Heaton, M. J., 1983, The adsorbtion of aquatic humic substances by two oxides of manganese: *Geochim. Cosmochim. Acta*, v. 47, p. 1393–1397.

Welte, E., 1956, Zur Konzentrationsmessung von Huminsäuren: *Z. Pflanzenernaehr. Dung. Bodenkd.*, v. 74, p. 219–227.

Wilson, S. A., and Weber, J. H., 1979, An EPR study of the reduction of vanadium(V) to vanadium(IV) by fulvic acid: *Chem. Geol.*, v. 26, p. 345–354.

M. Schnitzer
Chemistry and Biology Research Institute
Agriculture Canada
Ottawa, Ontario, Canada

24

REACTIONS OF HUMIC SUBSTANCES WITH METALS AND MINERALS

Humic substances, probably the most widely distributed organic carbon-containing materials in terrestrial and aquatic environments, are dark-colored, acidic, partly aromatic, hydrophilic, chemically complex, polyelectrolyte-like materials which range in molecular weights from a few hundred to several thousand. Among their most important characteristics is their ability to interact with metal ions, oxides, hydroxides, minerals, and organics to form associations of widely differing chemical and biological stabilities. Many of these reactions are affected by (1) the concentration of the humic material, (2) the pH, and (3) the metal and salt concentration. It is these parameters which also affect the shapes and sizes of humic acid and fulvic acid particles, which change from rigid spherocolloids at concentrations > 3.5 g/liter, pH 3.5 and lower, and at electrolyte concentrations of >0.01 M, to flexible, linear polyelectrolytes at the lower concentrations of humic materials and electrolytes and higher pH, that normally prevail in soils and inland waters.

Interactions of humic materials with inorganics are discussed under the following headings: the formation of water-soluble simple and mixed ligand complexes, sorption and desorption, dissolution of minerals, and adsorption on external mineral surfaces and in clay interlayers.

INTRODUCTION

Humic substances are the most widely distributed organic carbon-containing materials on the earth's surface, occurring in soils, fresh waters, and in the sea. They constitute between 70 and 80 percent of the organic matter in most inorganic soils and are formed from the chemical and biological degradation of plant and animal residues and from synthetic activities of microorganisms. The remaining 20 to 30 percent of the organic matter in soils consists mainly of protein-like materials, polysaccharides, fatty acids, and alkanes.

One of the most striking characteristics of humic substances is their ability to interact with metal ions, oxides, hydroxides, minerals, and organic chemicals, including toxic pollutants, to form water-soluble and water-insoluble associations of widely differing chemical and biological stabilities. These interactions have been described as ion exchange, surface adsorption, chelation, peptization, and coagulation reactions (Schnitzer, 1978), and it is likely that they affect many reactions that occur in soils and waters. For example, fulvic acid (FA) at any pH and humic acid (HA) at pH >6.5, can form water-soluble complexes with many organics and inorganics in competition with hydrolysis reactions and so change their solubilities. On the other hand, HA at pH <6.5 is water insoluble but can sorb organics and inorganics which may bring about their accumulation on the organic surfaces.

To provide a better understanding of how humic substances interact with metals and minerals, it may be appropriate to discuss first the current state of our knowledge of the chemistry of HAs and FAs. Thus, during the first part of the chapter the principal types of humic materials (i.e., HA, FA, and humin) are defined and a number of their analytical characteristics are described. I shall then discuss information obtained on the chemical structure of HAs and FAs by methods used in organic chemistry, by ^{13}C NMR spectroscopy and by colloid chemical methods. The second part of the chapter focuses on metal- and mineral–HA and –FA interactions.

DEFINITIONS

Based on their solubility in alkali and acid, humic substances are usually partitioned into the following three main fractions: (1) HA, which is soluble in dilute alkali but is coagulated by acidification of the alkaline extract; (2) FA, which is that humic fraction that remains in solution when the alkaline extract is acidified (i.e., it is soluble in both dilute alkali and dilute acid), and (3) humin, which is that humic fraction that cannot be extracted from the soil by dilute base or acid. The separation scheme described above and in Figure 24-1 has been accepted by most soil scientists and specialists investigating humic materials.

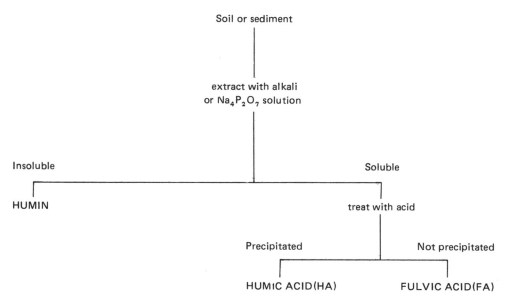

Fig. 24-1   Extraction and fractionation of humic substances.

Analytical Characteristics of Humic Substances

The elemental composition and functional group content of a HA extracted from a Mollisol Ah horizon and a FA extracted from the Bh horizon of a Spodosol are shown in Table 24-1. These two humic materials have been investigated in our laboratory over a period of many years, so these materials are emphasized in this chapter.

A more detailed analysis of the data shows that (1) the HA contains more C but less O than does the FA; (2) the HA contains more H, N, and S than the FA; (3) the total acidity and COOH content of the FA are appreciably higher than those of the HA;

TABLE 24-1 Analytical Characteristics
of HA and FA

	HA	FA
Element (%)		
C	56.4	50.9
H	5.5	3.3
N	4.1	0.7
S	1.1	0.3
O	33.9	44.8
Functional groups (meq/g)		
Total acidity	6.6	12.4
COOH	4.5	9.1
Phenolic OH	2.1	3.3
Alcoholic OH	2.8	3.6
Quinonoid C=O	2.5	0.6
Ketonic C=O	1.9	2.5
OCH$_3$	0.3	0.1
E_4/E_6	4.3	7.1

Source: Schnitzer and Skinner (1974).

(4) the FA is richer in phenolic OH, alcoholic OH, and ketonic C=O groups than the HA, but the latter contains more quinonoid C=O groups per unit weight; and (5) about 90 percent of the O in the HA can be accounted for in functional groups, but all of the O in the FA is similarly distributed. The E_4/E_6 ratio (ratio of absorbances at 465/665 nm) of the FA is considerably higher than that of the HA, which means that the FA has a lower particle or molecular weight than the HA. From data published in the literature (Schnitzer and Khan, 1972; Schnitzer 1978), it appears that the analytical characteristics of humins are similar to those of HAs.

Chemical Structure of HAs and FAs

Chemical degradation studies The oxidative degradation of HAs, FAs, and humins produces aliphatic, phenolic, and benzenecarboxylic acids in addition to *n*-alkanes and *n*-fatty acids (Schnitzer, 1977). The most abundant aliphatic degradation products are *n*-fatty acids (especially the *n*-C$_{16}$ and *n*-C$_{18}$ acids) and di- and tricarboxylic acids (Fig. 24-2). Major phenolic acids produced include those with between one and three OH groups and between one and five CO$_2$H groups per aromatic ring (Fig. 24-3). Prominent benzenecarboxylic acids are the tri, tetra, penta, and hexa forms (Fig. 24-4). The structures shown in Figures 24-2, 24-3, and 24-4 constitute the major "building block" of HAs, FAs, and humins.

The major types of chemical structures that make up the HA and FA are summarized in Table 24-2. The HA contains approximately equal proportions of aliphatic and phenolic structures, but a greater percentage of benzencarboxylic structures or structures producing benzenecarboxylic acids on oxidation. By contrast, the FA contains more phenolic than aliphatic and benzenecarboyxlic structures. Both the HA and FA

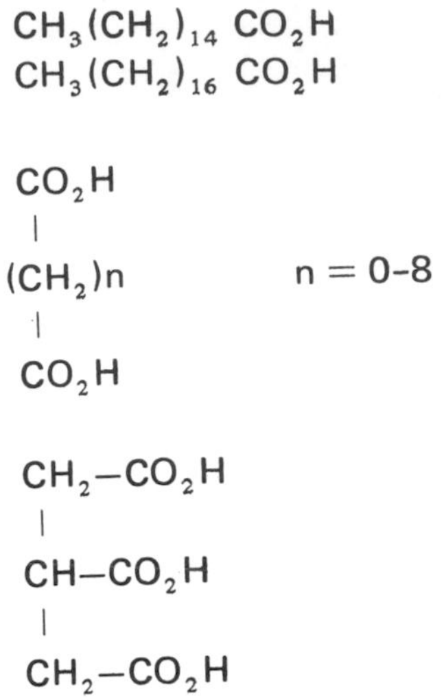

CH$_3$(CH$_2$)$_{14}$ CO$_2$H
CH$_3$(CH$_2$)$_{16}$ CO$_2$H

CO$_2$H
|
(CH$_2$)n   n = 0-8
|
CO$_2$H

CH$_2$—CO$_2$H
|
CH—CO$_2$H
|
CH$_2$—CO$_2$H

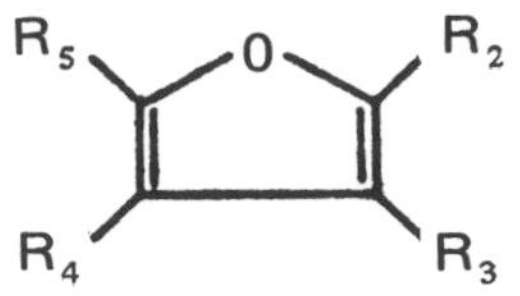

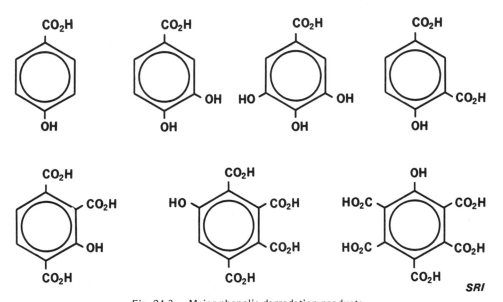

Fig. 24-2 Major aliphatic degradation products.

Fig. 24-3 Major phenolic degradation products.

contain approximately equal proportions of aliphatic structures and have similar aromaticities.

Thus the HA and FA have similar chemical structures, except that the FA is richer in phenolic but poorer in benzenecarboxylic structures than the HA.

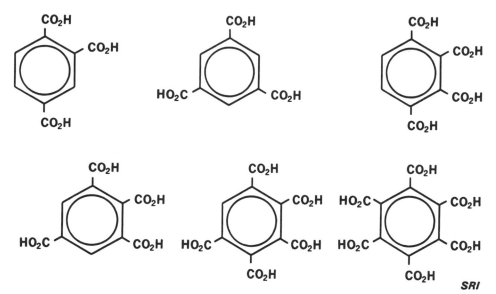

Fig. 24-4   Major benzene carboxylic degradation products.

TABLE 24-2   Major Chemical Structures
in HA and FA

Major Product	HA (%)	FA (%)
Aliphatic	24.0	22.2
Phenolic	20.3	30.2
Benzenecarboxylic	32.0	23.0
Total	76.3	75.4
Aromaticity[a]	69	71

[a] $\left(\dfrac{\text{phenolic} + \text{benzenecarboxylic}}{\text{total}}\right) \times 100.$

Source: Schnitzer (1977).

^{13}C NMR spectra of HAs and FAs Figure 24-5a shows the ^{13}C NMR spectrum of the HA dissolved in 0.5 M NaOD. The spectrum is well resolved and exhibits a number of distinct peaks in the aliphatic (0 to 105 ppm) and aromatic (105 to 165 ppm) regions, and near 180 ppm due to C in COOH groups. The chemical shifts at 16.5, 21.1, 25.0, 27.1, and 31.3 ppm are probably due to aliphatic carbons in alkyl chains; the peak at 16.5 ppm is probably due to terminal CH$_3$ groups, while the peaks at 31.3 ppm is characteristic of (CH$_2$)n in long alkyl chains, although other alkyl carbons may also occur at this chemical shift (Hatcher *et al.*, 1980; Lindeman and Adams, 1971). The resonance at 40.2 ppm is difficult to assign; it probably includes contributions from both alkyl

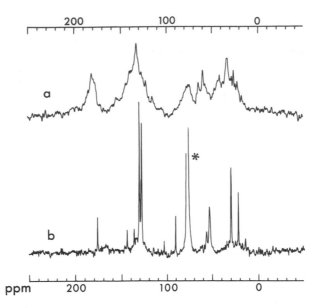

Fig. 24-5 ^{13}C NMR spectrum: (a) HA; (b) methylated HA.*, Solvent peak.

carbons and amino acid carbons (Breitmaier and Voelter, 1978). In the region 50 to 105 ppm, aliphatic carbons substituted by oxygen and nitrogen are usually observed. The peaks at 52.6 and 58.5 ppm may be due to $-OCH_3$, although the methoxyl content of the HA is only 1.0 percent. The HA contains 1.2 percent amino-N, so that some of the peaks in the region 50 to 70 ppm arise from amino acids. The HA contains 5.7 percent carbohydrates, which produce signals in the regions 60 to 65 (C_6), 70 to 80 (C_2-C_5 ring carbons), and 90 to 105 (anomeric carbons) ppm (Hatcher *et al.*, 1980; Breitmaier and Voelter, 1978), so that the peaks at 63.5, and the broader resonances at 73.4 and 105.0 ppm, are probably due to carbohydrates. The aromatic region shows a relatively sharp maximum at 131.8 ppm. This resonance is due to aromatic ring carbons in which the ring is not substituted by strong electron donors such as oxygen or nitrogen. The small peak at 155 ppm indicates the presence of O- and N-substituted aromatic C (phenolic OH or aromatic NH_2). The broad peak at 180 ppm is due to C of COOH groups.

The ^{13}C NMR spectrum of the methylated HA is shown in Figure 24-5b. Methylation increased the OCH_3 content from 0.3 percent in the unmethylated HA to 16.5 percent in the methylated HA. The methylated HA was soluble in $CDCl_3$, while the unmethylated HA had to be dissolved in NaOD.

The ^{13}C NMR spectrum of the methylated HA (Fig. 24-5b) exhibits distinct peaks at 13.9, 21.6, and 29.3 ppm due to aliphatic alkyl carbons. The well-defined resonances at 53.0 and 56.1 ppm arise from OCH_3 groups of aromatic esters and ethers, respectively (Mikita *et al.*, 1981). The small peaks in the region 57 to 62 ppm suggest the presence of aliphatic methyl ethers. But there are no indications of the presence of methylated amino groups [$R-N(CH_3)_2$] and aliphatic esters [signals at 45.6 and 51.2, respectively (Mikita et al., 1981)] in the methylated HA. It is likely that aliphatic COOH and amino groups are not sufficiently acidic to be methylated by diazomethane. Thus signals in the region 50 to 65 ppm indicate that the methylated HA contains aromatic COOH, phenolic OH, and aliphatic OH groups. Integration of the peak areas shows the aromatic COOH groups account for approximately 53 percent of the oxygen-containing functional groups, phenolic OH groups for 23 percent, and aliphatic OH groups for 22 percent. The spectrum in Figure 24-5b demonstrates the significant presence of phenolic OH groups. Integration of peak areas at 53.0 and 56.1 ppm indicates a OCH_3 content of 15.5 percent, which compares well with 16.5 percent OCH_3 obtained by the chemical (Zeisel) method. An interesting feature of Figure 24-5b is the group of sharp signals at 127.6, 128.2, 129.9, and 130.5 ppm, characteristic of aromatic carbons, including those substituted by car-bomethoxyl ($COOCH_3$) groups (Bremser et al., 1979). In unmethylated forms, these would be the benzenecarboxylic acid-type structures considered major "building blocks" of HAs and FAs (Schnitzer, 1978). Other smaller sharp peaks at 132.1, 136.1, and 143.8 ppm are likely due to carbon-substituted carbons in aromatic or heterocyclic rings (Verheyen et al., 1982). The broad peak at 166.9 ppm in Figure 24-5b arises from carbonyl carbons of aromatic methyl ester groups (Bremser et al., 1979), while the sharp signal at 175.9 ppm may represent a small amount of unmethylated COOH.

The ^{13}C NMR spectrum of the FA dissolved in NaOD is shown in Figure 24-6a. The spectrum consists of a number of aliphatic resonances in the region 20 to 50 ppm, fol-lowed by signals from amino acids and carbohydrates between 50 and 85 ppm. The presence of aromatic carbons is indicated by a broad region of intensity with a maximum at 130 to 133 ppm. The occurrence of several different types of COOH carbons is sug-

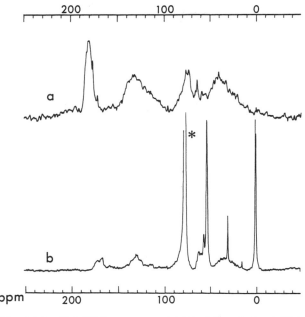

Fig. 24-6 ^{13}C NMR spectrum: (a) FA; (b) methylated FA.*, Solvent peak.

gested by signals at 171.2, 176.2, and near 181.0 ppm. Fewer sharp signals are observed in the ^{13}C NMR spectrum of the FA (Fig. 24-6a) than in that of the HA (Fig. 24-5a), possibly because of more H bonding in and greater structural complexity of the FA.

The spectrum of the methylated FA (Fig. 24-6b), which is 100 percent soluble in CDCl$_3$, is dominated by aliphatic signals at 14.2 and 29.8 ppm and by resonances at 52.9 ppm (OCH$_3$ of aromatic COOCH$_3$ groups), 56.3 (OCH$_3$ of aryl alkyl ethers), and 59.1 ppm and 62.6 ppm (OCH$_3$ of aliphatic OH groups). Integrations of peak areas between 50 and 65 ppm indicate that aromatic COOCH$_3$ groups account for 65 percent of the oxygen-containing functional groups, phenolic OH groups for 21 percent, and aliphatic OH groups for 14 percent. While chemical analysis shows a total OCH$_3$ content of 27.2 percent, area integration at 53.0 and 56.3 ppm indicates 24.0 percent OCH$_3$. In addition to these resonances, the spectrum in Figure 24-6b exhibits aromatic carbon signals at 112.6, 115.0, and in the 130.0-ppm region, while COOH-C resonances appear at 166.4 and 171.5 ppm.

The combination of chemical and ^{13}C NMR spectroscopic methods shows that all COOH groups in these materials are attached to aromatic structures. This finding has significant implications in studies on metal complexing by HA and FA. ^{13}C NMR can also be used to determine the relative concentrations of COOH and phenolic OH groups, which agree with results from chemical analyses.

Another significant observation is that following methylation and sometimes after acid hydrolysis, ^{13}C NMR spectra of HAs and FAs reveal fine chemical structure with sharp signals in the region 127 to 131 ppm, which appear to arise from benzenecarboxylic acid–type structures, major "building blocks" of humic materials. It seems that the building blocks are masked or coated by proteins, polysaccharides, alkanes, fatty acids, and so on, which are not chemically bonded to the humic materials but associated by H bonds. Once these have been removed, the building blocks become unmasked and can be detected by ^{13}C NMR spectroscopy.

Colloid chemical characteristics of HAs and FAs To obtain information on molecular sizes, shapes, and weights of HAs and FAs, Ghosh and Schnitzer (1980) did surface pressure and viscosity measurements at different pHs and varying concentrations of humic materials and neutral salts. One example of their data is presented in Table 24-3, which shows the effect of pH on the molecular characteristics of the FA. $\overline{Mn}$ is the

TABLE 24-3   Effect of pH on Molecular Characteristics of FA

pH	$\overline{Mn}$	A_0 (m²/mg)	$(\overline{R^2})^{1/2}$ (A)	$\overline{Mv}$
2.0	4270	0.044	302.8	9720
3.5	1180	0.030	233.9	2580
6.5	1020	0.024	210.0	2290
9.5	1080	0.026	226.6	2450

Source: Ghosh and Schnitzer (1980).

number-average molecular weight. A_0 is the molecular area, $(\overline{R^2})^{1/2}$ is the end-to-end separation, and Mv is the viscosity-average molecular weight. Note that at pH 2.0, the molecular weight of the FA (both $\overline{M}n$ and $\overline{M}v$) is four times as high as that at pH 6.5 and 9.5. At pH 2.0, four molecules of FA appear to combine to form an aggregate. The molecular area (A_0) and the end-to-end separation are also significantly greater at pH 2.0 than at the higher pH values. Ghosh and Schnitzer (1980) note that the three parameters controlling the molecular characteristics of HAs and FAs are the concentration of the humic material, the pH of the system, and the ionic strength of the medium. They conclude that HAs and FAs are rigid, uncharged colloids at: (1) high sample concentrations (>3.5 to 5.0 g/liter) (2) low pH (<3.5), and (3) electrolyte concentrations of 0.05 M and higher. But HAs and FAs are flexible, linear polyelectrolytes at (1) low sample concentrations (<3.5 g/liter), (2) pH >3.5, and (3) ionic strength <0.05 M. The different molecular configurations are summarized pictorially in Figure 24-7. In soil solutions and fresh waters, where normally both the humic and salt concentrations would be expected to be low and the pH >3.0, humic substances should occur as flexible, linear polyelectrolytes. Additional support for this view comes from transmission electron micrographs of HA and FA (Stevenson and Schnitzer, 1982) which show (Fig. 24-8) that in dilute aqueous solutions HAs and FAs form flat elongated, multibranched filaments, 20 to 100 nm in width.

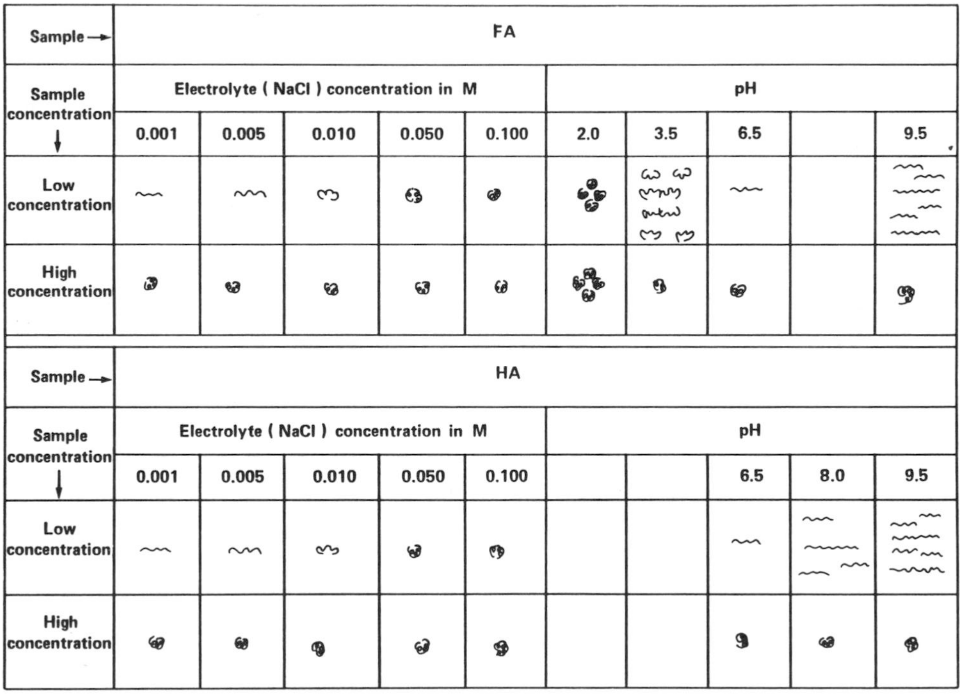

Fig. 24-7 Macromolecular HA and FA configurations at different pHs and electrolyte concentrations. (From Ghosh and Schnitzer, 1980.)

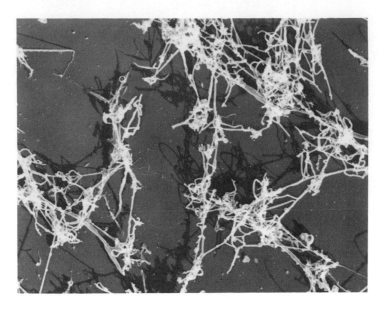

Fig. 24-8 Transmission electron micrograph of a 0.01 percent HA solution.

Concepts of the Chemical Structure of Humic Materials

From the information summarized in this paper we can conclude that between 50 and 60 percent of the weights of HAs and FAs consist of aromatic structures heavily substituted by COOH and OH groups. Whether the phenolic and benzenecarboxylic acids isolated as chemical degradation products exist in HAs and FAs in these forms or in more complex structures remains to be resolved. Colloid-chemical methods show that the shapes and sizes of HA and FA particles or molecules are strongly affected by the concentration of the humic material, the pH, and the ionic strength. At high sample concentration and ionic strength and at low pH, HA and FA tend to aggregate into larger particles through hydrogen bonding and possibly van der Waals forces. As the pH increases, these forces become weaker, and also because of increasing dissociation of CO_2H and phenolic OH groups, aggregates separate and begin to repel each other, so that the particles become smaller. In dilute aqueous solutions, between pH 3.5 and 9.0, HAs and FAs are flexible, linear polyelectrolytes. Because the processes of aggregation and dispersion are reversible, HAs and FAs must have the type of structure that can accommodate such changes. A partial chemical structure for FA that is in harmony with many experimental observations has been proposed by Schnitzer (1978) (Fig. 24-9). Each of the compounds that make up the structure has been isolated from FA with and without chemical degradation. Bonding between the building blocks is by hydrogen bonds, which makes the structure flexible, permits the building blocks to aggregate and disperse reversibly, and also allows the FA to react with inorganic and organic soil constituents either via oxygen-containing functional groups on the large external or internal surfaces, or by trapping them in internal voids.

Thus HAs and FAs are not single molecules such as benzoic or salicylic acid, but

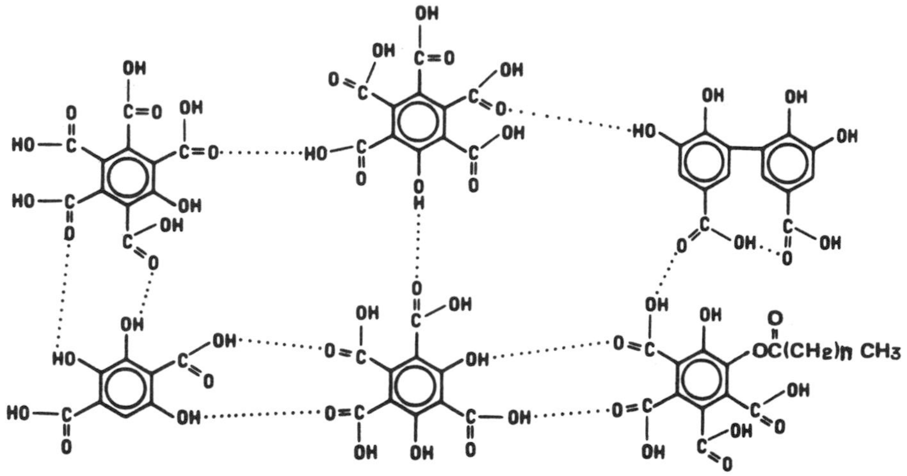

Fig. 24-9   Partial structure of FA (From Schnitzer, 1978.)

associations of molecules of microbiological and plant origin held together primarily by hydrogen bonding. The molecular makeup and reactivity of HAs and FAs will affect their interactions with inorganics and organics in soils and waters. It is for these reasons that I have attempted to present some background information on the chemical structure of HAs and FAs.

MAJOR METAL- AND MINERAL-HUMIC INTERACTIONS

Reactions between metals, minerals, and humic substances can occur via one or more of the following reaction mechanisms: (1) formation of water-soluble simple metal complexes, (2) formation of water-soluble mixed ligand complexes, (3) sorption on and desorption from water-insoluble HAs and metal-humate complexes, (4) dissolution of minerals, (5) adsorption on external mineral surfaces, and (6) adsorption in clay interlayers.

Formation of Water-Soluble Simple Metal Complexes

Reactions in water near pH 7 between di- and trivalent metal ions and HA or FA are likely to proceed by either one or more, or simultaneously all four, of the reaction mechanisms shown in Figure 24-10, taking divalent metal ion M^{2+} as an example. According to reaction (1), one CO_2H group reacts with one metal ion to form an organic salt or monodentate complex. Equation (2) describes a reaction in which one CO_2H and one adjacent OH group react simultaneously with the metal ion to form a bidentate complex or chelate. According to equation (3), two adjacent CO_2H groups interact simultaneously with the metal ion to also form a bidentate chelate. Equation (4) shows a situation in which the metal ion M^{n+} is linked to the FA, in addition to electrostatic bonding, also

Major metal–organic reaction mechanisms

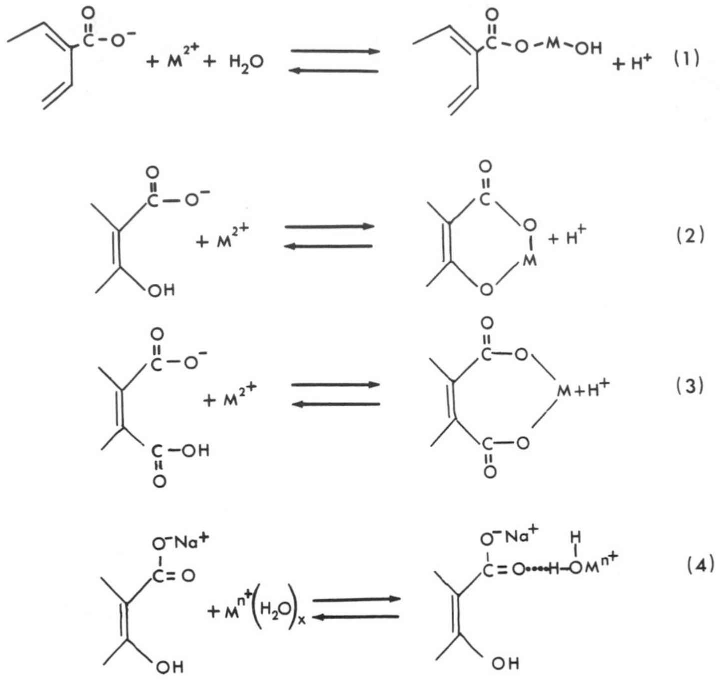

Fig. 24-10 Major metal–HA and metal–FA reaction mechanisms.

through a water molecule in its primary hydration shell via a hydrogen bond to a C=O group. This type of interaction is of special importance when the cation has a high solvation energy and so retains its primary hydration shell.

Equations (2) and (3) describe the formation of strong inner-sphere complexes, whereas equation (4) refers to the formation of a weaker outer-sphere complex.

Stability constants of water-soluble metal-HA and metal-FA complexes have been determined by a number of workers. The subject has been reviewed recently by Stevenson (1982). A number of serious problems have been encountered in the analysis and interpretation of data. Probably the most serious obstacle to progress in this area is our lack of adequate knowledge of the chemical structure and dissociation behavior of HAs and FAs.

Mixed Ligand Complexes

The formation of metal-FA-phosphate complexes was first described by Levesque and Schnitzer (1967). It is likely that in soils an appreciable portion of the total P exists in the form of such complexes, but it is difficult to demonstrate this because of the low P content of soils. The formation and stability of mixed ligand complexes of the type Cu^{2+}-FA-secondary ligand (Y) has been studied by Manning and Ramamoorthy (1973). Secondary ligands Y investigated were citrate, tartrate, salicylate, phosphate, NTA (nitrilotriacetate), aspartate, and glycine. In neutral and weakly acidic solutions, mixed complexes predominated over simple complexes. Values of equilibrium constants for mixed complexes with citrate, phosphate, and NTA were particularly high compared to simple complexes. If phosphate functions in the same way as other oxyanions, the relatively high concentration of HCO_3^- and SO_4^{2-} in some soil solutions should lead to the formation of mixed Cu^{2+}-FA-HCO_3 and Cu^{2+}-FA-HSO_4 complexes. There is increasing evidence that mixed complexes are important in the transport of trace metal ions in the soil and in the uptake of cations and anions by plant roots. Furthermore, the formation of mixed complexes will prevent the precipitation of metal ions by hydrolysis at elevated pH values and will also interfere with the precipitation of insoluble metal phosphates, sulfates, or carbonates. More information on the occurrence and properties of mixed ligand complexes in soils is needed.

Sorption and Desorption

One of the major characteristics of HAs is their ability to sorb inorganic and organic substances. I shall confine this discussion to the sorption of metal ions. In a recent investigation, Kerndorff and Schnitzer (1980) examined the interaction of HA with a solution containing equimolar concentrations of 11 different metal ions. They report that the efficiency of sorption of metal ions on HA increases with rises in pH and HA concentrations and decrease in metal concentration.

At pH 2.4, the order of sorption is

$$Hg > Fe > Pb > Cu = Al > Ni > Cr = Zn = Cd = Co = Mn$$

At pH 3.7, the order is

$$Hg > Fe > Al > Pb > Cu > Cr > Cd = Zn = Ni = Co = Mn$$

At pH 4.7, the order is

$$Hg = Fe = Pb = Cu = Al = Cr > Cd > Ni = Zn > Co > Mn$$

At pH 5.8, the order is

$$Hg = Fe = Pb = Al = Cr = Cu > Cd > Zn > Ni > Co > Mn$$

Hg(II) and Fe(III) are always sorbed most readily by HA, while Co and Mn are sorbed least readily. The different metal ions appear to compete for active sites (COOH and phenolic OH groups) on the HA. The sorption mechanism is complex. Not only do the 11 metal ions plus H^+ ions (a total of 12 ions) interact with the HA; but they also interact with each other. The mechanisms involved are likely to include ion exchange, coprecipitation, and the formation of inner-sphere and outer-sphere complexes. The affinities of the 12 ions for sorption on HA do not correlate with their atomic weights, atomic numbers, and crystal and hydrated ionic radii.

Figure 24-11 shows a family of curves that model the sorption of the metal ions on HA. The equation for these curves is

$$y = \frac{100}{1 + \exp\left[-(A + BX)\right]}$$

where y = percent metal sorbed by HA
   X = mg HA
A and B = empirical constants

The effects of varying constants A and B are shown in Figure 24-11. If A and B are large, the slopes are steep and the curves are close to the ordinate (curves V, U, T, S, and R). But if A and B are small, the curve is less steep and far from the ordinate (curve N). Medium values for A and B yield curves Q, P, and O.

Curves V to R in Figure 24-11 are representative of sorption on HA at pH 4.7 of Fe, Hg, Cr, and Al; curves Q to O of the sorption of Pb and Cu; and curve N of the sorption of Cd, Ni, Zn, Co, and Mn.

If the pH of system increases, the slopes increase because of a rapid increase in B, while A changes little or decreases. If the metal concentration increases, B tends to remain constant while A decreases, so that the slopes of the curves change little.

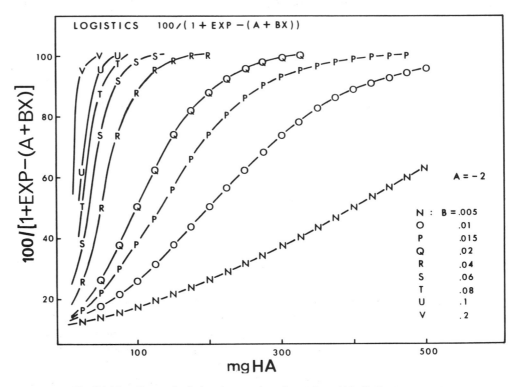

Fig. 24-11   Curves depicting the sorption of metals on HA. Ordinate, percent metal sorbed; abscissa, mg HA; A is constant, equaling −2. (From Kerndorff and Schnitzer, 1980.)

REACTIONS OF HUMIC SUBSTANCES WITH METALS AND MINERALS

The metal ions sorbed by HA can subsequently be desorbed by water-soluble FA. Similarly, FA will desorb metal ions sorbed on inorganic soil constituents such as clays and hydrous oxides. Thus FA can effectively change the sorption, desorption, and precipitation characteristics of metals.

Dissolution of Minerals

In view of their ability to complex mono-, di, tri-, and tetravalent metal ions, FAs and at pH >6.5 also HAs can attack and degrade minerals to form water-soluble and water-insoluble metal complexes. Thus the weathering of minerals in soils and sediments is often enhanced by the action of naturally occurring humic materials, especially water-soluble FA. The latter, because of its abundance in soils, mobility, and ability to complex metal ions and interact with silica, may increase the concentrations of these soil constituents in aqueous solutions to levels that are far in excess of their normal solubilities. In this manner, aqueous FA solutions may not only bring about the dissolution or degradation of existing minerals, but also lead to the synthesis of new minerals by permitting the complexed and dissolved metals and silica to form new combinations. Conversely, active surface of inorganic soil constituents may catalyze either the degradation or the synthesis of HAs and FAs.

Dilute FA solution is much more efficient in dissolving Fe, Al, and Mg from chlorites than is distilled water at the same pH under identical conditions (Table 24-4). Table 24-5 presents similar data for three micas. We found that 0.025 percent FA was relatively more efficient in degrading soil minerals than was 0.1 and 0.2 percent FA (Fig. 24-12), indicating that very dilute FA solutions have considerable activity in this regard. One point of special interest is illustrated by the data in Tables 24-4 and 24-5: minerals rich in Fe are more susceptible to attack by FA than are those free of Fe or containing little Fe. Thus the complexing of Fe by humic materials appears to have an adverse effect on the structural stability of Fe-rich minerals, so that along with Fe, other major constituent elements, such as Mg, Al, K, and Si, are also brought into solution. Senesi et al. (1977) combined ESR spectroscopy with chemical treatments to obtain information on oxidation states and site symmetries of iron bound by HA and FA. At least two,

TABLE 24-4   Dissolution of Metals (Fe, Al, Mg) from Chlorites by 0.2 Percent FA Solution after 360 Hours of Shaking at pH 2.5

	Percent of Sample Dissolved by:	
	H_2O	0.2% FA
Leuchtenbergite (Fe-poor)	2	4
Thuringite (Fe-rich)	6	26

Source: Kodama and Schnitzer (1973).

TABLE 24-5 Dissolution of Micas by 0.2 Percent FA
(pH 2.5)

Mineral	Element Dissolved	Percent Dissolved after 710 Hours
Biotite	Fe	11.5
(Fe-rich)	Al	14.5
	Mg	17.0
	K	18.5
	Si	14.0
Phlogopite	Al	8.0
(Fe-poor)	Mg	8.6
	K	8.0
	Si	9.0
Muscovite	Al	2.3
(Fe-poor)	K	1.8
	Si	0.6

Source: Schnitzer and Kodama (1976).

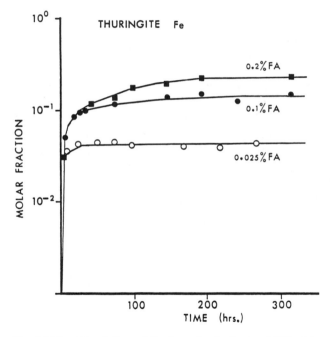

Fig. 24-12   Dissolution of Fe from thuringite by 0.025, 0.1, and 0.2 percent (w/v) FA solution.

possibly three, different binding sites for iron were found to occur in humic materials: (1) Fe^{3+} was strongly bound and protected by tetrahedral and/or octahedral coordination; this form of iron exhibited considerable resistance to complexation by known sequestering agents and reduction; and (2) Fe^{3+} was adsorbed on external surfaces of humic materials, weakly bound octahedrally, and easily complexed and reduced. These observations were made on laboratory-prepared Fe-HA and Fe-FA complexes and on a Fe-rich layer taken from a soil also rich in organic matter.

Adsorption on External Mineral Surfaces

The extent of adsorption of humic materials on mineral surfaces depends on the physical and chemical characteristics of the surface, the pH of the system, and its water content. One can visualize the formation of a wide range of mineral–humic associations, involving chemical bonding with widely differing strengths. Another mechanism of considerable importance is hydrogen bonding, the occurrence of which is clearly indicated in IR spectra of HA- and FA-mineral complexes (Schnitzer and Khan, 1972). These reactions are likely to involve H and O of CO_2H and OH groups in HA and FA and O and H on mineral surfaces and edges. It is probable, as has been mentioned above, that cations with high solvation energies on mineral surfaces react via water bridges with HA and FA functional groups. Van der Waals forces may also contribute to the adsorption of humic substances on mineral surfaces. Kodama and Schnitzer (1974) report high adsorption of FA on sepiolite surfaces. Sepiolite has a channel-like structure formed by the joining of edges of long and slender talc structures. In untreated sepiolite the channels are occupied by bound and/or zeolitic water, which apparently can be displaced by undissociated FA.

Adsorption in Clay Interlayers

The evidence currently available shows that the interlayer adsorption of FA by expanding clay minerals is pH dependent, being greatest at low pH, and no longer occurring at pH >5.0 (Schnitzer and Kodama, 1966) (Fig. 24-13). Adsorbed FA cannot be displaced

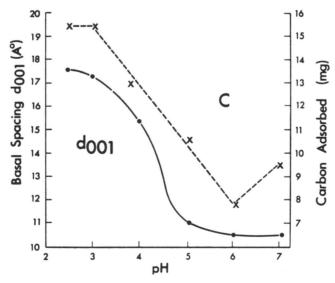

Fig. 24-13   Effect of pH on d_{001} of Na-montmorillonite and adsorption of FA (percent C X 2) on the clay. (From Schnitzer and Kodama, 1966.)

from clay interlayers by leaching with 1 M NaCl; an inflection occurs in the adsorption–pH curve near the pH corresponding to the pH of the acid species of FA (Schnitzer and Kodama, 1966). On the basis of these criteria, the adsorption could be classified as a "ligand-exchange" reaction (Greenland, 1971). This type of complex appears to be similar to an inner-sphere complex. In this type of reaction the anion is thought to penetrate the coordination shell of the dominant cation in the clay and displace water coordinated to the dominant cation in the clay interlayer.

The ease with which water can be displaced will depend on the affinity for water of the dominant cation with which the clay is saturated and also on the degree of dissociation of the FA. Since the latter is very low at low pH, interlayer adsorption of FA is greatest at low pH levels, as Schnitzer and Kodama (1966) have observed. Concurrently, the FA can dissolve a proportion of the dominant cation in the clay by forming a soluble complex and replacing the removed cation by H^+. If this process continues over long periods, the FA will eventually degrade the clay structure.

CONCLUSION

Humic substances, probably the most widely distributed organic carbon containing materials in terrestrial and aquatic environments, are dark brown, partly aromatic, acidic, hydrophilic, polyelectrolyte-like materials. Among their most important characteristics is their ability to interact with metal ions, oxides, hydroxides, minerals, and organics to form associations of widely differing chemical stabilities. These interactions include the formation of water-soluble simple and mixed ligand complexes, sorption and desorption, dissolution of minerals, and adsorption on external mineral surfaces and in clay interlayers.

REFERENCES

Breitmaier, E., and Voelter, W., 1978, ^{13}C NMR spectroscopy, p. 247–260 *in* Ebel, H. (Ed.), *Monographs in Modern Chemistry*, 2nd ed., Vol. 5; Verlag Chemie, Weinheim, West Germany.

Bremser, W., Ernst, L., Franke, B., Gerhards, R., and Hardt, A., 1979, *Carbon-13 NMR Spectral Data:* Verlag Chemie, New York (microfiche).

Ghosh, K., and Schnitzer, M., 1980, Macromolecular structures of humic substances: *Soil Sci.*, v. 129, p. 266–276.

Greenland, D. J., 1971, Interactions between humic and fulvic acids and clays: *Soil Sci.*, v. 111, p. 34–41.

Hatcher, P. G., Rowan, R., and Mattingly, M. A., 1980, ^{1}H and ^{13}C NMR of marine humic acids: *Org. Geochem.*, v. 2, p. 77–85.

Kerndorff, H., and Schnitzer, M., 1980, Sorption of metals on humic acid: *Geochim. Cosmochim. Acta*, v. 44, p. 1701–1708.

Kodama, H., and Schnitzer, M., 1973, Dissolution of chlorite minerals by fulvic acid: *Can J. Soil Sci.*, v. 53, p. 240–243.

____ , and Schnitzer, M., 1974, Adsorption of fulvic acid by non-expanding clay minerals: *10th Int. Congr. Soil. Sci.* (Moscow) *Trans. II*, p. 51–56.

Levesque, M., and Schnitzer, M., 1973, Dissolution of chlorite minerals by fulvic acid: *Can J. Soil Sci.*, v. 53, p. 240–243.

Lindeman, L. P., and Adams, J. Q., 1971, Carbon-13 nuclear magnetic resonance spectrometry: chemical shifts for the paraffins through C_9: *Anal. Chem.*, v. 43, p. 1245–1251.

Manning, P. G., and Ramamoorthy, S., 1973, Equilibrium studies of metal ion complexes of interest to natural waters. VII: *J. Inorg. Nucl. Chem.*, v. 35, p. 1577–1581.

Mikita, M., Steelink, C., and Wershaw, R. L., 1981, Carbon-13 enriched nuclear magnetic resonance methods for the determination of hydroxyl functionality in humic substances: *Anal. Chem.*, v. 53, p. 1715–1717.

Schnitzer, M., 1977, Recent findings on the characterization of humic substances extracted from soils from widely differing climatic zones, p. 117–132 *in Soil Organic Matter Studies II:* International Atomic Energy Agency, Vienna.

____ , 1978, Humic substances: chemistry and reactions: p. 1–64 *in* Schnitzer, M., and Khan, S.U. (Eds.), *Soil Organic Matter:* Elsevier, Amsterdam.

____ , and Khan, S. U., 1972, *Humic Substances in the Environment:* Marcel Dekker, New York.

____ , and Kodama, H., 1966, Montmorillorite: effect of pH on its adsorption of a soil humic compound: *Science*, v. 153, p. 70–71.

____ , and Kodama, H., 1976, The dissolution of micas by fulvic acid: *Geoderma*, v. 15, p. 381–389.

____ , and Skinner, S. I. M., 1974, The peracetic acid oxidation of humic substances: *Soil Sci.*, v. 118, p. 322–331.

Senesi, N., Griffith, S. M., Schnitzer, M., and Townsend, M. G., 1977, Binding of Fe^{3+} by humic materials: *Geochim. Cosmochim. Acta*, v. 41, p. 969–976.

Stevenson, F. J., 1982, *Humus Chemistry:* Wiley, New York.

Stevenson, I. L., and Schnitzer, M., 1982, Transmission electron microscopy of extracted fulvic and humic acids: *Soil Sci.*, v. 133, p. 179–185.

Verheyen, T. V., Johns, R. B., and Blackburn, D. T., 1982, Structural investigations of Australian coal. II. A ^{13}C NMR study of humic acids from Victorian brown coal lithotypes: *Geochim. Cosmochim. Acta*, v. 46, p. 269–277.

E. Michael Perdue
School of Geophysical Sciences
Georgia Institute of Technology
Atlanta, Georgia

Charles R. Lytle
Chemistry Department
Pacific University
Forest Grove, Oregon

25

CHEMICAL EQUILIBRIUM MODELING OF METAL COMPLEXATION BY HUMIC SUBSTANCES

The bioavailability and/or toxicity of a trace metal in aqueous solution is very sensitive to the "speciation" of the metal (i.e., whether it occurs primarily as the free ion, inorganic complexes, organic complexes, etc.). The interactions of trace metals with humic substances directly affect metal speciation and availability and therefore the potential applicability of geobotanical and/or biogeochemical methods of mineral exploration in a given soil environment. Two related topics are addressed: (1) the theoretical and experimental approaches toward describing metal-humus interactions in a quantitative manner, and (2) a basic overview of the effects of metal speciation on the "availability" of a metal to biological systems. These considerations lead to the prediction that geobotanical and biogeochemical methods of mineral exploration are probably most useful in soils with low humus content.

INTRODUCTION

In the field of exploration geology, an emerging interest in the use of both living and dead organic matter for mineral exploration has necessitated a more careful evaluation of the nature of the interaction of metals with humus (humic and fulvic acids). The underlying assumption of both geobotanical and biogeochemical methods of exploration is that plants growing in the vicinity of an ore deposit will tend to accumulate unusually high levels of the element of interest in their tissues. Elevated levels of metal may lead to morphological changes in affected plants, even selecting for the survival of resistant "indicator" species. The ability of humus to complex many of the metals of economic interest will undoubtedly need to be considered inasmuch as the bioavailability of a metal can be profoundly modified by complexation.

The complexation of trace metals by humic substances has been the subject of extensive research. Soil scientists have sought to understand mechanisms of uptake of metals by plants and toxic effects of some metals to plants. These same interests have subsequently led environmental chemists and biologists to study metal complexation by humus in order to understand the effects of trace metals on the growth of phytoplankton, the primary producers in natural waters. To the geochemist, it is apparent that humus may be intimately involved in the mobilization, transport, and deposition of trace metals of economic interest. All these groups of scientists have contributed in some measure to our current level of understanding of metal-humus interactions in aqueous environments.

As in any other field of endeavor, initial efforts to describe metal-humus interactions in a quantitative fashion were based on very simple theoretical models. In retrospect, it is now apparent that such complex systems cannot be described by simple models that work well for pure substances. In this paper, two commonly held erroneous perceptions about metal-humus complexation will be examined and more recent and theoretically sound approaches to these problems will be presented. The two subjects are (1) the measurement of total ligand concentration in humus, and (2) the determination of stability "constants" for metal complexation by humus. At the outset it must be stated that no attempt will be made to describe the uptake of elements that occur primarily in anionic forms (Mo, W, As, Se, etc.) and that no particular emphasis will be placed on trace metals of economic interest. The principles outlined in this chapter are applicable to cationic metal species in general.

EFFECTS OF COMPLEXATION ON METAL UPTAKE BY MICROORGANISMS AND PLANTS

Several of the papers presented at this colloquium addressed the question of how organisms are affected by trace metals. With only one or two exceptions, little or no attention was given to the solution chemistry of metals. We would simply like to point out the potential for experimental ambiguity that can result from neglecting the solubilities and complexation equilibria of metals. Then we would like to thrust our chemical perspective into an admittedly alien field: growth dynamics of organisms. We are, however, fully aware of the fact that our colleagues in biology can justifiably accuse us of oversimplification.

A great many of the metals form rather insoluble hydroxides, carbonates, phosphates, and so on. In experimental studies that involve the addition of a metal to some growth medium (either defined or natural), meaningful results are not readily obtained if the solubility limits have been exceeded. Approximate solubility limits can be estimated from available thermodynamic data (Lindsay, 1979) for any pH and CO_2 pressure. For example, at pH 7 and a CO_2 pressure of $10^{-3.5}$ atm, the solubility of Cu(II) is limited to about 0.75 mg/liter. The use of higher levels of added Cu(II) would be chemically meaningful only if other complexing ligands were present in solution at levels sufficient to prevent precipitation of copper hydroxide or copper carbonate.

The question of how the growth of organisms is influenced by complexation of trace metals has been addressed by soil and aquatic scientists. The addition of chelating agents can lead to either growth stimulation or growth inhibition. Historically, the former observation is more common and has been rationalized in two ways: (1) that complexation increases the bioavailability of an essential trace metal or (2) that complexation decreases the toxicity of a toxic trace metal (Jackson and Morgan, 1978). Some trace metals are essential micronutrients at low concentrations, becoming toxic at higher concentrations. From a purely chemical point of view, a complexing agent is expected to bind such a trace metal regardless of whether that metal is in a toxic or nontoxic concentration range. It seems highly unlikely that chelation would simultaneously increase the availability and decrease the toxicity of such a metal. Neither of the above explanations can account for the occasionally reported growth inhibition by chelating agents.

In several recent studies, it has been suggested that complexation generally decreases metal availability to organisms, both for toxic and essential metals. Sunda and coworkers (Sunda and Guilliard, 1976; Sunda and Lewis, 1978; Sunda et al., 1978) have clearly shown that the toxicities of copper and cadmium to aquatic organisms are a function of the free aquo ion concentrations, not of total or dissolved metal concentrations. In other words, the "available" concentration of a trace metal is the concentration of the free, uncomplexed ion. This nomenclature is somewhat misleading, because the free ion is in chemical equilibrium with all its complexed forms. It is quite useful to consider the analogous process of proton binding. Just as the free hydrogen ion concentration (or $pH = -\log[H^+]$) determines the extent to which ligand sites on a cell surface can be protonated, the free metal ion concentration (or $pM = -\log[M^{z+}]$) determines the extent of metal binding to cell surfaces that are in equilibrium with the external solution.

Petersen (1982) determined the toxicities of copper and zinc to *Scenedesmus quadricauda* and found that cell growth rate was reduced by 50 percent either when pZn was less than 5.0 or when pCu was less than 9.0. If zinc was added to a solution that already contained a small amount of copper, growth rates were reduced at surprisingly low levels of added zinc. A careful analysis of the chemical speciation of the nutrient medium revealed that the addition of zinc resulted in displacement of copper from Cu-EDTA complexes. The suppression of growth rate was due to *copper* toxicity rather than toxicity of zinc (the added metal). This study clearly illustrates the difficulties facing the laboratory investigator who wishes to determine the relative toxicities of trace metals. It is not possible to increase the concentration of one metal without simultaneously increasing the free ion concentrations of all other metals in the medium. A biological response to the addition of a metal to a nutrient medium could be the result of increased free ion concentrations of any of the metal ions in the solution. The same problem will exist in studies of trace metal nutrient requirements. The "availability" of a nutrient will be increased by addition of another metal that can displace the nutrient metal from "unavailable" complexes.

The question of whether complexation can increase the availability of a trace metal has been addressed by Jackson and Morgan (1978), who concluded that, with the exception of highly specific ionophores (biochemicals that are produced by an organism for the sole purpose of transporting a metal across cell membranes), complexing agents do not generally increase metal availability to aquatic organisms. They demonstrated that in several documented cases where the addition of chelating agents stimulated the growth of phytoplankton, suppression of trace metal toxicity was the most probable explanation for the observed results.

The success of geobotanical and biogeochemical methods of mineral exploration is contingent on the extent to which levels of metal in an organism reflect levels of that metal in the immediate environment. As a first approximation, factors that favor a rapid rate of uptake of a metal can be presumed to favor high levels of metal in the organism. A general relationship between growth rate and metal concentration must consider nutrient limitation at very low levels of metal and toxicity at high metal concentrations. Assuming a saturatable uptake mechanism that can be described by Monod kinetics and a toxic effect arising from the binding of the metal to a ligand on the cell surface, the following very simplistic model can be written:

$$V = V_{max} \left(\frac{[M]}{K_{uptake} + [M]} \right) \left(\frac{1}{1 + K_{toxic}[M]} \right) \tag{1}$$

where K_{uptake} is a half-saturation constant that equals the nutrient concentration of metal required to produce a growth rate (V) that equals one-half of the maximum possible growth rate (V_{max}), and K_{toxic} is the formation constant for binding the metal to a cellular ligand site that is thus inactivated for necessary metabolic functions. Although this theoretical model is highly simplistic, it will give some insight into how organisms respond to increased levels of trace metals and how that response is affected by metal–humus complexation.

As the free metal ion concentration increases from very low concentrations (high pM) to higher concentrations (low pM), the general effect on growth rate can be predicted from Eq. (1) if the K_{uptake} and K_{toxic} values are known. For illustrative purposes, Figure 25-1 has been constructed using log $K_{uptake} = -14$ and log $K_{toxic} = +6$. This plot bears a superficial resemblance to traditional plots based on total metal concentration; however, there is no fundamental chemical justification for the latter plots. The appropriate independent variable is the free metal ion concentration.

The first term in Eq. (1) gradually increases to unity as [M] exceeds K_{uptake}. Assuming that significantly higher levels of metal are required to produce a toxic effect, growth rates may be optimum over a range of free metal concentrations. At some higher concentration of metal, other essential binding sites on a cell may be deactivated by binding the metal. The last term in Eq. (1) will decrease from unity to near zero as K_{toxic}[M] exceeds 1.

There will be cases where K_{uptake} is so small that the metal could not be considered to be a nutrient and cases where K_{toxic} is so small that the metal would not exhibit a toxic effect even at high concentrations. The region of maximum growth may cover only a very small range of [M] values, in which case an organism would probably not survive the stress of high metal concentrations. Conversely, the region of maximum growth may be wide enough to allow a metal-resistant species to flourish when other organisms are killed.

To predict whether or not an organism will concentrate high levels of metal in its tissues, the free metal ion concentration must be known or reasonably estimated from other chemical data, such as total metal concentration, humus concentration, pH, and so on. The model described later in this chapter is an initial effort to provide a reasonable

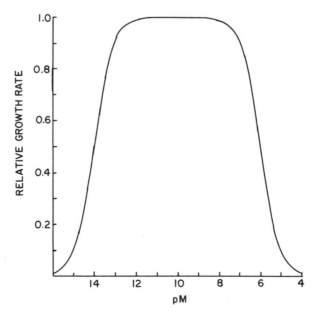

Fig. 25-1   Simulated growth curve for an organism with $K_{uptake} = 10^{-14}$ and $K_{toxic} = 10^{6}$.

CHEMICAL EQUILIBRIUM MODELING OF METAL COMPLEXATION

basis for such calculations. Perhaps the most direct approach toward modeling the effects of humus on metal uptake should start with the mass balance equation for the metal:

$$C_m = [M] + \Sigma\,[M(OH)_x] + \Sigma\,[M(CO_3)_x] + [M\text{-clay}] + [M\text{-humus}] \qquad (2)$$

To achieve a particular free metal concentration, the total metal concentration must be sufficient to satisfy the complexation capacity of the soil solution. Soils with higher pH, clay, or humus content buffer the free metal concentration, thus attenuating the uptake of metal by a plant. In such cases it would seem likely that the elevated total metal concentrations in the soil might be more readily detected. Geobotanical and biogeochemical methods will probably be more effective in environments with minimal soil development (i.e., soils that do not contain high levels of either clays or humus).

MATHEMATICS OF MULTILIGAND MIXTURES

In recent years, largely through the efforts of Gamble and coworkers (1970a, 1980) and MacCarthy and coworkers (1976, 1977, 1979), the more important mathematical properties of multiligand mixtures have been presented. Perdue and Lytle (1983a,b) have extended these concepts and reviewed the mathematical details of determination of stability "constants" from laboratory data. Only a brief overview of the mathematics will be presented in this chapter.

The complexation of a metal (M) by a single ligand (L_i) at constant pH can be described by a conditional stability constant (K_i). Letting $[L_i]$ represent the sum of the concentrations of all ligand species that are not bound to the metal,

$$M + L_i = ML_i \qquad (3)$$

$$K_i = \frac{[ML_i]}{[M]\,[L_i]} \qquad (4)$$

Historically, it was simply assumed that an average stability constant for a complex mixture of ligands could be written in a similar fashion.

$$\overline{K} = \frac{\Sigma\,[ML_i]}{[M]\,(\Sigma[L_i])} = \frac{C_m - [M]}{[M]\,(C_l - C_m + [M])} \qquad (5)$$

The total stoichiometric concentrations of metal and ligand are given by C_m and C_l, respectively. The attractiveness of this approach is undoubtedly enhanced by the fact that $\overline{K}$ can be directly calculated from experimental data. However, if Eqs. (4) and (5) are combined and all terms in the summations are divided by $[L_r]$, the concentration of an arbitrarily selected reference ligand, $\overline{K}$ can be expressed as

$$\overline{K} = \frac{\Sigma K_i([L_i]/[L_r])}{\Sigma([L_i]/[L_r])} \qquad (6)$$

The average stability "constant" ($\overline{K}$) is a weighted average of many K_i values and cannot be a constant unless all the weighting factors $([L_i]/[L_r])$ are constants. Because the ligands probably have a range of K_i values, L_i and L_r will not remain in a constant ratio as metal is added to a ligand mixture. For example, if the weakest ligand in the mixture is chosen as the reference ligand, then all the $([L_i]/[L_r])$ values will be greatest at very low levels of metal and will steadily decrease as metal is added. It follows that $\bar{K}$ *values must be functions rather than constants*, and that $\bar{K}$ values will decrease as metal is titrated into a ligand mixture [see MacCarthy and Smith, (1979) for a detailed development of these equations and conclusions].

Thus far in the discussion, it has been assumed that C_l values are known. Unless C_l values have been determined, $\bar{K}$ values cannot be calculated. Furthermore, any error in the determination of C_l leads directly to an error in $\bar{K}$ values.

DETERMINATION OF TOTAL LIGAND CONCENTRATIONS

It has been widely recognized that there is some ambiguity in reported total ligand concentrations (e.g., Gächter *et al.*, 1973; Reuter and Perdue, 1977; Weber, 1983; Langford *et al.*, 1983). Reasons that have been cited include competitive effects from spectator metals and pH effects. It has also been noted that higher apparent complexation capacities (mmol/g) are obtained with solutions of higher humus concentration and it has been recommended that "complexation capacities" of natural waters should not even be determined (Langford *et al.*, 1983).

It has been implied that the difficulties encountered in determinations of total ligand concentrations (C_l) are due to the complexity of humus. However, at least one of the observations, that of the dependence of complexation capacity (mmol/g) on the weight concentration of humus, has a much simpler explanation. Even for a pure ligand, the equilibrium $M + L \rightleftharpoons ML$ must shift to the left in dilute solutions, making it more difficult to saturate the ligand with metal. This phenomenon is illustrated in Figure 25-2 for a pure ligand with a stability constant for metal complexation of $\log K = 4$. At the low ligand concentrations that are typical of natural waters, only a small fraction of the potential binding capacity of the ligand is realized. At C_l values of 0.001 M or higher, the total ligand concentration would be more accurately determined. Of course, the problem would be less important for a ligand with a very large $\log K$ value. In humic substances, a wide range of $\log K$ values is possible, so it is reasonable to presume that complexation "capacities" that were determined at low humus concentrations are probably invalid.

The prerequisites for accurate determination of complexation capacity, high humus concentration and a large excess of metal ion, cannot be met because humus is precipitated from solution under those conditions (MacCarthy and Mark, 1976; Saar and Weber, 1982; Langford *et al.*, 1983). Although other approaches toward the determination of metal complexation capacity will be examined, this experimental limitation may ultimately be insurmountable. Should that be the case, complexation capacities may have to be estimated from elemental composition and total acidity data, with the understanding that equilibrium "constants" cannot then be unambiguously determined.

There are several more mathematically based methods for extracting total ligand concentrations from metal complexation data. One of these methods, the Scatchard plot, will be discussed in the next section, where the inapplicability of this method to multi-

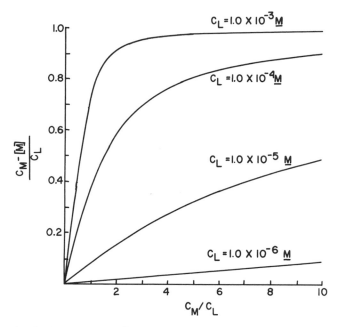

Fig. 25-2 Simulated "complexation capacity" experiment for a ligand with $K = 10^4$ at selected C_L values.

ligand systems will be demonstrated. The other method involves the analysis of a plot of bound metal versus log (free metal) as exemplified in Figure 25-3 for a pure ligand with log $K = 4$. The inflection point in this plot corresponds to log $[M] = -\log K$ and a bound

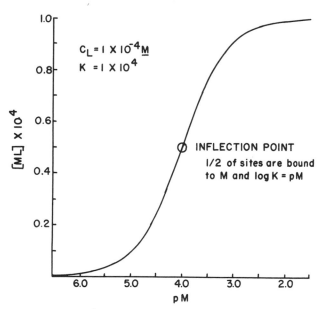

INFLECTION POINT
1/2 of sites are bound
to M and log K = pM

Fig. 25-3 Semilogarithmic plot of metal complexation data with $K = 10^4$ and $C_L = 10^{-4}$ M.

metal concentration that equals one-half of the complexation capacity. Even in complex ligand mixtures, the complexation capacity of a ligand mixture can be estimated by this method. The superiority of this method relative to the Scatchard plot for complex biological materials has recently been discussed (Klotz, 1982).

DETERMINATION OF STABILITY "CONSTANTS"

Despite the fact that the mathematical fallacy of assigning stability "constants" to a complex ligand mixture such as humus has been clearly established for more than 10 years (see the previous discussion), the current literature is rich with examples of such efforts to describe metal complexation by humus. In 1975, the use of the Scatchard equation (Scatchard, 1949; Scatchard et al., 1957) for analysis of metal–humus complexation data was suggested (Mantoura and Riley, 1975). Since that time, this equation has been widely applied in similar studies (e.g., Guy and Chakrabarti, 1976; Bresnahan et al., 1978; Sposito and Holtzclaw, 1979; Hirata, 1981; Sohn and Hughes, 1981; Sposito, 1981; McKnight et al., 1984; Tuschall and Brezonik, 1983; Alberts and Giesy, 1983).

If mathematical reality is temporarily suspended, it is easy to see why such a large group of scientists could have found the Scatchard method to be useful. Consider the typical Scatchard plot of metal–humus complexation data given in Figure 25-4, noting in particular how well the data points are fitted by the calculated curve. To fit these data, it is necessary to combine Eqs. 4 and 5 with a ligand mass balance equation to obtain

$$\Sigma\,[ML_i] = \Sigma\,\frac{C_i K_i [M]}{1 + K_i [M]} \tag{7}$$

As it is usually employed, the Scatchard method involves the use of Eq. (7) to calculate bound metal concentrations for each known [M] value. The summation includes only enough terms to adequately fit the data (usually two terms). For each term used in Eq. (7), there are two fitting parameters (C_i and K_i), so when two terms are used in Eq. (7), there is a total of four fitting parameters that can be determined by nonlinear regression techniques. The fitted curve in Figure 25-4 was obtained with this so-called two-component discrete ligand model. Undeniably, the experimental data are accurately represented by this equation.

If the erroneous premise that average stability "constants" are formally analogous to K_i values for discrete ligands is assumed to be true, it would be natural to assume that humus contains two "classes" of binding sites, whose concentrations and stability constants were obtained by fitting data to Eq. (7). On the other hand, if the nonconstant nature of average stability "constants" is recognized, the fitting parameters in Eq. (7) must be regarded as nothing more than empirical curve-fitting parameters that have no chemical significance. Accordingly, it is highly risky to suggest that humic substances have two "classes" of binding sites and any speculation about the chemical nature of those hypothetical binding sites is clearly unwarranted.

To elaborate further on the conclusions of the preceding paragraph, the arguments that were presented earlier concerning the nonconstant nature of $\overline{K}$ values should be recalled. Because any change in C_m or C_l would change the value of the weighting factors

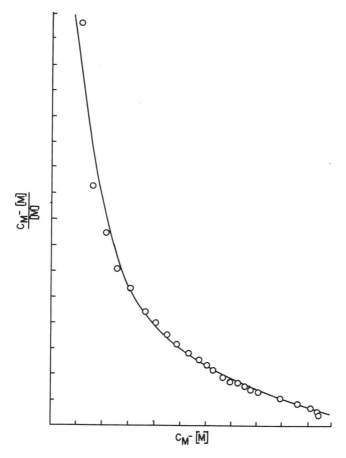

$$\frac{C_M - [M]}{[M]}$$

$$C_M - [M]$$

Fig. 25-4 Two-component scatchard fit of copper–humus complexation data at pH 5.0.

in Eq. (6), $\overline{K}$ values must decrease continuously as metal is added to a ligand mixture. If the ligand mixture could be subdivided into two "classes" of ligands, the average stability "constant" for each class of ligands would also decrease continuously as progressively weaker ligands in each "class" are titrated with metal. In other words, the average stability "constant" for a single class of ligands is also a function that varies during a titration. Returning to the two-component Scatchard equation [Eq. (7), truncated to two terms], we have

$$\Sigma [ML_i] = \frac{C_1 K_1 [M]}{1 + K_1 [M]} + \frac{C_2 K_2 [M]}{1 + K_2 [M]} \tag{8}$$

If, in fact, C_1, K_1, C_2, and K_2 are determined by fitting data that span a finite range of $\Sigma [ML_i]$ and $[M]$ values, it is implied that those fitting parameters are fixed constants over the range of the experimental data. Therefore, these *fitting constants* are not equivalent to the stability *functions* that might pertain to two classes of ligands. It also follows that the concentrations of the two hypothesized "classes" of ligands are not chemically meaningful parameters, but rather they too are simply curve-fitting parameters.

The indiscriminate use of the Scatchard equation in multiligand mixtures has also occurred frequently in the biochemical literature. Recent publications in that literature have clearly demonstrated the nature of the problem (Klotz, 1982; Peters and Pingoud, 1982). Similarly, several recent papers have clearly demonstrated that this equation is not applicable to metal–humus complexation equilibria (e.g., Gamble et al., 1980; Langford et al., 1983; Dempsey and O'Melia, 1983; Perdue and Lytle, 1983a,b; Shuman et al., 1983). All these authors have suggested that humus appears to consist of a very large number of nonidentical ligands that span a wide range of log K_i values. New approaches to the modeling of metal–humus complexation equilibria are clearly needed. One possible alternative is discussed in the next section.

CONTINUOUS DISTRIBUTION MODEL FOR METAL COMPLEXATION

The fact that metal binding by humus is accompanied by the release of hydrogen ions from the humus indicates that the same functional groups are involved in both proton binding and metal binding. Although much remains to be learned about the acid-base properties of humus, all the current evidence points to the presence of a very large number of nonidentical proton binding sites. To gain some insight into the range of pK_a values that is likely to exist for a complex mixture of organic acids, it is useful to consider the pK_a values that have been reported for simple organic acids. A frequency histogram of pK_a values was constructed for all the organic acids containing only C, H, and O that were listed in *Lange's Handbook of Chemistry*, 12th ed. (Dean, 1979). A total of 484 pK_a values for carboxylic acids, phenols, and other acids were included. The acids ranged from monoprotic to hexaprotic, so the observed pK_a distribution (see Fig. 25-5) is the result of statistical, electrostatic, and structural variations.

To a good approximation, the observed distribution of acidic functional groups can be described as normal distribution of relatively acidic groups with a mean pK_a value of 4 and another less well-defined distribution of groups with a mean pK_a value of about 10. This type of continuous distribution of functional groups is intuitively a reasonable model of the acidic groups in humus, as previously recognized by Posner (1964, 1966).

Perdue and Lytle (1983a,b) have extended this model to include metal-binding sites. The model assumes that a continuous distribution of binding sites is present in humus. If it is further assumed that only a single normal distribution of ligands is present, Eq. (7) can be modified to obtain

$$\Sigma [ML_i] = \frac{C_l}{\sigma\sqrt{2\pi}} \int_{-\infty}^{+\infty} \left(\frac{K[M]}{1 + K[M]}\right) \exp\left(-\frac{1}{2}\left[\frac{\mu - \log K}{\sigma}\right]^2\right) d \log K \qquad (9)$$

where σ is the standard deviation of log K values around the mean value (μ). Thus bound metal (calculated from experimental data as $C_m - [M]$) is a nonlinear function of [M] and the fitting parameters to be determined by nonlinear regression analysis of experi-

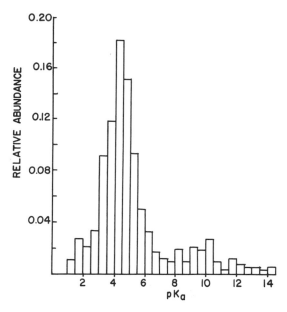

Fig. 25-5   Frequency histogram of 484 pK_a values of organic acids that contain only C, H, and O.

mental data are the total ligand concentration (C_l) and the mean (μ) and standard deviation (σ) of the distribution of log K values. The overall procedure is described more fully by Perdue and Lytle (1983b).

Lytle (1982) recently studied the binding of Cu(II) by humus at concentrations of both metal and ligand that approximate natural water conditions. The experimental procedures and data processing methods of Lytle (1982) are quite complex and will be mentioned only briefly in this chapter. Titrations were conducted at pH values of 5.0, 5.5, 6.0, and 6.5. The complexation capacity of the humus was estimated to be 5.7 mmol/g, using a semilogarithmic plot (e.g., see Fig. 25-3). The variation of log $\bar{K}$ with the C_l/C_m ratio for titrations at all pH values is given in Figure 25-6.

The normal distribution model was used to describe the variation in bound Cu(II) as a function of [Cu^{2+}]. Because total ligand concentrations (C_l) were estimated from the complexation capacity data, only μ and σ values were determined by nonlinear regression analysis of experimental data. The calculated average values are given in Table 25-1 and a typical fitted data set (pH 6.0) is given in Figure 25-7. The ordinate is given as the fraction of occupied ligand sites to emphasize the fact that only very little of the complexation capacity of the humus was reacted. Only a few strong binding sites were occupied by Cu(II). Accordingly, the log $\bar{K}$ values are quite high, reflecting the dominance of a few strong binding sites in the ligand mixture. The weighting factor used in the nonlinear regression analysis only partially compensated for the tendency of regression techniques to more closely fit large values of the dependent variable, so the agreement between experimental and calculated values is not as good at very low [Cu] values.

The results in Table 25-1 indicate a slight shift of μ toward larger values with increasing pH. This is expected, inasmuch as the conditional stability constants (K_i) of the

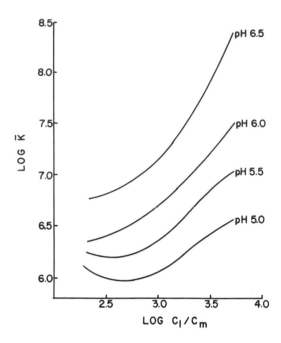

Fig. 25-6 Dependence of experimental stability functions for copper-humus complexation on ligand-metal ratios at selected pH values.

TABLE 25-1 Fitting Parameters for the Distribution Model of Copper Complexation by Aquatic Humus at pH 5.0–6.5.

pH	Mean (μ)	Std. Dev. (σ)
5.0	3.81	1.52
5.5	3.81	1.62
6.0	3.98	1.62
6.5	4.01	1.80

individual metal–ligand complexes will tend to increase with increasing pH. The slight increase in σ values could result from variable increases in K_i values; that is, the K_i values of ligands that are already extensively deprotonated at pH 5.0 will not increase at higher pH, whereas the K_i values of more fully protonated ligands will increase substantially. In any case, the marked variation in $\bar{K}$ values that is evident in Figure 25-6 can be attributed to only minor variations in the μ and σ values of a continuous distribution of ligands.

For comparative purposes, an attempt was made to fit the data to the two-component Scatchard equation; however, because the fraction of occupied ligand sites was so low, only one term in Eq. (7) was needed to fit the data. In fact, neither nonlinear regression nor a simplex method was successful in fitting the data to the two-component equation. The results of the one-component fits were so bad that they will not even be given in this chapter. It is sufficient to say simply that if the wrong equation is used, meaningless results are obtained.

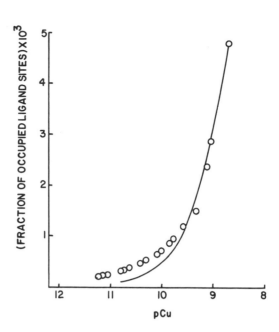

Fig. 25-7 Normal distribution model fit of copper-humus complexation data at pH 6.0.

CONCLUSION

The inherent difficulties involved in determination of the complexation capacity of humus have been examined. The experimental prerequisites of high ligand concentration and excess metal concentration appear to be mutually exclusive, indicating that accurate experimental determinations of complexation capacity may not be possible. Errors in C_l values yield concomitant uncertainties in computed stability functions.

A continuous distribution model has been used to describe metal-humus complexation equilibria. This approach avoids the conceptual errors of so-called discrete models that treat humus as a mixture of one or two simple ligands. The equations presented in this chapter can be used to predict the extent of complexation of a metal cation by humus in the absence of significant competing equilibria. An extension of the continuous distribution model to multimetal systems would be highly desirable for modeling metal-humus complexation in soil solutions.

If the field of geobotanical or biogeochemical mineral exploration is to gain wide acceptance, greater attention must be paid to the solution chemistry of metals (not just complexation equilibria but also solubility equilibria) and to the fundamental chemistry that is involved in metal uptake by organisms. In the field of environmental chemistry, most recent evidence tends to support the hypothesis that complexation generally decreases metal availability and decreases metal toxicity, both effects being attributed to a decrease in free metal ion concentration in the presence of a chelating agent.

It seems reasonable to speculate that the "signal-to-noise ratio" of geobotanical and biogeochemical mineral exploration methods will be highly dependent on the overall complexation capacity of a soil, all other things being equal. A greater probability of success probably exists in those environments where the soil has little complexation capacity to buffer the effects of elevated levels of total metal concentration. The soils in such environments should have minimal amounts of clays and humus. In soils of higher clay and humus contents, free metal ion concentrations will be far lower than total dissolved metal and less metal should be incorporated into growing plants. The standard soil analysis methods are able to detect elevated levels of metal in soil, whether or not the metal is complexed. Therefore, those methods are probably warranted in soils of high clay or humus content.

REFERENCES

Alberts, J. J., and Giesy, J. P., 1983, Conditional stability constants of trace metals and naturally occurring humic materials: application in equilibrium models and verification with field data, p. 333–348 *in* Christman, R. F., and Gjessing, E. T. (Eds.), *Aquatic and Terrestrial Humic Materials:* Ann Arbor Science, Ann Arbor, Mich.

Bresnahan, W. T., Grant, C. L., and Weber, J. H., 1978, Stability constants for the complexation of copper(II) ions with water and soil fulvic acids measured by an ion selective electrode: *Anal. Chem.*, v. 50, p. 1675–1679.

Dean, J. A. (Ed.), 1979, *Lange's Handbook of Chemistry, 12th ed.:* McGraw-Hill, New York, Chap. 5, p. 17–41.

Dempsey, B. A., and O'Melia, C. R., 1983, Proton and calcium complexation of four fulvic acid fractions, p. 239–273 *in* Christman, R. F., and Gjessing, E. T. (Eds.), *Aquatic and Terrestrial Humic Materials*, Ann Arbor Science, Ann Arbor, Mich.

Gächter, R., Lum-Shue-Chan, K., and Chau, Y. K., 1973, Complexing capacity of the nutrient medium and its relation to inhibition of algal photosynthesis by copper: *Schweiz. Zt. Hydrol.*, v. 35, p. 252–260.

Gamble, D. S., 1970, Titration curves of fulvic acid: the analytical chemistry of a weak acid polyelectrolyte: *Can. J. Chem.*, v. 48, p. 2662–2669.

———, Schnitzer, M., and Hoffman, I., 1970, Copper(II) fulvic acid chelation equilibrium in 0.1 M KCl at 25.0C: *Can. J. Chem.*, v. 48, p. 3197–3204.

———, Underdown, A. W., and Langford, C. H., 1980, Copper(II) titration of fulvic acid ligand sites with theoretical potentiometric, and spectrophotometric analysis: *Anal. Chem.*, v. 52, p. 1901–1908.

Guy, R. D., and Chakrabarti, C. L., 1976, Studies of metal–organic interactions in model systems pertaining to natural waters: *Can. J. Chem.*, v. 54, p. 2600–2611.

Hirata, S., 1981, Stability constants for the complexes of transition-metal ions with fulvic and humic acids in sediments measured by gel filtration: *Talanta*, v. 28, p. 809–815.

Jackson, G. A., and Morgan, J. J., 1978, Trace metal–chelator interactions and phytoplankton growth in seawater media: theoretical analysis and comparison with reported observations: *Limnol. Oceanogr.*, v. 23, p. 268–282.

Klotz, I. M., 1982, Numbers of receptor sites from Scatchard graphs: facts and fantasies: *Science*, v. 217, p. 1247–1249.

Langford, C. H., Gamble, D. S., Underdown, A. W., and Lee, S., 1983, Interaction of metal ions with a well characterized fulvic acid, p. 219–237 *in* Christman, R. F., and Gjessing, E. T. (Eds.), *Aquatic and Terrestrial Humic Materials:* Ann Arbor Science, Ann Arbor, Mich.

Lindsay, W. L., 1979, *Chemical Equilibria in Soils:* Wiley, New York, 449 p.

Lytle, C. R., 1982, The Copper Complexation Properties of Dissolved Organic Matter from the Williamson River, Oregon: Ph.D. dissertation, Portland State University, 193 p.

MacCarthy, P., 1977b, An interpretation of stability constants for soil organic matter-metal ion complexes under Schubert's conditions: *J. Environ. Sci. Health*, v. A12, p. 43–59.

——., and Mark, H. B., Jr., 1976, An evaluation of Job's method of continuous variations as applied to soil organic matter–metal ion interactions: *Soil Sci. Soc. Am. J.*, v. 40, p. 267–276.

——, and Smith, G. C., 1979, Stability surface concept: a quantitative model for complexation in multiligand mixtures: *in* Jenne, E. A. (Ed.), *Chemical Modeling in Aqueous Systems*, American Chemical Society Symp. Ser. 93, Washington, D.C.

Mantoura, R. F. C., and Riley, J. P., 1975, The use of gel filtration in the study of metal binding by humic acids and related compounds: *Anal. Chim. Acta*, v. 78, p. 193–200.

McKnight, D. M., Feder, G. L., Thurman, E. M., Wershaw, R. L., and Westall, J. C., 1983, Complexation of copper by aquatic humic substances from different environments: *in* Wildung, R. E. and Jenne, E. A. (Eds.), *Biological Availability of Trace Metals*, Elsevier, New York, p. 65–76.

Perdue, E. M., and Lytle, C. R., 1983a, A critical examination of metal–ligand complexation models: application to defined multiligand mixtures, p. 295–313 *in* Christman, R. F., and Gjessing, E. T. (Eds.), *Aquatic and Terrestrial Humic Materials*, Ann Arbor Science, Ann Arbor, Mich.

——, and Lytle, C. R., 1983b, Distribution model for binding of protons and metal ions by humic substances: *Environ. Sci. Technol.*, v. 17, p. 654–667.

Peters, F., and Pingoud, V. A., 1982, A critical interpretation of experiments on binding of peptide hormone to specific receptors by computer modeling: *Biochim. Biophys. Acta*, v. 714, p. 442–447.

Petersen, R., 1982, Influence of copper and zinc on the growth of a freshwater alga, *Scenedesmus quadricauda:* the significance of chemical speciation: *Environ. Sci. Technol.*, v. 16, p. 443–447.

Posner, A. M., 1964, Titration curves of humic acid: *Proc. 8th Int. Congr. Soil Sci., Part II*, Bucharest, Romania.

——, 1966, The humic acids extracted by various reagents from a soil. Part I. Yield, inorganic components, and titration curves: *J. Soil Sci.*, v. 17, p. 65–78.

Reuter, J. H., and Perdue, E. M., 1977, Importance of heavy metal–organic matter interactions in natural waters: *Geochim. Cosmochim. Acta*, v. 41, p. 325–334.

Saar, R. A., and Weber, J. H., 1982, Fulvic acid: modifier of metal-ion chemistry: *Environ. Sci. Technol.*, v. 16, p. 510A–517A.

Scatchard, G., 1949, The attractions of proteins for small molecules and ions: *Ann. N.Y. Acad. Sci.*, v. 51, p. 660–672.

——, Coleman, J. S., and Shen, A. L., 1957, Physical chemistry of protein solutions: VII. The binding of some small anions to serum albumin: *J. Am. Chem. Soc.*, v. 79, p. 12–20.

Shuman, M. S., Collins, B. J., Fitzgerald, P. J., and Olson, D. L., 1983, Distribution of stability constants and dissociation rate constants among binding sites on estuarine copper–organic complexes: rotated disk electrode studies and an affinity spectrum analysis of ion-selective electrode and photometric data, p. 349–370 *in* Christman, R. F., and Gjessing, E. T. (Eds.), *Aquatic and Terrestrial Humic Materials*, Ann Arbor Science, Ann Arbor, Mich.

Sohn, M. L., and Hughes, M. C., 1981, Metal ion complex formation constants of some sedimentary humic acids with Zn(II), Cu(II), and Cd(II): *Geochim. Cosmochim. Acta*, v. 45, p. 2393–2399.

Sposito, G., 1981, Trace metals in contaminated waters: *Environ. Sci. Technol.*, v. 15, p. 396–403.

——, and Holtzclaw, K. M., 1979, Copper (II) complexation by fulvic acid extracted from sewage sludge as influenced by nitrate versus perchlorate background ionic media: *Soil Sci. Soc. Am. J.*, v. 43, p. 47–51.

Sunda, W., and Guillard, R. R. L., 1976, The relation between copper ion activity and the toxicity of copper to phytoplankton: *J. Mar. Res.*, v. 34, p. 511–529.

Sunda, W. G., and Lewis, J. M., 1978, Effect of complexation by natural organic ligands on the toxicity of copper to a unicellular alga: *Monochrysis lutheri. Limnol. Oceanogr.*, v. 23, p. 870–876.

Sunda, W., Engel, D. W., and Thuotte, R. M., 1978, Effect of chemical speciation on the toxicity of cadmium to grass shrimp: importance of free cadmium ion: *Environ. Sci. Technol.*, v. 12, p. 409–413.

Tuschall, J. R., and Brezonik, P. L., 1983, Complexation of heavy metals by aquatic humus: a comparative study of five analytical methods, p. 275–294 *in* Christman, R. F., and Gjessing, E. T. (Eds.), *Aquatic and Terrestrial Humic Materials*, Ann Arbor Science, Ann Arbor, Mich.

Weber, J. H., 1983, Metal ion speciation studies in the presence of humic material, p. 313–331 *in* Christman, R. F., and Gjessing, E. T. (Eds.), *Aquatic and Terrestrial Humic Materials* Ann Arbor Science, Ann Arbor, Mich.

INDEX

445

element concentration in roots and stems of, 147
LANDSAT satellites, 100, 113–16
Lanthanum (La), 120
Larix laricina, 135
Larrea tridentata (creosote bush [*Covillea tridentata*]), 86–88
Lateral spread of roots, 84, 86
Leach, aqua regia, 151
Lead (Pb), 2, 3, 5, 137, 207
  alternative sample types for prospecting for, 119
  complexes, 53
  derived from normal and anomalous parent materials, 50
  differential uptake of, 118
  distribution, 198–99
  in Eagle Bluff, 41
  extraction of available, 51–52
  herbarium material in exploration for, 125
  in horsetails, 104
  in igneous and sedimentary rocks, 48, 49
  indicator plants for, 106
  in Milan mine, 188
  plant communities used to prospect for, 109
  plant organs concentrating, 147
  pollen and, 6–7
  relative concentrations of, 55, 120
  root localization of, 54
  seasonal variation in levels of, 56
  solubility and mobility of, 49, 57
  statistics, 195–98
  toxicity, 35
Leather leaf (*Chamaedaphne calyculata*), 134, 135
Leave(s), 24, 73, 117, 139, 141, 165
  ash weights in, 22
  biogeochemical prospecting and, 119
  chromium trioxalate and, 68
  gold content of, 155
  metal concentration in, 19–20
  rapid growth of, 92
  tungsten in, 143
  See also specific plants
Lecanora cenisia Ach., 42
Lecythis ollaria, 65, 66
Ledum groenlandicum (labrador tea), 134, 135, 140, 143, 147

Legume genus, 65
Leguminosae, 105, 106
Leptospermum scoparium, 68
Lethal concentrations, 16
Liatris punctata, 87
Lichen, 42, 119, 135
Limitations, salt, 94
Limits, tolerance, 39
Limonium suffruticosum, 105
Lindernia damblonii, 105
Lisbon Valley, 205
Lisle Valley, 154–56
Lithium (Li), 120
Lithogeochemical samples, 171–73
Local indicators, 104
Lodgepole pine (*Pinus contorta*), 119, 152, 204
Low-barrier system, 117, 118
Lowland tropical areas, 94
Lowland tundra, 94
Low-molecular-weight compounds, 65–71
Lusitanicum, 67
Lychnia alpina, 105, 125
Lychnis alpina var. *serpentinicola*, 106

Magnesium (Mg), 120
Maine, *See* Parmachenee sulfide exploration
Maize, 94
Man fern, 154
Manganese (Mn), 2
  accumulation in relation to evolutionary status, 67
  herbarium material in exploration for, 125
  hydrous oxides of, 35
  indicator plants for, 106
  relative concentrations of, 55, 120
  soil acidity and, 53
  soil drainage and, 53
  solubility and mobility of, 49
Maple, 185
Marine flora, 174
Massive sulfide mineralization, 185
Maximum lateral length of roots, 86–89
Maximum rooting depths, 86–89
Maytenus bureauvianus, 106
Medium-barrier system, 118
Melaleuca
  erocifolia, 154
  sheathiana, 103
Membrane protection, 76
Mercury (Hg), 2, 6

in igneous and sedimentary rocks, 48, 49
plant tolerance to, 70
Metabolic adaptations to metal stress, 64–73
  enzyme changes and, 72–73
  high-molecular-weight compounds and, 65–71
  low-molecular-weight compounds and, 65–71
Metal exclusion, 21, 28, 62–64
Metalliferous soils, 61
Metallophytes, 36
Metallothionein, 61, 62, 71–72, 76
Metal-rich exudates, 119
Metal(s)
  accumulation. *See* Accumulation, metal
  availability, 39, 44, 51
  concentration
    abnormal, 16–19
    background, 39
    heavy, 26
    in leaves and stems, 19–20
    lethal, 16, 17
    soil, 16
    tissue, 16
  essential, 36
  exclusion, 62–64
  heavy, 25, 26
  immobilization of, 61, 62
  nonessential trace, 15
  in plant tissue, 20–28
  sequestering of, 62
  in soils, 48–51
  toxicity to. *See* Metal stress; Toxicity
  uptake of. *See* Uptake
Metal stress, 14, 34, 36–40, 61–81, 110
  airborne biogeophysical measurement of, 112, 207
  cellular adaptations to, 73–76
  cell membrane permeability and, 76
  cytoplasmic sensitivity and, 73–76
  metabolic adaptations to, 64–73
    enzyme changes and, 72–73
    high-molecular-weight compounds and, 71–72
    low-molecular-weight compounds and, 65–71
  metal exclusion and, 62–64
  plant communities and, 38, 40, 101–4
  response to, 36–38

Statistical procedures, computer–assisted, 101
Stems, 24
  lead concentrations in, 147
  metal concentration in, 19–20
  See also specific plants
Stereochlaena cameronii, 63
Stillwater Complex, biogeochemical exploration of, 171–82
  sample preparation and analysis, 174–80
  sample types in, 171–74
Storage tissues, 24, 28
Stress
  geochemical, 107–10
  metal, *See* Metal stress
  nonmetal, 14
Stringy bark, 154, 155
Strontium (Sr)
  absorption of, 23
  relative concentration of, 26, 120
Sugar maple (*Acer saccharum*), 185
Sulfur (S), 25
SURTRACE system, 99, 119, 121–24, 126
Sutera fodina, 68
Sycamore, 91
Synergisms, metal, 44

Tallgrass prairie, 85, 91
Tamarack, 134
Tamarix pentandra, 88
Taproots, 84–85, 200
  development of, 88
Tasmania, gold prospecting in, 151–58
  analytical techniques for, 151–54
  gold content of plants and, 154–57
Tea tree, 154
Tellurium (Te), 2, 6
Temperate desert, 91–92
Temperate forest, 91
Tephrosia, 106, 109
Terrestrial mosses, biogeochemical prospecting and, 119
Thallium (Tl), 64
Thermographic imagery, 113
Thistle, 152
Thlaspi, 107
  alpestre, 69
  calaminare, 107
  rotundifolium ssp. *cepaeifolium*, 106
Threshold, toxic, 17–18, 23, 26, 37

Thunbergia, 108
Tilia americana (basswood), 185
Timmins–Porcupine greenstone belts, 165–66
Tin (Sn), 203
  extreme accumulation of, 64
  phytotoxicity, 35
  relative concentrations of, 120
Tissue plant
  age, 20
  barrier, 28
  barrier-free, 28
  chronological age, 25
  metal content of, 16, 20–28
  multielement analysis of, 28
  nonbarrier, 28
  physiological, 25
  storage, 24, 28
  tolerance mechanisms and composition of, 14
  woody, 22, 28
Tissue concentration–dose response curve, 26–28
Titanium (Ti)
  normalization to, 121
  relative concentrations of, 120
Tobacco, 71
Tolerance, 24, 26, 28, 34, 62
  active, 24
  biochemical, 19, 20
  to environmental stress, 34
  limits, 39
  nonessential elements and, 23
  physiological adaptations and, 20
  plant tissue composition and, 14
  plateau, 36, 37
  salinity, 19
  strategies, 19–20
  zone, 17–18, 26, 27
  See also Adaptations to metal stress; Metal stress; specific elements and plants
Tomato, 71
Toxicity, 26
  acute, 37
  chronic, 37
  gold-cyanide formation and, 68
  metal, 35–36, 39
  nonessential elements and, 23
  parameters of toxic threshold, 17–18, 23, 26, 37
  unit, 17, 18
  plant community and, 37

See also Metal stress; specific elements and plants
Toxic metals, available, 39
Toxic zone, 17–18, 27
Trace metals, nonessential, 15
Trachypogon spicatus, 63
Tracing, boulder, 3
Translocation, 64
Transpiration stream, 24
Transport
  of copper, 73
  restricted, 62–64
Tree sap, 99, 119
Triodia pungens, 63
Triumfetta, 108
Tropical areas, 91, 94, 95
Trunks, 134, 143–44
Trunkwood, 155
Trymalium ledifolium, 103
Tundra, 91, 94
Tungsten (W), 136, 160, 203
  in Canadian boreal forests, 134, 135, 142–43
  differential uptake of, 118
  extreme accumulation of, 64
  in Georgetown District, 204
  INAA determination of, 162
  plant organs concentrating, 147
  relative concentration of, 26
Twigs, 139, 154, 171, 174, 184, 185, 188, 195, 200
  gold content of, 155
  lead concentrations in, 147
  tungsten in, 143
  uranium in, 143–44, 146
  See also specific plants

Uapaca robynsii, 108
Ulmus americana (white elm), 185
Ultraviolet aerial photography, 112–13
Unit toxicity, 17, 18
Universal indicators, 104
Uptake, 52–59
  capacity, 21
  of essential elements, 23
  of minerals, 89
  passive, 117
  passive, 21–23, 26, 117
  plant factors influencing, 52–57
  soils factors in, 53–54
Uranium (U), 20, 117, 136, 205
  alternative sample types for prospecting for, 119
  in Canadian boreal forests, 134, 135, 142–47
  differential uptake of, 118
  extreme accumulation of, 64

Uranium *(cont.)*
 indicator plants for, 106
 plant communities used to
  prospect for, 109

V. ledocteanus, 106
Vaccinium myrtilloides (blueberry), 134, 135, 147
Vanadium (V), 205
 absorption of, 15
 extreme accumulation of, 64
 relative concentrations of,
  55, 120
Variability
 genetic, 20
 in root form, 83
Variation, seasonal, 138
Vascular plants, 151
Vegetation
 aerial sampling and analysis
  of, 119–24
 halophytic, 34
 rain forest, 89–91
 reflectance patterns of,
  110–12
 remote sensing of, 44, 100,
  110–16
 See also specific plants
Vegetation zonation, 33–46
 metal stress and, 36–40
  communities response to,
   38–40
  species response to, 36–38
 multifaceted nature of, 40–43
Vernonia cinerea, 106
Vigna radiata, 66
Viola calaminaria, 107
Violaceae, 106
Viscaria alpina, 104
Visible spectrum aerial photography, 112–13

Waterlogged horizons, 94
Wattle, 157
Weight, ash and dry, 22
Westringia cephalantha, 103

White birch (*Fraxinus ameri-
  cana*), 185
White elm (*Ulmus ameri-
  cana*), 185
White pine (*Pinus strobus*), 185
White spruce (*picea glauca*),
  135, 141, 184, 185, 188
Wild lilac, 154
Willow (*Salix* sp.), 134, 135,
  140, 143, 152
Wood, biogeochemical pros-
  pecting and, 119
Woody tissue, 22, 28

Xanthoparmelia wyomingica,
  42
Xerophyllum tenax (beargrass),
  203
Xerophyta, 108

Yarrow, 55
Yellow birch *(betula alle-
  ghaniensis)*, 185
Yellow Pine District, 203
Yield–dose response curve,
  16–18, 26
Yttrium (Y), 120
Yorkshire Fog, 55, 56
Yucca glauca, 87

Zinc-malate, 61
Zinc–organic acids, 69–70
Zinc (Zn), 2, 3, 61, 63, 117,
  174, 189, 204, 206, 207
 agronomic plant analysis and,
  15
 alternative sample types for
  prospecting for, 119
 Betula spp. accumulation of,
  55
 in cytoplasms, 72, 73
 derived from normal and
  anomalous parent mate-
  rials, 50
 distribution, 199

in Eagle Bluff, 41
as essential element, 23
estimation of availability of,
  52
free cations of, 51
herbarium material in exploration for, 125
in igneous and sedimentary
  rocks, 48, 49
indicator plants for, 106–7
in jack pine, 134
in Milan mine, 188
in Panachenee, 194, 195, 199
plant communities used to
  prospect for, 109
plant organs concentrating,
  147
pollen and, 6–7
relative concentrations of, 24,
  55, 74–75, 120
roots and, 54
seasonal variation in levels of,
  56
soil acidity and, 53
statistics, 195
in Stillwater Complex,
  176–77
in sulfide deposits, 184
tolerance to, 70, 76
toxicity of, 35, 208
transfer across cytoplasms,
  69–70
uptake of trace elements and,
  53
Zirconium (Zr), 120
Zn^{2+}, 51
Zonation, vegetation
 metal stress and, 36–40
  communities response to,
   38–40
  species response to, 36–38
 multifaceted nature of, 40–43
Zone
 deficient, 16, 27
 tolerance, 17–18, 26, 27
 toxic, 16–18, 27
Zygophyllum dumosum, 87

Index to Part II:
Microorganisms

Conjugation, 266, 267
Contamination, soil bacteria and heavy-metal, 352
Copper (Cu), 218, 332, 352
  –binding agent, penicillamine as, 345–46
  penicillin, inactivating effect on, 344
  percent organic matter vs., 340–41
  pH and toxicity of, 243
  resistance, 343
  seawater and toxicity of, 249
  See also Poorman Creek porphyry copper deposit
Copy number control, 277–78
Correlation analysis, 335–36, 338, 339
Cost, 211, 239
Cotter Basin Prospect, 347
Cross-resistance, 284
Crustal abundance, 210
Cycloheximide, 333

Deoxyribonucleotides, 269
Diagenetic processes, 231
Direct genetic selective pressure, 340
Disk method for multiple antibiotic resistance, 304
Dispersion patterns, ore-associated geochemical, 329
DNA, 266–68, 340. *See also* Genetic mechanisms

E_h, 243–44
Ecotoxicology, 240
EDTA, 248–50
Egg-yolk agar, 334, 347
Electron acceptors, terminal, 217
Elements, indicator, 211, 329
Endospores, 330
Enrichment of elements, 210–11. *See also* Natural metal enrichments, influence on penicillin resistance.
Enumerating bacteria
  B. cereus, 334
  methods of, 215–16
Environmental factors. *See* Physicochemical environmental factors
Enzyme, resistance involving substitute, 276–77
Eriogonum ovalifolium, 331
Erythromycin, resistance to, 287, 304

Estuarine environment, 283–300
  metal-resistant bacteria as index of potential metal mobilization, 286–93
  multiresistant bacteria, 284–86, 287, 289
  plasmid-mediated resistance in, 298–99
  seasonal variation in population of resistant bacteria, 294–95
  total metal concentration and TVC in, 296–98
Ethane, 215
  C-labeled, 216
  –oxidizing bacteria, 215
Ethylene, 215
Ethylenediaminetetraacetic acid (EDTA), 248–50
Eukaryotes, 266
Exchange, gene, 267
Expression, gene, 269–70
Exudates, organic, 249

Fecal samples, analysis for antibiotic-resistant bacteria, 304, 305–6
Feed lots, 307
Fungi
  copper as inhibitor of, 340–41
  over Cotter Basin Prospect, 347
  penicillin producing, 344, 345
  spores, 341

Gallionella ferruginea, 227
Gas deposits, detection of, 214–17
Gaseous hydrocarbon-oxidizing bacteria, 214-15, 217
Gas uptake method, 216
Geiger–Muller counting tube, 216
Gelatin, 251
Gene theory, 343
Genetic mechanisms, 265–81
  adaptation, 312, 313, 329
  gene amplification, 277–78
  gene exchange, modes of, 267
  gene expression, 269–70
  genetic elements of bacteria, 266–67
  metal tolerance by, 231
  recombination, 267, 268–69
  resistance to heavy metal and antibiotics, 254, 266, 270–79, 340
  selective pressure, direct, 340

ways to kill bacterial cells, 270
Gentamicin, resistance to, 287, 288, 290–93
Geologic time, 342–43
Geomicrobiological prospecting, 213–20
  for metallic and nonmetallic elements, 217–19
  for petroleum, 214–17
Glucose, anaerobic digestion of, 215
Gold (Au), 331, 346
Gram-negative and gram-positive bacteria, 352
Gram reaction, 333
Growth
  factors, 251
  inhibition by metals, 222, 229–30
  pellicle of, 215
  requirements, 225–26
  stimulation by metals, 222, 229, 230

Heavy-metal concentration antibiotic resistance and. *See under* Antibiotic resistance
  in fresh water, 223
  in some rock types, 223–24
Heavy-metal contamination, soil bacteria and, 352
Heavy-metal indicator microbe, 347
Heavy-metal penicillamine complexes, 346
Heavy-metal resistance, 218–19, 284–93
  coevolution with antibiotic resistance, 330
  genetic mechanisms, 254, 270–78, 340
  in high-altitude watershed, 301–10
  as index of potential metal mobilization, 286–93
  mineral exploration, use in, 312–27, 330
  multiresistant bacteria, 284–86, 289, 307, 308
  penicillin production and, 343–46
  plasmid-determined, 271–78, 313, 342
  plasmid-mediated, in aquatic bacteria, 298–99
  seasonal variation, 294–95, 352
  selection pressure for, 307

Simple organic substances, mineral solubilization by, 385–86
Soil
  profiles, analysis of, 370–75
  sample preparation and analysis of mull, 358–59
Solubilization of metals, humic substances and, 378–407
  exploration geochemistry and, 403–4
  investigation of, 387–98
    electrophoresis studies, 392–93
    polarographic studies, 389–92
    potentiometric titration technique, 387–89
    reaction kinetics, 395–98
    X-ray diffraction, 394–95, 396
  metal transport, 378, 387, 401–3
  mobile phase in weathering cycle, 398–400
  orebody weathering, 401
  solubilization of minerals, constraints on, 379–81
    experimental arrangements, 381–83
    experimental studies of, 379–86
    in presence of excess carbon dioxide, 384–85
    results, 383–84
    by simple organic substances, 385–86
Sorption of metal on humic substances, 421–23
Speciation, 429. *See also* Complexation, metal–humus

Spectrographic method, 359, 370, 371
Spectroscopy, ^{13}C NMR, 413–16
Sphagnum spp. (moss), 380
Spruce, 360
Stability "constant"
  determination of, 436–38
  for multiligand mixtures, 433–34, 439–40
  for single ligand, 433
  of water–soluble metal–HA and –FA complexes, 420
Stibnite, Idaho, 359, 365–70
Strontium (Sr), 357, 372–74
Subalpine fir (*Abies lasiocarpa*), 365, 371, 373
Subalpine forest, 370

Tea tree (*Melaleuca ericifolia*), 380
Tin (Sn), 357, 365, 372–74
Titanium (Ti), 357, 372–74
Titration, potentiometric, 387–89
Total ligand concentration, determination of, 434–36, 439
Toxicity of metals, 429, 430
Trace metals, complexation by humic substances. *See* Complexation, metal–humus
Transport, metal, 378, 387, 401–3
Tungsten (W), 365
Twigs, 371, 373

Uranium (U), 354, 387, 389, 391, 392

Vaccinium myrtillus L. (myrtle blueberry), 371
Vanadium (V), 354, 372–74

Water–soluble simple metal complexes, formation of, 419–20
Weathering
  cycle, humic–metal association as mobile phase in, 398–400
  orebody, 401, 402–3
White topped stringy-bark (*Eucalyptus delegatensis*), 380

X-ray diffraction, 394–95, 396

Yttrium (Y), 357, 373, 374

Zinc (Zn), 371
  concentration, 372–74
  leaching of, 398, 399–400
  in mull ash, 357, 360, 365, 370, 372
  potentiometric titration curves for, 388–89
  solubilization, 386
  sorption and desorption by HA, 421, 422
  toxicity, 431
  X-ray diffraction, 394
Zirconium (Zr), 357, 372–74